Some modern methods of organic synthesis

W. CARRUTHERS

Chemistry Department
University of Exeter

SECOND EDITION

Cambridge University Press

Cambridge
London New York Melbourne

Published by the Syndics of the Cambridge University Press
The Pitt Building, Trumpington Street, Cambridge CB2 1RP
Bentley House, 200 Euston Road, London NW1 2DB
32 East 57th Street, New York, NY 10022, USA
296 Beaconsfield Parade, Middle Park, Melbourne 3206, Australia

First published 1971
Second edition 1978

Printed in Great Britain at The Spottiswoode Ballantyne Press by
William Clowes & Sons Limited, London, Colchester and Beccles

Library of Congress Cataloguing in Publication Data

Carruthers, W.

Some modern methods of organic synthesis.

(Cambridge texts in chemistry and biochemistry)

Includes bibliographical references and index.

1. Chemistry, Organic – Synthesis. I. Title.

II. Series.

QD262.C33 1978 547′.2 77–77735

ISBN 0 521 21715 6 hard covers 2nd edition
(ISBN 0 521 08145 9 hard covers 1st edition)
ISBN 0 521 29241 7 paperback 2nd edition
(ISBN 0 521 09643 X paperback 1st edition)

Contents

Preface to the first edition

This book is addressed principally to advanced undergraduates and to graduates at the beginning of their research careers, and aims to bring to their notice some of the reactions used in modern organic syntheses. Clearly, the whole field of synthesis could not be covered in a book of this size, even in a cursory manner, and a selection has had to be made. This has been governed largely by consideration of the usefulness of the reactions, their versatility and, in some cases, their selectivity.

A large part of the book is concerned with reactions which lead to the formation of carbon–carbon single and double bonds. Some of the reactions discussed, such as the alkylation of ketones and the Diels–Alder reaction, are well established reactions whose scope and usefulness has increased with advancing knowledge. Others, such as those involving phosphorus ylids, organoboranes and new organometallic reagents derived from copper, nickel, and aluminium, have only recently been introduced and add powerfully to the resources available to the synthetic chemist. Other reactions discussed provide methods for the functionalisation of unactivated methyl and methylene groups through intramolecular attack by free radicals at unactivated carbon–hydrogen bonds. The final chapters of the book are concerned with the modification of functional groups by oxidation and reduction, and emphasise the scope and limitations of modern methods, particularly with regard to their selectivity.

Discussion of the various topics is not exhaustive. My object has been to bring out the salient features of each reaction rather than to provide a comprehensive account. In general, reaction mechanisms are not discussed except in so far as is necessary for an understanding of the course or stereochemistry of a reaction. In line with the general policy in the series references have been kept to a minimum. Relevant reviews are noted but, for the most part, references to the original literature are given only for points of outstanding interest and for very recent work. Particular reference is made here to the excellent book by H. O. House, *Modern Synthetic Reactions* which has been my guide at several points and on which I have tried to build, I fear all too inadequately.

I am indebted to my friend and colleague, Dr K. Schofield, for much helpful comment and careful advice which has greatly assisted me in writing the book.

W. CARRUTHERS

26 October 1970

Preface to the second edition

The general plan of this second edition follows that of the first edition, but the opportunity has been taken to bring the content up to date as far as possible. A considerable amount of new material has been included to take account of advances in knowledge and of new synthetic methods which have come into use since publication of the first edition. The increasing application of organic derivatives of sulphur, selenium and silicon in synthesis and improvements in the methods for selective alkylation of ketones and for reversing the polarity of functional groups ('umpolung') are among the subjects discussed more fully in this edition. To prevent the book from becoming too big, some material of less immediate interest which appeared in the first edition has been excised – it is hoped without detriment to the usefulness of the book. My aim, as before, has been to bring out the salient features of the reactions rather than to provide a comprehensive account. I have supported the discussion of new reactions by numerous references.

I am indebted to Dr I. Fleming for helpful correspondence about the reactions of organosilanes.

W. CARRUTHERS

January 1977

Nomenclature of olefins

Olefin configurations are expressed as *cis* or *trans*, using the priority rules formulated for the *E* (entgegen, *trans*) and *Z* (zusammen, *cis*) convention (Blackwood, Gladys, Loening, Petrarca and Rush, 1968).

1 Formation of carbon–carbon single bonds

In spite of the fundamental importance in organic synthesis of the formation of carbon–carbon single bonds there are comparatively few general methods available for effecting this process, and fewer still which proceed in good yield under mild conditions. Many of the most useful procedures involve carbanions, themselves derived from organometallic compounds, or from compounds containing 'activated' methyl or methylene groups. They include reactions which proceed by attack of the carbanion on a carbonyl or conjugated carbonyl group, as in the Grignard reaction, the aldol and Claisen ester condensations and the Michael reaction, and other reactions, with which this chapter will be largely concerned, which involve nucleophilic displacement at a saturated carbon atom, as in the alkylation of ketones and the coupling reactions of some organometallic compounds.

1.1. Alkylation: importance of enolate anions

It is well known that certain unsaturated groups attached to a saturated carbon atom render hydrogen atoms attached to that carbon relatively acidic, so that the compound can be converted into an anion on treatment with an appropriate base. Table 1.1, taken from House (1965), shows the pK_a values for some compounds of this type and for some common solvents and reagents.

The acidity of the C—H bonds in these compounds is due to a combination of the inductive electron-withdrawing effect of the unsaturated groups and resonance stabilisation of the anion formed by removal of a proton (1.1). Not all groups are equally effective in 'activating' a neighbouring CH_2 or CH_3; nitro is the most powerful of the common groups and thereafter the series follows the approximate order $—NO_2 >$ $—COR > —SO_2R > —CO_2R > —CN > —C_6H_5$. Two activating groups reinforce each other, as can be seen by comparing diethyl malonate

TABLE 1.1 *Approximate acidities of active methylene compounds and other common reagents*

Compound	pK_a	Compound	pK_a
$CH_3CO_2\underline{H}$	5	$C_6H_5COC\underline{H}_3$	19
$C\underline{H}_2(CN)CO_2C_2H_5$	9	$CH_3COC\underline{H}_3$	20
$C\underline{H}_2(CO.CH_3)_2$	9	$C\underline{H}_3SO_2C\underline{H}_3$	~23
$C\underline{H}_3NO_2$	10	$C\underline{H}_3CO_2C_2H_5$	~24
$CH_3COC\underline{H}_2CO_2C_2H_5$	11	$CH_3CO_2^-$	~24
$C\underline{H}_2(CO_2C_2H_5)_2$	13	$C\underline{H}_3CN$	~25
$CH_3O\underline{H}$	16	$C_6H_5N\underline{H}_2$	~30
$C_2H_5O\underline{H}$	18	$(C_6H_5)_3C\underline{H}$	~40
$(CH_3)_3CO\underline{H}$	19	$C\underline{H}_3SOC\underline{H}_3$	~40

(Acidic hydrogen atoms are underlined)

H. O. House, *Modern synthetic reactions*, copyright 1972, W. A. Benjamin, Inc., Menlo Park, California.

(p$K_a \approx 13$) with ethyl acetate (p$K_a \approx 24$). Acidity is also increased slightly by electron-withdrawing substituents, and decreased by alkyl groups, so that diethyl methylmalonate, for example, has a slightly less acidic C—H group than diethyl malonate itself.

By far the most important activating groups in synthesis are the carbonyl and carboxylic ester groups. Removal of a proton from the

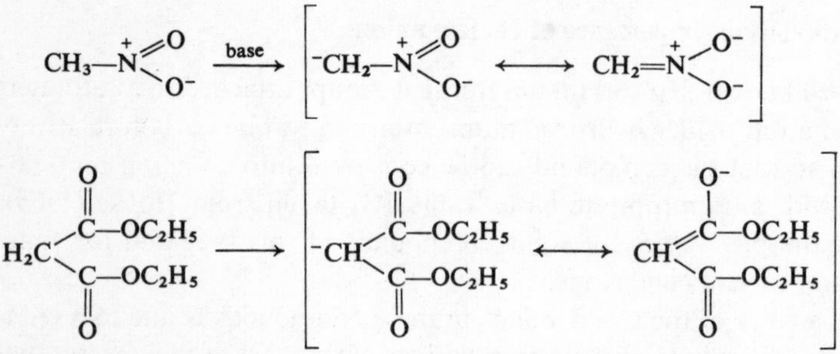

$$CH_3{-}\overset{+}{N}{\overset{\displaystyle =O}{\underset{\displaystyle O^-}{}}} \xrightarrow{\text{base}} \left[{}^-CH_2{-}\overset{+}{N}{\overset{\displaystyle =O}{\underset{\displaystyle O^-}{}}} \longleftrightarrow CH_2{=}\overset{+}{N}{\overset{\displaystyle O^-}{\underset{\displaystyle O^-}{}}} \right] \tag{1.1}$$

$$H_2C{\overset{\displaystyle C(=O){-}OC_2H_5}{\underset{\displaystyle C(=O){-}OC_2H_5}{}}} \longrightarrow \left[{}^-CH{\overset{\displaystyle C(=O){-}OC_2H_5}{\underset{\displaystyle C(=O){-}OC_2H_5}{}}} \longleftrightarrow CH{\overset{\displaystyle =C(O^-){-}OC_2H_5}{\underset{\displaystyle C(=O){-}OC_2H_5}{}}} \right]$$

α-carbon atom of a carbonyl compound with base gives the corresponding enolate anion, and it is these anions which are involved in base-catalysed condensation reactions of carbonyl compounds, such as the aldol condensation, and in bimolecular nucleophilic displacements (alkylations) (1.2). The enolate anions must be distinguished from the enols themselves, which are always present in equilibrium with the

$$R{-}CH_2{-}CO{-}R' \xrightleftharpoons{\text{base (slow)}} R{-}\bar{C}H{-}CO{-}R' \longleftrightarrow R{-}CH{=}C(O^-){-}R'$$

$$-C(O^-){=}CH \quad >C{=}O \rightleftharpoons -C(=O){-}CH{-}C{-}O^- \tag{1.2}$$

$$-C(O^-){=}CH \quad -C{-}X \longrightarrow -C(=O){-}CH{-}C{-} + X^-$$

$$R{-}CH_2{-}CO{-}R' \rightleftharpoons R{-}CH{=}C(OH){-}R' \tag{1.3}$$

carbonyl compound in presence of acidic or basic catalysts (1.3). The enols are concerned in certain acid-catalysed condensations of carbonyl compounds. Most monoketones and esters contain only small amounts of enol (<1 per cent) at equilibrium, but with 1,2- and 1,3-dicarbonyl compounds much higher amounts of enol (>50 per cent) may be present.

The formation of the enolate anion results from an equilibrium reaction between the carbonyl compound and the base. A competing equilibrium involves the enolate anion and the solvent. Thus, with diethyl malonate in solvent SolH in presence of base B^-, we have

$$\begin{aligned} CH_2(CO_2C_2H_5)_2 + B^- &\rightleftharpoons {}^-CH(CO_2C_2H_5)_2 + BH \\ {}^-CH(CO_2C_2H_5)_2 + \text{SolH} &\rightleftharpoons CH_2(CO_2C_2H_5)_2 + \text{Sol}^-, \end{aligned} \tag{1.4}$$

and to ensure an adequate concentration of the enolate anion at equilibrium clearly both the solvent and the conjugate acid of the base must be much weaker acids than the active methylene compound. The correct choice of base and solvent is thus of great importance if the subsequent alkylation, or other, reaction is to be successful. Reactions must normally be effected under anhydrous conditions since water is a much stronger acid than the usual activated methylene compounds and, if present, would instantly protonate any carbanion produced. Another point of importance is that the solvent must not be a much stronger acid than the conjugate acid of the base, otherwise the equilibrium

$$B^- + \text{SolH} \rightleftharpoons BH + \text{Sol}^- \tag{1.5}$$

will lie too far to the right and lower the concentration of B^-. For example, sodamide can be used as base in liquid ammonia or in benzene, but, obviously, not in ethanol. Base–solvent combinations commonly used to convert active methylene compounds into the corresponding anions include sodium methoxide, sodium ethoxide and sodium or potassium t-butoxide in solution in the corresponding alcohol, or as suspensions in ether, benzene or dimethoxyethane. Potassium t-butoxide is a particularly useful reagent, since it is a poor nucleophile and its solutions in different solvents have widely different basic strengths; it is most active in solution in dry dimethyl sulphoxide (Pearson and Buehler, 1974). Metallic sodium or potassium, or sodium hydride, in suspension in benzene, ether or dimethoxyethane, sodamide in suspension in an inert solvent or in solution in liquid ammonia, and solutions of sodium or potassium triphenylmethyl in ether or benzene are also widely used with the less 'active' compounds.

Recently the lithium salts of secondary amines, particularly lithium diisopropylamide and lithium 2,2,6,6-tetramethylpiperidide have been finding increasing application (see Olofson and Dougherty, 1973). These amide bases are non-nucleophilic and have the advantage that they are soluble in non-polar, even hydrocarbon, solvents. The insolubility of the bases listed above, in such solvents, seriously limits their usefulness.

1.2. Alkylation of relatively acidic methylene groups

In order to effect a reasonably rapid reaction it is, of course, necessary to have a high concentration of the appropriate carbanion. Because of their relatively high acidity (see Table 1.1) compounds in which a C—H bond is activated by a nitro group or by two or more carbonyl, ester or cyano groups can be converted largely into their anions with an anhydrous alcoholic solution of a metal alkoxide, such as sodium ethoxide or potassium t-butoxide. An alternative procedure is to prepare the enolate in benzene or ether, using finely divided sodium or potassium metal or sodium hydride, which react irreversibly with compounds containing active methylene groups with formation of the metal salt and evolution of hydrogen. β-Diketones can often be converted into their enolates with alkali metal hydroxides or carbonates in aqueous alcohol or acetone.

Much faster alkylation of enolate anions can often be achieved in dimethylformamide, dimethyl sulphoxide, 1,2-dimethoxyethane or hexamethylphosphoramide than in the usual protic solvents. This appears to be due to the fact that the former solvents do not solvate the enolate anion and thus do not diminish its reactivity as a nucleophile.

At the same time they are able to solvate the cation, separating it from the cation–enolate ion pair and leaving a relatively free enolate ion which would be expected to be a more reactive nucleophile than the ion pair (Parker, 1962). Reactions effected with aqueous alkali as base are often improved in the presence of a phase-transfer catalyst such as a tetra-alkylammonium salt (cf. Makosza and Jończyk, 1976).

Alkylation of enolate anions is readily effected with alkyl halides or other alkylating agents. Both primary and secondary alkyl, allyl or benzyl halides may be used successfully, but with tertiary halides poor yields of alkylated product often result because of competing dehydro-halogenation of the halide. It is often advantageous to proceed by way of the toluene-*p*-sulphonate or methanesulphonate rather than a halide. The sulphonates are excellent alkylating agents, and can usually be obtained from the alcohol in a pure condition more readily than the corresponding halides. Epoxides have also been used as alkylating agents, generally reacting at the less substituted carbon atom. Attack of the enolate anion on the alkylating agent takes place by a bimolecular nucleophilic displacement (S_N2) process and thus results in inversion of configuration at the carbon atom of the alkylating agent.

$$p\text{-}CH_3C_6H_4SO_2O\text{-(steroid, }H_3C\text{, H)} \xrightarrow[C_2H_5OH]{CH_2(CO_2C_2H_5)_2,\ C_2H_5ONa} \text{(steroid, }H_3C\text{, H)}\text{-}CH(CO_2C_2H_5)_2 \quad (1.6)$$

With secondary and tertiary allylic halides or sulphonates, reaction of an enolate anion may give mixtures of products formed by competing attack at the α- and γ-positions (1.7).

$$C_2H_5\text{—}\underset{\displaystyle Cl}{CH}\text{—}CH{=}CH_2 \xrightarrow[C_2H_5ONa,\ C_2H_5OH]{CH_2(CO_2C_2H_5)_2}$$

$$C_2H_5\text{—}\overset{\displaystyle CH(CO_2C_2H_5)_2}{CH}\text{—}CH{=}CH_2 + \underset{(10\%\ \text{of product})}{C_2H_5CH{=}CHCH_2CH(CO_2C_2H_5)_2} \quad (1.7)$$

Alkylation of active methylene compounds with αω-polymethylene dihalides, and intramolecular alkylation of ω-haloalkylmalonic esters provides a useful method for synthesising three- to seven-membered rings. Non-cyclic products are frequently formed at the same time by competing intermolecular reactions and conditions have to be carefully chosen to suppress their formation (1.8).

$$Br(CH_2)_5Br + CH_2(CO_2C_2H_5)_2 \xrightarrow[C_2H_5OH]{C_2H_5ONa} \text{(cyclohexane-1,1-diyl)}(CO_2C_2H_5)_2 + (C_2H_5O_2C)_2CH(CH_2)_5CH(CO_2C_2H_5)_2 \quad (1.8)$$

A difficulty sometimes encountered in the alkylation of active methylene compounds is the formation of unwanted dialkylated products. During the alkylation of diethyl sodiomalonate, the monoalkyl derivative formed initially is in equilibrium with its anion as indicated in the first equation of (1.9). In ethanol solution, dialkylation does not take place to any appreciable extent because ethanol is sufficiently acidic to reduce the concentration of the anion of the alkyl derivative, but not that of the more acidic diethyl malonate itself, to a very low value.

$$RCH(CO_2C_2H_5)_2 + \bar{C}H(CO_2C_2H_5)_2 \rightleftharpoons R\bar{C}(CO_2C_2H_5)_2 + CH_2(CO_2C_2H_5)_2 \quad (1.9)$$

$$R\bar{C}(CO_2C_2H_5)_2 + C_2H_5OH \rightleftharpoons RCH(CO_2C_2H_5)_2 + C_2H_5O^-$$

However, replacement of ethanol by an inert solvent favours dialkylation, and dialkylation also becomes a more serious problem with the more acidic alkylcyanoacetic esters, and in alkylations with very reactive compounds such as allyl or benzyl halides or sulphonates. It has been said that alkylation of β-dicarbonyl compounds by reaction of the thallium(I) salts with an alkyl iodide or bromide, leads to high yields of monoalkylated product without any dialkylation, but this has been questioned (Hooz and Smith, 1972). Dialkylation may, of course, be effected deliberately if required by carrying out two successive operations, using either the same or a different alkylating agent in the two steps. It is often found that active methylene compounds with a secondary or tertiary alkyl substituent in the position adjacent to the activating group undergo further alkylation only with difficulty. This is partly due to increased steric hindrance to approach of the base for proton abstraction and partly, in the case of carbonyl compounds at any rate, to steric interference with the attainment of a transition state for proton removal that allows continuous overlap of the p-orbitals concerned (1.10). This difficulty may be overcome by use of a stronger base in a less acidic

$$(CH_3)_2CH{-}C(H)(CO_2C_2H_5){-}C(OC_2H_5){=}O \longrightarrow \text{[base removes H; orbital overlap with C=O]} \longrightarrow (CH_3)_2CH(H_5C_2O_2C)C{=}C(OC_2H_5)O^- \quad (1.10)$$

solvent, such as potassium t-butoxide in t-butanol, or, if the choice is available, by introducing the branched-chain substituent last.

Under ordinary conditions aryl or vinyl halides do not react with enolate anions, although aryl halides with strongly electronegative substituents in the *ortho* and *para* positions do. 2,4-Dinitrochlorobenzene, for example, with ethyl cyanoacetate gives ethyl (2,4-dinitrophenyl)cyanoacetate in 90 per cent yield. Aryl halides may also react with enolates under more vigorous conditions. Reaction of bromobenzene with diethyl malonate, for example, takes place readily in presence of an excess of sodium amide in liquid ammonia, to give diethyl phenylmalonate in 50 per cent yield. The reaction is not a direct nucleophilic displacement, however, but takes place by an elimination–addition sequence in which benzyne is an intermediate (1.11). Similar reactions can be effected intramolecularly and provide a good route to

$$C_6H_5Br \xrightarrow[\text{liq. } NH_3]{NaNH_2} \text{benzyne} \xrightarrow{\bar{C}H(CO_2C_2H_5)_2} \bar{C}_6H_4CH(CO_2C_2H_5)_2 \longrightarrow C_6H_5\bar{C}(CO_2C_2H_5)_2 \xrightarrow{H_3O^+} C_6H_5CH(CO_2C_2H_5)_2 \quad (1.11)$$

some cyclic systems (Gilchrist and Rees, 1969) (1.12). Vinyl derivatives of active methylene compounds can be obtained indirectly from alkylidene derivatives by moving the double bond out of conjugation as illustrated in (1.13). Kinetically controlled alkylation of the delocalised anion takes place at the α-carbon atom to give the $\beta\gamma$-unsaturated compound directly. A similar course is followed in the kinetically controlled protonation of such anions, but now rapid reaction takes place at the

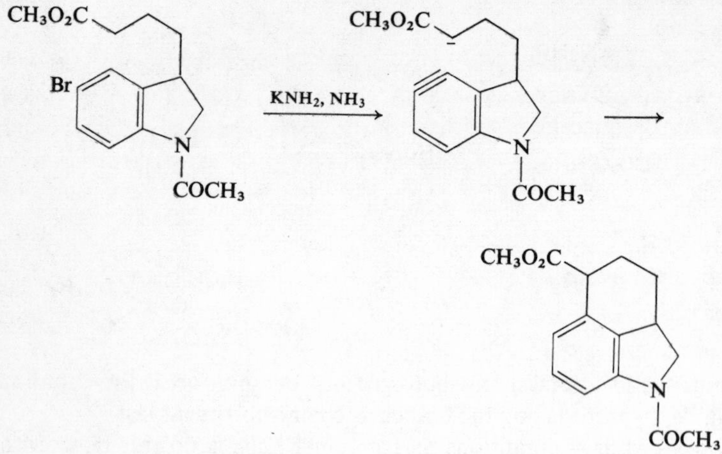

CH_3O_2C Br N $COCH_3$ $\xrightarrow{KNH_2,\ NH_3}$ CH_3O_2C N $COCH_3$ $\longrightarrow$

(1.12)

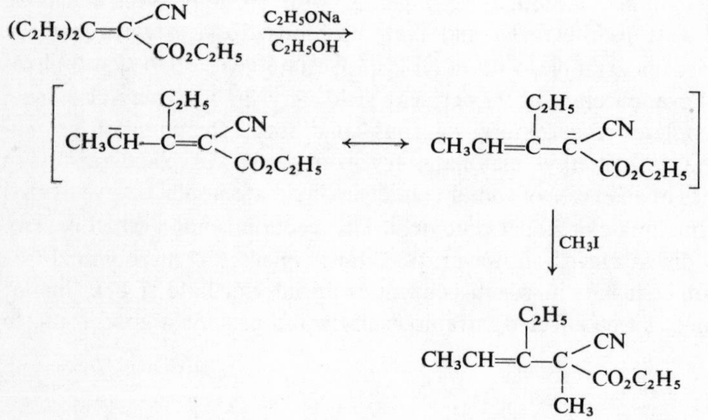

CH_3O_2C N $COCH_3$

$$(C_2H_5)_2C{=}C\begin{matrix}CN\\CO_2C_2H_5\end{matrix}\ \xrightarrow[C_2H_5OH]{C_2H_5ONa}$$

$$\left[CH_3\bar{C}H{-}\underset{}{\overset{C_2H_5}{C}}{=}C\begin{matrix}CN\\CO_2C_2H_5\end{matrix} \longleftrightarrow CH_3CH{=}\overset{C_2H_5}{C}{-}\bar{C}\begin{matrix}CN\\CO_2C_2H_5\end{matrix}\right] \quad (1.13)$$

$$\downarrow CH_3I$$

$$CH_3CH{=}\overset{C_2H_5}{C}{-}\underset{CH_3}{C}\begin{matrix}CN\\CO_2C_2H_5\end{matrix}$$

oxygen to give an enol which, under neutral or weakly acidic conditions, isomerises more slowly to the $\beta\gamma$-unsaturated compound.

A wasteful side-reaction which frequently occurs in the alkylation of 1,3-dicarbonyl compounds is the formation of the *O*-alkylated product. Thus, reaction of the sodium salt of cyclohexan-1,3-dione with butyl bromide gives 37 per cent of 1-butoxycyclohexen-3-one and only 15 per cent of 2-butylcyclohexan-1,3-dione. In general, however, *O*-alkylation competes significantly with *C*-alkylation only with reactive methylene compounds in which the equilibrium concentration of enol is relatively high, as in 1,3-dicarbonyl compounds and phenols. Phenols, of course, generally undergo predominant *O*-alkylation.

Alkylation of malonic ester, cyanacetic ester and β-keto esters is useful in synthesis because the alkylated products on hydrolysis and decarboxylation afford carboxylic acids and ketones. From alkylated malonic or cyanacetic esters substituted acetic acids are obtained, and alkylated acetoacetic esters give substituted acetones (1.14). Alkaline hydrolysis of β-keto esters is often complicated by competing attack of

$$C_2H_5CH(CH_3)CH(CO_2C_2H_5)_2 \xrightarrow[\text{boil}]{H_2O,\ KOH} C_2H_5CH(CH_3)CH(COOK)_2 \xrightarrow[\text{boil}]{H_2O,\ H_2SO_4} C_2H_5CH(CH_3)CH_2CO_2H \quad (1.14)$$

$$C_4H_9CH(CO_2C_2H_5)COCH_3 \xrightarrow[NaOH]{5\%\ aq.} C_4H_9CH(CO_2Na)COCH_3 \xrightarrow[\text{boil}]{H_2O,\ H_2SO_4} C_4H_9CH_2COCH_3$$

hydroxide ion on the ketone group, leading to fission products. This is particularly liable to occur when the α-position is disubstituted; where the molecule contains an α-hydrogen atom, the carbonyl group is protected from attack by enolisation (1.15). These cleavage reactions can be avoided

2-benzyl-2-ethoxycarbonylcyclopentanone $\xrightarrow{H_2O,\ NaOH}$ (attack of $\bar{O}H$ on C=O) tetrahedral intermediate ($\bar{O}$, OH, $CH_2C_6H_5$, $CO_2C_2H_5$) → ring-opened $HO(O=)C(CH_2)_3\bar{C}(CH_2C_6H_5)CO_2C_2H_5$ → $HOOC(CH_2)_3CH(CH_2C_6H_5)CO_2H$ (90%) (1.15)

2-benzyl-2-ethoxycarbonylcyclopentanone $\xrightarrow{48\%\ HBr,\ reflux}$ 2-benzylcyclopentanone ($CH_2C_6H_5$) (80%)

by effecting the hydrolysis and decarboxylation under acid conditions, or by using benzyl, t-butyl or tetrahydropyranyl esters in place of ethyl esters. Benzyl esters are readily cleaved by hydrogenolysis (see p. 408) and acid-catalysed cleavage of t-butyl or tetrahydropyranyl esters takes place easily.

Enolate anions derived from β-dicarbonyl compounds can also be acylated by reaction with an acid chloride or acid anhydride in an inert solvent. These reactions are synthetically useful for the preparation of 'mixed' β-keto esters which would be obtainable only with difficulty by a Claisen ester condensation using two different esters (1.16).

$$CH_3COCH_2CO_2C_2H_5 \xrightarrow[C_6H_6]{Na} CH_3CO\bar{C}HCO_2C_2H_5$$

$$\downarrow C_6H_5COCl, \text{ benzene} \qquad (1.16)$$

$$C_6H_5COCH_2CO_2C_2H_5 \xleftarrow[H_2O,\ NH_4Cl]{NH_4OH} CH_3COCH(COC_6H_5)CO_2C_2H_5$$

1.3. γ-Alkylation of 1,3-dicarbonyl compounds

Alkylation of a 1,3-diketone at one of the 'flanking' methyl or methylene groups instead of at the doubly activated CH_2 does not usually take place to any significant extent under ordinary conditions. It can be accomplished selectively and in good yield by way of the corresponding *dianion*, itself prepared from the diketone and two equivalents of sodium or potassium amide in liquid ammonia, by reaction with one equivalent of alkylating agent (Harris and Harris, 1969). Thus, acetylacetone is converted into 2,4-nonanedione in 82 per cent yield (1.17), and benzoylacetone gives 1,6-diphenyl-1,3-pentanedione in 77 per cent yield by reaction of the dianion with benzyl chloride. Keto acids and triketones can also be obtained, by reaction of the dianions with carbon dioxide or with esters.

With unsymmetrical diketones, which could apparently give rise to two different dianions, it is found in practice that in most cases only one is formed, and a single alkylation product results. Thus, with 2,4-hexanedione, alkylation at the methyl group greatly predominates over that at the methylene group, and 2-acetylcyclohexanone and 2-acetylcyclopentanone are both alkylated exclusively at the methyl group. In general, the ease of alkylation follows the order $C_6H_5CH_2— > CH_3— > —CH_2—$.

$$CH_3COCH_2COCH_3 \xrightarrow[\text{liq. } NH_3]{\text{2 equivs. } KNH_2} CH_3CO\bar{C}HCO\bar{C}H_2 \longleftrightarrow CH_3C(O^-){=}CH{-}C(O^-){=}CH_2$$

$$\xrightarrow[\text{(2) } H_3O^+]{\text{(1) } C_4H_9Br\text{, ether}} CH_3COCH_2COCH_2C_4H_9 \quad (82\%)$$

$$CH_3COCH_2CHO \longrightarrow CH_3COCH{=}CH\bar{O}\overset{+}{Na} \xrightarrow{KNH_2,\ NH_3} CH_2{=}C(O^-){-}CH{=}CH{-}O^- \xrightarrow[\text{(2) } H_3O^+]{\text{(1) } C_4H_9Br} C_4H_9CH_2COCH_2CHO \quad (72\%) \qquad (1.17)$$

The reaction can be applied equally well to β-keto aldehydes and β-keto esters, and provides a useful alternative route to 'mixed' Claisen ester condensation products. Reactions with β-keto aldehydes are generally effected by treating the monosodium salt of the aldehyde with alkali amide, to prevent self-condensation of the aldehyde (1.17). With β-keto esters it is advantageous to generate the dianion by reaction with two equivalents of lithium diisopropylamide, thus avoiding wasteful aminolysis of the ester by sodium amide. The dianions produced give γ-alkylated products in high yield with a wide range of alkylating agents (Huckin and Weiler, 1974).

Dianions derived from β-keto sulphoxides are also alkylated at the γ-carbon atom. The auxiliary sulphoxide group can be removed after alkylation by pyrolysis (see p. 99) or reduction, to give αβ-unsaturated ketones or α-alkylated ketones (1.18) (Russell and Ochrymowycz, 1969; Grieco and Pogonowski, 1974, 1975). Reduction of the substituted β-keto sulphoxides obtained by condensation of esters with the anion of dimethylsulphoxide and subsequent alkylation, provides a good route for the conversion of esters into a wide variety of ketones (Gassmann and Richmond, 1966) (1.19). In many reactions β-keto sulphones, prepared by reaction of the methylsulphonyl carbanion with esters, or by oxidation of the sulphoxides, are superior to the β-keto sulphoxides,

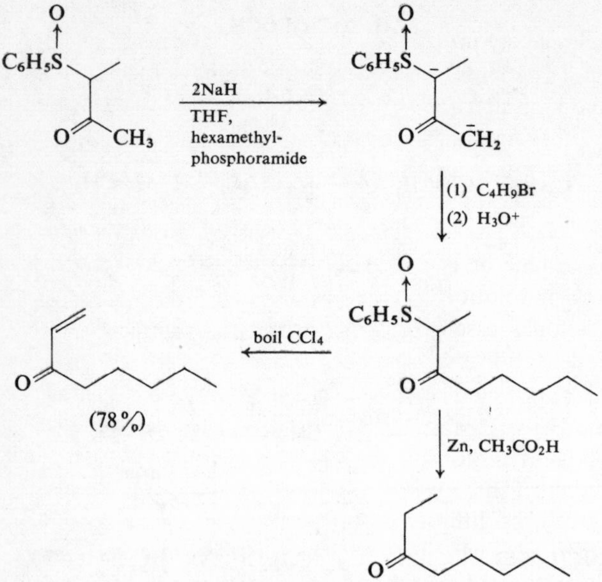

$C_6H_5S(\rightarrow O)$–CH(CH_3)–CO–CH_3 $\xrightarrow[\text{THF, hexamethylphosphoramide}]{2NaH}$ $C_6H_5S(\rightarrow O)$–$\bar{C}$(CH_3)–CO–$\bar{C}H_2$

(1) C_4H_9Br
(2) H_3O^+

boil CCl_4

(78%)

Zn, CH_3CO_2H

(1.18)

$$C_6H_5CO_2C_2H_5 \xrightarrow[\text{CH}_3\text{SOCH}_3,\ \text{then H}_3\text{O}^+]{\text{CH}_3\text{SOCH}_2^-\ \text{Na}^+} C_6H_5COCH_2SOCH_3$$

(1) 2NaH, dimethylformamide
(2) CH_3I

$$C_6H_5COCH(CH_3)_2 \xleftarrow{\text{Zn, CH}_3\text{CO}_2\text{H}} C_6H_5COC(CH_3)_2SOCH_3 \quad (1.19)$$

particularly where subsequent oxidations or reductions are to be effected elsewhere in the molecule (House and Larson, 1968).

1.4. Alkylation of ketones

Alkylation of monofunctional carbonyl compounds, aldehydes, ketones and esters, is more difficult than that of the 1,3-dicarbonyl compounds discussed above. As can be seen from Table 1.1, a methyl or methylene group which is activated by only one carbonyl, ester or cyano group requires a stronger base than sodium ethoxide or methoxide to convert it into the enolate anion in high enough concentration to be useful for subsequent alkylation. Alkali metal salts of tertiary alcohols, such as t-butanol or t-amyl alcohol, in solution or suspension in the corresponding alcohol or in an inert solvent, have been used with success, but suffer from the disadvantage that they are not sufficiently basic to convert the ketone completely into the enolate anion, thus allowing the possibility of an aldol condensation between the anion and unchanged carbonyl compound. An alternative procedure, which largely obviates this difficulty, is to use a much stronger base which will convert the compound completely into the anion. Typical bases of this type are sodium and potassium amide, sodium hydride, lithium diisopropylamide and sodium and potassium triphenylmethyl, in such solvents as ether, benzene, dimethoxyethane or dimethylformamide. The alkali metal amides are often used in solution in liquid ammonia. Although these bases can convert ketones essentially quantitatively into their enolate anions, aldol condensation may again be a difficulty with sodium hydride or sodamide in inert solvents, because of the insolubility of the reagents. Formation of the anion takes place only slowly in the heterogeneous reaction medium and both the ketone and the enolate ion are present at some point. This difficulty does not arise with the lithium dialkylamides, such as lithium diisopropylamide or lithium 2,2,6,6-tetramethylpiperidide, which are soluble in non-polar solvents, and these bases are being used increasingly for the generation of enolate anions.

O; H_5C_6; (1) NaH, benzene, reflux; (2) CH_2Br; O; H_5C_6; (88%)

CO_2CH_3; $LiN(iso\text{-}C_3H_7)_2$; THF, −78°C; CO_2CH_3; Br

(90%); CO_2CH_3; repeat using C_2H_5Br; CO_2CH_3 (1.20)

Intramolecular alkylation of ketones can be used to prepare cyclic compounds, as in the synthesis of 9-methyl-1-decalone, and in the key step in the synthesis of the sesquiterpene hydrocarbon seychellene (1.21).

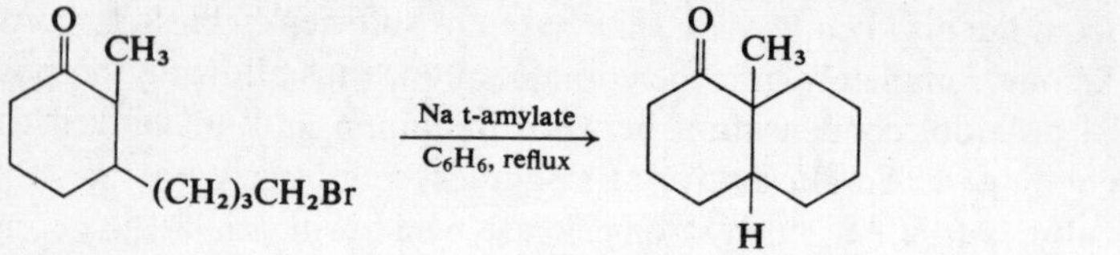

Na t-amylate
C_6H_6, reflux

$\bar{C}H_2SOCH_3$
CH_3SOCH_3

(90%)

(1.21)

Tos = H_3C–C_6H_4–SO_2

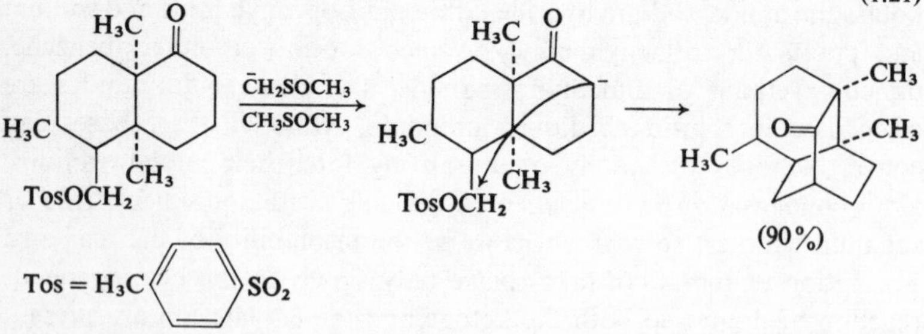

A common side-reaction in the direct alkylation of ketones is the formation of di- and poly-alkylated products through interaction of the original enolate with the monoalkylated product (1.22). This difficulty

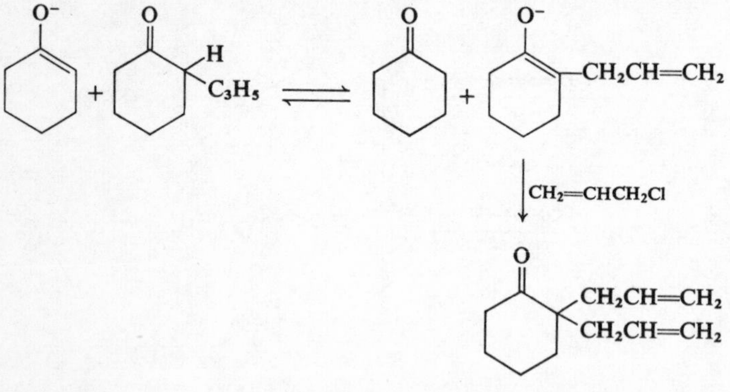

$CH_2{=}CHCH_2Cl$

(1.22)

can be avoided to some extent by adding a solution of the enolate in dimethoxyethane to a large excess of the alkylating agent. The enolate may thereby be rapidly consumed before equilibration with the alkylated ketone can take place. Nevertheless, formation of polysubstituted products is a serious problem in the direct alkylation of ketones and often results in decreased yields of the desired monoalkyl compound.

Alkylation of symmetrical ketones and of ketones which can enolise in one direction only can, of course, give only one mono-*C*-alkylated product. *O*-Alkylation usually does not take place to any appreciable extent with simple ketones. With unsymmetrical ketones, however, two different monoalkylated products may be formed by way of the two possible structurally isomeric enolates (1.23).

$$\mathrm{RCH_2COCH_2R'} \underset{}{\overset{\text{base}}{\rightleftharpoons}} \mathrm{RCH{=}C(O^-){-}CH_2R'} + \mathrm{RCH_2C(O^-){=}CHR'} \xrightarrow{\text{AlkX}} \mathrm{RCH(Alk.)COCH_2R'} + \mathrm{RCH_2COCH(Alk.)R'} \quad (1.23)$$

If one of the isomeric enolate anions is stabilised by conjugation with another group such as cyano, nitro or ethoxycarbonyl, then for all practical purposes only this stabilised anion is formed and alkylation takes place at the position activated by both groups. Even an α-phenyl or an α-vinyl group provides sufficient stabilisation of the resulting anion to direct substitution into the adjacent position (1.24).

$$\mathrm{C_6H_5CH_2COCH_3} \xrightarrow[\text{dimethoxy-ethane}]{(\mathrm{C_6H_5})_3\mathrm{CK}} \mathrm{C_6H_5CH{=}C(O^-)CH_3} + \mathrm{C_6H_5CH_2C(O^-){=}CH_2} \xrightarrow{\mathrm{CH_3I}} \underset{(93\%\text{ of product})}{\mathrm{C_6H_5CH(CH_3)COCH_3}} + \underset{(<1\%\text{ of product})}{\mathrm{C_6H_5CH_2COCH_2CH_3}} \quad (1.24)$$

Alkylation of unsymmetrical ketones bearing only α-alkyl substituents, however, generally leads to mixtures containing both α-alkylated products. The relative amounts of the two products depend on the structure of the ketone and may also be influenced by experimental factors such as the nature of the cation and the solvent (see Table 1.2). In the presence of the free ketone or an equivalent proton donor such as a protic solvent, equilibration of the enolate ions can take place (1.25). Thus, if the enolate is prepared by slow addition of the base to the ketone, or if an excess of ketone remains, the equilibrium mixture of enolate anions is obtained. Slow addition of the ketone to an excess of a strong base in an aprotic solvent, on the other hand, leads to the kinetic mixture of enolates; under these conditions the ketone is converted essentially quantitatively into the anion, and equilibration does not occur. The composition of mixtures of enolate ions formed under

TABLE 1.2 *Composition of mixtures of enolate anions generated from the ketone and a triphenylmethyl metal derivative in dimethoxyethane* (House, 1967)

Ketone, cation and reaction conditions	Enolate anion composition %	
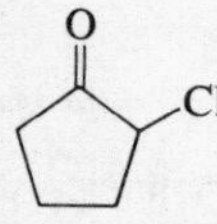 2-methylcyclopentanone	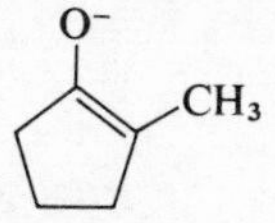	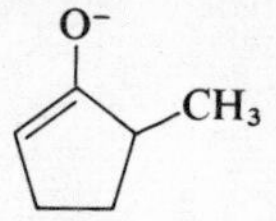
K^+ (kinetic control)	55	45
K^+ (equil. control)	78	22
Li^+ (kinetic control)	28	72
Li^+ (equil. control)	94	6
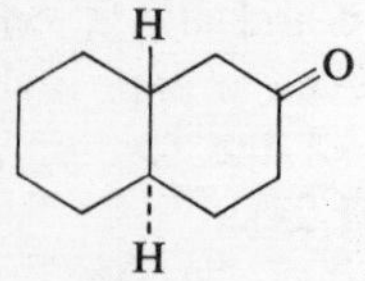 trans-2-decalone	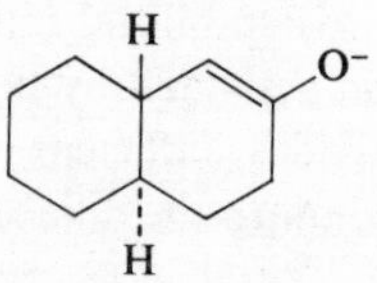	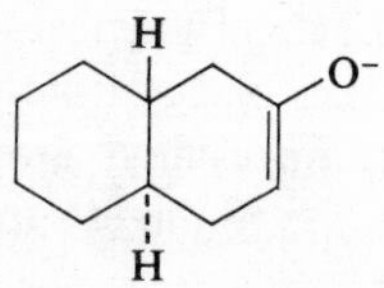
Li^+ (kinetic control)	13	87
Li^+ (equil. control)	53	47
$C_4H_9CH_2COCH_3$	$C_4H_9CH{=}C(O^-){-}CH_3$	$C_4H_9CH_2C(O^-){=}CH_2$
K^+ (kinetic control)	46	54
K^+ (equil. control)	58	42
Li^+ (kinetic control)	30	70
Li^+ (equil. control)	87	13

conditions of kinetic control differs substantially from that of mixtures formed under equilibrium conditions. In general, enolate mixtures formed under kinetic conditions contain more of the less highly substituted enolate than the equilibrium mixture, indicating that the less

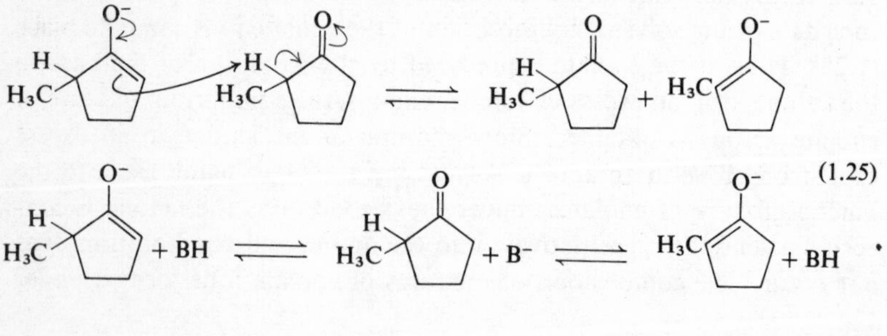

$$\text{enolate} + BH \rightleftharpoons \text{ketone} + B^- \rightleftharpoons \text{enolate} + BH \qquad (1.25)$$

hindered α-protons are removed more rapidly by the strong base (House, 1967). This is especially true of five- and six-membered cyclic ketones, where the more highly substituted enolate is favoured at equilibrium. However, whichever method is used, mixtures of both structurally isomeric enolates are generally obtained, and mixtures of products result on alkylation. Di- and tri-alkylation products may also be formed by equilibration of the monoalkylated product with the original anion (see p. 14) and it is not always easy to isolate the pure monoalkylated derivative in good yield from the resulting complex mixtures.

A number of methods have been used to improve selectivity in the alkylation of unsymmetrical ketones and to reduce the amount of polyalkylation. One widely used procedure is to introduce temporarily an activating group at one of the α-positions to stabilise the corresponding enolate anion; this group is removed later after the alkylation has been effected. Common activating groups used for this purpose are the ethoxycarbonyl, ethoxyoxalyl and formyl groups. Thus, to prepare 2-methylcyclohexanone from cyclohexanone the best procedure is to go through the 2-ethoxycarbonyl derivative, which is easily obtained from the ketone by reaction with ethyl carbonate, or by condensation with diethyloxalate followed by decarbonylation. Conversion into the enolate ion with a base such as sodium ethoxide takes place exclusively at the doubly activated position. Methylation with methyl iodide, and removal of the β-ketoester group with acid or base gives 2-methylcyclohexanone, free from polyalkylated products.

$$\text{cyclohexanone} \xrightarrow[C_2H_5OH]{(CO_2C_2H_5)_2,\ NaOC_2H_5} \text{2-}COCO_2C_2H_5 \xrightarrow[\text{borax}]{\text{heat}} \text{2-}CO_2C_2H_5$$

$$\text{2-}CO_2C_2H_5 \xrightarrow[(2)\ CH_3I]{(1)\ NaOC_2H_5,\ C_2H_5OH} \text{2-}CH_3\text{-2-}CO_2C_2H_5 \xrightarrow[HCl]{\text{boil}} \text{2-}CH_3 \qquad (1.26)$$

Another useful technique is to block one of the α-positions by introduction of a removable substituent which *prevents* formation of the corresponding enolate. A widely used method is acylation with ethyl formate and transformation of the resulting formyl or hydroxymethylene substituent into a group that is stable to base, such as an enamine, an enol ether or an enol thioether (1.27). An example of this procedure is shown below in the preparation of 9-methyl-1-decalone from *trans*-1-decalone. Direct alkylation of this compound gives mainly the 2-alkyl derivative (1.28). A useful alternative procedure is alkylation of the

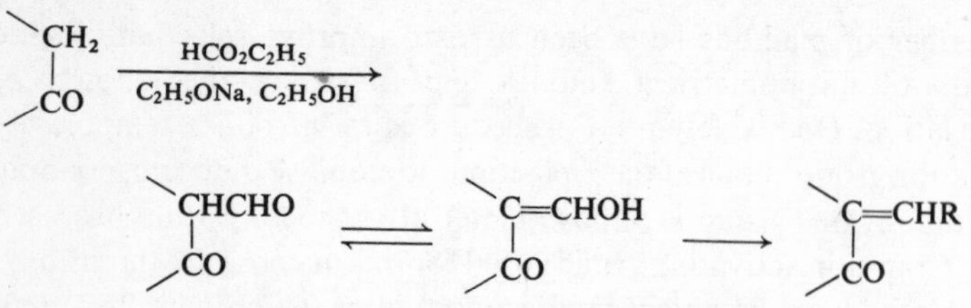

CH_2–CO $\xrightarrow[C_2H_5ONa,\ C_2H_5OH]{HCO_2C_2H_5}$ CHCHO–CO ⇌ C=CHOH–CO ⟶ C=CHR–CO (1.27)

$R = —N(CH_3)C_6H_5$, $—OCH(CH_3)_2$, or $—SC_4H_9$

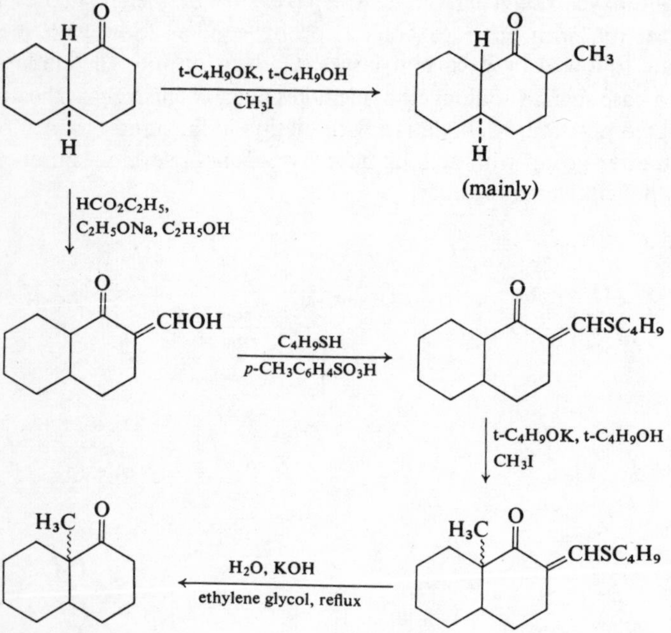

$\xrightarrow[CH_3I]{t\text{-}C_4H_9OK,\ t\text{-}C_4H_9OH}$ (mainly)

$HCO_2C_2H_5$, C_2H_5ONa, C_2H_5OH

(1.28)

CHOH $\xrightarrow[p\text{-}CH_3C_6H_4SO_3H]{C_4H_9SH}$ $CHSC_4H_9$

$t\text{-}C_4H_9OK$, $t\text{-}C_4H_9OH$, CH_3I

H_3C … $CHSC_4H_9$ $\xrightarrow[\text{ethylene glycol, reflux}]{H_2O,\ KOH}$ H_3C

(78% mixture of *cis*- and *trans*)

dianion prepared from the formyl derivative with potassium amide (p. 11).

Specific enolate anions may also be obtained from unsymmetrical ketones by reaction of the structurally specific enol acetates with two equivalents of methyl-lithium in dimethoxyethane (House and Trost, 1965). Under these conditions the enol acetate is converted into the corresponding lithium enolate and there is no isomerisation to the alternative enolate, as long as care is taken to ensure the absence of proton donors such as an alcohol or excess of ketone. Reaction with an alkyl halide then affords a specific monoalkylated ketone, accompanied, usually, by some dialkylated product (1.29). A disadvantage of this

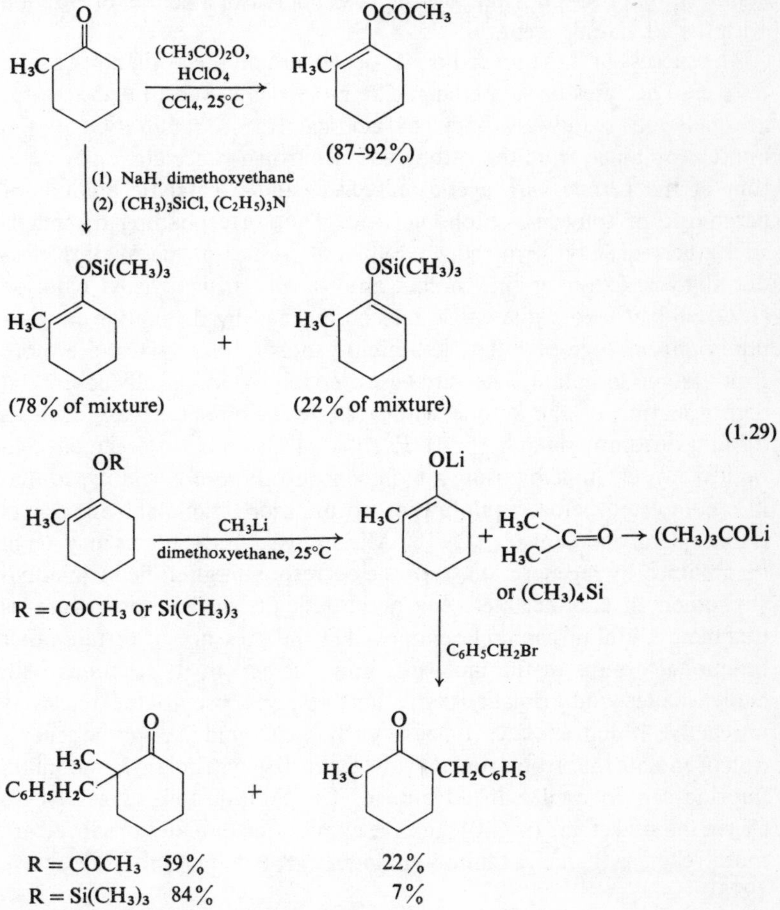

O
H_3C
$(CH_3CO)_2O$, $HClO_4$
CCl_4, 25°C
$OCOCH_3$
H_3C
(87–92%)
(1) NaH, dimethoxyethane
(2) $(CH_3)_3SiCl$, $(C_2H_5)_3N$

$OSi(CH_3)_3$
H_3C
+
$OSi(CH_3)_3$
H_3C
(78% of mixture)
(22% of mixture)

(1.29)

OR
H_3C
CH_3Li
dimethoxyethane, 25°C
OLi
H_3C
+ $(H_3C)_2C{=}O$ → $(CH_3)_3COLi$
or $(CH_3)_4Si$
R = $COCH_3$ or $Si(CH_3)_3$
$C_6H_5CH_2Br$

O
H_3C
$C_6H_5H_2C$
+
O
H_3C
$CH_2C_6H_5$

R = $COCH_3$ 59% 22%
R = $Si(CH_3)_3$ 84% 7%

procedure is that t-butoxide ion is formed as by-product and may bring about unwanted side-reactions, including formation of polyalkylated products. To overcome this difficulty the corresponding trimethylsilyl ethers are often used instead of the enol acetates (Stork and Hudrlik, 1968); reaction of the purified ethers with methyl-lithium generates the specific lithium enolates along with inert tetramethylsilane (see also Benkley and Heathcock, 1975). It is rarely possible to obtain completely selective alkylation of the enolates formed by either method, because as soon as some monoalkyl ketone is formed in the reaction mixture it can bring about equilibration of the original enolate. This problem is lessened by using the covalent *lithium* enolates which give maximum stabilisation of the enolate, consistent with a reasonable rate of reaction with the alkylating agent.

The success of these procedures is dependent on the availability of the pure enol acetates or silyl ethers. The more highly substituted acetates are generally readily available by acid-catalysed equilibration of the mixture obtained from the ketone with isopropenyl acetate, or by reaction of the ketone with acetic anhydride and a catalytic amount of perchloric or toluene-*p*-sulphonic acid. The corresponding trimethylsilyl ethers usually form the predominant isomer in the mixture produced by reaction of the enolate anions with trimethylsilyl chloride (1.29) and in favourable cases may be purified by distillation or gas–liquid chromatography. The less highly substituted enolates are more troublesome to obtain. They are best prepared by kinetically controlled deprotonation of the ketone with a sterically hindered base such as lithium diisopropylamide (p. 4). Even so, alkylation has to be effected rapidly, for example by using a high concentration of reactants, so that it is completed before equilibration to the more stable enolate occurs (House, Gall and Olmstead, 1971). Alternatively, the enolates may again be obtained by regeneration from the corresponding purified trimethylsilyl ethers or enol acetates. A general difficulty in these procedures is that methyl-lithium cannot be employed in the presence of certain other functional groups in the molecule, and, further, in its reactions with enol acetates and trimethylsilyl ethers it gives rise to the relatively unreactive lithium enolates. It has recently been found that for the generation of enolate ions from trimethylsilyl ethers benzyltrimethylammonium fluoride can be used with advantage. The fluoride ion serves well to cleave the silyl ethers (p. 310), and the ammonium enolates produced are more reactive than the lithium analogues (Kuwajima and Nakamura, 1975).

Silyl enol ethers can be used in another way to achieve specific α-methylation of aldehydes and ketones. Cyclopropanation of the silyl enol ethers of saturated aldehydes and ketones by the Simmons–Smith procedure (p. 77) and alkaline hydrolysis of the resulting silyloxocyclopropanes leads to the α-methyl derivative of the carbonyl compound (Conia and Girard, 1973 (1.30)). The method can also be used to alkylate αβ-unsaturated ketones. With simple cyclohexenones the α′-methyl derivative is obtained, but in the steroid series either the α- or α′-derivative can be made depending on how the silyl enol ether is prepared. This procedure is more specific than direct methylation of the cross-conjugated dienolates formed from the enones with lithium amide bases (p. 24), and leads exclusively to monomethylated ketone (Girard and Conia, 1974).

$$(CH_3)_2CHCHO \xrightarrow[(C_2H_5)_3N,\ \text{dimethylformamide}]{(CH_3)_3SiCl} (CH_3)_2C{=}CHOSi(CH_3)_3 \xrightarrow{CH_2I_2,\ Zn\text{–}Ag} \text{H}_3\text{C},\text{H}_2\text{C}\text{-cyclopropane}(\text{H})(OSi(CH_3)_3) \xrightarrow{NaOH,\ CH_3OH} (CH_3)_3CCHO$$

$$\text{cyclohexenone} \xrightarrow[(CH_3)_3SiCl,\ THF]{LiN(C_2H_5)_2} \text{OSi(CH}_3)_3\text{-cyclohexadiene} \xrightarrow{CH_2I_2,\ Zn\text{–}Ag} \text{OSi(CH}_3)_3\text{-bicyclic cyclopropane} \xrightarrow{NaOH,\ CH_3OH} \text{6-methylcyclohex-2-enone} \quad (1.30)$$

A promising new method for regioselective alkylation of unsymmetrical ketones proceeds from the corresponding oximes (Kofron and Yeh, 1976). Lithium derivatives of oximes are formed preferentially at the α-carbon atom which is *syn* to the oxygen, presumably as a result of chelate formation. Reaction with an alkyl halide then occurs only on that side to give a specific α-alkyl derivative. Directed aldol condensations are also possible. Thus dilithiocyclohexanone oxime and acetone gave only 2-isopropylidenecyclohexanone in 48 per cent yield, after cleavage of the first-formed oxime.

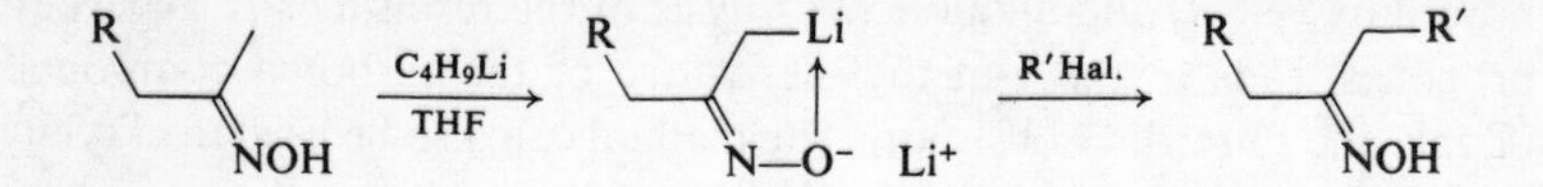

Specific enolate ions can also be obtained by reduction of $\alpha\beta$-unsaturated ketones, or of α-bromo or α-acetoxy ketones, with lithium in liquid ammonia (cf. p. 446). Alkylation then affords an α-alkyl derivative of the *saturated* ketone, which may not be the same as that obtained by direct base-catalysed alkylation of the saturated ketone. For example, alkylation of 2-decalone in presence of base generally leads to 3-alkyl derivatives, but by proceeding from the corresponding 1(9)-enone the 1-alkyl compound can be obtained readily (Stork, Rosen, Golman, Coombs and Tsuji, 1965) (1.31). The success of this method depends on the fact that in liquid ammonia the alkylation step

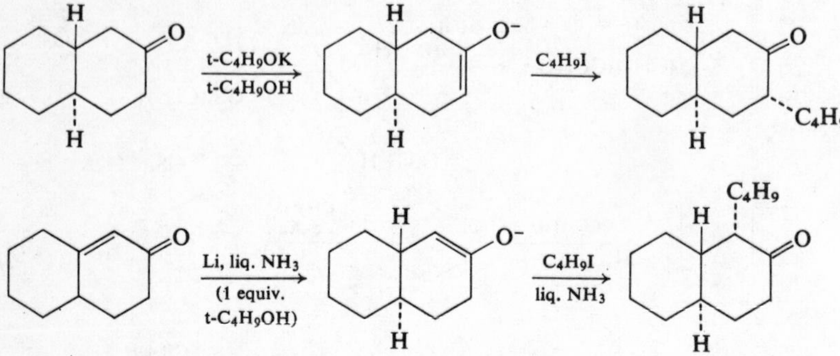

(1.31)

is faster than equilibration of the initially formed enolate anion with the more stable isomer. Again it is essential to use the *lithium* enolates; with the sodium and potassium salts, equilibration takes place and mixtures of alkylated products are obtained.

Alkylation of $\alpha\beta$-unsaturated ketones with alkali metal salts of tertiary alcohols as base follows a different course to give either the α-alkyl-$\alpha\beta$-unsaturated ketone or the $\alpha\alpha$-dialkyl-$\beta\gamma$-unsaturated ketone, depending on the reaction conditions, although in many cases mixtures of the two products are obtained (Conia, 1963; Ringold and Malhotra, 1962). The reaction has been widely used in the synthesis of α-alkylated

(1) $t\text{-}C_4H_9OK$, $t\text{-}C_4H_9OH$; (2) CH_3I added rapidly → (65%) + (3%)

(1.32)

(1) $t\text{-}C_4H_9OK$, $t\text{-}C_4H_9OH$; (2) CH_3Cl added over 20 h → (40%) + (12%)

$$-\overset{|}{C}H-CH=CH-\overset{O}{\overset{\|}{C}}- \xrightleftharpoons{\text{base (slow)}} \left[-\overset{-}{\underset{|}{C}}-CH=CH-\overset{O}{\overset{\|}{C}}- \longleftrightarrow -\underset{|}{C}=CH-\overset{-}{C}H-\overset{O}{\overset{\|}{C}}-\right] \xrightarrow{CH_3I}$$

$$\left[-\overset{-}{\underset{|}{C}}-CH=\underset{CH_3}{\underset{|}{C}}-\overset{O}{\overset{\|}{C}}- \longleftrightarrow -\underset{|}{C}=CH-\overset{-}{\underset{CH_3}{\underset{|}{C}}}-\overset{O}{\overset{\|}{C}}-\right] \xrightleftharpoons{\text{base (fast)}} -\underset{|}{C}=CH-\underset{CH_3}{\underset{|}{C}H}-\overset{O}{\overset{\|}{C}}-$$

$$\downarrow H^+ \qquad\qquad \downarrow CH_3I$$

$$-\underset{|}{C}H-CH=\underset{CH_3}{\underset{|}{C}}-\overset{O}{\overset{\|}{C}}- \qquad -\underset{|}{C}=CH-\underset{CH_3}{\underset{|}{\overset{H_3C}{\overset{|}{C}}}}-\overset{O}{\overset{\|}{C}}- \qquad (1.33)$$

enones in the steroid series, and for the introduction of *gem*-dimethyl groups in the synthesis of natural products (1.32).

Reaction proceeds by initial formation of the delocalised anion which undergoes kinetically controlled alkylation at the α-position (compare p. 7) to give the monoalkyl-$\beta\gamma$-unsaturated ketone. The α-proton in this compound is removed by reaction with base more readily than the γ-proton in the starting material, since it is activated by the carbonyl group and an ethylenic double bond. The resulting anion, in the presence of excess alkylating agent, is again alkylated at the α-position to give the $\alpha\alpha$-dialkyl-$\beta\gamma$-unsaturated ketone. If further alkylation does not occur, however, the thermodynamically more stable $\alpha\beta$-unsaturated ketone gradually accumulates (1.33). In accordance with this scheme it is found that dialkylation is diminished by slow addition of the alkylating agent or by use of a less reactive alkylating agent (for example, methyl chloride instead of methyl iodide in the example), thus allowing isomerisation of the $\beta\gamma$-unsaturated ketone to the less acidic $\alpha\beta$-isomer to take place before alkylation.

A disadvantage of this procedure is that it generally gives mixtures of products, particularly in experiments aimed at the monoalkylated compound. A much better route to the latter through metalloenamines derived from cyclohexylimines or *N*,*N*-dimethylhydrazones has been described by Stork and Benaim (1971) (1.34). Dialkylation does not occur because transfer of a proton from the monoalkylated compound to the metalloenamine is a slow process.

CH_3 $C_6H_{11}N$ — slightly less than 1 equiv. of $LiN(iso\text{-}C_3H_7)_2$, THF, 8 h → Li^+ $C_6H_{11}\bar{N}$ CH_3

CH_3I ↓

CH_3 $C_6H_{11}N$ CH_3 — CH_3COOH, H_2O, reflux → CH_3 O CH_3 (90%)

(1.34)

Alkylation of $\alpha\beta$-unsaturated ketones can also be brought about at the α'-position by way of the kinetically preferred dienolate anions formed with lithium diisopropylamide or potassium t-butoxide in an aprotic solvent (1.35) (Stork and Benaim, 1971; Lee, McAndrews, Patel and Reusch, 1973; Nedelec, Gasc and Bucourt, 1974; see also p. 21). This reaction has been exploited in a useful route to 4-alkyl-2-cyclohexenones from cyclohexan-1,3-diones (Stork and Danheiser, 1973).

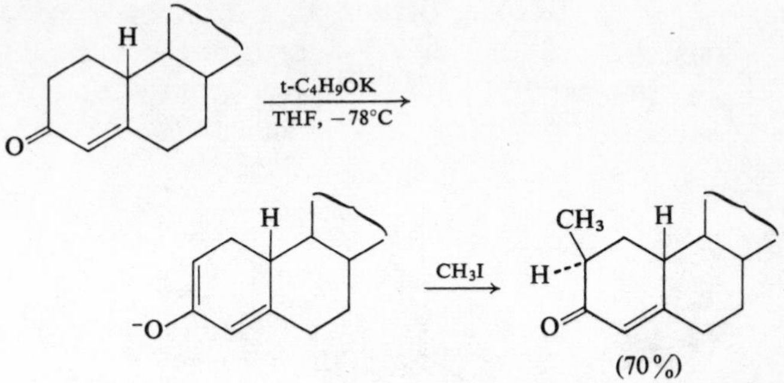

t-C_4H_9OK
THF, $-78°C$

CH_3I

(70%)

(1.35)

$\alpha\beta$-Unsaturated ketones can be converted into β-alkyl derivatives of the corresponding saturated ketone by conjugate addition of Grignard reagents or organocopper reagents, as discussed on p. 66. The enolates formed as intermediates in these reactions may be alkylated further to give $\alpha\beta$-disubstituted ketones. In the following example intramolecular alkylation is used to prepare the *cis*-9,10-dimethyldecalin ring system present in some sesquiterpenes.

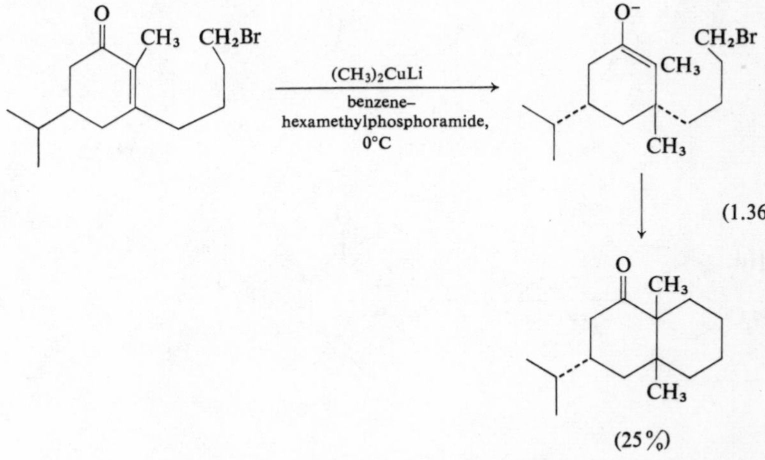

$(CH_3)_2CuLi$
benzene–hexamethylphosphoramide, 0°C

(25%)

(1.36)

Specific α,α-disubstituted ketones can be obtained by way of the butylthiomethylene derivates described on p. 18. Reduction with lithium–ammonia solution containing two equivalents of an alcohol affords a methyl substituted enolate which can be alkylated directly with primary alkylating agents. Cyclohexanone, for example, gave 2,2-dimethylcyclohexanone in 82 per cent yield (Coates and Sowerby, 1971).

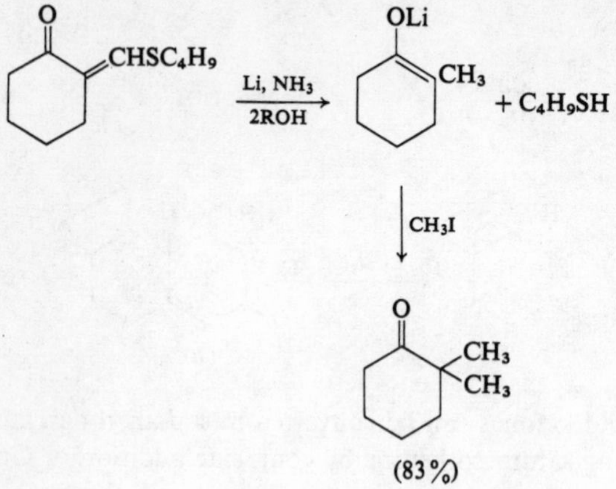

CHSC$_4$H$_9$ Li, NH$_3$ 2ROH OLi CH$_3$ + C$_4$H$_9$SH

CH$_3$I

CH$_3$ CH$_3$ (83%)

CHSC$_4$H$_9$ (CH$_3$)$_2$CuLi ether OM CH$_3$I dimethoxy ethane

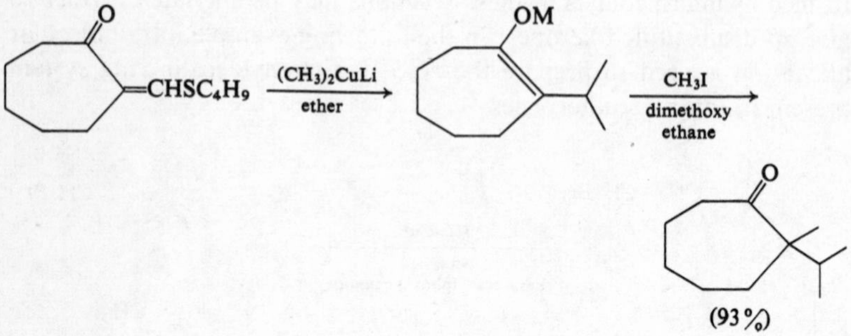

(93%)

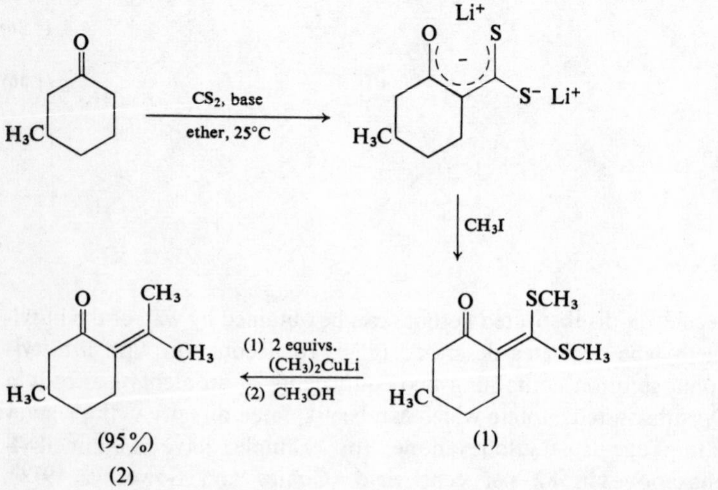

H$_3$C CS$_2$, base ether, 25°C Li$^+$ S S$^-$ Li$^+$ H$_3$C

CH$_3$I

CH$_3$ CH$_3$ H$_3$C (95%) (2) (1) 2 equivs. (CH$_3$)$_2$CuLi (2) CH$_3$OH SCH$_3$ SCH$_3$ H$_3$C (1) (1.37)

Alternatively, reaction with lithium dialkylcuprates affords α-substituted ketones with branched α-substituents. The copper enolates formed as intermediates in this reaction are very readily alkylated with primary alkyl halides to give α,α-dialkylated ketones (1.37) (Coates and Sandefur, 1974).

In a related reaction, α-dithiomethylene ketones (as 1), which are readily available from ketones and carbon disulphide (as exemplified in (1.37)), are converted into 2-isopropylidene or 2-t-butyl ketones by reaction with two or three equivalents of lithium dimethylcuprate. Pulegone (2), for example, was obtained from 3-methylcyclohexanone in 95 per cent yield (Corey and Chen, 1973).

The stereochemistry of the product obtained in the alkylation of cyclic ketones is important in synthesis but is not always easy to predict. If the product still contains a hydrogen atom attached to the carbon which has been alkylated, equilibration under the alkaline conditions of the reaction generally leads to formation of the more stable isomer, irrespective of the initial direction of attack on the enolate anion. Thus, in the methylation of cholestanone, the product is the 2-α-methyl compound in which the methyl group has the more stable equatorial conformation.

If the alkylated product has no hydrogen atom at the alkylated position, then the product is usually that formed by attack on the less hindered side of the enolate anion. This is particularly the case with relatively rigid ketones which cannot easily adopt different conformations. Hence, methylation of norcamphor (3) with sodium triphenylmethyl and methyl iodide leads almost entirely to the *exo* methyl derivative, even though equilibration studies show that the *endo* isomer is (slightly) more stable. Further alkylation of this methyl derivative with methylpentenyl chloride then gives the *endo*-methyl-*exo*-methylpentenyl derivative, again by attack of the alkylating agent from the less hindered *exo* side of the enolate ion (1.38). Reversing the order of the alkylations leads to the epimeric product, showing that the stereochemistry is controlled by the direction of attack on the enolate ion and not by the size of alkyl group.

If the enolate anion is reasonably flexible, both conformational factors and steric hindrance in the anion may influence the steric course of the alkylation, and it is often difficult to predict what the stereochemical result in a particular case will be. Mixtures of products often result.

Stereoelectronic effects have been invoked to rationalise the stereochemistry of alkylation of some cyclohexanone derivatives (House,

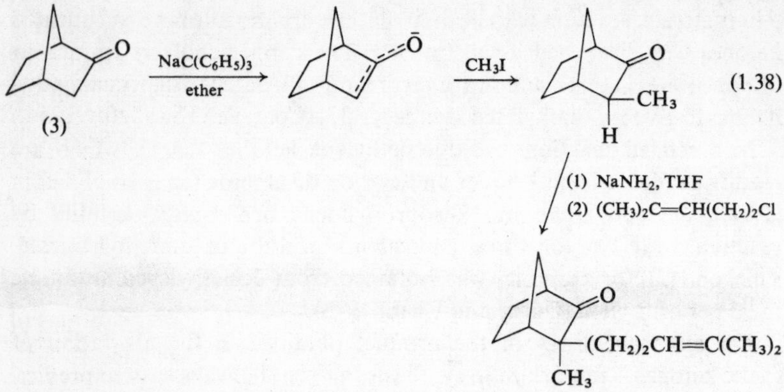

(3) $\xrightarrow[\text{ether}]{NaC(C_6H_5)_3}$ $\xrightarrow{CH_3I}$ (1.38)

(1) $NaNH_2$, THF
(2) $(CH_3)_2C{=}CH(CH_2)_2Cl$

$-(CH_2)_2CH{=}C(CH_3)_2$ CH_3

1967). Alkylation of an enolate anion should be energetically most favourable when the developing carbonyl group remains in a plane perpendicular to the developing carbon–carbon bond between the enolate and the alkylating agent. With this geometry the stabilising interaction of the π-orbital of the carbonyl group with the *p*-orbital at the α-carbon continues at a maximum while bonding with the alkylating agent is in progress (1.39). The formation of *cis*- and *trans*-4-t-butyl-2-ethylcyclo-

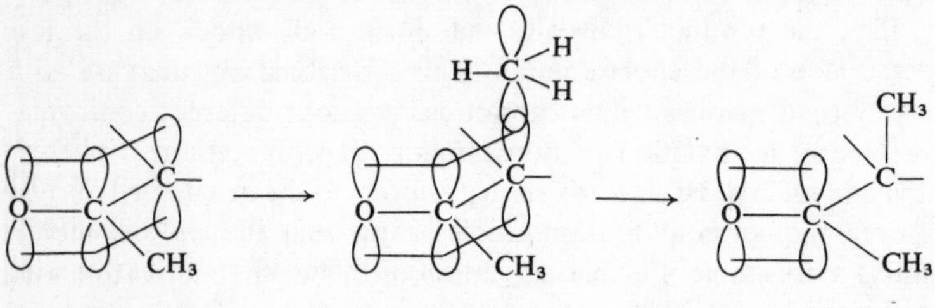

(1.39)

hexanone by alkylation of the enolate anion derived from 4-t-butyl-cyclohexanone is considered to take place by way of the partial chair and partial twisted boat transition states shown (1.40), in which the ethyl group is introduced perpendicular to the plane of the $>C{=}O$ bond. If a cyclohexanone derivative already has an alkyl substituent at the α-position, further alkylation at the same position takes place predominantly from the axial side, unless this would produce an unfavourable 1,3-diaxial interaction between alkyl substituents (House and Umen, 1973). Of course, if one side of the enolate is substantially more hindered than the other, alkylation will take place from the less hindered side as in (1.41).

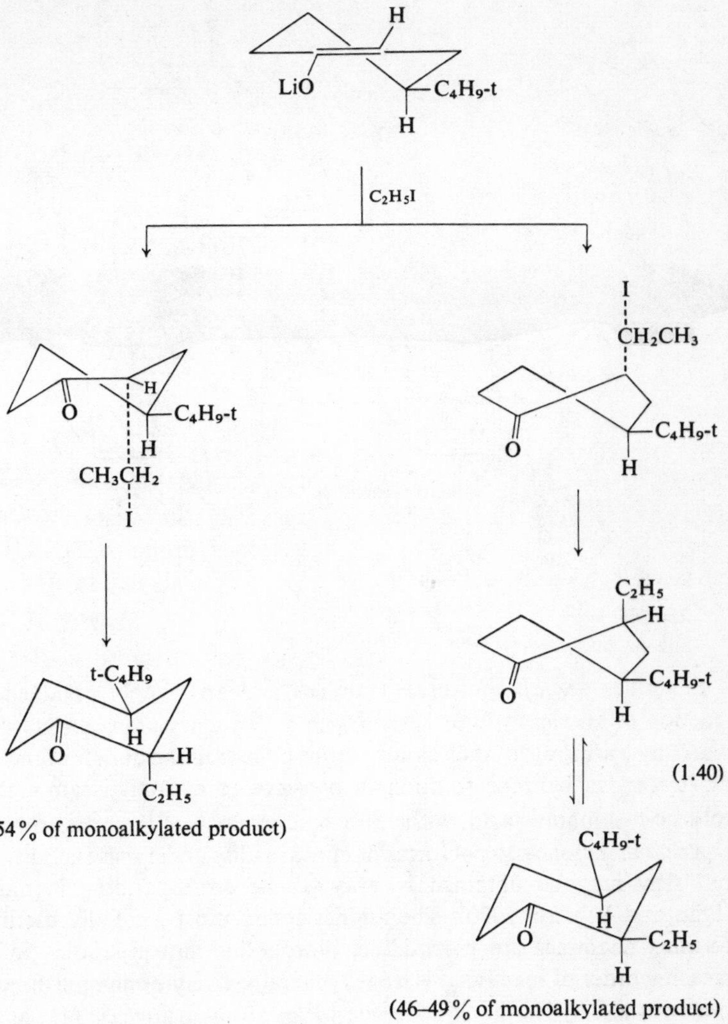

H
LiO
C_4H_9-t
H

C_2H_5I

H
O
C_4H_9-t
H
CH_3CH_2
I

I
CH_2CH_3
C_4H_9-t
O
H

C_2H_5
H
C_4H_9-t
O
H

t-C_4H_9
O
H
H
C_2H_5

(1.40)

(51–54 % of monoalkylated product)

C_4H_9-t
H
O
C_2H_5
H

(46–49 % of monoalkylated product)

1.5. The enamine and related reactions

The enamine reaction, originally interpreted by Robinson (1916) and reintroduced by Stork and his co-workers (Stork, Brizzolana, Landesman, Szmuszkovicz and Terrell, 1963; Szmuszkovicz, 1963; Cook, 1968), provides a valuable alternative method for the selective alkylation and acylation of aldehydes and ketones.

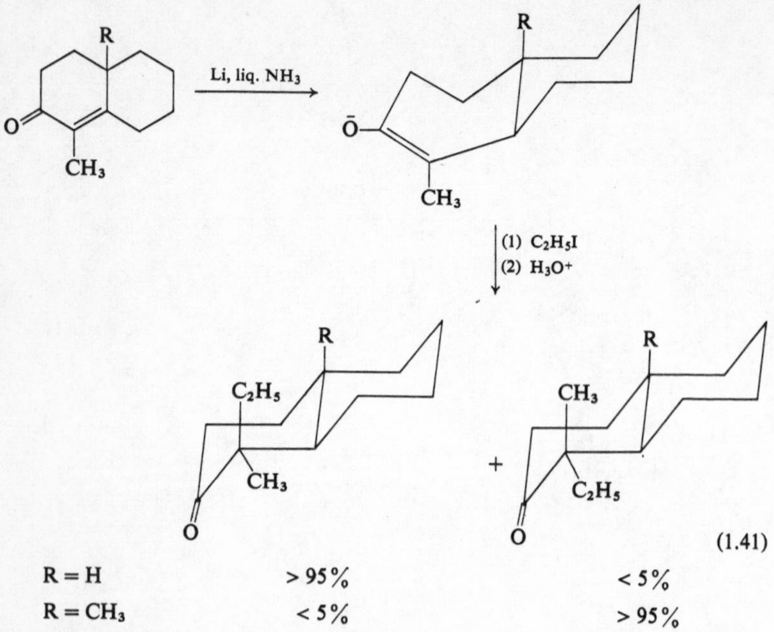

Li, liq. NH_3

(1) C_2H_5I
(2) H_3O^+

(1.41)

R = H	> 95%	< 5%
R = CH_3	< 5%	> 95%

Enamines are $\alpha\beta$-unsaturated amines and are simply obtained by reaction of an aldehyde or ketone with a secondary amine in presence of a dehydrating agent such as anhydrous potassium carbonate or, better, by heating in benzene solution in presence of a catalytic amount of toluene-*p*-sulphonic acid, with azeotropic removal of the water formed. Open chain ketones do not form enamines readily under these conditions, but they may be obtained by way of the corresponding ketimines (Pfau and Ribière, 1970). The amines found most generally useful in forming enamines are pyrrolidine, morpholine and piperidine in decreasing order of reactivity. Ketones generally give the enamine directly, but with aldehydes the *N*-analogue of an acetal is formed first and is converted into the enamine on distillation (1.42). All the steps of the reaction are reversible, and enamines are readily hydrolysed by water to reform the carbonyl compound. All reactions of enamines must therefore be conducted under strictly anhydrous conditions, but once reaction has been effected the modified carbonyl compound is easily liberated from the product by addition of water to the reaction mixture.

The usefulness of enamines in synthesis is due to the fact that there is some negative charge on the β-carbon atom which can therefore act as

$$RCH_2COR^1 + HNR^2R^3 \rightleftharpoons RCH_2C(R^1)(OH)NR^2R^3$$

$$RCH_2C(R^1)(OH)NR^2R^3 \rightleftharpoons RCH_2C(R^1){=}\overset{+}{N}R^2R^3\ OH^-$$

$$RCH_2C(R^1){=}\overset{+}{N}R^2R^3\ OH^- \rightleftharpoons RCH{=}C(R^1){-}NR^2R^3 \qquad (1.42)$$

$$\overset{|}{R}CHCHO + 2HNR^1R^2 \rightleftharpoons \overset{|}{R}CHCH(NR^1R^2)_2 \xrightarrow{\Delta} \overset{|}{R}C{=}CH{-}NR^1R^2$$

a nucleophile in reactions with alkyl and acyl halides and with electrophilic olefins (1.43). Reaction with alkyl halides, for example, leads irreversibly to *C*-alkylated and *N*-alkylated products. Subsequent hy-

$$>C{=}\overset{|}{C}{-}\ddot{N}< \;\longleftrightarrow\; >\bar{C}{-}\overset{|}{C}{=}\overset{+}{N}< \qquad (1.43)$$

drolysis of the *C*-alkylated iminium salt gives the alkylated ketone; the *N*-alkylated product is usually water soluble and unaffected by the hydrolysis (1.44).

$$RCH{=}C(R^1){-}\ddot{N}R^2R^3 \longleftrightarrow R\bar{C}H{-}C(R^1){=}\overset{+}{N}R^2R^3$$

$$\xrightarrow{CH_3I,\ C_6H_6,\ \text{reflux}} RCH{=}C(R^1){-}\overset{+}{N}(CH_3)R^2R^3\ I^- + RCH(CH_3){-}C(R^1){=}\overset{+}{N}R^2R^3\ I^- \qquad (1.44)$$

$$\xrightarrow{\text{hydrolysis}} RCH(CH_3){-}C(R^1){=}O$$

This procedure has a number of advantages over direct base-catalysed alkylation of aldehydes and ketones. Since no base or other catalyst is required, there is less tendency for wasteful self-condensation reactions of the carbonyl compound, and even aldehydes can be alkylated and acylated in good yield. A valuable feature of the enamine reaction is that in the alkylation of unsymmetrical ketones the product of reaction at the *less* substituted α-carbon atom is formed in preponderant amount, in contrast to base-catalysed alkylation, which usually gives a mixture of products. For example, reaction of the pyrrolidine enamine of 2-methylcyclohexanone with methyl iodide gives 2,6-dimethylcyclohexanone almost exclusively. This selectivity derives from the fact that the enamine from an unsymmetrical ketone consists mainly of the isomer in which the double bond is directed toward the less substituted carbon. The 'more substituted' enamine is destabilised by steric inhibition of the resonance involving the nitrogen lone pair and the double bond, caused by interference of the substituent (CH_3— in the example) (1.45) and the α-methylene group of the amine.
leads to high yields of alkylated aldehydes and ketones even with simple aliphatic alkyl halides (Stork and Dowd, 1963, 1974). It is found that many *imines* formed from aliphatic primary amines and enolisable

$$H_3C\text{-cyclohexanone} + \text{pyrrolidine} \xrightarrow[C_6H_6]{p\text{-}CH_3C_6H_4SO_3H} \text{enamine (85\% of mixture)} + \text{enamine (15\% of mixture)} \qquad (1.45)$$

Alkylation of enamines with alkyl halides generally proceeds in only poor yield because the main reaction is *N*-alkylation rather than *C*-alkylation. Good yields of alkylated products are obtained using reactive halides such as benzyl or allyl halides, and it is believed that in these cases there is migration of the substituent group from nitrogen to carbon. This may take place in some cases by an intramolecular pathway, resulting in rearrangement of allyl substituents or, in other cases, by dissociation of the *N*-alkyl derivative followed by irreversible *C*-alkylation. Fortunately, a modification of the original procedure provides a method which aldehydes and ketones undergo essentially complete enolisation on treatment with one equivalent of ethylmagnesium bromide in boiling tetrahydrofuran. The magnesium salts so formed react readily with primary and secondary alkyl halides to give, after hydrolysis, high yields of monoalkylated carbonyl compound. Even aldehydes are readily alkylated by this method (1.46). Allylic halides give products without rearrangement of the allyl group, and the new alkyl substituent is again introduced at the less substituted α-carbon of an unsymmetrical ketone.

$$CH_3CH(CH_3)CHO + (CH_3)_3CNH_2 \xrightarrow[\text{reflux}]{C_6H_6,} CH_3CH(CH_3)CH{=}NC(CH_3)_3$$

$$\xrightarrow[\text{reflux}]{C_2H_5MgBr,\ THF,} CH_3C(CH_3){=}CH{-}N(MgBr)C(CH_3)_3 \xrightarrow[(2)\ H_2O]{(1)\ C_6H_5CH_2Cl} \underset{(80\%)}{CH_3C(CH_3)(CH_2C_6H_5)CHO} \qquad (1.46)$$

A reaction of this type has been exploited in an asymmetric synthesis of 2-alkylcyclohexanones using the optically active amine (4). Reaction with cyclohexanone and treatment of the resulting imine with lithium diisopropylamide gives the chelated lithio derivative (5), which is

$$\text{cyclohexanone} + \underset{(4)}{H_2N{-}CH(C_6H_5)CH_2OCH_3} \xrightarrow{C_6H_6} \text{cyclohexylidene}{=}N{-}CH(C_6H_5)CH_2OCH_3$$

$$\xrightarrow[-20°C]{LiN(iso\text{-}C_3H_7)_2,\ THF,} \left[\underset{(5)}{\text{cyclohexenyl}{-}N(Li{\leftarrow}O(CH_3))CH(C_6H_5)CH_2}\right] \xrightarrow[(2)\ H_3O^+]{(1)\ RHal.} \text{2-R-cyclohexanone} \qquad (1.47)$$

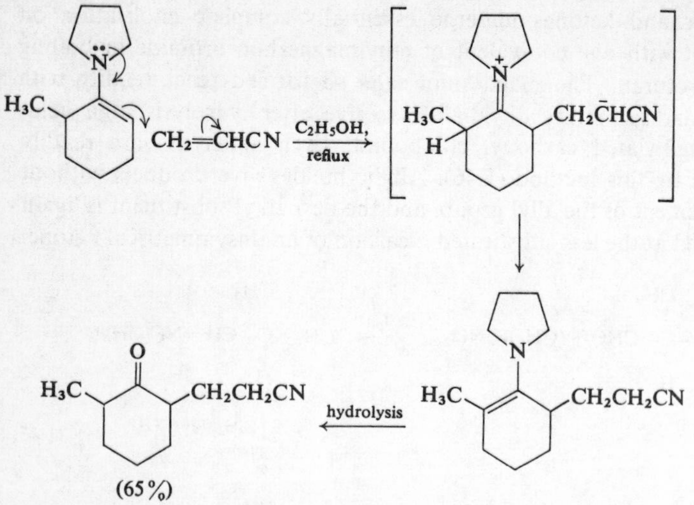

H_3C $CH_2{=}CHCN$ $\xrightarrow{C_2H_5OH,\ reflux}$ H_3C H N^+ $CH_2\bar{C}HCN$

(1.48)

O H_3C CH_2CH_2CN $\xleftarrow{hydrolysis}$ N H_3C CH_2CH_2CN

(65%)

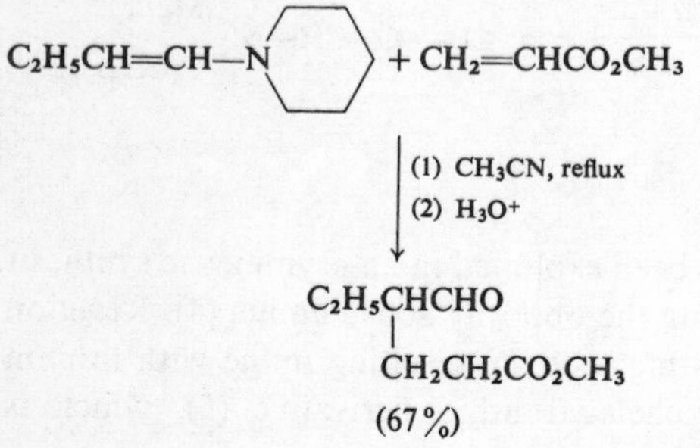

$C_2H_5CH{=}CH{-}N$ + $CH_2{=}CHCO_2CH_3$

(1) CH_3CN, reflux
(2) H_3O^+

C_2H_5CHCHO
$CH_2CH_2CO_2CH_3$
(67%)

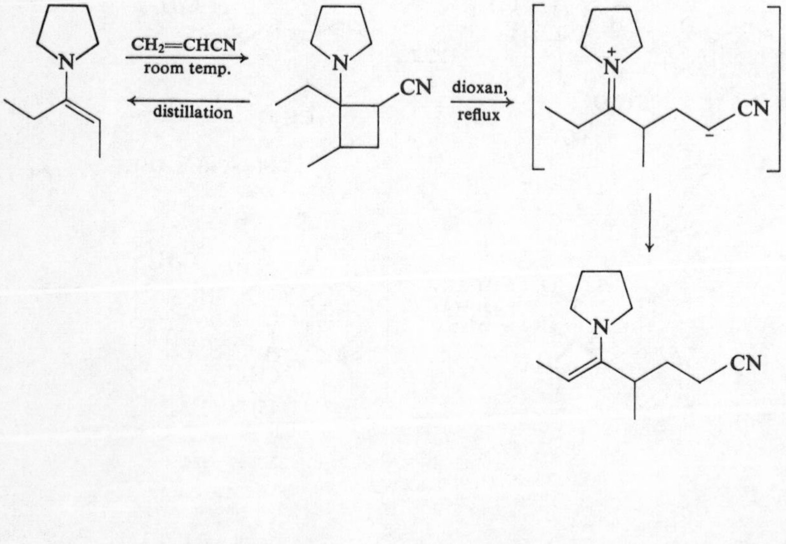

N $\xrightarrow{CH_2{=}CHCN,\ room\ temp.}$ N CN $\xrightarrow{dioxan,\ reflux}$ N^+ CN

$\xleftarrow{distillation}$

N CN

alkylated preferentially on the less hindered face. Subsequent hydrolysis affords 2-alkylcyclohexanones of high optical purity. Alkylation with ethyl iodide, for example, led to (*R*)-2-ethylcyclohexanone with, it is claimed, complete optical purity (Meyers, Williams and Druelinger, 1976).

Alkylation of enamines of aldehydes and ketones can also be effected with electrophilic olefins, such as $\alpha\beta$-unsaturated ketones, esters or nitriles to give high yields of monoalkylated carbonyl compounds, and the sequence provides a useful alternative to base-catalysed Michael addition. In these reactions *N*-alkylation is reversible and good yields of *C*-alkylated products are usually obtained. Reaction again takes place at the less substituted α-carbon atom (1.48).

According to Fleming and Harley-Mason (1964) reaction of enamines with acrylonitrile and some other electrophilic olefins forms cyclobutane derivatives which may be isolated from reaction at room temperature but which are converted into substituted enamines on warming. With acrylic aldehydes and $\alpha\beta$-unsaturated ketones, a dipolar addition product is first formed which affords the substituted enamine on heating, but at room temperature leads, reversibly, to a dihydropyran (Colonna, Fattuta, Risaliti and Russo, 1970).

Enamines also react readily with acid chlorides or anhydrides to give products which, on hydrolysis, afford β-diketones or β-ketoesters. Reaction at the nitrogen atom is again reversible, and good yields of *C*-acylated products are obtained. The morpholine enamine of cyclohexanone, for example, and heptanoyl chloride give 2-heptanoylcyclohexanone in 75 per cent yield. In these reactions triethylamine is often added to neutralise the hydrogen chloride formed which would otherwise combine with the enamine; alternatively, two equivalents of enamine may be used (1.49).

$$\underset{R^2CO{-}Cl}{R{-}\overset{H}{\overset{|}{C}}{=}\overset{R^1}{\overset{|}{C}}{-}\ddot{N}\langle} \longrightarrow R{-}\underset{R^2CO}{\underset{|}{\overset{H}{\overset{|}{C}}}}{-}\overset{R^1}{\overset{|}{C}}{=}\overset{+}{N}\langle\; Cl^- \quad (:N(C_2H_5)_3)$$

$$\downarrow$$

$$\underset{R^2CO}{\underset{|}{RCHCOR^1}} \xleftarrow{\text{hydrolysis}} R{-}\underset{R^2CO}{\underset{|}{C}}{=}\overset{R^1}{\overset{|}{C}}{-}N\langle \qquad (1.49)$$

Endocyclic enamines, prepared by oxidation of piperidines or pyrrolidines, or from acyclic imines (Evans and Domeier, 1974), are useful for the synthesis of more complex nitrogen heterocyclic systems, as in the alkaloids (Wenkert, 1968). The alkaloid mesembrine and the carbocyclic skeleton of the hasubanan alkaloids, for example, were obtained as shown in (1.50) (Evans, Bryan and Wahl, 1970; Stevens and Wentland, 1968).

OCH3 OCH3 N CH3 — O, ethylene glycol, 119°C → Ar N+ CH3 O− ⇌ Ar N+ CH2 O− ↓ CH3O OCH3 N H3C H O (56%) mesembrine

(1.50)

N CH3 — O, CH3CN, CH3CO2H, 80°C → O N CH3 (32%)

A promising alternative procedure described recently by Corey and Enders (1976) starts not from an imine or an enamine but from the dimethylhydrazone of an aldehyde or ketone. These compounds, on reaction with lithium diisopropylamide are readily converted into lithium derivatives which can be alkylated with alkyl halides, epoxides and carbonyl compounds. At the end of the sequence the dimethylhydrazone grouping is removed by oxidation with sodium periodate, liberating the alkylated aldehyde or ketone. Reactions take place easily under mild conditions and with high positional- and stereo- selectivity. Generally alkylation occurs at the less substituted α-position of an unsymmetrical ketone, and with cyclohexanone derivatives *axial* methylation is highly favoured. Thus, methyl pentyl ketone was converted into ethyl pentyl ketone in 95 per cent yield, and 2-methylcyclohexanone gave trans-2,6-dimethylcyclohexanone (1.51). With epoxides, γ-hydroxycarbonyl compounds and thence, by oxidation, 1,4-dicarbonyl compounds are obtained in high yield, and reaction with aldehydes and ketones leads to β-hydroxycarbonyl compounds, formed, in effect, by a 'directed' aldol condensation (1.51) (see also p. 52).

2-methylcyclohexanone N,N-dimethylhydrazone ($NN(CH_3)_2$, H_3C) $\xrightarrow[\text{(2) } CH_3I]{\text{(1) } LiN(iso\text{-}C_3H_7)_2,\ THF}$ 2,6-dimethyl hydrazone ($NN(CH_3)_2$, H_3C, CH_3) $\xrightarrow[CH_3OH,\ H_2O]{NaIO_4}$ 2,6-dimethylcyclohexanone (H_3C, O, CH_3)

(95%; 97% *trans*)

$$C_3H_7C(CH_3){=}NN(CH_3)_2 + CH_3CH{=}CHCHO \longrightarrow CH_3CH{=}CHCHOHCH_2COC_3H_7$$

(95% overall)

(1.51)

1.6. Reactions of α-thio carbanions

Because of its empty 3*d*-orbitals, which it is generally believed can interact with filled *p*-orbitals on an adjacent carbanion, or simply because of its electronegativity (compare Streitwieser and Williams, 1975), sulphur can stabilise a neighbouring carbanion, and carbanions derived from sulphur compounds are playing an increasingly important role in synthesis. The sulphur atom may be present in a sulphide, a sulphoxide or a sulphone, and may be removed from the product after reaction, if desired, by reduction (hydrogenolysis, pp. 455, 489), by hydrolysis to a carbonyl compound, or by elimination to give an olefin (see p. 99). Controlled rearrangement of the alkylated product before removal of the sulphur extends the scope of the synthetic sequence in many cases.

Alkyl sulphides are apparently not acidic enough to be converted easily into anions under experimentally useful conditions; the presence of another activating group, such as a second sulphur atom or a double bond, is generally desirable for convenient reaction. Thus, allyl sulphides (allyl *phenyl* sulphides are usually employed) are readily converted by butyl-lithium into carbanions which are alkylated with active alkylating agents such as methyl iodide or allyl or benzyl halides, mainly at the position α to the sulphur atom, although in some cases γ-alkylation is also observed. Reduction of the products with lithium and ethylamine removes the sulphur substituent to give the α-alkylated olefin. The sequence provides an excellent method for coupling allyl groups; squalene, for example, was synthesised from farnesyl bromide and farnesyl phenyl sulphide in 70 per cent yield. Coupling of two different

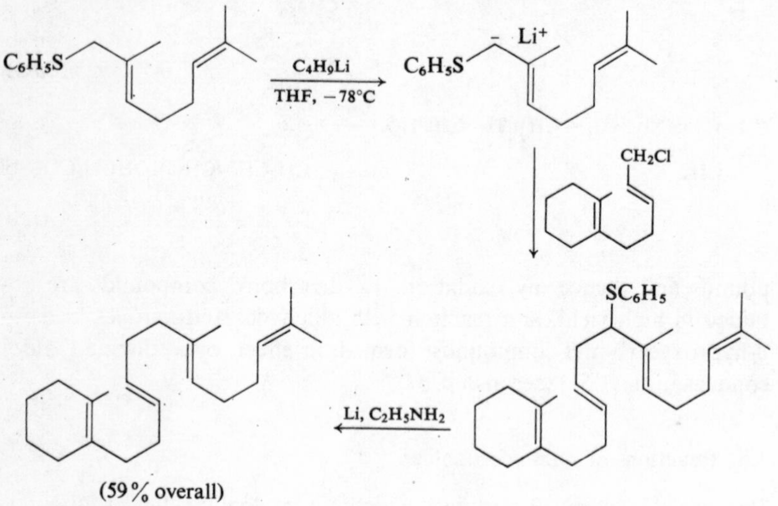

C_6H_5S — $\xrightarrow[\text{THF, }-78^\circ\text{C}]{C_4H_9Li}$ C_6H_5S $^-$ Li^+

CH_2Cl

SC_6H_5

$\xleftarrow{\text{Li, }C_2H_5NH_2}$

(59% overall)

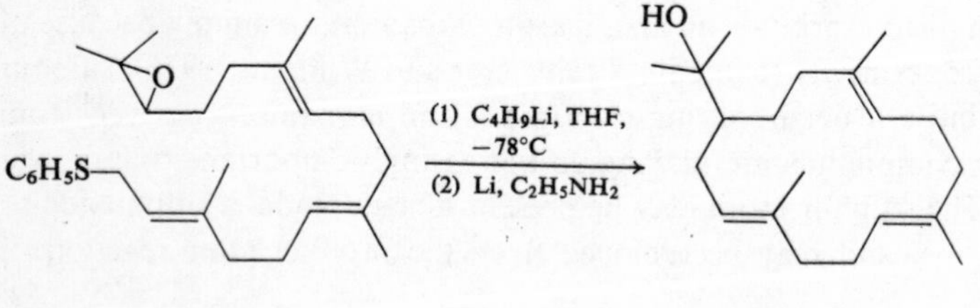

C_6H_5S — O — $\xrightarrow[\text{(2) Li, }C_2H_5NH_2]{\text{(1) }C_4H_9Li\text{, THF, }-78^\circ\text{C}}$ HO (1.52)

allyl groups is easily effected, as in (1.52). A practical advantage is that no dialkylation products are formed because the monoalkylated compound is not sufficiently acidic to form an anion under the reaction conditions. Helpfully, the stereochemistry of the double bond in the allylic sulphides is preserved in the products (Bielmann and Ducep, 1971). Alkylation of the allyl vinyl sulphide (6) may be combined with thio-Claisen rearrangement to give, after hydrolysis of the product, a γ-keto aldehyde as in (1.53) (see p. 158). In this sequence the anion (7) is equivalent to the specific enolate anion $^{-}CH_2COCH_2CH_2CHO$.

C_2H_5O (6) $\xrightarrow[\text{THF, }-78^\circ\text{C}]{s\text{-}C_4H_9Li}$ C_2H_5O (7) $\xrightarrow{C_5H_{11}Br}$ C_5H_{11} S C_2H_5O

heat in aqueous dimethoxy-ethane

C_5H_{11} CHO (66%) $\longleftarrow$ [C_5H_{11} H S OC_2H_5]

(1.53)

α-(Phenylthio)- ketones and -aldehydes are also useful in synthesis. They are easily formed by sulphenylation of the corresponding enolates (see p. 100) and are alkylated at the carbon bearing the sulphur atom. The alkylated products can be reductively cleaved with lithium and ethylamine, to give the α-alkylcarbonyl compound, or oxidised to the sulphoxides, which on elimination afford the α-alkyl-αβ-unsaturated carbonyl derivative (Coates, Pigott and Ollinger, 1974).

A promising route to cyclic α-(phenylthio)ketones by Diels–Alder addition of 2-methoxy-3-phenylthio-1,3-butadiene to dienophiles has been described recently by Trost and Bridges (1976). The α-phenylthio-ketone is produced initially in a masked form, allowing further selective transformations to be effected. Carvone (8), for example, was synthesised from methyl vinyl ketone as shown in (1.54). Selective alkylations may also be effected and, in a new reaction, diosphenols are produced regio-specifically by oxidation of the keto sulphide with lead tetra-acetate.

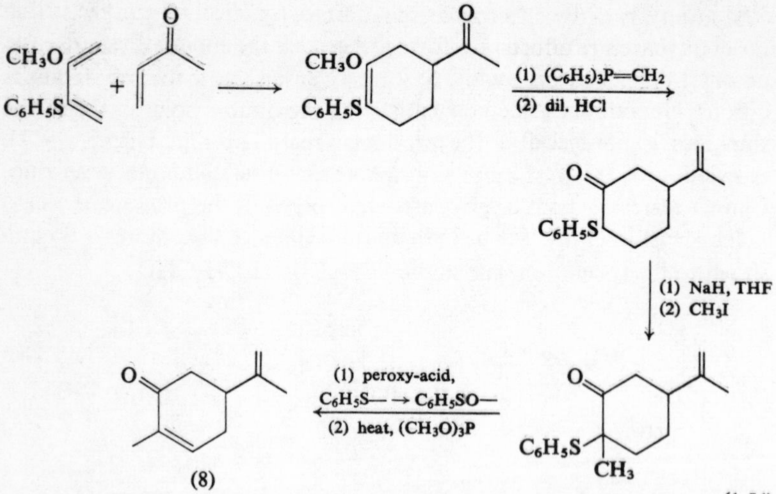

CH_3O C_6H_5S + → CH_3O C_6H_5S —(1) $(C_6H_5)_3P{=}CH_2$; (2) dil. HCl→ C_6H_5S —(1) NaH, THF; (2) CH_3I→ C_6H_5S CH_3 —(1) peroxy-acid, $C_6H_5S\text{—} \rightarrow C_6H_5SO\text{—}$; (2) heat, $(CH_3O)_3P$→ (8)

(1.54)

The reaction of the sulphur ylid diphenylsulphonium cyclopropylid with non-conjugated aldehydes and ketones to form oxaspiropentanes, which are readily converted into cyclobutanones, is discussed on p. 122. A related reagent is 1-lithiocyclopropyl phenyl sulphide (1.55, 9), which is readily formed from cyclopropyl phenyl sulphide and butyl-lithium. It reacts easily with aldehydes and ketones to give adducts which rearrange to cyclobutanones under acidic conditions. In contrast to the cyclopropylid the lithio derivative reacts with $\alpha\beta$-unsaturated carbonyl compounds exclusively at the carbonyl group. Both reagents

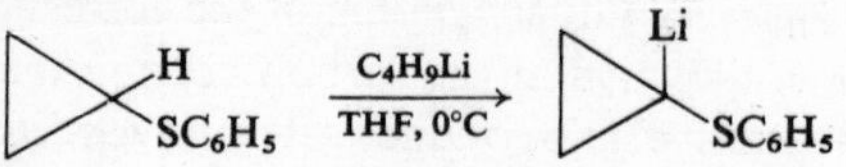

H SC_6H_5 —C_4H_9Li, THF, 0°C→ Li SC_6H_5 (9)

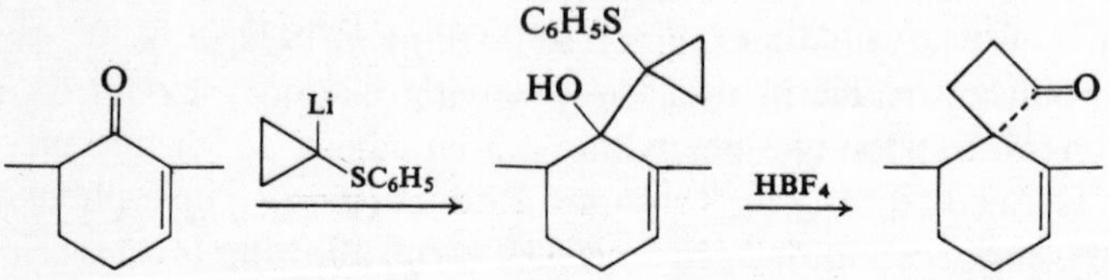

O —Li SC_6H_5→ C_6H_5S HO —HBF_4→ O

(1.55)

are valuable in synthesis because of the variety of useful transformations which can be effected with the cyclobutanones and with the intermediate oxaspiropentanes and hydroxy sulphides (Trost, 1974, 1975).

Thus, dehydration of the hydroxy sulphides with thionyl chloride at room temperature affords vinylcyclopropanes which rearrange to cyclopentenyl sulphides on pyrolysis. The alkenyl sulphide group in these products represents a masked carbonyl group, and hydrolysis affords a cyclopentanone derivative (1.56). Oxaspiropentanes, formed from diphenylsulphonium cyclopropylid can equally well be used as starting materials in this route to five-membered rings.

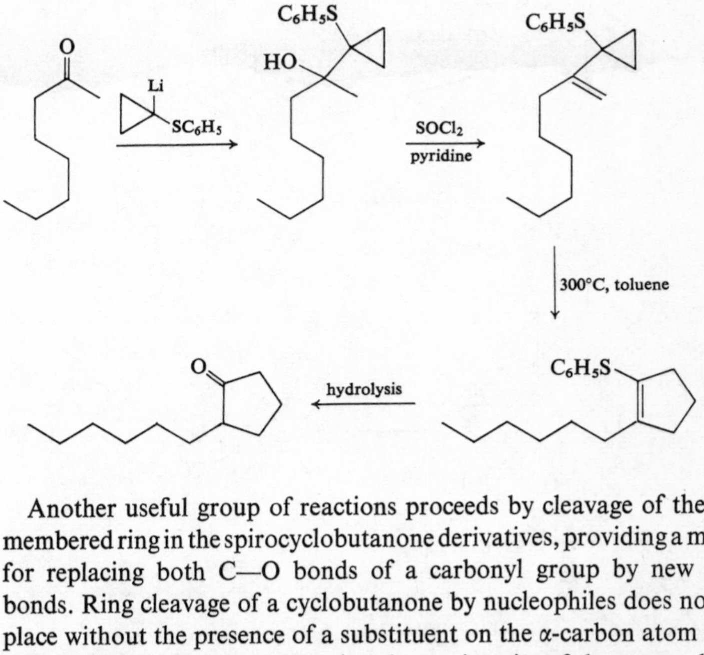

(1.56)

Another useful group of reactions proceeds by cleavage of the four-membered ring in the spirocyclobutanone derivatives, providing a method for replacing both C—O bonds of a carbonyl group by new C—C bonds. Ring cleavage of a cyclobutanone by nucleophiles does not take place without the presence of a substituent on the α-carbon atom which can stabilise the resultant negative charge, in spite of the strong driving force in the relief of ring strain. Bromine atoms provide sufficient stabilisation and bromination, followed by ring opening with alkali, allows the transformations exemplified in (1.57), among others (Trost, Bogdanowicz and Kern, 1975). This procedure has some limitations. For example, it would be unsuitable with compounds containing olefinic bonds elsewhere in the molecule, and in some cases reaction of the dibromo ketone with methoxide results in ring contraction rather than ring cleavage (1.58).

Better results are obtained in many cases with the trimethylene dithiane derivatives (as 1.59, 10) prepared from the ketone by reaction of the α-acetoxymethylene derivative with trimethylene dithiotosylate (1.59).

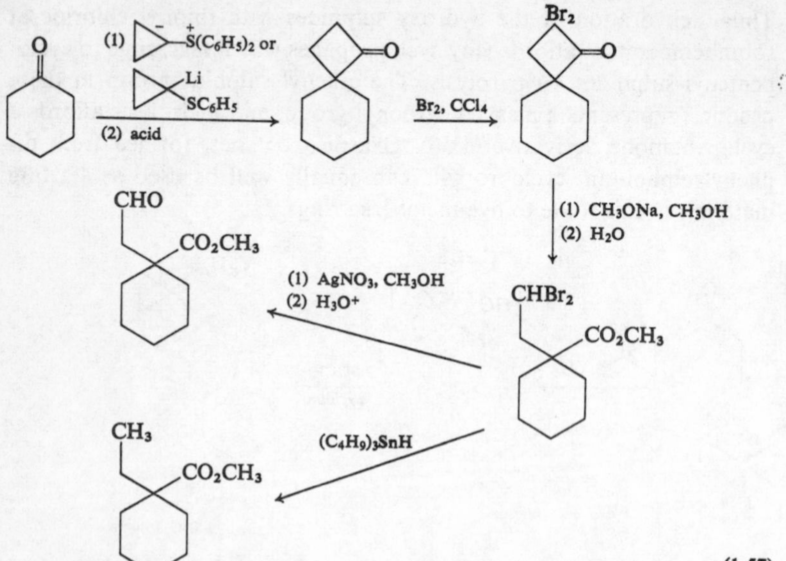

(1) $\text{cyclopropyl}^{-}\,\overset{+}{S}(C_6H_5)_2$ or $\text{1-lithio-1-(phenylthio)cyclopropane}$ (Li, SC_6H_5)
(2) acid

Br_2, CCl_4

Br_2

(1) CH_3ONa, CH_3OH
(2) H_2O

$CHBr_2$

CO_2CH_3

(1) $AgNO_3, CH_3OH$
(2) H_3O^+

CHO

CO_2CH_3

$(C_4H_9)_3SnH$

CH_3

CO_2CH_3

(1.57)

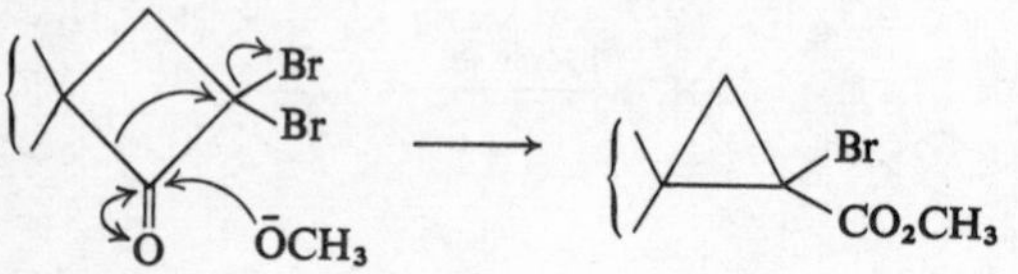

Br

Br

O

$\bar{O}CH_3$

Br

CO_2CH_3

(1.58)

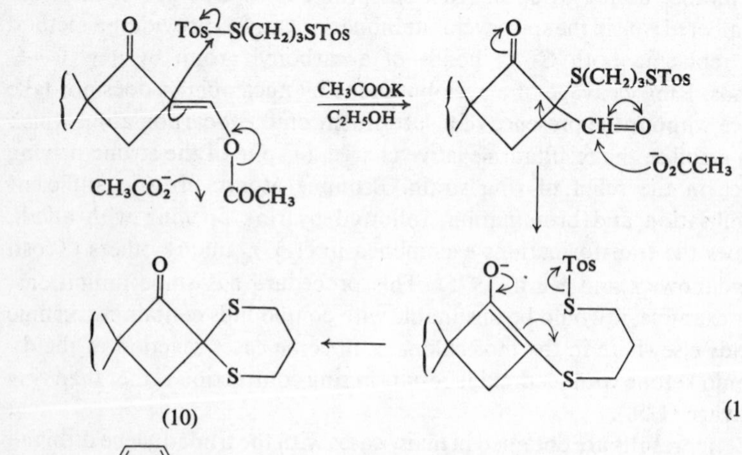

O

Tos—$S(CH_2)_3STos$

O

$CH_3CO_2^-$

$COCH_3$

CH_3COOK
C_2H_5OH

O

$S(CH_2)_3STos$

CH=O

$\bar{O}_2CCH_3$

O^-

Tos

S

S

O

S

S

(10)

(1.59)

Tos = H_3C–C_6H_4–SO_2

These derivatives are cleaved easily with methanolic sodium methoxide, and since sulphur is not a good leaving group ring contraction is avoided (Trost, Preckel and Leichter, 1975). An application of this sequence in the synthesis of methyl podocarpate is shown in (1.60). Ring cleavage

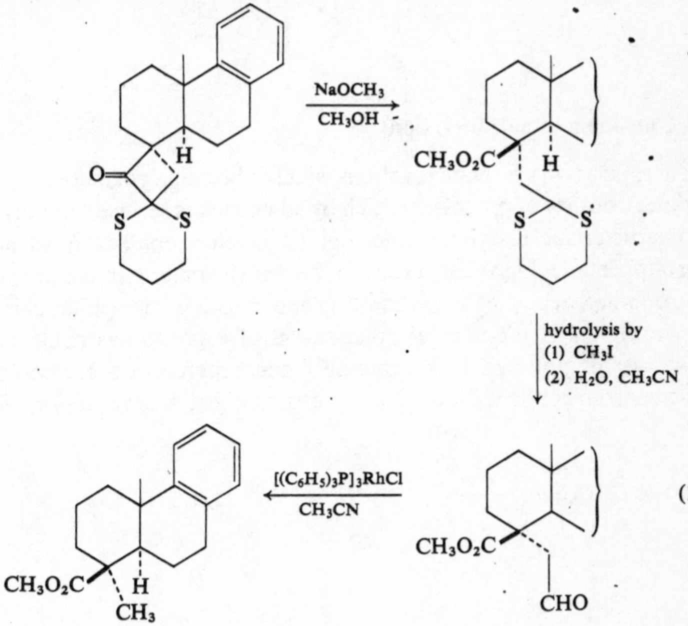

$NaOCH_3$, CH_3OH

CH_3O_2C H S S

hydrolysis by (1) CH_3I (2) H_2O, CH_3CN

$[(C_6H_5)_3P]_3RhCl$, CH_3CN

CH_3O_2C H CH_3 CH_3O_2C CHO

(1.60)

and hydrolysis of the dithiane, followed by decarbonylation of the aldehyde with tris(triphenylphosphine)chlororhodium (p. 429), generates the *gem*-methylcarboxylic ester group, a common structural feature in natural products. The group (and also other *gem* dialkyl groups produced by this sequence of reactions) is formed with high stereoselectivity because the addition of sulphonium cyclopropylids and of lithiocyclopropyl phenyl sulphides to carbonyl groups and the rearrangement of the products to spirocyclobutanones are both highly stereoselective reactions. In general, spiroannelation of a cyclic ketone with the sulphonium cyclopropylid gives the spirocyclobutanone with the carbon–carbonyl bond on the more hindered face of the starting ketone, whereas spiroannelation with lithiocyclopropyl phenyl sulphide gives the cyclobutanone with the carbon–carbonyl bond on the less hindered side. In a variation of the procedure described above, reaction of the dithiane derivative of the cyclobutanone with methyl-lithium, followed by ring cleavage with base, unmasking of the carbonyl group and aldol condensation, provides a route for the conversion below.

1.7. Umpolung (dipole inversion)

A general class of synthetic reactions which is becoming of increasing importance consists of processes which by some means reverse temporarily the characteristic reactivity, nucleophilic or electrophilic, of an atom or group. One such process results in the transformation of the normally electrophilic carbon of a carbonyl group into a nucleophilic carbon. This inversion of the normal polarisation of a group has come to be known as 'umpolung'. It is commonly encountered in the laboratory in the benzoin condensation which results, in effect, from addition of the

$$C_6H_5CHO \xrightarrow{CN^-} C_6H_5CH(O^-)(CN) \longrightarrow C_6H_5C^-(OH)(CN)\ \textbf{(11)}$$

$$\textbf{(11)} \xrightarrow{C_6H_5CHO} C_6H_5C(OH)(CN)\text{—}C(O^-)(H)\text{—}C_6H_5 \longrightarrow C_6H_5\overset{O}{\overset{\|}{C}}\text{—}CHOHC_6H_5 \qquad (1.61)$$

benzoyl anion, $C_6H_5\bar{C}{=}O$, to the carbonyl group of benzaldehyde, the intermediate (11) behaving as a 'benzoyl anion equivalent'. The process is important in nature in reactions catalysed by thiamine coenzyme in which acyl anion equivalents, $R\text{—}\bar{C}{=}O$, are generated with the aid of the thiazole ring of the thiamine molecule. For example in the transketolase reaction, the intermediate ($12 \equiv X\text{—}\bar{C}{=}O$) goes to the ketol (13) by reaction with an aldehyde. This sequence has been emulated in the laboratory using *N*-laurylthiazolium bromide as catalyst (Tagaki and Hara, 1973; Stetter and Kuhlmann, 1974).

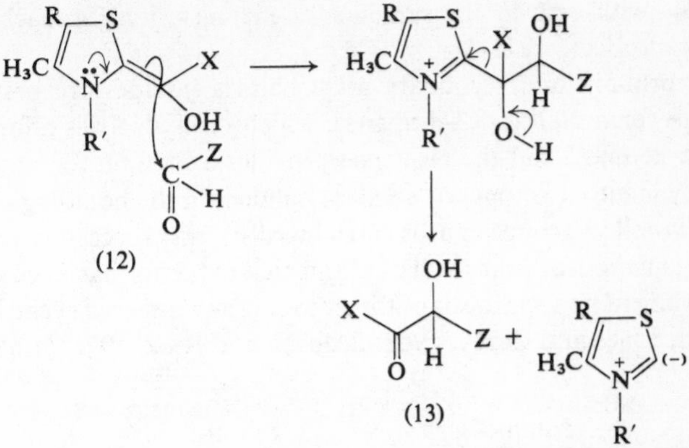

(12)

(13)

(1.62)

A number of other methods for the generation of acyl anion equivalents from aldehydes have been developed for use in the laboratory. One proceeds from cyanohydrins, as in the benzoin condensation. Aldehyde cyanohydrins, protected as the acetals with ethyl vinyl ether, are readily transformed into anions by treatment with lithium diisopropylamide. Reaction with an alkyl halide gives the protected cyanohydrin of a ketone from which the parent compound is easily liberated by successive treatment with dilute sulphuric acid and dilute sodium hydroxide (Stork and Maldonado, 1974).

In another widely used route (Corey and Seebach, 1965; Seebach and Corey, 1975), the key step is the conversion of the carbonyl group into a 1,3-dithiane, which, because of the stabilising effect of the two electronegative sulphur atoms, is easily converted into the corresponding carbanion at C-2, i.e. at the carbon of the original carbonyl group, by reaction with butyl lithium (1.63). The resulting lithium compounds are stable in solution at low temperatures, and undergo the whole range of reactions shown by other organolithium compounds. After reaction the 1,3-dithiane system can be reconverted into the carbonyl group by

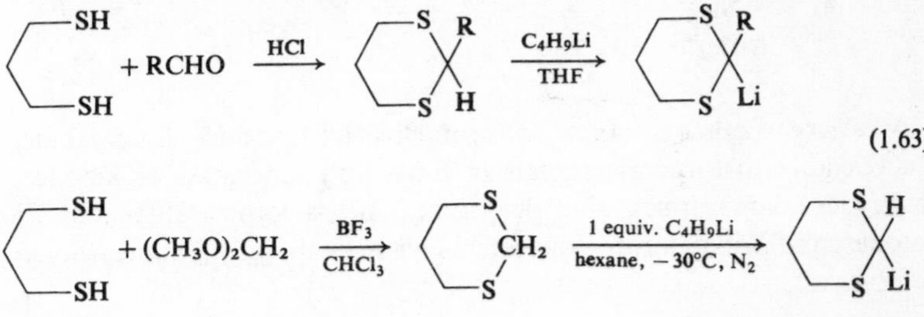

$$+ RCHO \xrightarrow{HCl} \xrightarrow[THF]{C_4H_9Li}$$

(1.63)

$$+ (CH_3O)_2CH_2 \xrightarrow[CHCl_3]{BF_3} \xrightarrow[\text{hexane, } -30°C,\ N_2]{\text{1 equiv. } C_4H_9Li}$$

hydrolysis with acid in the presence of mercuric ion (Seebach, 1969; Seebach and Beck, 1971).

Thus, primary and secondary alkyl halides (iodides are best) react readily to form 2-alkyl-1,3-dithianes, which on hydrolysis afford aldehydes or ketones, and the sequence provides a method for converting an aldehyde into a ketone, or an alkyl halide into the homologous aldehyde. Two alkyl groups can be introduced by two successive reactions without isolation of intermediates, and this sequence has been applied in a convenient new synthesis of three- to seven-membered cyclic ketones (Seebach, Jones and Corey, 1968; Seebach and Beck, 1971) (1.64).

SH SH $\xrightarrow[HCl]{C_6H_5CHO}$ S S (H, C_6H_5) $\xrightarrow[(2)\ D_2O]{(1)\ C_4H_9Li}$ S S (D, C_6H_5) $\xrightarrow{\text{hydrolysis}}$ $C_6H_5C(=O)D$

S S (CH_3, Li) $\xrightarrow[(2)\ \text{hydrolysis, } H_2O,\ HgCl_2]{(1)\ C_6H_5CH_2Br}$ $C_6H_5CH_2COCH_3$

S S (H, Li) $\xrightarrow[THF,\ -50°C]{Cl(CH_2)_3I}$ S S (H, $(CH_2)_3Cl$) $\xrightarrow{1\ \text{equiv. } C_4H_9Li,\ -70°C}$ S S (spiro-cyclobutane) $\xrightarrow[H_2O,\ HgCl_2]{\text{hydrolysis}}$ O= (cyclobutanone) (60%) (1.64)

A variety of other reactions has been effected. Epoxides, for example, react readily to form mercaptals of β-hydroxy aldehydes or ketones, aldehydes and ketones give derivatives of α-hydroxy aldehydes or ketones, and acid chlorides and esters give mercaptals of 1,2-dicarbonyl compounds.

A limitation on the use of dithianes for carbonyl umpolung is that although the anions react readily with alkylating agents they do not add to Michael acceptors, reacting for example, with $\alpha\beta$-unsaturated aldehydes and ketones at the carbonyl group. A more versatile reagent is ethyl ethylthiomethyl sulphoxide (14), the anion of which undergoes both alkylation with halides and conjugate addition to enones, allowing the formation of a wide range of aldehydes and ketones and 1,4-dicarbonyl compounds (Richman, Herrmann and Schlessinger, 1973).

C_2H_5S / $C_2H_5S(\rightarrow O)$ (14) $\xrightarrow[\text{THF, 20°C}]{LiN(iso\text{-}C_3H_7)_2}$ Li^+ anion $\xrightarrow{C_2H_5I}$ C_2H_5S / $C_2H_5S(\rightarrow O)$, H

$\downarrow$ $Li(iso\text{-}C_3H_7)_2$ (1.65)

(1) methyl vinyl ketone, (2) hydrolysis ← C_2H_5S / $C_2H_5S(\rightarrow O)$ anion $\xrightarrow[(2)\ H_3O^+,\ Hg^{2+}]{(1)\ C_3H_7Br}$ O= ketone (90%)

The corresponding methyl methylthiomethyl sulphoxide has also been used, notably for the preparation of cyclic compounds. 3-Cyclopentenone, for example, was obtained from *cis*-1,4-dichloro-2-butene and the anion of the sulphoxide in 60 per cent yield (Ogura, Yamashita, Suzuki and Tsuchihashi, 1974).

Another useful reagent is bis(phenylthio)methane. Successive alkylation of the derived anion with an alkyl halide and an aldehyde or ketone and treatment of the resulting carbinol with trifluoroacetic acid provides a convenient route to ketones. Hydrolysis of the intermediate dithioacetals is easier in this case than in the dithiane route (Blatcher, Grayson and Warren, 1976) (1.66).

Metallated enol ethers also are versatile and efficient acyl anion equivalents (Baldwin, Höfle and Lever, 1974). The reagents are easily prepared by action of t-butyl-lithium on an enol ether in tetrahydrofuran at low temperature. Most work so far has been done with the parent

$CH_2{=}C(OCH_3)Li \equiv CH_3\overset{O}{C}^{(-)} \xrightarrow{E^+} CH_3\overset{O}{\overset{\|}{C}}E$

compound, methoxyvinyl-lithium, but a few monosubstituted enol ethers have also been used. The crotyl anion equivalent,

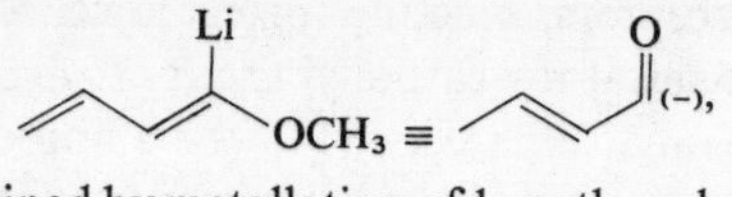

Li OCH$_3$ ≡ O (−),

for example, is obtained by metallation of 1-methoxybutadiene. Reaction of the lithium compounds with electrophiles (alkyl halides, aldehydes and ketones) gives, initially, vinyl ethers which may be further elaborated or converted by mild hydrolysis with acid into the corresponding carbonyl compounds. Reaction with alkyl halides leads to ketones and aldehydes and ketones give α-ketols; even hindered compounds such as oestrone methyl ether react readily (1.66). $\alpha\beta$-Unsaturated compounds react

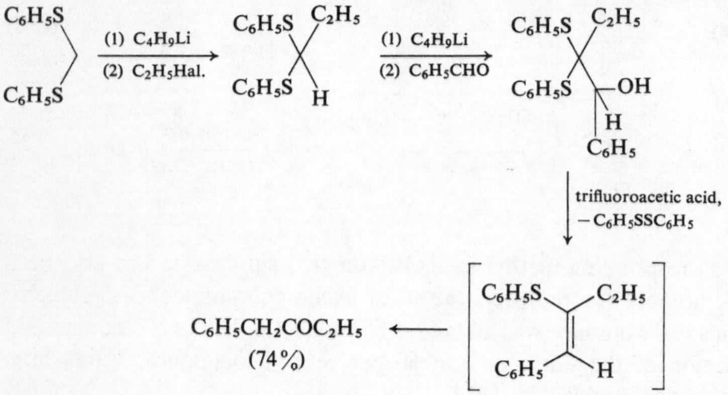

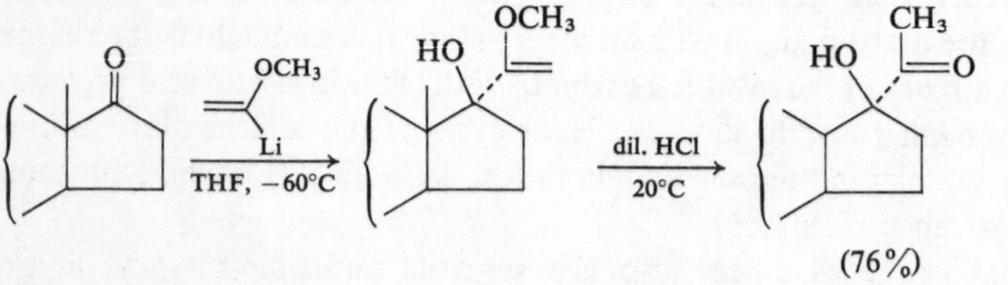

C_6H_5S, C_6H_5S —(1) C_4H_9Li, (2) $C_2H_5Hal.$→ C_6H_5S, C_2H_5, C_6H_5S, H —(1) C_4H_9Li, (2) C_6H_5CHO→ C_6H_5S, C_2H_5, C_6H_5S, OH, H, C_6H_5

↓ trifluoroacetic acid, $-C_6H_5SSC_6H_5$

$C_6H_5CH_2COC_2H_5$ (74%) ← [C_6H_5S, C_2H_5, C_6H_5, H]

O, OCH$_3$, Li, THF, −60°C → HO, OCH$_3$ —dil. HCl, 20°C→ HO, CH$_3$, O (1.66)

(76%)

entirely at the carbonyl group, as they do with the lithiodithianes, but the present method has the advantage, if the 1,2-addition products are required, that the adducts are much more easily hydrolysed than in the dithiane route. Conjugate addition can be achieved, if desired, by first converting the lithio derivatives into the corresponding cuprate reagents (see p. 67) (Boeckman, Bruza, Baldwin and Lever, 1975). 1-Alkylthio-vinyl-lithium reagents have also been prepared, and give similar results in reactions with allyl halides and aldehydes and ketones (Oshima, Shimoji, Takahashi, Yamamoto and Nozaki, 1973).

Another group of versatile synthetic intermediates are the alkylidene dithianes or ketene thioacetals (15).

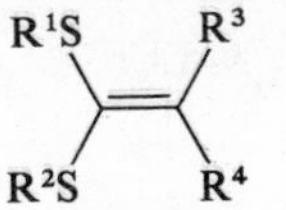

R^1S R^3 R^2S R^4 (15)

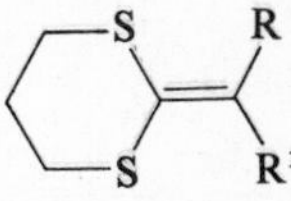

S R S R^1 (16)

A number of methods can be used to prepare compounds of this type. The valuable derivatives (16), for example, are easily obtained from commercially available 1,3-dithiane, using the Peterson reaction (p. 318) (Jones, Lappert and Szary, 1973), and undergo a variety of useful reactions. The double bond is easily reduced by protonation followed by hydride transfer to give a product which, on hydrolysis, completes the conversion of RCHO into RCH_2CHO. Alkyl groups attached to the double bond are readily metallated by action of butyl-lithium or lithium diisopropylamide to give the corresponding allyl anions. Reaction of these with various electrophiles takes place predominantly at the carbon adjacent to the sulphur atoms, giving products which are hydrolysed to $\alpha\beta$-unsaturated ketones. In this sequence, the allyl anions serve as masked $\alpha\beta$-unsaturated acyl anions (Seebach, Kolb and Gröbel,

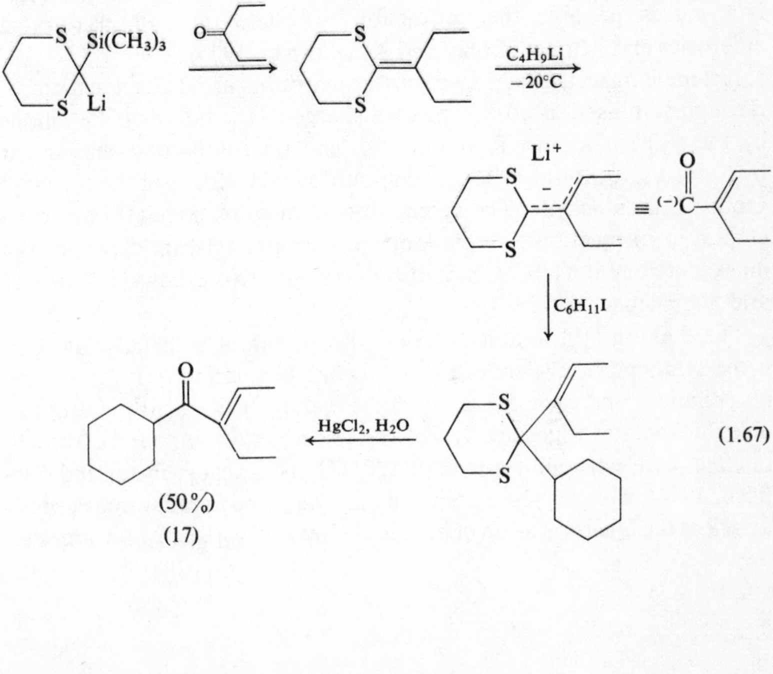

S $Si(CH_3)_3$ S Li → S S $\xrightarrow[-20°C]{C_4H_9Li}$

Li^+ S S ≡ $^{(-)}$C O

$C_6H_{11}I$

O (50%) (17) $\xleftarrow{HgCl_2, H_2O}$ S S (1.67)

1974; Corey and Kozikowsky, 1975). Thus, the ketone (17) was obtained in 50 per cent overall yield from cyclohexyl iodide and 3-pentanone, as shown in (1.67). An alternative route to the alkylidene dithianes (16)

SCH_3 / $SCH_3 \rightarrow O$ $\equiv$ $^+CH_2CHO$

(18)

CH_3O_2C $^-$ Li^+ + (SCH_3, $SCH_3 \rightarrow O$) $\longrightarrow$ CH_3O_2C ... Li^+ $^-$ SCH_3, $SCH_3 \rightarrow O$

H_3O^+, CH_3CN

CH_3O_2C ... CHO

(1.68)

from esters by reaction with bis(dimethylaluminium)-1,3-propanedithiolate makes possible the conversion of esters into $\alpha\beta$-unsaturated aldehydes and ketones (Corey and Kozikowski, 1975).

Ketene thioacetals act as electron acceptors in Michael reactions, through stabilisation of the negative charge in the intermediate anions by the sulphur atoms. Reaction with enolate anions and subsequent hydrolysis gives 1,4-dicarbonyl compounds in which one of the carbonyl groups is an aldehyde. The ketene thioacetal sulphoxide (18) has been employed in reactions of this kind, in which it acts, in effect, as the masked carbonium ion, $^+CH_2CHO$ (Herrmann, Kieczykowski, Romaret and Schlessinger, 1973).

The dianion (19), obtained from 2-propenethiol with butyl-lithium, is the synthetic equivalent of the homoenolate anion (20). It reacts with alkyl halides and carbonyl compounds mainly at the γ-carbon atom to give thioenolates which are hydrolysed to carbonyl compounds, formed, in effect by electrophilic attack on (20) (Geiss, Seuring, Pieter and Seebach, 1974). This sequence complements the more familiar nucleophilic attack at the β-carbon atom of acrylic aldehyde and other $\alpha\beta$-unsaturated

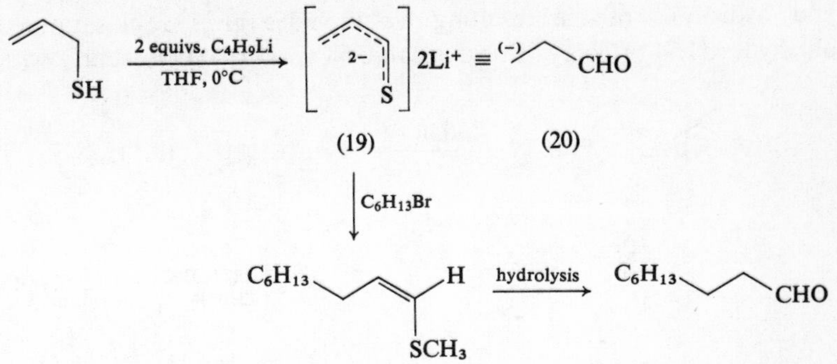

(19) (20)

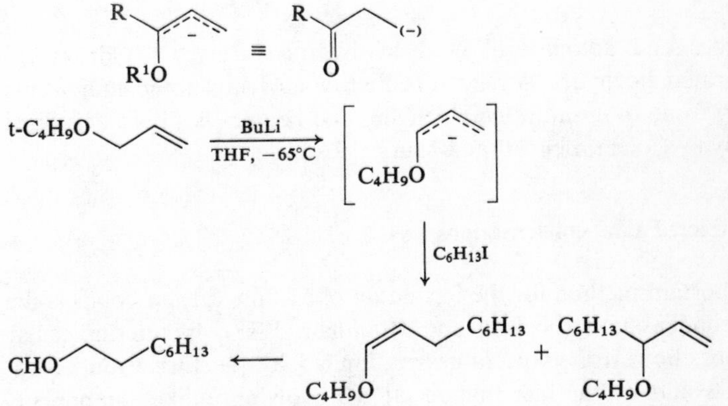

(1.69)

(90% of product) (10% of product)

aldehydes. More generally, β-acylcarbanion equivalents are obtained by deprotonation of allyl ethers. The anions thus obtained, by proper choice of experimental conditions, react with alkyl halides at the γ-carbon atom giving enol ethers which are hydrolysed to carbonyl compounds (1.69) (Evans, Andrews and Buckwalter, 1974; Still and Macdonald, 1974).

Another useful anion, bis(methylthio)allyl anion (21), which is easily generated from epichlorohydrin (Erickson, 1974), acts as the synthetic equivalent of the β-formylvinyl anion (22). Alkylation with alkyl halides

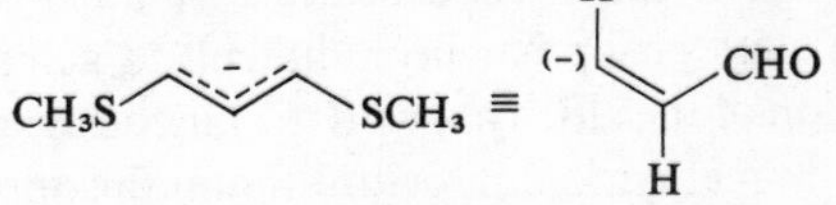

(1.70)

(21) (22)

and hydrolysis of the resulting bis-sulphides gives $\alpha\beta$-unsaturated aldehydes (1.71) (Corey, Erickson and Noyori, 1971). Reaction with

$$\mathrm{CH_3S\text{-}CH{=}CH{-}CH^{-}\text{-}SCH_3\ Li^+} \xrightarrow[\text{THF, }-78^\circ\text{C}]{\mathrm{C_5H_{11}Br}} \mathrm{C_5H_{11}CH(SCH_3)CH{=}CHSCH_3} \xrightarrow{\mathrm{HgCl_2,\ H_2O,\ CH_3CN}} \underset{trans}{\mathrm{C_5H_{11}CH{=}CHCHO}}\ (84\%) \tag{1.71}$$

aldehydes and ketones followed by hydrolysis gives γ-hydroxy-$\alpha\beta$-unsaturated aldehydes. A related route to β-acylvinyl anion equivalents directly from $\alpha\beta$-unsaturated aldehydes and ketones has been described recently by Cohen, Bennett and Mura (1976).

1.8. Directed aldol condensations

An important method for the formation of carbon–carbon bonds is the aldol condensation (Nielson and Houlihan, 1968). In intermolecular reactions, however, its usefulness is limited by the formation of by-products and by the fact that reactions involving unlike components frequently give mixtures of aldols. Various methods have been adopted to circumvent these difficulties and to bring about 'directed' aldol condensations, particularly in reactions between aldehydes and ketones, usually proceeding from the preformed enolate or enol ether of one of the components. In one method condensation of a preformed lithium enolate with an aliphatic or aromatic aldehyde in ether or dimethoxyethane, in the presence of zinc chloride, gives a single aldol product in high yield; here the intermediate β-keto alkoxide derivative is trapped as a metal chelate in the aprotic reaction solvent and protected from further reaction (Auerbach, Crumrine, Ellison and House, 1974). Methyl pentyl ketone, for example, was converted into the aldol (22) in 80 per cent yield (1.72). Directed aldol condensation of aldehydes with methyl ketones at the methyl group has been difficult to achieve, but it can be effected by reaction of the aldehyde with the kinetically generated enolate obtained from the methyl ketone with lithium diisopropylamide (1.72) (Stork, Kraus and Garcia, 1974; see also Kuwajima, Sato, Arai and

$$C_4H_9CH{=}C(OCOCH_3){-}CH_3 \xrightarrow[\text{ether}]{CH_3Li} C_4H_9CH{=}C(OLi){-}CH_3$$

$$\downarrow \text{(1) } C_6H_5CHO,\ ZnCl_2,\ \text{dimethoxyethane, 0°C; (2) } H_2O,\ NH_4Cl$$

$$C_4H_9CH(COCH_3)CHOHC_6H_5 \quad (80\%)$$

(22)

$$C_3H_7COCH_3 \xrightarrow[\text{THF, } -78°C]{LiN(iso\text{-}C_3H_7)_2} C_3H_7C(OLi){=}CH_2 \qquad (1.72)$$

$$\downarrow \text{(1) } C_3H_7CHO; \text{ (2) } H_3O^+$$

$$C_3H_7COCH_2CHOHC_3H_7 \quad (65\%)$$

Minami, 1976). Another useful and versatile method was found by Mukaiyama, Banno and Narasaka (1974). Silyl enol ethers of both aldehydes and ketones (see p. 20) react with aldehydes and ketones under mild conditions in the presence of titanium tetrachloride to give specific aldols in good yields. Enol ethers derived from unsymmetrical ketones react specifically at the carbon atom at the terminus of the double bond. The reaction is thought to proceed as in (1.73). Thus, condensation

$$R^1R^2C{=}C(R^3)OSi(CH_3)_3 + R^4R^5C{=}O \xrightarrow[CH_2Cl_2]{TiCl_4} \left[R^3C(O{-}Si(CH_3)_3){=}CR^2R^1 \cdots Cl{-}TiCl_3 \cdots O{=}CR^4R^5 \right]$$

$$\downarrow -(CH_3)_3SiCl$$

$$R^3C(=O)CR^1R^2C(OH)R^4R^5 \xleftarrow{H_2O} \left[R^3C(=O\cdots TiCl_3){-}CR^1R^2{-}C(R^4R^5){-}O{-} \right] \qquad (1.73)$$

of isobutyraldehyde and phenylacetaldehyde gave the aldol (23) in 86 per cent yield, and in another useful application high yields of primary alcohols are obtained from reaction of the silyl ethers with formaldehyde, without the by-products commonly encountered. In reactions of aldehyde enol ethers which contain a β-hydrogen atom, dehydration of derived aldols may take place under the reaction conditions to give αβ-unsaturated carbonyl compounds.

$$(CH_3)_2CHCHO \longrightarrow (CH_3)_2C{=}C\begin{smallmatrix}OSi(CH_3)_3\\ H\end{smallmatrix} \xrightarrow[(2)\ H_2O]{(1)\ C_6H_5CH_2CHO,\ TiCl_4,\ CH_2Cl_2,\ -78°C} C_6H_5CH_2CHOHC(CH_3)_2CHO \quad (86\%)$$

(23) (1.74)

1.9. Allylic alkylation of olefins

Alkylation of a methyl or methylene group adjacent to a carbon–carbon double bond can sometimes be effected, and conditions can be arranged to give replacement of hydrogen by an electrophilic or nucleophilic substituent. Metallation of olefins with very strong bases generates allylic metallo derivatives which behave as typical organometallic compounds. Thus, reaction of limonene with the complex of butyl-lithium and tetramethylethylenediamine results in selective metallation at C-10, giving an allyl-lithium species which undergoes the whole range of reactions shown by other organolithium compounds. Reaction with 1-bromo-3-methyl-2-butene, for example, gave β-bisabolene (Crawford, Erman and Broaddus, 1972) (1.75). By reaction with butyl-lithium in the presence of potassium t-butoxide, arylalkanes and olefins are converted into *potassium* derivatives which can be alkylated with alkyl halides or epoxides, and carbonated by carbon dioxide. Reaction takes

C_4H_9Li, C_6H_{14} → Li → $BrCH_2CH{=}C(CH_3)_2$, −40°C →

(1.75)

place without change in the stereochemistry of the olefinic double bond, and preferentially at an allylic methyl group (Schlosser, Hartmann and David, 1974).

The strongly basic conditions employed in the above procedures severely limits their application. An alternative mild procedure has been described recently by Trost (1975) and involves activation of the allylic position of an olefin by formation of a π-allylpalladium complex, followed by condensation of this ambident electrophile with a polarisable (soft) anion. The complexes are easily prepared from a variety of olefins

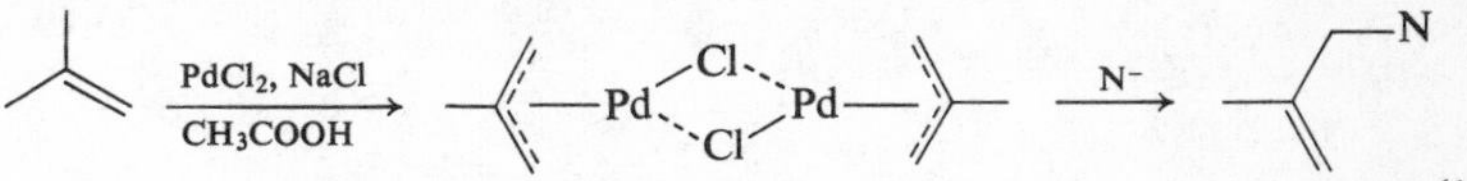

$PdCl_2$, NaCl / CH_3COOH ; N^-

(1.76)

by action of palladium chloride and sodium chloride in acetic acid. Depending on the structure of the olefin, a mixture of complexes may be obtained from which the individual components may be separated by fractional crystallisation or by chromatography, but in all cases reaction involves deprotonation from the allylic site adjacent to the more substituted end of the double bond. Reaction of the complexes with polarisable

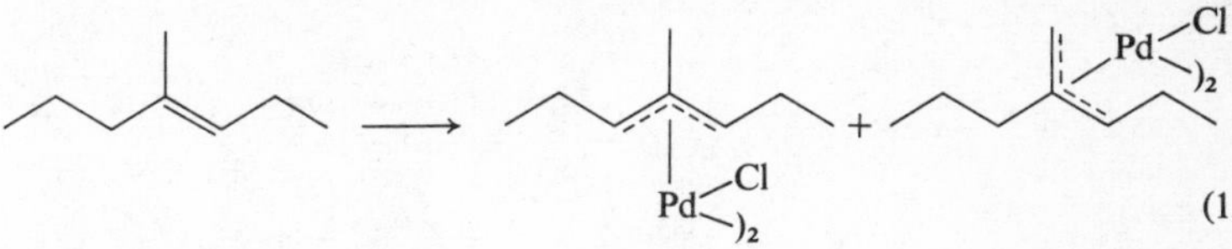

(1.77)

carbanions, such as the anion of diethyl malonate or of methyl methylsulphonylacetate, in the presence of a phosphorus ligand (for example, triphenylphosphine or hexamethylphosphoramide) leads preferentially to alkylation, under appropriate conditions, at the less substituted carbon of the complex. Harder anions, such as those in methyl-lithium, methylmagnesium iodide or lithium dimethylcuprate, do not react. Equation (1.78) shows how the sequence can be used to prepare $\alpha\beta,\gamma\delta$-dienoic acids, and to convert an allylic methyl group into an ethyl group (Trost and Fullerton, 1973). Carbonyl groups do not interfere with the reaction, and $\alpha\beta$-unsaturated ketones and esters can be alkylated at the γ-position by reaction of the derived π-allyl palladium complexes with enolate anions; that is they can be made to react as (+)⌒⌒C(=O)R, providing another example of umpolung (see p. 44) (Jackson and Strauss, 1975).

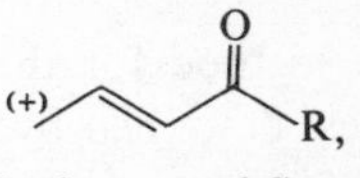

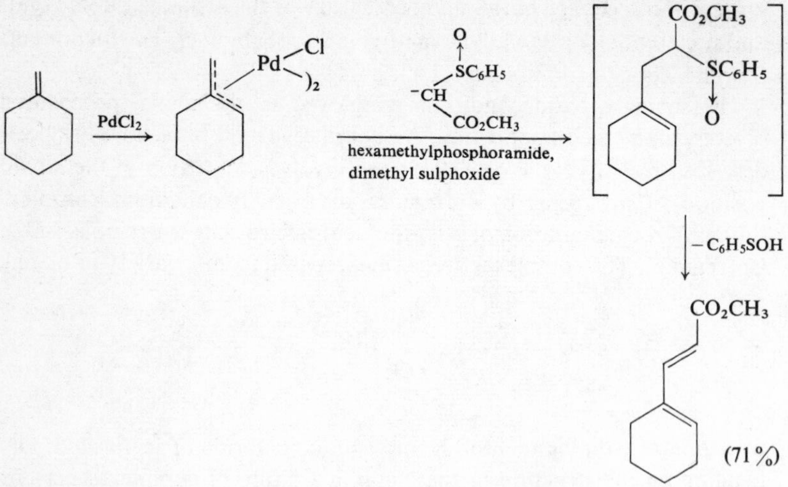

$PdCl_2$

$^{-}CH(SOC_6H_5)(CO_2CH_3)$

hexamethylphosphoramide, dimethyl sulphoxide

CO_2CH_3

SC_6H_5

O

$-C_6H_5SOH$

CO_2CH_3

(71%)

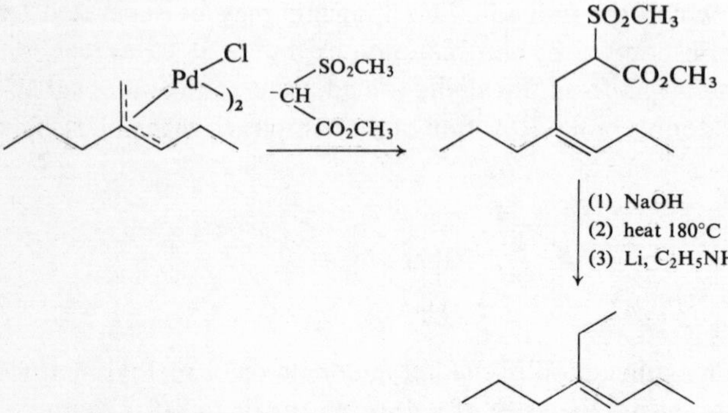

$^{-}CH(SO_2CH_3)(CO_2CH_3)$

SO_2CH_3

CO_2CH_3

(1) NaOH
(2) heat 180°C
(3) Li, $C_2H_5NH_2$

(1.78)

The mechanism of these alkylations is not completely understood, but it is thought to involve ionic complexes of the type

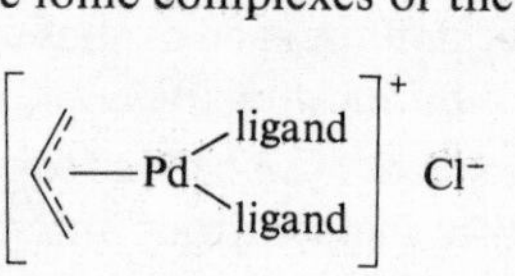

$$\left[(\text{allyl})\text{—Pd}\begin{matrix}\text{ligand}\\ \text{ligand}\end{matrix} \right]^{+} Cl^{-}$$

1.10. The dihydro-1,3-oxazine synthesis of aldehydes and ketones

A versatile method for the synthesis of aldehydes and ketones from alkyl halides, described by Meyers and his co-workers (1973), is illustrated in (1.79) in general terms. The starting material is a dihydro-1,3-

oxazine derivative (available commercially or by a simple preparation from 2-methyl-2,4-pentandiol and a nitrile), the 'activated' alkyl group of which is readily metallated by treatment with butyl-lithium at −80°C. The resulting anions react rapidly with a wide variety of alkyl halides (bromides and iodides) to give high yields of substituted dihydro-oxazines which are smoothly reduced in quantitative yield with sodium borohydride to the tetrahydro derivatives. These compounds are readily cleaved in dilute acid to give the aldehyde. By this method

O N CH_2X $\xrightarrow[\text{THF, } -80°C]{C_4H_9Li}$ O N CHLi(X) $\xrightarrow{\text{RHal.}}$ O N CHR(X)

$\downarrow$ $NaBH_4$, THF, H_2O, C_2H_5OH

(1.79)

RCHCHO(X) $\xleftarrow[\text{oxalic acid}]{H_2O}$ O N H, H, CHR(X)

halides are converted into aldehydes with two more carbon atoms, and the sequence thus complements the synthesis from lithio-1,3-dithiane described above (p. 45) which gives an aldehyde with one more carbon. A useful feature of the reaction is that by reduction with borodeuteride instead of borohydride C-1 labelled aldehydes can be obtained. By appropriate choice of starting materials a variety of different kinds of aldehydes and ketones can be synthesised. Thus, successive alkylation with two molecules of alkyl halide leads to $\alpha\alpha$-disubstituted aldehydes, and with one molecule of an $\alpha\omega$-dihalide alicyclic aldehydes can be obtained (Politzer and Meyers, 1971). Aldehydes and ketones afford precursors of $\alpha\beta$-unsaturated aldehydes, and with epoxides γ-hydroxy-aldehydes are formed. Hydrolysis at the dihydro-oxazine stage leads to carboxylic acids instead of aldehydes.

The reaction has been extended to the synthesis of ketones, as illustrated by the following synthesis of 1-phenyl-3-pentanone (Meyers and Smith, 1970) (1.80). The oxazine synthesis of aldehydes and ketones complements the method using organoboranes (p. 294) which is difficult to apply in some cases because of the complexity of the olefins required.

$(CH_2)_2C_6H_5$ $\xrightarrow{CH_3I}$ $\overset{+}{N}$ $(CH_2)_2C_6H_5$ CH_3 I^-

$\downarrow C_2H_5MgBr$ (1.80)

$C_6H_5(CH_2)_2COC_2H_5$ (78%) $\xleftarrow{H_3O^+}$ $(CH_2)_2C_6H_5$ C_2H_5 N CH_3

S N C_2H_5 $\xrightarrow[(2)\ CH_3I]{(1)\ C_4H_9Li,\ THF,\ -78°C}$ S N CH CH_3 CH_3 $\xrightarrow[(2)\ C_6H_{13}I]{(1)\ C_4H_9Li}$ S N CH_3 C C_6H_{13} CH_3

$\downarrow$ Al–Hg, moist ether

CHO (81%) $\xleftarrow[H_2O,\ CH_3CN]{HgCl_2}$ S N H CH_3 C C_6H_{13} CH_3

A limitation of the dihydro-oxazine synthesis of aldehydes is that, in general, it is suitable only for the preparation of monosubstituted acetaldehydes. An alternative procedure using 2-alkylthiazolines which can be used to prepare di- and tri-substituted acetaldehydes has been described recently by Meyers and Durandetta (1975).

Another difficulty is that dihydro-1,3-oxazines are not always readily accessible, and an alternative procedure using 2-alkyl-4,4-dimethyl-2-oxazolines, which are easily prepared by heating carboxylic acids with 2-amino-2-methyl-1-propanol, is often more convenient (Meyers,

OH NH_2 $\xrightarrow[100–150°C]{RCO_2H}$ O N R (1.81)

Temple, Nolen and Mihelich, 1974; Meyers and Mihelich, 1976). The lithio derivatives, prepared with butyl-lithium, react with alkyl halides giving 2-alkyloxazolines which may be hydrolysed to homologated acetic acids. Dialkylation leads to α,α-dialkylacetic acids. By reaction with

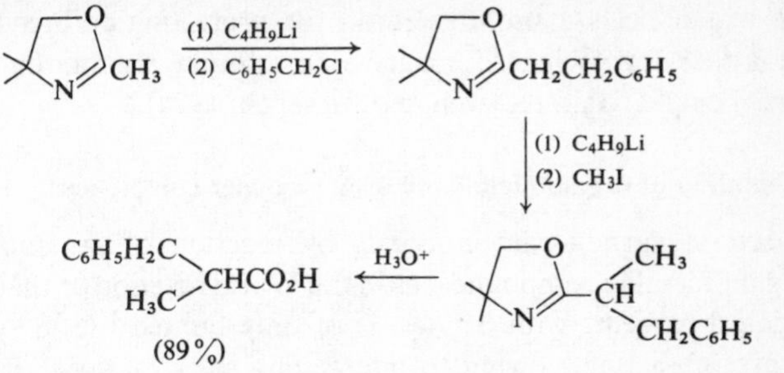

(1) C_4H_9Li (2) $C_6H_5CH_2Cl$ → $CH_2CH_2C_6H_5$; (1) C_4H_9Li (2) CH_3I; H_3O^+ → $C_6H_5H_2C(H_3C)CHCO_2H$ (89%)

(1.82)

aldehydes and ketones, followed by hydrolysis, $\alpha\beta$-unsaturated acids are formed, and epoxides give γ-butyrolactones.

An interesting development is the use of the chiral oxazoline (23) to effect asymmetric synthesis of α-alkylcarboxylic acids. High optical yields are obtained, and either enantiomer of the alkylated acids can be prepared by selecting the order of introduction of the substituent groups (Meyers, Knaus, Kamata and Ford, 1976). Even higher optical yields

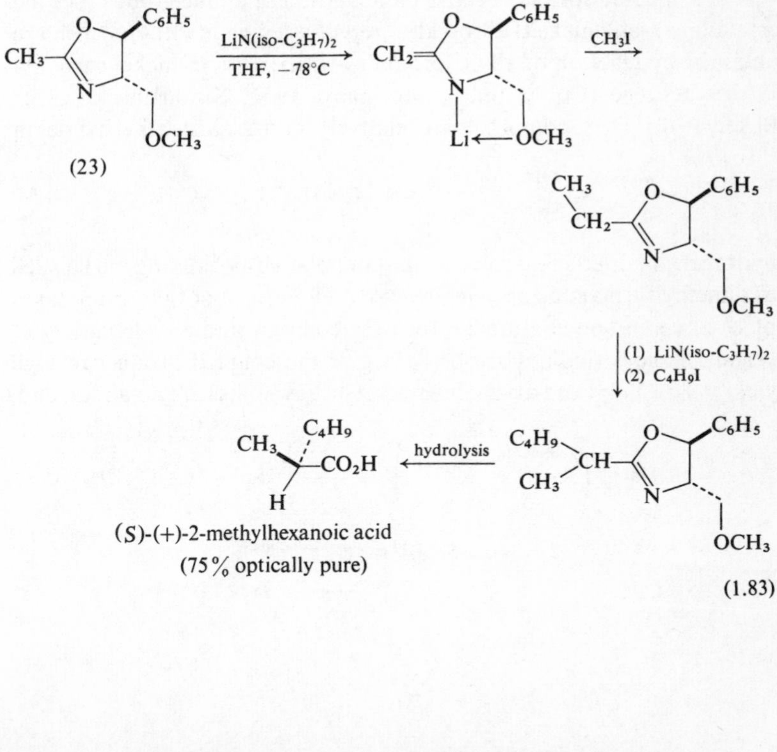

(23) $\xrightarrow{\text{LiN(iso-}C_3H_7)_2,\ \text{THF},\ -78°C}$ → CH_3I → (1) LiN(iso-$C_3H_7)_2$ (2) C_4H_9I → hydrolysis →

(*S*)-(+)-2-methylhexanoic acid
(75% optically pure)

(1.83)

of 3-alkylalkanoic acids have been obtained by conjugate addition of organolithium compounds to 2-alkenyl analogues of (23) (Meyers and Whitten, 1975).

The oxazoline ring is in effect a masked carboxyl group, and conversion into an oxazoline is a useful method for protecting carboxyl groups against attack by Grignard reagents or lithium aluminium hydride (Meyers, Temple, Haidukewych and Mihelich, 1974).

1.11. Coupling of organonickel and organocopper complexes

Formation of carbon–carbon bonds by reaction of organic halides with organometallic compounds, as in the Wurtz reaction or the reaction of Grignard reagents with halides, is of only limited use in synthesis. Yields are often poor, owing to intervening side-reactions such as α- and β-eliminations, and in 'mixed' reactions, as with RHal. and R′M, substantial amounts of RR and R′R′ are formed besides the desired RR′ because of fast halogen–metal interchange. Improved yields of coupled products can be obtained from Grignard reagents and organohalides in the presence of transition metal complexes (Tamao, Sumitani and Kumada, 1972) or thallium(I) salts (McKillop, Elsom and Taylor, 1968).

A new method for the selective combination of unlike groups uses the crystalline π-allylnickel(I) bromides, represented as in (1.84), which are obtained by reaction of allylic bromides with excess of nickel carbonyl in dry benzene (Corey and Semmelhack, 1967; Semmelhack, 1972; Baker, 1973). These complexes are relatively inert towards alkyl halides in

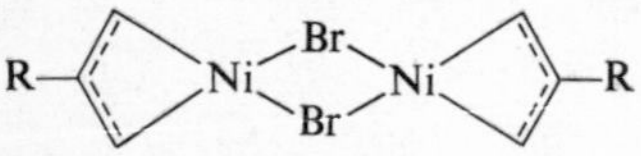

R–(π-allyl)Ni(μ-Br)₂Ni(π-allyl)–R $R = H, CH_3, CO_2C_2H_5$, etc.   (1.84)

hydrocarbon or ethereal solvents, but in polar co-ordinating media such as dimethylformamide or *N*-methylpyrrolidone, a smooth reaction takes place at room temperature between the complex and a wide variety of organic halides (iodides are best) to give the coupled product in high yield (1.85). Vinyl and aryl halides react just as well as alkyl halides, and,

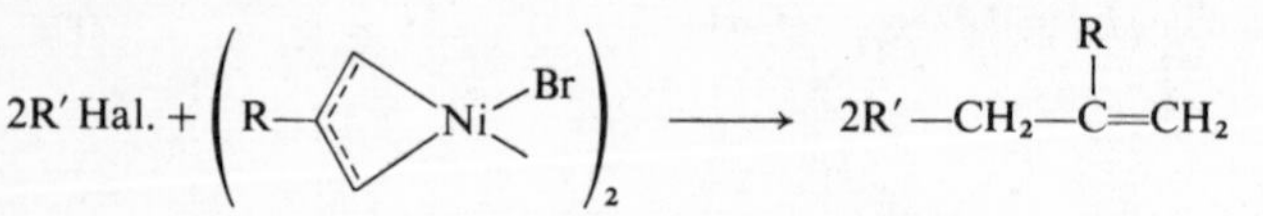

$$2R'\text{Hal.} + \left(R\text{–}(\pi\text{-allyl})\text{Ni}\text{–Br}\right)_2 \longrightarrow 2R'\text{–}CH_2\text{–}\overset{\displaystyle R}{\overset{|}{C}}{=}CH_2 \quad (1.85)$$

$R' = $ 4-hydroxycyclohexyl, $R = CH_3$ ⟶ 88%

$R' = C_6H_5$ $R = CH_3$ ⟶ 98%

helpfully, hydroxyl and carbonyl groups do not interfere with the reaction. Carbonyl groups do react with the complexes, but more slowly and at higher temperatures. With dihalides, disubstitution products can be obtained with the appropriate quantity of the complex, but, with chloro-iodides, selective reaction can be effected at the iodo group affording a route to halides useful for further chain extension or other reaction. A useful reagent is the 2-methoxyallyl nickel complex (24) which can be used to prepare substituted acetones as illustrated in (1.86) (Hegedus and Stiverson, 1974).

Br–C₆H₄–Br + (2-methylallyl)Ni(Br)$)_2$ $\xrightarrow[25°C]{\text{dimethylformamide}}$ $CH_2{=}C(CH_3)CH_2{-}C_6H_4{-}CH_2C(CH_3){=}CH_2$ (85%)

CH_3O–(allyl)Ni(Br)$)_2$ (24) $\xrightarrow{\text{RHal.}}$ $RCH_2C(OCH_3){=}CH_2$ $\xrightarrow{H_3O^+}$ RCH_2COCH_3 (1.86)

The complex from αα-dimethylallyl bromide generally undergoes preferential coupling at the primary rather than the tertiary position. In cases where substituents at C-1 and C-3 of the allyl group allow the possibility of geometrically isomeric products both isomers are usually produced (1.87).

CH_3I + (CH_3, CH_3-allyl)Ni(Br)$)_2$ ⟶ $CH_3{-}CH_2{-}CH{=}C(CH_3){-}CH_3$ (90%)

(1.87)

$C_6H_{11}I$ + (CH_3, R-allyl)Ni(Br)$)_2$ ⟶ $C_6H_{11}{-}CH_2{-}CH{=}C(CH_3)R$

(40% *cis*, 60% *trans*)

$R = (CH_3)_2C{=}CHCH_2CH_2{-}$

Exactly how the new carbon–carbon bond is formed in these reactions is still uncertain, but it is thought that a complex formed by the route shown in (1.88) may be involved.

$$\text{(π-allyl)Ni(Br)(L)} + RX \rightleftharpoons \text{(allyl)Ni(Br)(L)(R–X)} \longrightarrow \text{(allyl)Ni(Br)(R)(X)} + L$$

L = e.g. dimethylformamide (1.88)

Allyl halides undergo a more complex reaction with π-allylnickel(I) complexes due to halogen–metal exchange and cannot be used in coupling reactions of the type described above, except where the allyl group in the halide and in the complex is the same. In other cases, mixtures of products are obtained (Corey, Semmelhack and Hegedus, 1968; but see Sato, Inou, Ota and Fujita, 1972).

The reaction has been usefully extended to the synthesis of cyclic 1,5-dienes, which are very easily obtained in remarkably high yield by reaction of $\alpha\omega$-bisallyl bromides with nickel carbonyl (1.89). The pro-

$$(CH_2)_n\begin{cases}CH{=}CHCH_2Br\\CH{=}CHCH_2Br\end{cases} \xrightarrow[\text{dimethylformamide}]{Ni(CO)_4} (CH_2)_n\begin{cases}CH{=}CHCH_2\\CH{=}CHCH_2\end{cases} \quad (1.89)$$

$n = 6$ (59%)
$n = 8$ (70%)
$n = 12$ (84%)

ducts consist very largely of the *trans*, *trans* cyclic dienes, irrespective of whether the starting material contained *cis* or *trans* double bonds.

With the bisallyl bromide where $n = 2$, however, the main reaction was the formation of a *six*- and not an eight-membered ring; 4-vinylcyclohexene was the predominant product. Similarly, for $n = 4$, the product consisted entirely of *cis*- and *trans*-1,2-divinylcyclohexane; no cyclodeca-1,5-diene was detected. Evidently, in these cases, six-membered ring formation is so much favoured over eight- and ten-membered ring formation that the usual strong preference for the joining of primary carbon atoms is overcome.

This cyclisation procedure, because it leads to cyclic 1,5-dienes, makes available a wide variety of cyclic structures which are not readily obtainable by any other method, and has already been used in the synthesis of a number of natural products. The important sesquiterpene hydrocarbon humulene, for example, was readily synthesised using this method (Corey and Hamanaka, 1967) and the ester (25) was converted into the macrolide (26) (Corey and Kirst, 1972).

A number of other organic reactions leading to the formation of new carbon–carbon bonds are promoted by copper and copper ions

$$\text{(25)} \xrightarrow[\text{N-methylpyrrolidone}]{Ni(CO)_4} \text{(26)}\ (70\%) \quad (1.90)$$

(Bacon and Hill, 1965). Some of these may involve the transient formation of organocopper species. A very useful reaction is the oxidative coupling of terminal acetylenes, which is easily effected by agitation of the ethynyl compound with an aqueous mixture of cuprous chloride in an atmosphere of air or oxygen (the Glaser reaction) or by reaction of the ethynyl compound with cupric acetate in pyridine solution. Good yields of coupled products can be obtained from substrates containing a variety of functional groups, and the reaction has been applied to the synthesis of cyclic compounds from αω-diacetylenes (Eglinton and McCrae, 1963), and as a key step in the synthesis of a variety of natural products, for example the fungal polyene, corticrocin (27) (1.91). The mechanism of the reaction is not entirely clear, but it

$$HO_2CCH{=}CH(CH_2)_2C{\equiv}CH \xrightarrow{Cu_2Cl_2,\ O_2,\ H_2O,\ NH_4Cl} HO_2CCH{=}CH(CH_2)_2C{\equiv}CC{\equiv}C(CH_2)_2CH{=}CHCO_2H\ (100\%) \xrightarrow{20\%\ aq.\ KOH} HO_2C(CH{=}CH)_6CO_2H\ \text{(27)} \quad (1.91)$$

$$HC{\equiv}C(CH_2)_2OCO(CH_2)_8C{\equiv}CH \xrightarrow[\text{pyridine}]{Cu(OCOCH_3)_2} \underset{O\text{———}CO}{(CH_2)_2(C{\equiv}C)_2(CH_2)_8}\ (88\%)$$

$$\text{o-}C_6H_4(C{\equiv}CH)_2 \xrightarrow[\text{pyridine}]{Cu(OCOCH_3)_2} \text{cyclic dimer with two } C{\equiv}C{-}C{\equiv}C \text{ bridges}\ (40\%)$$

seems to be generally held that it proceeds by an initial ionisation step, facilitated probably by cuprous ion complex formation, followed by a one-electron transfer involving cupric copper and subsequent dimerisation of the radical formed. There is no doubt, however, that the reaction provides one of the easiest methods of forming carbon–carbon bonds and ranks with the Ullmann, Kolbe and acyloin reactions for molecular duplication.

The direct coupling of terminal acetylenes discussed above is obviously not so well suited for the synthesis of unsymmetrical products by coupling of two different acetylenes. Happily, this difficulty has been overcome by the discovery that, in presence of a base and a catalytic amount of cuprous ion, terminal acetylenes react rapidly with 1-bromoacetylenes with elimination of hydrogen bromide, to form the unsymmetrical diyne in high yield. The reaction is believed to take the course shown in (1.92). The second step regenerates the cuprous ion

$$\begin{aligned} RC{\equiv}CH + Cu^+ &\longrightarrow RC{\equiv}CCu + H^+ \quad \text{(fast)} \\ RC{\equiv}CCu + BrC{\equiv}CR' &\longrightarrow R(C{\equiv}C)_2R' + CuBr \end{aligned} \qquad (1.92)$$

which is best kept at low concentration to avoid self-coupling of the bromoalkyne. The addition of a base facilitates the reaction by removing the liberated acid and assisting in the solution of the cuprous derivative. This valuable reaction has already been used to synthesise a variety of unsymmetrical diynes, including a number of naturally occurring polyacetylenic compounds (1.93).

$$HO_2CCH_2C{\equiv}CH + BrC{\equiv}CC_6H_5 \xrightarrow[20^\circ C,\ C_2H_5NH_2]{Cu^+,\ H_2O} HO_2CCH_2C{\equiv}CC{\equiv}CC_6H_5 \quad (82\%)$$

$$CH_3(CH_2)_2C{\equiv}CBr + HC{\equiv}CCH{=}CHCH_2OH \xrightarrow{Cu^+,\ CH_3OH,\ H_2O,\ 20^\circ C} CH_3(CH_2)_2(C{\equiv}C)_2CH{=}CHCH_2OH \quad (84\%) \qquad (1.93)$$

1.12. Reactions of lithium organocuprates

A more general procedure for coupling unlike organohalides which is not restricted to the coupling of allylic with non-allylic halides as in the reactions discussed above, makes use of the now widely employed lithium organocuprates, which are generally represented as LiR_2Cu (Posner,

1975*b*). These reagents are more stable and more reactive than the well-known organocopper(I) reagents. They are prepared *in situ* and not isolated, most conveniently by reaction of two equivalents of an organolithium compound with cuprous iodide in ether. Aryl, alkenyl and primary alkyl cuprates are readily obtained by this procedure. Secondary and tertiary alkyl cuprates are best obtained from the corresponding lithium compound and an ether-soluble derivative of cuprous iodide such as the complex with tributylphosphine or with dimethylsulphide.

$$CH_3Li + CuI \xrightarrow{\text{ether}} CH_3Cu \xrightarrow{CH_3Li} Li(CH_3)_2Cu$$

$$2(CH_3)_3CLi + CuI(C_4H_9)_3P \xrightarrow{\text{ether}} Li[(CH_3)_3C]_2CuP(C_4H_9)_3 \quad (1.94)$$

The composition of the reagent solutions is not well defined. The state of aggregation of the complexes is uncertain, but it seems probable that in many the organic ligands are bonded to tetrahedral clusters of four metal atoms.

The complexes are excellent reagents for the specific replacement of iodine or bromine in a variety of organic substrates by alkyl, alkenyl or aryl groups. Reactions take place readily at or below room temperature to give high yields of coupled products. Ketonic carbonyl groups react only slowly under the conditions employed; aldehydes, however, undergo ordinary carbonyl addition reactions at temperatures above about –90°C. It has recently been found that the reagent prepared by addition of methyl-lithium to lithium dimethylcuprate, possibly $Li_2(CH_3)_3Cu$, reacts readily with cyclohexanones giving the axial alcohols in high yield (Still and Macdonald, 1976). Some typical reactions are shown in (1.95). Reaction with alkyl halides takes place with inversion of configuration at the carbon atom attacked. In contrast, substitution of halogen in an alkenyl halide proceeds with retention of the configuration of the double bond. Likewise, alkenylcuprates react with retention of the geometry of the double bond. Both points are exemplified in the last example given in (1.95), which also serves to emphasise the high selectivity of the reagents; neither the ester nor the lactone group was attacked. Since alkenyl free radicals are easily interconverted, these results suggest that alkenyl radicals are not involved in these reactions. Little is yet known about the mode of formation of the new carbon–carbon bond in any of the reactions.

Lithium organocuprates undergo a variety of other synthetically useful reactions besides those described above (Carruthers, 1973).

$$C_{10}H_{21}I + Li(CH_3)_2Cu \xrightarrow[0°C,\ 6\ h]{ether} C_{11}H_{24} \quad (90\%)$$

$$I(CH_2)_{10}CO_2H \xrightarrow[ether,\ -40°C]{Li(C_2H_5)_2Cu} C_{12}H_{25}CO_2H \quad (70\%)$$

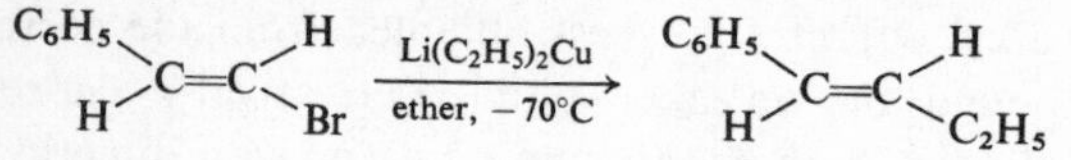

(65%)

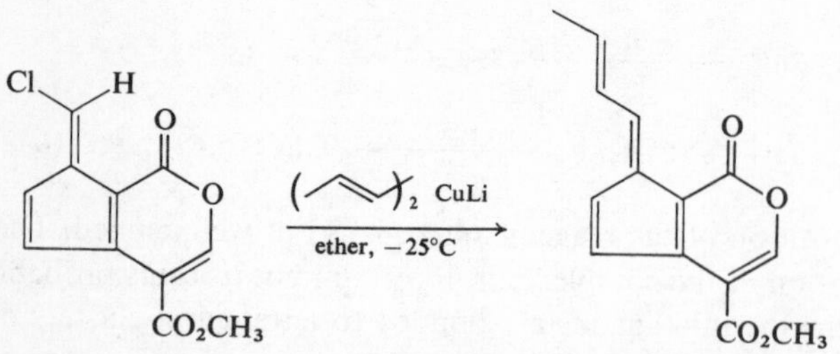

(1.95)

With acid chlorides, ketones are formed in high yield under mild conditions. Epoxides form alcohols and with $\alpha\beta$-unsaturated epoxides *trans* allylic alcohols are formed stereoselectively by 1,4-addition (1.96). In another useful application $\alpha\beta$-epoxy ketones are converted into α-alkyl-β-hydroxy ketones and thence into α-alkyl-$\alpha\beta$-unsaturated ketones, by reaction of cuprates with the corresponding epoxy ketoximes (Corey, Melvin and Haslanger, 1975).

1,4-Addition is also a feature of the reaction of organocuprates with $\alpha\beta$-unsaturated ketones (1.96) and $\alpha\beta$-acetylenic esters (see p. 131) (Posner, 1975*a*). Conjugate addition to $\alpha\beta$-unsaturated ketones gives β-substituted derivatives of the corresponding saturated ketone. The reactions are highly specific; non-conjugated carbonyl groups do not react under the conditions employed and $\alpha\beta$-unsaturated (olefinic) acids and esters are generally unaffected. Conjugate addition of Grignard reagents to $\alpha\beta$-unsaturated ketones and esters in presence of catalytic amounts of copper(I) salts has been known and used for many years (Kharasch and Tawney, 1941; Munch-Petersen, 1958), but in the former cases better results are obtained with the organocuprates. The last example in (1.96) illustrates a useful route to 1,4-diketones and γ-keto esters by conjugate addition of di(α-methoxyvinyl)cuprate to $\alpha\beta$-unsaturated ketones; in this sequence the cuprate reagent is acting, in effect, as a masked acyl anion, CH_3CO^- (Chavdarian and Heathcock, 1975; Boeckman, Bruza, Baldwin and Lever, 1975).

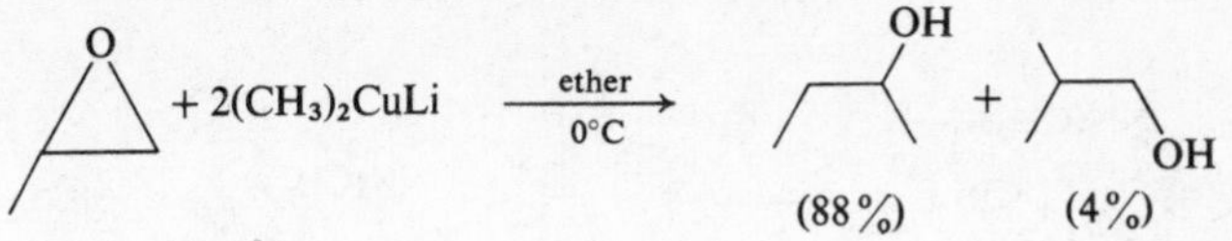

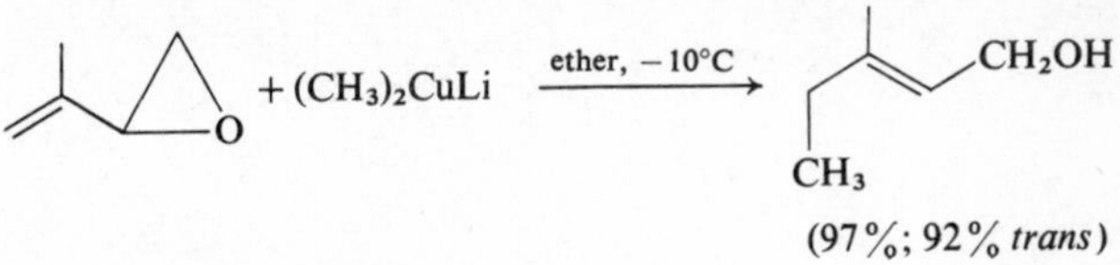

O + 2$(CH_3)_2CuLi$ —ether, 0°C→ OH (88%) + OH (4%)

+ $(CH_3)_2CuLi$ —ether, −10°C→ CH_2OH CH_3 (97%; 92% *trans*)

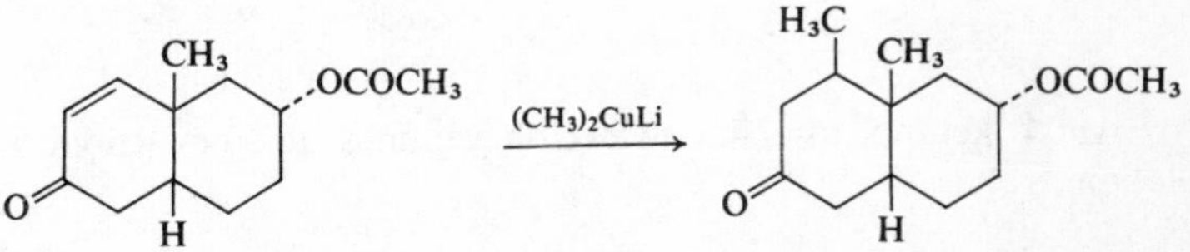

CH_3 OCOCH$_3$ O H —$(CH_3)_2CuLi$→ H_3C CH_3 OCOCH$_3$ O H

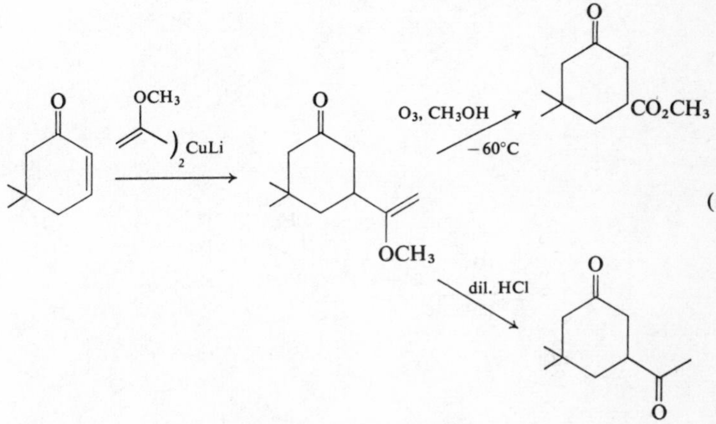

O OCH$_3$ CuLi 2 → O OCH$_3$ —O_3, CH_3OH, −60°C→ O CO_2CH_3

—dil. HCl→ O O

(1.96)

Another variation which leads to 1,5-dicarbonyl compounds has been described recently by Corey and Enders (1976). Dimethylhydrazones of ketones are readily converted into α-lithio derivatives with lithium diisopropylamide selectively at the less alkylated carbon atom (cf. p. 36). Reaction of the lithio derivatives with the cuprous iodide–isopropyl sulphide complex gives cuprate reagents which add conjugatively to αβ-unsaturated ketones and esters to give products which are converted into 1,5-dicarbonyl compounds after oxidative removal of the hydrazone group. The sequence allows a reversal of the usual mode of addition of

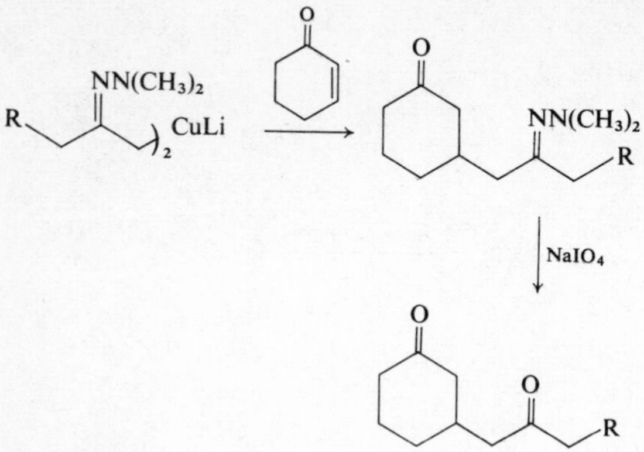

$$\text{(1.97)}$$

methyl vinyl ketone to 2-methylcyclohexanones in the Robinson annelation.

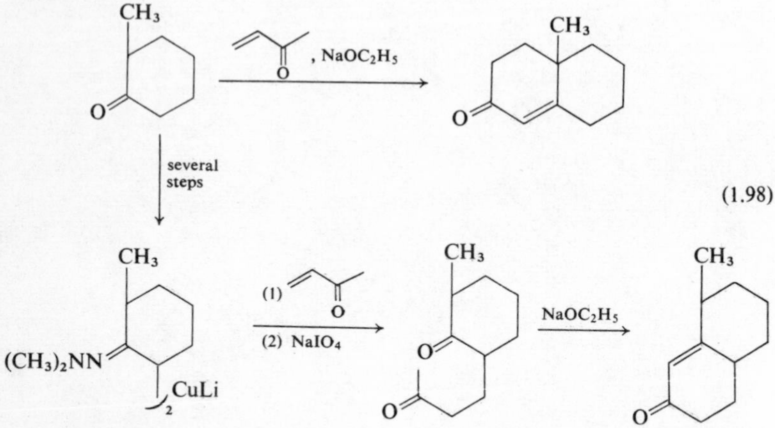

$$\text{(1.98)}$$

The mechanism of the transfer of alkyl groups from organocuprates to the β-position of conjugated ketones is still uncertain. One suggestion is that reaction takes place by one-electron transfer from the organocopper(I) species to the ketone to form an anion radical, followed by transfer of an alkyl radical from the metal to the β-carbon atom. Experimental evidence based on polarographic reduction potentials of a series of carbonyl compounds supports the first step of this proposed mechanism (House and Umen, 1972; House and Weeks, 1975). The enolate anions (28) produced initially in these reactions can be alkylated further by reaction with alkyl halides. Thus, 3-butyl-2-methyl-

$$\mathrm{R{-}CH{=}CH{-}C({=}O){-}R + CH_3{-}\dot{C}u^{I-}(ligand)\ Li^+ \longrightarrow R{-}\dot{C}H{-}CH{=}C(O^-){-}R\ Li^+ + CH_3{:}\ Cu^{II}(ligand) \longrightarrow R{-}CH(CH_3){-}CH{=}C(O^-){-}R\ Li^+\ (28) + ligand{-}\dot{C}u^{I}} \quad (1.99)$$

cyclohexanone was obtained from cyclohexenone in 84 per cent yield by successive reaction with lithium dibutylcuprate and methyl iodide. Intramolecular alkylation is also possible as in the synthesis of the *cis*-decalone (29), a precursor of the sesquiterpene valerane (Posner, Sterling, Whitten, Lentz and Brunelle, 1975). Intramolecular aldol condensations can also be effected, giving cyclic aldols with high stereoselectivity (Näf, Decorzant and Thommen, 1975).

$$\xrightarrow[0^\circ C]{(CH_3)_2CuLi} \quad (1.100)$$

CH$_2$Br, CH$_3$, O$^-$, CH$_3$

(29)

The factors controlling the stereochemistry of addition of organo-cuprates to cyclic conjugated ketones are not completely understood. Mixtures of geometrical isomers are often produced, but generally one isomer predominates, formed by approach of the reagent in a direction perpendicular to the plane of the enone system. With bicyclic enones, both steric and stereoelectronic factors play a part and the solvent may also have an effect.

A disadvantage of the straightforward reactions of organocuprates described above is that in many cases an excess of reagent has to be employed, and in the conjugate addition to enones at any rate only one of the two organo groups in the cuprate takes part in the reaction; the other is effectively wasted. To overcome these difficulties a number of 'mixed' organocuprates, R_rR_tCuLi, have been developed in which the group R_r is tightly bound to the copper and only R_t is transferred. One such reagent has pentynyl as the residual bound group, R_r. It reacts readily with cyclohexenones, transferring only the other group, R_t, to give β-substituted cyclohexanones in excellent yield (Corey and Beames, 1972) (1.101). The mixed complexes Li[CN—Cu—R],

$$C_3H_7C{\equiv}CCu \xrightarrow[\text{hexamethyl-phosphoramide, } -78°C]{RLi} [C_3H_7C{\equiv}CCuR]^- Li^+$$

↓ (cyclohexenone)

(1.101)

(> 95% for R = C_4H_9, t-C_4H_9 or vinyl) → 3-R-cyclohexanone

formed by treating cuprous cyanide with one equivalent of an alkyllithium are also very effective. They react readily with alkyl halides and $\alpha\beta$-unsaturated ketones to give products formed by transfer of R in high yield (Gorlier, Hamon, Levisalles and Wagnon, 1973).

Other promising reagents are the phenylthio derivatives Li[C_6H_5S—Cu—R], which are obtained by treatment of C_6H_5SCu with an organolithium compound. They also react readily with halides and by conjugate addition to $\alpha\beta$-unsaturated ketones; they are particularly effective for the transfer of secondary and tertiary alkyl groups, which is not always easy with ordinary organocuprates (Posner, Whitten and Sterling, 1973; Piers and Nakamura, 1975, Grieco, Chia-Lin, and Majetich, 1976; Posner and Whitten, 1976).

$$C_6H_5SLi + CuI \xrightarrow[25°C]{THF} C_6H_5SCu$$

$$C_6H_5SCu \xrightarrow{(CH_3)_3CLi,\ THF,\ -60°C} [C_6H_5S{-}Cu{-}C(CH_3)_3]Li$$

$$[C_6H_5S{-}Cu{-}C(CH_3)_3]Li \xrightarrow[-60°C]{C_6H_5COCl} C_6H_5COC(CH_3)_3\ (84\%)$$

Cyclohex-2-enone $+ \left[C_6H_5S{-}Cu{-}C(=CH_2)CH(OC_2H_5)_2\right]Li \xrightarrow{ether,\ -40°C}$ 3-[1-(diethoxymethyl)vinyl]cyclohexanone, $CH(OC_2H_5)_2$ (88%) (1.102)

1.13. Synthetic applications of carbenes and carbenoids

A carbene is a neutral intermediate containing bivalent carbon, in which a carbon atom is covalently bonded to two other groups and has two valency electrons distributed between two non-bonding orbitals. If the two electrons are spin-paired the carbene is a singlet; if the spins of the electrons are parallel it is a triplet (Kirmse, 1964; Hine, 1964).

$$\begin{matrix} A \diagdown \\ \quad C: \\ B \diagup \end{matrix}$$

A singlet carbene is believed to have a bent sp^2 hybrid structure in which the paired electrons occupy the vacant sp^2-orbital. A triplet carbene may be either a bent sp^2 hybrid with an electron in each unoccupied orbital, or a linear sp hybrid with one electron in each of the unoccupied p-orbitals. Structures intermediate between the last two are also possible (1.103). The results of experimental observations and molecular orbital calculations indicate that many carbenes have a non-linear triplet ground state. Exceptions are the dihalogenocarbenes and carbenes with oxygen, nitrogen and sulphur atoms attached to the

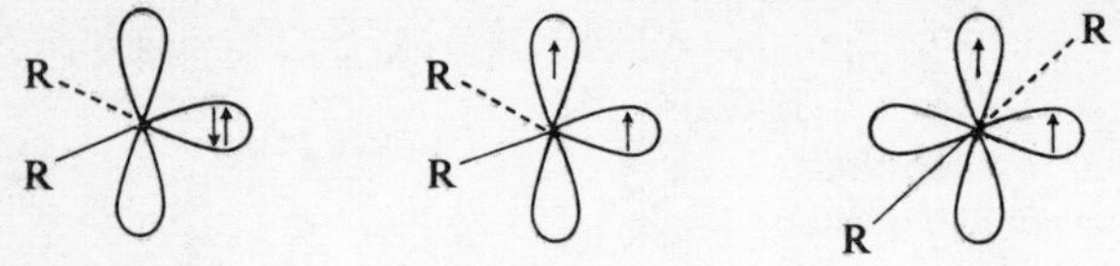

lowest singlet triplet triplet (1.103)

bivalent carbon, which are probably singlets. The singlet and triplet states of a carbene do not necessarily show the same chemical behaviour (Bethell, 1969). For example, addition of singlet carbenes to olefinic double bonds to form cyclopropane derivatives is much more stereoselective than addition of triplet carbenes.

A variety of methods is available for the generation of carbenes, but for synthetic purposes they are usually obtained by thermal or photolytic decomposition of diazoalkanes, or by α-elimination of hydrogen halide from a haloform or of halogen from a *gem*-dihalide by action of base or a metal (Jones and Moss, 1972). In many of these latter reactions it is doubtful whether a 'free' carbene is actually formed. It seems more likely that in these reactions the carbene is complexed with a metal or held in a solvent cage with a salt, or that the reactive intermediate is, in fact, an organometallic compound and not a carbene (see Kobrich, 1972; Cardin, Cetinkaya, Doyle and Lappert, 1973). Such organometallic or complexed intermediates which, while not 'free' carbenes, give rise to products expected of carbenes are usually called carbenoids (1.104). Carbenes produced by photolysis of diazoalkanes are highly energetic species and indiscriminate in their action, and photolysis is not, therefore, a good method for generating alkylcarbenes for synthesis. Thermal decomposition of diazoalkanes often produces a less energetic, and more selective, carbene, particularly in the presence of copper powder

$$RCHN_2 \xrightarrow{h\nu} [RCH{:}] + N_2$$

$$N_2CHCO_2C_2H_5 \xrightarrow{\Delta} [{:}CHCO_2C_2H_5] + N_2$$

$$CHCl_3 \xrightarrow{B^-} BH + {:}CCl_3^- \longrightarrow {:}CCl_2 + Cl^- + BH$$

$$R_2CBr_2 + R'Li \longrightarrow R_2CBrLi + R'Br \xrightarrow{?} [R_2C{:}] + LiBr \qquad (1.104)$$

or copper salts. Copper–carbene complexes are probably involved in these reactions (see p. 77). Another convenient and widely used route to alkylcarbenes is the thermal or photolytic decomposition of the lithium or sodium salts of toluene-*p*-sulphonylhydrazones. The diazoalkane is first formed and decomposes under the reaction conditions (1.105). Ketocarbenes and alkoxycarbonylcarbenes are usually produced by heating or photolysing diazoketones and diazoesters.

$$\text{cyclopropyl–CH{=}NNHSO_2C_6H_4CH_3} \xrightarrow[\text{diglyme, 150°C}]{CH_3ONa} \text{cyclopropyl–CHN_2} \longrightarrow \text{cyclopropyl–CH:} \quad (1.105)$$

Carbenes in general are very reactive electrophilic species. Their activity depends to some extent on the method of preparation, on the nature of the substituent groups R and R′ in $R{-}\ddot{C}{-}R'$, and also on the presence or absence of certain metals or metallic salts (see p. 76). Carbenes undergo a variety of reactions, including insertion into C—H bonds, addition to olefinic and acetylenic bonds, and skeletal rearrangements, some of which are useful in synthesis. Insertion into C—H bonds is not of general synthetic value because mixtures are often produced. Methylene itself attacks primary, secondary and tertiary C—H bonds indiscriminately, although other carbenes may be more selective. With alkylcarbenes intramolecular insertion into C—H bonds is the preferred course of reaction. The major product is usually an olefin, formed by insertion at the β-C—H bond; insertion at the γ-C—H bond gives a cyclopropane as a second product. In general, no intermolecular reactions are observed when intramolecular insertion is possible (1.106). Only in favourable cases where the possibilities of different reactions are limited by geometric factors are insertion reactions of carbenes synthetically

$$(CH_3)_2CHCHN_2 \longrightarrow [(CH_3)_2CHCH\!:]$$

$$\downarrow \quad (1.106)$$

$$\underset{(75\%)}{CH_3{-}C(CH_3){=}CH_2} + \underset{(25\%)}{CH_3{-}CH\langle(CH_2)_2\rangle}$$

useful. For example, camphor toluene-*p*-sulphonylhydrazone, when heated with sodium methoxide in diglyme, gives tricyclene by intramolecular insertion of the derived carbene. A similar reaction was used by Corey, Chow and Scherrer (1957) in a key step in their synthesis of α-santalene (1.107).

Br (1) H_2NNH_2, C_2H_5OH (2) HgO, CH_3OH, reflux → [Br :] → Br (68%)

(1.107)

Probably the most useful application of carbenes in synthesis is in the formation of three-membered rings by addition to olefinic and acetylenic bonds. This is a common reaction of all carbenes which do not undergo intramolecular insertion. Particularly useful synthetically are the halocarbenes, which are readily obtained from a variety of precursors (p. 72). Generated in presence of an olefin they give rise to halocyclopropanes which are valuable intermediates for the preparation of cyclopropanes, allenes, and ring-expanded products (Parham and Schweizer, 1963) (1.108). Addition of halocarbenes to olefins is a stereospecific *cis* reaction, but this is not necessarily the case with other carbenes. Intramolecular additions are also possible. The toluene-*p*-sulphonylhydrazones of *αβ*-unsaturated aldehydes and ketones, for example, on reaction with sodium methoxide in aprotic media at 160–220°C yield alkyl substituted cyclopropenes, presumably by way of the corresponding alkenylcarbene. *gem*-Dihalocyclopropenes are obtained by reaction of dihalocarbenes with disubstituted acetylenes. They have been used to prepare cyclopropenones.

Skell attributed the stereospecificity of the reactions with dihalogenocarbenes to the fact that these carbenes have a singlet ground state which allows concerted addition of the carbene to the olefin, since the two new σ-bonds of the cyclopropane can be formed without changing the spin of any of the electrons involved. Addition of triplet carbenes, on the other hand, would be stepwise, going through a triplet diradical intermediate, with the possibility of rotation about one of the bonds before spin inversion and closure to the cyclopropane could take place. It is now believed that the addition of dihalogenocarbenes is one of a general class of non-linear, cheletropic reactions which are symmetry-

$$(CH_3)_2C{=}C(CH_3)CH{=}NNHTos \xrightarrow[\text{diglyme, } 160\text{–}220\,^\circ C]{CH_3ONa,} \left[(CH_3)_2C{=}C(CH_3)\ddot{C}H\right] \longrightarrow \text{1,3,3-trimethylcyclopropene} \ (72\%)$$

$$H_3C\text{–}CH{=}CH\text{–}CH_3 \xrightarrow[t\text{-}C_4H_9OK \ [:CBr_2]]{CHBr_3} \text{1,1-dibromo-2,3-dimethylcyclopropane} \ (79\%) \xrightarrow[\text{ether}]{LiAlH_4} \text{1,2-dimethylcyclopropane}$$

$$\text{6,6-dibromobicyclo[3.1.0]hexane} \ (OH_2, \ Ag^+) \xrightarrow[H_2O]{AgNO_3} \text{2-bromocyclohex-2-en-1-ol} \qquad (1.108)$$

allowed and concerted. A necessary feature of such reactions is that both electrons must be in the same orbital, as they are in singlet carbenes.

Addition of carbenes to aromatic systems to form ring-expanded products is a valuable reaction in synthesis. Methylene itself, formed by photolysis of diazomethane, adds to benzene to form cycloheptatriene in 32 per cent yield; a small amount of toluene is also formed by an insertion reaction. Better yields of cycloheptatriene are obtained in the presence of copper salts (p. 76). One of the oldest known carbene reactions is the addition of ethoxycarbonylcarbene, from diazoacetic ester, to benzene to form a mixture of cycloheptatrienylcarboxylic esters. Under the conditions of the reaction the intermediate norcaradiene undergoes a Cope rearrangement to give the ring expanded product. In the reaction with polycyclic aromatic hydrocarbons the norcaradiene can often be isolated (Dave and Warnhoff, 1970; Marchand and Brockway, 1974). Dichlorocarbene also adds to some aromatic compounds. It is insufficiently reactive to attack benzene, but it adds readily to more reactive aromatic compounds. In most cases ring expansion proceeds spontaneously. Pyrroles and indoles, for example, in neutral solution, are readily converted into pyridines and quinolines (1.109). Under basic conditions substitution products are also formed in a Reimer–Tiemann type reaction (Jones and Rees, 1969). The well-known Reimer–Tiemann reaction, of course, proceeds by electrophilic attack of dichlorocarbene on a phenoxide anion.

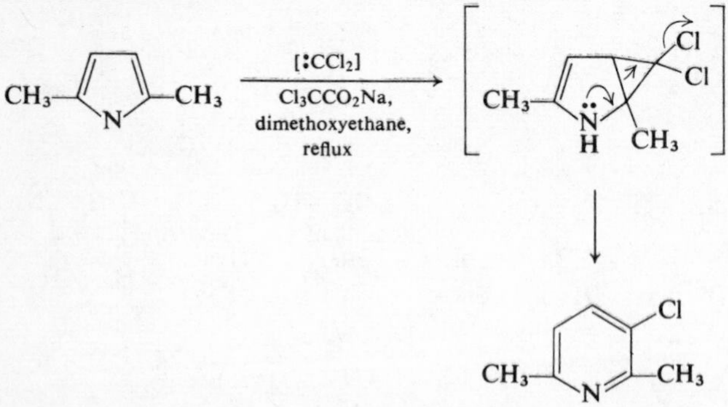
CH₃
CH₃
N
[:CCl₂]
Cl₃CCO₂Na,
dimethoxyethane,
reflux
CH₃
N
H
CH₃
Cl
Cl
Cl
CH₃
N
CH₃

(1.109)

The composition of the mixture of products obtained from reactions of carbenes is profoundly altered by the presence of certain transition metals, notably copper and its salts. Under these conditions the intermediates obtained, for example, by decomposition of diazo compounds are more selective than 'free' carbenes. Insertion reactions are suppressed and higher yields of addition products are obtained in reactions with olefins and aromatic compounds (Müller, Kessler and Zeeh, 1966). Thus, benzene reacts readily with diazomethane in the presence of cuprous chloride to form cycloheptatriene in 85 per cent yield. The reaction is general for aromatic systems, substituted benzenes giving a mixture of the corresponding substituted cycloheptatrienes. Related reactions of cyclic and acyclic olefins produce cyclopropanes in good yield and with complete retention of configuration. Intramolecular addition to olefinic double bonds also takes place readily in presence of cuprous salts, and this was exploited by Corey and Achiwa (1969) in a

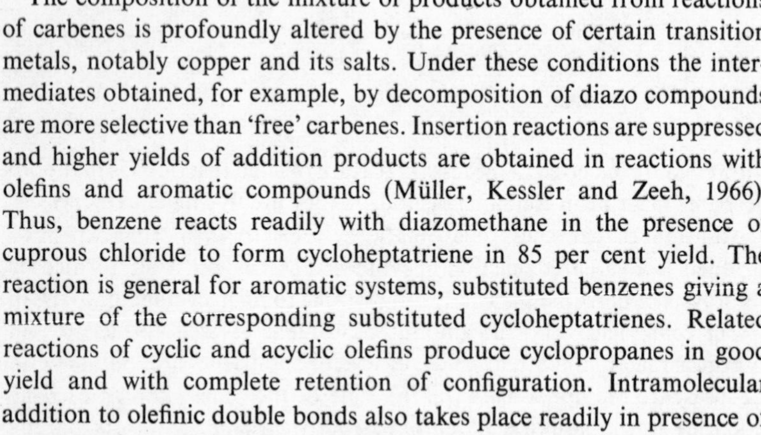
CH
NNH₂
MnO₂
CH₂Cl₂, 0°C
N₂CH
Cu₂I₂

(30)

(1.110)

neat synthesis of the hydrocarbon sesquicarene (30) from *cis,trans*-farnesal (1.110). Unsaturated diazoketones and diazoacetic esters likewise form intramolecular addition products in the presence of cuprous salts and reactions of this type have been used to prepare bridged bicyclic and 'cage' ketones (1.111). In the absence of copper the main reaction of diazoketones is the Wolff rearrangement (p. 79).

$$CH_2{=}CH(CH_2)_3COCl \xrightarrow{CH_2N_2} CH_2{=}CH(CH_2)_3COCHN_2 \xrightarrow{Cu,\ 80^\circ C} \text{bicyclic ketone } (50\%) \quad (1.111)$$

$$N_2CHCO_2CH_2CH{=}CHCH_3 \xrightarrow[\text{cyclohexane, boil}]{Cu_2O} \text{bicyclic lactone } (CH_3, H) \ (66\%)$$

It is very unlikely that 'free' carbenes are involved in any of these catalysed reactions, and cyclopropane formation is believed to involve a copper–carbene–olefin complex similar to that invoked in the Simmons–Smith reaction (1.113). The view that the reactive intermediate is some kind of carbene–copper complex gains strong support from kinetic studies and from the observation that catalysed decomposition of ethyl diazoacetate with optically active copper complexes in presence of styrene gave a mixture of cyclopropane derivatives which was optically active (Moser, 1969; Cowell and Ledwith, 1970).

Related to these copper-catalysed reactions of diazoalkanes and diazoketones is the valuable Simmons–Smith reaction (Wendisch, 1971; Simmons, Cairns, Vladuchick and Hoiness, 1973), which is widely used for the synthesis of cyclopropane derivatives from olefins by reaction with methylene iodide and zinc–copper or, better, zinc–silver couple (Denis and Conia, 1972). This is a versatile reaction and has been applied with success to a wide variety of olefins. Many functional groups are unaffected, making possible the formation of cyclopropane derivatives. Dihydrosterculic acid, for example, was obtained from methyl oleate in 51 per cent yield (1.112). The reaction is stereospecific and takes place

$$CH_3(CH_2)_7CH{=}CH(CH_2)_7CO_2CH_3 \xrightarrow[\text{(2) } H_2O,\ NaOH]{\text{(1) } CH_2I_2,\ Zn\text{–}Cu,\ \text{ether}} \text{cyclopropane: } CH_3(CH_2)_7,\ H,\ H,\ (CH_2)_7CO_2H \quad (51\%) \tag{1.112}$$

by *cis* addition of methylene to the less hindered side of the double bond. The reactive intermediate is believed to be an iodomethylenezinc iodide complex which reacts with the olefin in a bimolecular process to give a cyclopropane and zinc iodide (1.113). A valuable feature of the

$$\text{olefin} + ICH_2ZnI \longrightarrow [\text{transition state}] \longrightarrow \text{cyclopropane} + ZnI_2 \tag{1.113}$$

reaction in synthesis is the stereochemical control exerted on the developing cyclopropane ring by a suitably situated hydroxyl group in the olefin. With allylic and homoallylic alcohols or ethers, the rate of the reaction is greatly increased and in five- and six-membered cyclic allylic alcohols the product in which the cyclopropane ring is *cis* to the hydroxyl group is formed stereospecifically. A methoxycarbonyl group in the α-position to the double bond has a similar directing effect (1.114). These effects are ascribed to co-ordination of oxygen to the zinc,

$$\xrightarrow[\text{ether}]{CH_2I_2,\ Zn\text{–}Cu} \tag{1.114}$$

followed by transfer of methylene to the nearer face of the adjacent double bond. In eight- and nine-membered cyclic allylic alcohols, reaction with methylene iodide and zinc–copper couple takes place stereospecifically to give products in which the cyclopropane ring is *trans* to the hydroxyl group. Presumably, in these reactions the conformation of the ring in the intermediate is such that it is the *trans* face of the double bond which is nearer to the co-ordinated reagent (Poulter, Friedrich and Winstein, 1969).

High yields of cyclopropanes can also be obtained from olefins and ethylidene iodide using diethylzinc as catalyst; zinc–copper couple was ineffective. Reaction is again stereospecific and with cyclic allylic and homoallylic alcohols gives the *cis,anti*-cyclopropane only (Nishimuru, Kawabata and Furukawa, 1969) (1.115).

$\xrightarrow[\text{isopropyl ether, reflux}]{CH_3CHI_2,\ Zn(C_2H_5)_2}$ (45%) (1.115)

Carbenes undergo a number of skeletal rearrangements, some of which are useful in synthesis. The most important of these is the Wolff rearrangement of diazoketones to ketenes, which is brought about by heat, light or by action of some metallic catalysts. This reaction is the key step in the well-known Arndt–Eistert method for converting a carboxylic acid into its next higher homologue (Bachman and Struve, 1942) (1.116). With cyclic diazoketones the rearrangement leads to

$$RCOCl \xrightarrow{CH_2N_2} RCOCHN_2 \longrightarrow [R\text{–}C(=O)\text{–}\ddot{C}H] \longrightarrow RCH{=}C{=}O \xrightarrow{R'OH} RCH_2CO_2R' \quad (1.116)$$

ring contraction, and this reaction has been widely used to prepare derivatives of strained small-ring compounds such as bicyclo[2,1,1]-hexane and benzocyclobutene (Meinwald and Meinwald, 1966) (1.117).

$\xrightarrow[h\nu]{CH_3OH}$ CO_2CH_3 (54%)

$\xrightarrow[h\nu]{H_2O,\ THF}$ CO_2H (20%) (1.117)

Migration of alkyl groups in alkyl- and dialkyl-carbenes does not occur easily, but cyclopropylcarbenes are exceptional and rearrange to cyclobutenes in high yield (Gutsche and Redmore, 1968; Bird, Frey and Stevens, 1967).

There is no clear evidence that a carbene is involved in any of these rearrangements. A concerted migration and expulsion of nitrogen is usually a valid alternative (1.118).

$$-\underset{\substack{|\\R}}{\overset{|}{C}}-\overset{-}{C}H-\overset{+}{N_2} \longrightarrow \;\rangle C{=}CHR + N_2 \qquad (1.118)$$

1.14. Some photocyclisation reactions

Few areas of organic chemistry have been more productive of new synthetic reactions in recent years than organic photochemistry (see, for example, Schönberg, 1968; Bryce-Smith, 1974). In general, absorption of light by an organic molecule can produce three types of activated molecule not accessible by normal thermal means, namely the electronically excited singlet and triplet states, and, often, a vibrationally 'hot' ground state. Each of these excited states may undergo different chemical reactions in proceeding back to the ground state (Neckers, 1967). The triplet excited state, which generally has a relatively long lifetime, is frequently encountered in photochemical reactions. Because of the high energy of the excited states photochemical reactions often lead to strained structures which would be difficult to obtain by thermal reactions.

Many photochemical transformations are of value in synthesis, and some are referred to elsewhere (see pp. 241, 380). Particularly useful are photoelectrocyclic reactions of the type (1.119), which lead to the

$$\text{(hexatriene)} \rightleftharpoons \text{(1,3-cyclohexadiene)} \qquad (1.119)$$

interconversion of acyclic conjugated trienes and 1,3-cyclohexadienes. The formation of an acyclic hexatriene from a 1,3-cyclohexadiene was first noted in the case of the light-induced formation of calciferol from ergosterol via precalciferol. More recent studies show that this ring cleavage reaction takes place with a number of other simple and complex 1,3-cyclohexadienes and may be fairly general (see Barton, 1959; Corey and Hortmann, 1963). The reaction is reversible and under appropriate conditions certain acyclic 1,3,5-trienes are converted into cyclohexadienes. Thus, irradiation of *trans,cis,trans*-2,4,6-octatriene in ether affords a stationary state containing 10 per cent of *trans*-1,2-dimethyl-3,5-cyclohexadiene. By far the most useful application of this reaction in synthesis is in the conversion of stilbene derivatives, in which two of the double bonds of the 'triene' are contained in benzene rings, into 4*a*,5*a*-dihydrophenanthrenes and thence into phenanthrenes (1.120). Stilbene

hν

oxidation

H

H

(1.120)

undergoes a rapid *cis–trans* isomerisation under the influence of ultraviolet light, and *cis*-stilbene, upon further irradiation, cyclises to give the dihydrophenanthrene. The dihydrophenanthrene has not been isolated but there is convincing evidence for its formation (Moore, Morgan and Stermitz, 1963; Cuppen and Laarhoven, 1972). In presence of mild oxidising agents such as oxygen or iodine it is readily converted into phenanthrene, and the sequence of reactions has been widely used to prepare a variety of phenanthrene derivatives containing alkyl, chloro, bromo, methoxy, phenyl, and carboxyl substituents from the appropriately substituted stilbene derivatives. For example, 1,2,7,8-tetramethyl phenanthrene was obtained from 2,2′,3,3′-tetramethylstilbene in 50 per cent yield by irradiation of a dilute solution in hexane in presence of iodine; a minor product in this reaction was 1,2,5-trimethylphenanthrene formed by cyclisation in the alternative direction with expulsion of a methyl group. Polycyclic aromatic hydrocarbons can also be obtained by photocyclisation of the appropriate stilbene analogues (Blackburn and Timmons, 1969; Laarhoven, Cuppen and Nivard, 1970; Tinnemans and Laarhoven, 1974). Thus 1-α-styrylnaphthalene is converted into chrysene and 2-α-styrylnaphthalene gives benzo[*c*]phenanthrene. In the latter case, cyclisation takes place at the 1-position of the naphthalene nucleus only. No benz[*a*]anthracene, formed by cyclisation at the 3-position was obtained. The convenience of this synthetic method compensates for the moderate yields obtained in some cases. The reaction has also been very useful for the preparation of helicenes and heterohelicenes (Dopper, Oudman and Wynberg, 1973; Martin, 1974).

Heterocyclic aromatic compounds can also be obtained by this route. Thus, 2-α-styrylpyridine gave 1-azaphenanthrene and 2-α-styrylthiophen cyclised to naphtho[2,1-*b*]thiophen. Azobenzene derivatives and Schiff's bases are cyclised to the corresponding benzocinnoline and phenanthridine derivatives on irradiation in strongly acid solution. The anil (31), for example, gave calycanine, a degradation product of the alkaloid calycanthine (1.121).

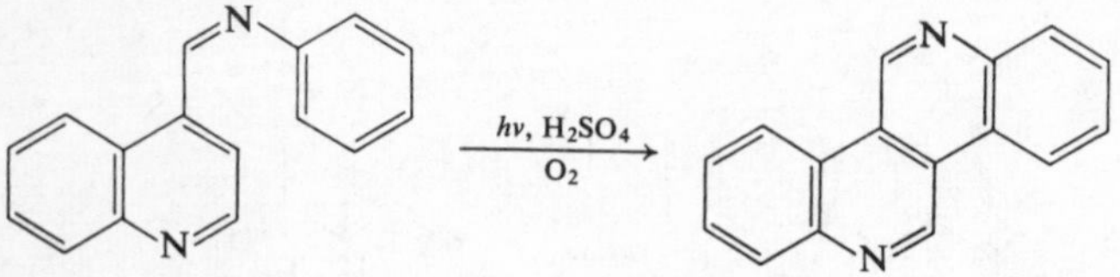

(31) $\xrightarrow[O_2]{h\nu,\ H_2SO_4}$ (53%)

(1.121)

CH_3O, CH_3O, $NCO_2C_2H_5$ $\xrightarrow[I_2]{h\nu,\ C_2H_5OH}$ CH_3O, CH_3O, $NCO_2C_2H_5$

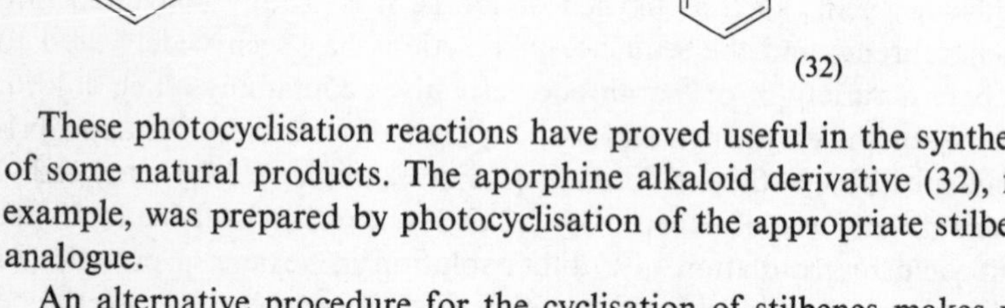

(32)

These photocyclisation reactions have proved useful in the synthesis of some natural products. The aporphine alkaloid derivative (32), for example, was prepared by photocyclisation of the appropriate stilbene analogue.

An alternative procedure for the cyclisation of stilbenes makes use of the ready photolysis of the carbon–iodine bond in iodoaromatic compounds. It had been found earlier that photolysis of iodoaromatic compounds in aromatic solvents, particularly benzene, was useful for the preparation of biphenyl and polyphenyl derivatives. In an intramolecular version of the reaction 2-iodostilbene gave phenanthrene in

90 per cent yield by photolysis in hexane solution. This procedure is particularly useful for the preparation of nitrophenanthrenes which cannot be obtained by irradiation of nitrostilbenes (cf. Kupchan and Wormsey, 1965). It seems likely that these reactions proceed by a radical mechanism and not by way of a dihydrophenanthrene-type of intermediate.

A number of alkaloids of the proaporphine, homoproaporphine and morphinan–dienone types have been prepared by photolysis of phenolic bromoisoquinolines in alkaline solution. (±)-Pronuciferine (33), for example, was obtained as shown in (1.122) (Kametani, Shibuya, Nakano and Fukumoto, 1971).

CH_3O CH_3O NCH_3 Br HO $\xrightarrow[CH_3OH,\ H_2O,\ NaOH]{h\nu}$ CH_3O CH_3O NCH_3 O (33) (1.122)

N-Benzoylenamines of cyclohexanones also cyclise stereoselectively on irradiation to form *trans*-octahydrophenanthridones (Ninomiya, Naito and Kiguchi, 1973), and this reaction also has been applied to the synthesis of some alkaloid structures. (±)-Crinan, for example, the basic ring system of the *Lycoris* alkaloids was synthesised. A related method was used to prepare protoberberine alkaloids (Ninomiya, Naito and Takasugi, 1975) (1.123).

Another reaction of practical value in synthesis is the cyclo-addition of olefins to double bonds to form four-membered rings. Most simple olefins absorb in the far ultraviolet and in the absence of sensitisers undergo mainly fragmentations and *trans–cis* isomerisation, but conjugated olefins which absorb at longer wavelengths form cyclo-addition compounds readily (Dilling, 1966; Eaton, 1968). Thus butadiene on irradiation in dilute solution with light from a high pressure mercury arc forms cyclobutene and bicyclo[1,1,0]butane. Many substituted acyclic 1,3-butadienes behave similarly and conjugated cyclic dienes also are often converted into cyclobutenes by direct irradiation. Thus, *trans,trans*-2,4-hexadiene gives *cis*-3,4-dimethylcyclobutene and the anhydride (34) forms the tricyclic compound (35), a reaction which was

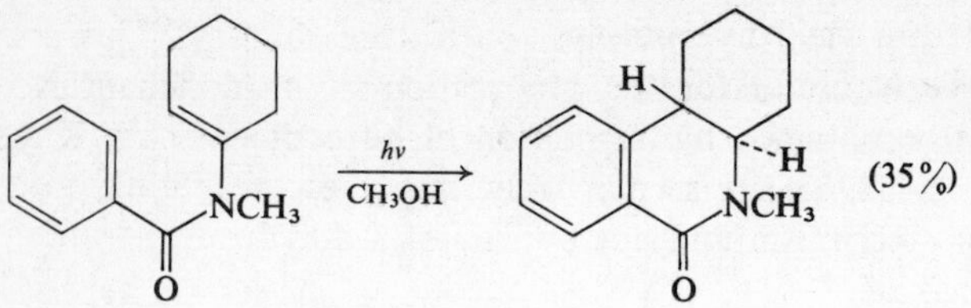

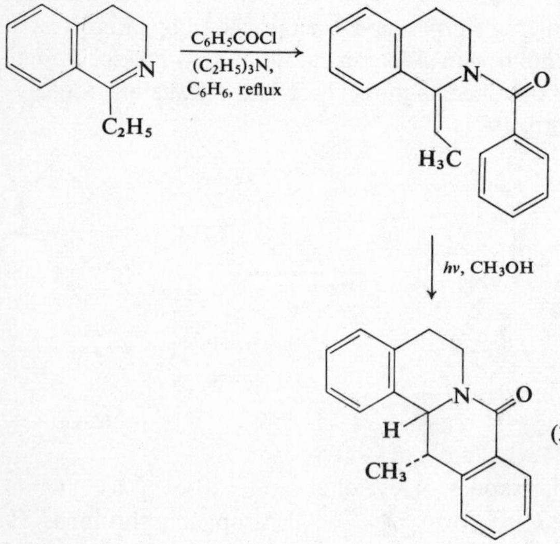

$h\nu$, CH_3OH (35%)

C_6H_5COCl, $(C_2H_5)_3N$, C_6H_6, reflux

$h\nu$, CH_3OH (1.123)

(30%)

exploited in the preparation of 'Dewar-benzene' (van Tamelen, Pappas and Kirk, 1971). These cyclisations, and the reverse ring-opening reactions, are stereospecific. The stereochemical course of the reactions is controlled by the symmetries of the engaged orbitals in reactants and products – in general terms by the number of electrons involved and the conditions, thermal or photochemical, of the experiment (Woodward and Hoffmann, 1970; Gilchrist and Storr, 1972).

In the presence of a sensitiser reaction may follow a different course. Thus butadiene itself on irradiation with a sensitiser dimerises to form,

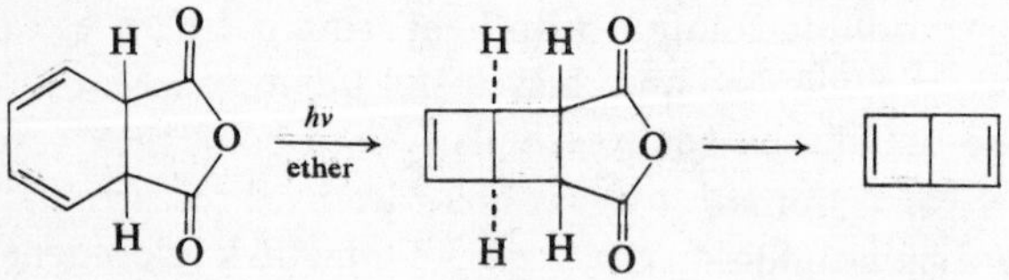

$h\nu$, ether (1.124)

(34) (35)

mainly, *trans*-1,2-divinylcyclobutane, and many other acyclic dienes behave in a similar way. Myrcene, on the other hand, which on direct irradiation in ether solution affords a mixture of the cyclobutene (36) and β-pinene, in presence of ketonic sensitisers gives only the bridged ring compound (37) (1.125). The sensitised reaction proceeds by way of a

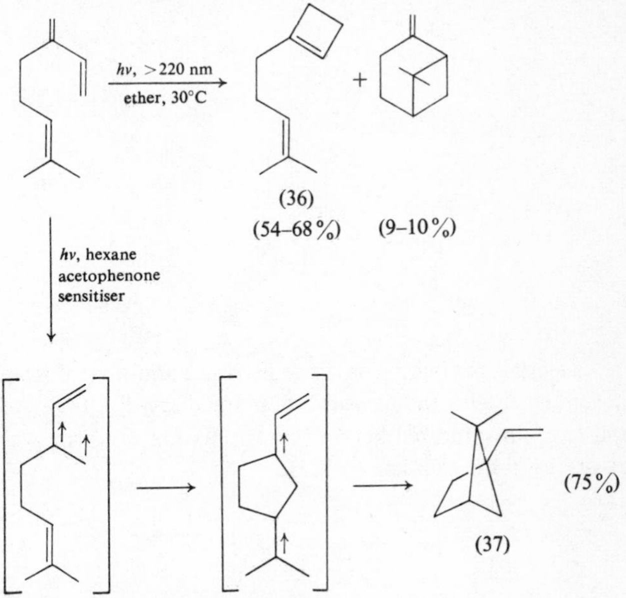

$h\nu$, >220 nm
ether, 30°C

(36)
(54–68%) (9–10%)

$h\nu$, hexane
acetophenone
sensitiser

(1.125)

(37) (75%)

triplet excited state. This state behaves as a diradical and the two bond-forming steps occur consecutively. It is found in such cases that, where a choice is available, the initial bond formation takes place to give a five-membered ring rather than a smaller or larger ring.

$\alpha\beta$-Unsaturated carbonyl compounds also readily undergo intramolecular cyclo-addition reactions. One of the first-recorded examples was the conversion of carvone into carvone camphor (38). Many other cases have since been observed. Thus 1,5-hexadien-3-one gives the highly strained bicyclo[2,1,1]hexan-2-one (1.126).

In general, these reactions are brought about by irradiation with light of wavelength greater than 300 nm. Reaction is thought to take place through a triplet excited state of the enone formed by intersystem crossing from the initial $n \rightarrow \pi^*$ excited singlet.

Intermolecular addition of olefins to $\alpha\beta$-unsaturated ketones is also of value in synthesis (Sammes, 1970; de Mayo, 1971). Thus, the first step

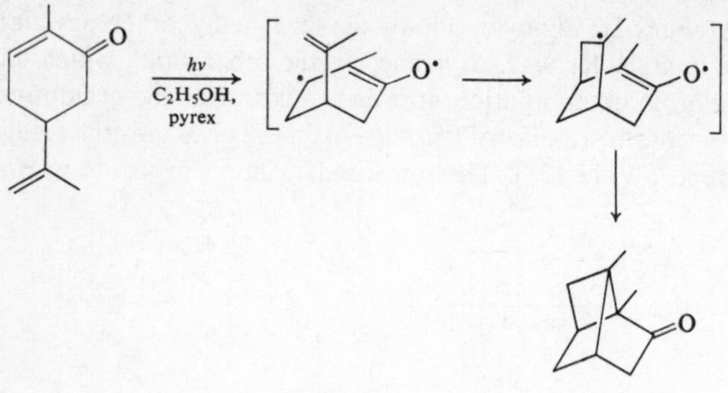

$h\nu$, C_2H_5OH, pyrex

(1.126)

(38)

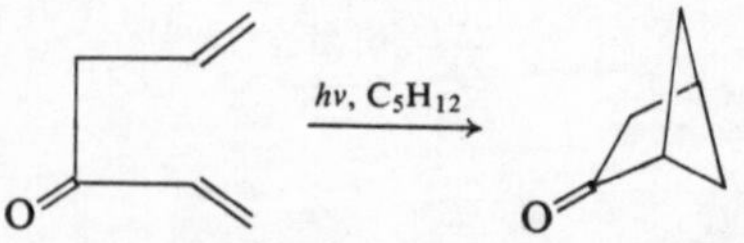

$h\nu$, C_5H_{12}

(30%)

in Corey's synthesis of caryophyllene involved addition of isobutene to cyclohexenone to give the *trans*-cyclobutane derivative (39). Acetylenes also add to $\alpha\beta$-unsaturated ketones on irradiation, to form cyclobutene derivatives, as in the example (1.127).

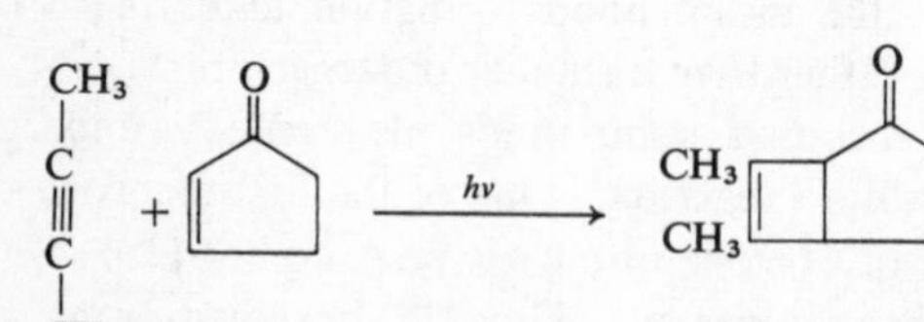

$h\nu$, >250 nm, C_5H_{12}

(26%)

(39)

(1.127)

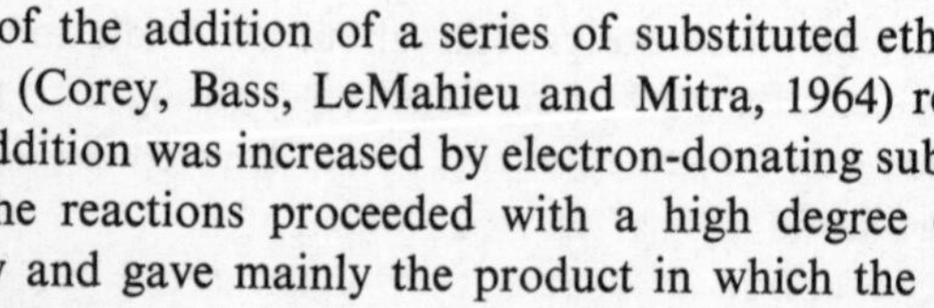

CH_3–C≡C–CH_3 + cyclopentenone $\xrightarrow{h\nu}$ (25%)

Study of the addition of a series of substituted ethylenes to cyclohexenone (Corey, Bass, LeMahieu and Mitra, 1964) revealed that the ease of addition was increased by electron-donating substituents on the olefin. The reactions proceeded with a high degree of orientational specificity and gave mainly the product in which the α-carbon of the enone was attached to the more nucleophilic carbon of the olefin.

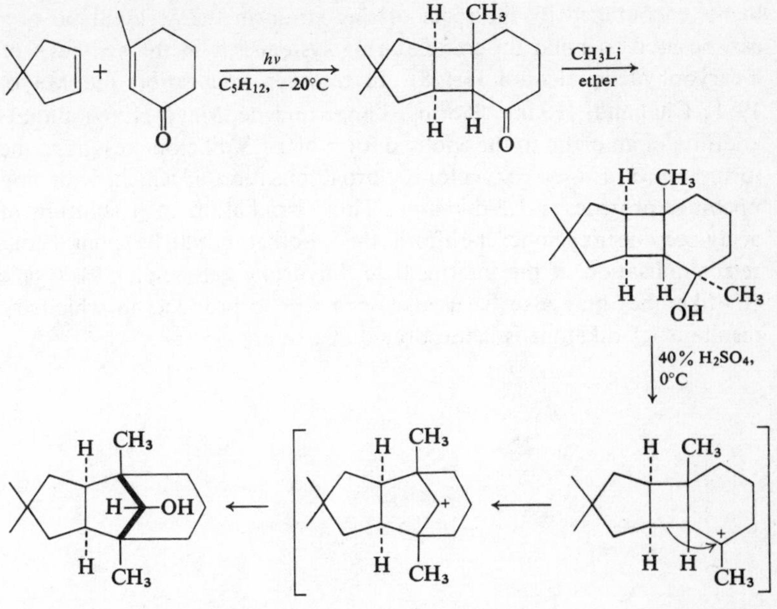

+ $\xrightarrow[C_5H_{12},\ -20°C]{h\nu}$ H CH_3 H H O $\xrightarrow[\text{ether}]{CH_3Li}$

H CH_3 H H CH_3 OH

40% H_2SO_4, 0°C

H CH_3 H–OH H CH_3 ← [H CH_3 + H CH_3 ← H CH_3 + H H CH_3]

(1.128)

Addition of *cis*- and *trans*-2-butene to cyclohexenone gave the same mixture of products in each case, suggesting that the reactions proceed in a stepwise manner through radical intermediates.

The synthetic usefulness of these cyclo-addition reactions extends beyond the immediate formation of cyclobutane derivatives. Rearrange-

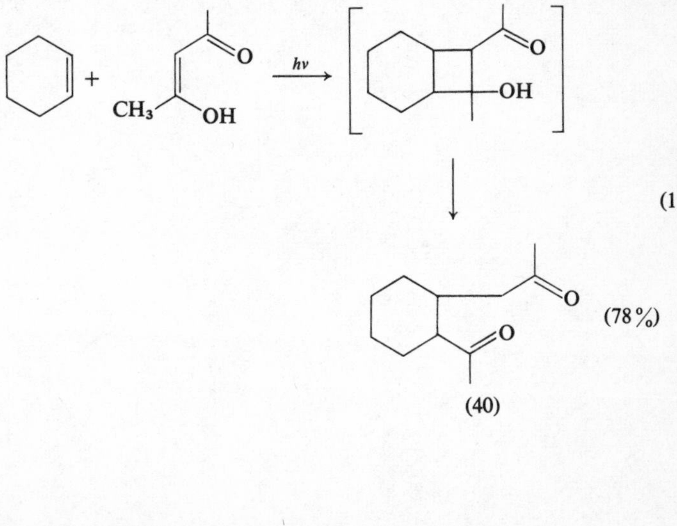

+ O CH_3 OH $\xrightarrow{h\nu}$ [O OH]

(1.129)

O O (78%)

(40)

ments encouraged by the relief of ring strain in the cyclobutane ring can be used to build up complex ring systems, as in the synthesis of α-caryophyllene alcohol (1.128). In another application (de Mayo, 1971; Challand, Hikino, Kornis, Lange and de Mayo, 1969) photo-addition of an olefin to the enolised form of a 1,3-diketone results in the formation of a β-hydroxyketone, retroaldolisation of which, with ring opening, provides a 1,5-diketone. Thus, irradiation of a solution of acetylacetone in cyclohexene affords the 1,5-diketone (40) by spontaneous retroaldolisation of the intermediate β-hydroxy ketone (1.129). Cyclic 1,3-diketones may also be used, giving rise to products in which the resultant 1,5-diketone is incorporated in a ring.

2 Formation of carbon–carbon double bonds

2.1. β-Elimination reactions

The formation of carbon–carbon double bonds is important in synthesis, not only for the obvious reason that the compound being synthesised may contain a double bond but also because formation of the double bond followed by reduction may be the most convenient route to a new carbon–carbon single bond.

One of the most commonly used methods for forming carbon–carbon double bonds is by β-elimination reactions of the type

$$\underset{\mathrm{H}\ \ \mathrm{X}}{-\overset{|}{\underset{|}{\mathrm{C}}}-\overset{|}{\underset{|}{\mathrm{C}}}-} \longrightarrow -\overset{|}{\mathrm{C}}=\overset{|}{\mathrm{C}}- + \mathrm{HX} \qquad (2.1)$$

where X = e.g. OH, OCOR, halogen, OSO_2Ar, $\overset{+}{N}R_3$, $\overset{+}{S}R_2$. Included among these reactions are acid-catalysed dehydrations of alcohols, solvolytic and base-induced eliminations from alkyl halides and sulphonates, and the well-known Hofmann elimination from quaternary ammonium salts. They proceed by both E1 and E2 mechanisms and synthetically useful reactions are found in each category (Saunders, 1964; Bunnett, 1969). Some examples are given in (2.2). These reactions,

$$\text{[bicyclic alcohol, HO]} \xrightarrow{PCl_5} \text{[bicyclic alkene]} \qquad (2.2)$$

$$CH_3CHBrCH_2CH_3 \xrightarrow[C_2H_5OH]{C_2H_5ONa} \underset{(81\%\ \text{of product})}{CH_3CH{=}CHCH_3} + \underset{(19\%\ \text{of product})}{CH_3CH_2CH{=}CH_2}$$

$$\underset{\overset{|}{^{+}S(CH_3)_2}}{CH_3CHCH_2CH_3} \xrightarrow[C_2H_5OH]{C_2H_5ONa} \underset{(26\%\ \text{of product})}{CH_3CH{=}CHCH_3} + \underset{(74\%\ \text{of product})}{CH_3CH_2CH{=}CH_2}$$

$$\underset{\overset{|}{^{+}N(CH_3)_3I^-}}{CH_3CH(CH_2)_2CH_3} \xrightarrow[130^\circ C]{H_2O,\ KOH} \underset{(98\%\ \text{of product})}{CH_2{=}CH(CH_2)_2CH_3} + \underset{(2\%\ \text{of product})}{CH_3CH{=}CHCH_2CH_3}$$

although often used, leave much to be desired as synthetic procedures. One disadvantage is that in many cases elimination can take place in more than one way, so that mixtures of products are obtained. Mixtures of geometrical isomers may also be formed (see below). The direction of elimination in unsymmetrical compounds is governed largely by the nature of the leaving group, but may be influenced to some extent by the experimental conditions. It is found in general that acid-catalysed dehydration of alcohols, and other E1 eliminations, as well as eliminations from alkyl halides and arylsulphonates with base, gives rise to the more highly substituted olefin as the principal product (the Saytzeff rule), whereas base-induced eliminations from quaternary ammonium salts and from sulphonium salts gives predominantly the less substituted olefin (the Hofmann rule). Exceptions to both rules are observed, however. If there is a conjugating substituent such as CO or C_6H_5 at one β-carbon atom, elimination will take place towards that carbon to give the conjugated olefin, irrespective of the method used. For example, (*trans*-2-phenylcyclohexyl)trimethylammonium hydroxide on Hofmann elimination gives 1-phenylcyclohexene exclusively; none of the 'expected' 3-phenylcyclohexene is detected. Another exception is found in the elimination of hydrogen chloride from 2-chloro-2,4,4-trimethylpentane, which gives mainly the terminal olefin; presumably in this case the intermediate leading to the expected Saytzeff product is destabilised by steric interaction between a methyl group and the t-butyl substituent (2.3). An additional disadvantage of acid-catalysed dehydration of

$$(CH_3)_3C{-}CH_2{-}C(CH_3)_2{-}Cl \xrightarrow{\text{base}} (CH_3)_3C{-}CH_2{-}C(CH_3){=}CH_2\ (81\%) + (CH_3)_3C{-}CH{=}C(CH_3){-}CH_3\ (19\%) \qquad (2.3)$$

alcohols, and other E1 eliminations, is that since they proceed through an intermediate carbonium ion, elimination is frequently accompanied by rearrangement of the carbon skeleton. Thus, if the α-hydroxy compound in the first equation of (2.2) is replaced by the β-isomer, dehydration is accompanied by ring contraction to give an isopropylidenecyclopentane derivative. Numerous other examples are known in the terpene series, as in the conversion of camphenilol into santene (2.4).

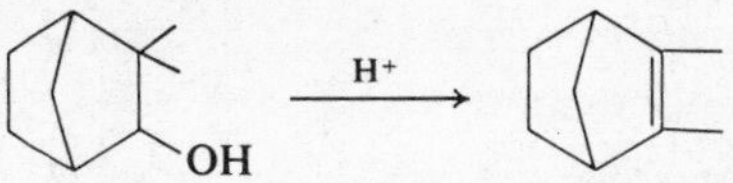

OH $\xrightarrow{H^+}$ (2.4)

The Hofmann reaction with quaternary ammonium salts and base-induced eliminations from alkyl halides and arylsulphonates are generally *anti* elimination processes, that is to say the hydrogen atom and the leaving group depart from opposite sides of the incipient double bond (Saunders, 1964; Cope and Trumbull, 1960) although examples involving *syn* elimination are known (Bailey and Saunders, 1970; DePuy, Naylor and Beckman, 1970; Saunders and Cockerill, 1973). This has been established in a number of ways. For example, reaction of ethanolic potassium hydroxide with *meso*-stilbene dibromide gave *cis*-bromostilbene whereas the (±)-dibromide gave the *trans*-stilbene. Again, the quaternary ammonium salts derived from *threo*- and *erythro*-1,2-diphenylpropylamine were found to undergo stereospecific elimination on treatment with sodium ethoxide in ethanol as shown in (2.5). In agreement

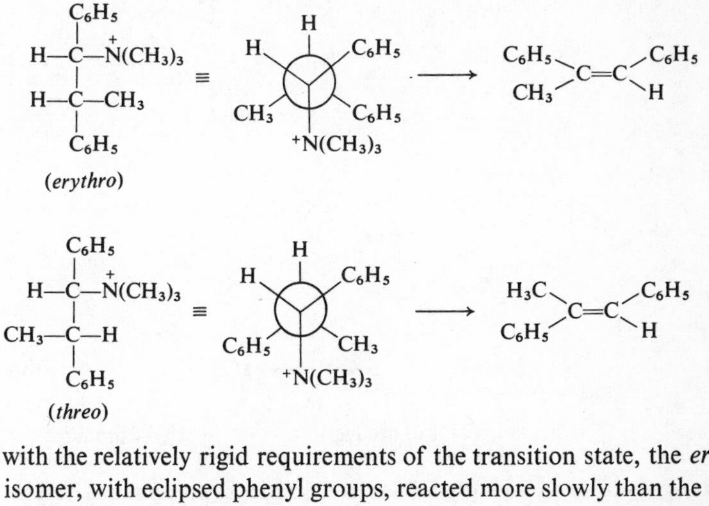

C_6H_5; H—C—$\overset{+}{N}(CH_3)_3$; H—C—CH_3; C_6H_5 ≡ H, C_6H_5, H, CH_3, C_6H_5, $^+N(CH_3)_3$ ⟶ C_6H_5(CH_3)C=C(C_6H_5)H

(*erythro*)

(2.5)

C_6H_5; H—C—$\overset{+}{N}(CH_3)_3$; CH_3—C—H; C_6H_5 ≡ H, C_6H_5, H, C_6H_5, CH_3, $^+N(CH_3)_3$ ⟶ $H_3C(C_6H_5)C=C(C_6H_5)H$

(*threo*)

with the relatively rigid requirements of the transition state, the *erythro* isomer, with eclipsed phenyl groups, reacted more slowly than the *threo*.

For stereoelectronic reasons elimination takes place most readily when the hydrogen atom and the leaving group are in the anti-periplanar arrangement (1), although a number of '*syn*' eliminations have been observed in which hydrogen and the leaving group are eclipsed (2) (2.6). In open-chain compounds the molecule can usually adopt a conformation in which H and R are anti-periplanar, but in cyclic systems this is not always so, and this may have an important bearing on the direction of

(1) (2) (2.6)

elimination in cyclic compounds. In cyclohexyl derivatives antiperiplanarity of the leaving groups requires that they be diaxial even if this is the less stable conformation, and on this basis we can understand why menthyl chloride (3) on treatment with ethanolic sodium ethoxide gives only 2-menthene (2.7), while neomenthyl chloride (4) gives a mixture of

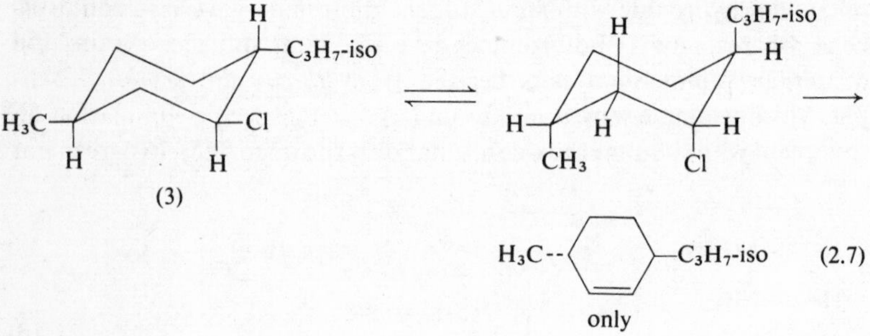
H
C_3H_7-iso
H_3C
Cl
H
H
H
C_3H_7-iso
H
H
H
H
CH_3
Cl
H_3C
C_3H_7-iso
only

(3) (2.7)

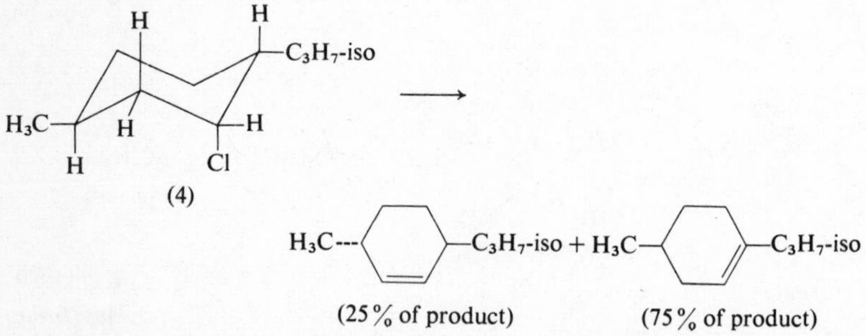
H
H
C_3H_7-iso
H_3C
H
H
H
Cl
H_3C
C_3H_7-iso +
H_3C
C_3H_7-iso
(25% of product)
(75% of product)

(4)

2- and 3-menthene in which the 'Saytzeff product' predominates. The elimination from menthyl chloride is much slower than that from neomenthyl chloride because the molecule has to adopt an unfavourable conformation with axial substituents before elimination can take place.

Syn eliminations are apparently more common than was formerly appreciated, particularly in reactions involving quaternary ammonium salts, where *syn* and *anti* eliminations often take place at the same time (Sicher, 1972). They are also observed in reactions of some bridged bicyclic compounds where *anti*-elimination is disfavoured by steric or conformationl factors (2.8), and in compounds where a strongly electron-attracting substituent on the β-carbon atom favours elimination in that direction, outweighing other effects (cf. Schlosser, 1972).

D, H, $\overset{+}{N}(CH_3)_3\overset{-}{O}H$, H —pyrolysis→ H, H (2.8)

In most cases, however, *anti* elimination is preferred.

In spite of the disadvantages acid-catalysed dehydration of alcohols and base-induced eliminations from halides and arylsulphonates are widely used in the preparation of olefins. The bases used in the latter reactions include alkali metal hydroxides and alkoxides, as well as organic bases such as pyridine and triethylamine. Superior results have been obtained using 1,5-diazabicyclo[3,4,0]non-5-ene (6) and the related 1,5-diazabicyclo-[5,4,0]undec-5-ene. Thus, the chloroester (5) could not be dehydrochlorinated with pyridine or quinoline but with the base (6) at 90°C the enyne was obtained in 85 per cent yield (2.9). Lithium chloride

$$CH{\equiv}CCH_2CH(Cl)(CH_2)_8CO_2CH_3 \; (5) + (6) \xrightarrow{90^\circ C} CH{\equiv}CCH{=}CH(CH_2)_8CO_2CH_3 \quad (2.9)$$

(or bromide) in dimethylformamide has also been found to be an effective reagent for dehydrohalogenation in some cases (Wendler, Taub and Kuo, 1960). Other methods are discussed by Schlosser (1972).

2.2. Pyrolytic *syn* eliminations

Another important group of olefin-forming reactions, some of which are useful in synthesis, are pyrolytic eliminations (DePuy and King, 1960; Saunders and Cockerill, 1973). Included in this group are the pyrolyses of carboxylic esters and xanthates, which provide valuable alternative methods for dehydration of alcohols without rearrangement, and pyrolysis of amine oxides. For the majority of cases these reactions are believed to take place in a concerted manner by way of a cyclic transition state and, in contrast to the eliminations discussed above, they are necessarily *syn* eliminations; i.e. the hydrogen atom and the leaving group depart from the same side of the incipient double bond (2.10).

$$\text{H–C–C–O–C(=O)R} \longrightarrow \text{–C=C–} + \text{HOCOR}$$

$$\text{H–C–C–O–C(=S)SR} \longrightarrow \text{–C=C–} + \text{HSCOSR} \quad (\text{COS} + \text{RSH}) \qquad (2.10)$$

$$\text{H–C–C–}\overset{+}{\text{N}}\text{R}_2\text{–}\ddot{\text{O}}^{-} \longrightarrow \text{–C=C–} + \text{HONR}_2$$

The *syn* character of these pyrolytic eliminations has been demonstrated in a number of ways. Thus, pyrolysis of the *erythro* and *threo* isomers of 1-acetoxy-2-deutero-1,2-diphenylethane gave in each case *trans*-stilbene, but the stilbene from the *erythro* compound retained nearly all its deuterium, whereas the stilbene from the *threo* compound had lost most of its deuterium. Either the hydrogen or the deuterium could be *syn* to the acetoxy group, but the preferred conformations (shown) are those in which the phenyl groups are as far removed from each other as possible (2.11). Similarly, the oxide of *cis*-1-phenyl-2-dimethylaminocyclohexane gave 3-phenylcyclohexene, elimination involving the *syn* and not the *anti* hydrogen even although the latter is activated by the phenyl group.

Pyrolysis of esters to give an olefin and a carboxylic acid is usually effected at a temperature of about 300–500°C and may be carried out by simply heating the ester if its boiling point is high enough, or by passing the vapour through a heated tube. Yields are usually good, and the absence of solvents and other reactants simplifies the isolation of the product. In practice acetates are nearly always employed, but esters of other carboxylic acids can be used. The reaction provides an excellent

$$\text{(erythro)}\ C_6H_5CH(OCOCH_3)CH(^2H)C_6H_5 \xrightarrow{\Delta} (C_6H_5)HC{=}C(^2H)(C_6H_5)$$

$$\text{(threo)}\ C_6H_5CH(OCOCH_3)CH(^2H)C_6H_5 \xrightarrow{\Delta} (C_6H_5)HC{=}CH(C_6H_5) \qquad (2.11)$$

method for preparing pure terminal olefins from primary acetates and, because it does not involve either acidic or basic reagents, it is especially useful for preparing highly reactive or thermodynamically disfavoured dienes and trienes. An example is the preparation of 4,5-dimethylene-cyclohexene, which is obtained from 4,5-diacetoxymethylcyclohexene without extensive rearrangement to *o*-xylene (2.12). With secondary and

$$CH_3CH_2CH_2CH_2OCOCH_3 \xrightarrow[N_2]{500°C} CH_3CH_2CH{=}CH_2 \quad (100\%)$$

$$\text{4,5-bis}(CH_2OCOCH_3)\text{cyclohexene} \xrightarrow[N_2]{500°C} \text{4,5-di}(=CH_2)\text{cyclohexene} \quad (47\%) \qquad (2.12)$$

tertiary acetates, where elimination can take place in more than one direction, mixtures of products are usually obtained.

The high temperature required for elimination is a disadvantage in these reactions, and in some cases the olefin formed may not be stable under the reaction conditions. Thus, pyrolysis of 1-cyclopropylethyl acetate affords mainly cyclopentene by rearrangement of the initially formed 1-vinylcyclopropane. In some cases also rearrangement of the ester may take place before elimination, leading to mixtures of products. This is especially liable to occur with allylic esters as in the example (2.13).

Pyrolysis of amine oxides (the Cope reaction) (Cope and Trumbull, 1960) and of methyl xanthates (the Chugaev reaction) (Nace, 1962) takes

$$CH_3CH(OCOCH_3)CH{=}CH(CH_2)_2CH_3 \rightleftharpoons CH_3CH{=}CHCH(OCOCH_3)(CH_2)_2CH_3 \xrightarrow{\Delta} CH_2{=}CHCH{=}CH(CH_2)_2CH_3 + CH_3CH{=}CHCH{=}CHCH_2CH_3 \quad (2.13)$$

place at much lower temperatures (100–200°C) than that of carboxylic esters, with the result that further decomposition of sensitive olefins can often be avoided in these reactions. On the other hand separation of the olefin from the other products of the reaction may be more troublesome, particularly in the Chugaev reaction, where sulphur-containing impurities have often to be removed by distillation from sodium. Pyrolysis of amine oxides generally takes place under particularly mild conditions, and this often allows the generation of a new olefinic bond without migration into conjugation with other unsaturated systems in the molecule, as in the synthesis of 1,4-pentadiene (2.14). If an allyl or benzyl group is attached to the nitrogen atom, rearrangement to give an *O*-substituted hydroxylamine may compete with elimination.

$$t\text{-}C_4H_9{-}C(CH_3)(H){-}OCSSCH_3 \xrightarrow{170°C} t\text{-}C_4H_9CH{=}CH_2 \quad (71\%)$$

$$CH_2{=}CH(CH_2)_3N(CH_3)_2{\rightarrow}O \xrightarrow{140°C} CH_2{=}CHCH_2CH{=}CH_2 \quad (61\%) + (CH_3)_2NOH \quad (2.14)$$

$$CH_2{=}CHCH_2N(C_2H_5)_2{\rightarrow}O \xrightarrow{150°C} (C_2H_5)_2NOCH_2CH{=}CH_2 \quad (59\%) + CH_2{=}CH_2$$

Like the base-induced eliminations discussed above, pyrolytic *syn* eliminations have the disadvantage that with derivatives of unsymmetrical secondary and tertiary alcohols and with unsymmetrical amine oxides, mixtures of products are formed. With open-chain compounds, the three methods give closely similar results. If there is a conjugating substituent in the β-position elimination takes place to give the conjugated olefin, but otherwise the composition of the product is determined mainly by the number of hydrogen atoms on each β-carbon. For example, pyrolysis of s-butyl acetate affords a mixture containing 57 per cent

$$CH_3-C(=O)-O\cdots H \quad \text{(Newman projections: } CH_3, CH_3, H, H \text{ and } CH_3, CH_3, H, H) \quad (2.15)$$

cis-2-butene (15%) *trans*-2-butene (28%)

1-butene and 43 per cent 2-butene, in close agreement with the 3:2 distribution predicted on the basis of the number of β-hydrogen atoms. Of the 2-butenes the *trans* isomer is formed in larger amount, presumably because there is less steric interaction between the two methyl substituents in the transition state which leads to the *trans* olefin (2.15). It is found in general with aliphatic esters, xanthates or amine oxides which could form either a *cis* or *trans* olefin on pyrolysis that the more stable *trans* isomer is formed in larger amount (2.16). In the third example in (2.16), the necessity for *syn* elimination ensures the formation of the *cis* olefin from the *threo* isomer and of the *trans* olefin from the *erythro* isomer of the amine oxide.

$$CH_3CH_2CH(OCSSCH_3)CH_2CH_2CH_3 \xrightarrow{250°C} CH_3CH_2CH{=}CHCH_2CH_3 \; (17\%\ cis)\ (33\%\ trans) + CH_3CH{=}CH(CH_2)_2CH_3 \; (16\%\ cis)\ (34\%\ trans) \quad (2.16)$$

$$CH_3CH_2CH(N(CH_3)_2{\rightarrow}O)CH_3 \xrightarrow[85\text{–}150°C]{91\%} CH_3CH{=}CHCH_3 \; (12\%\ cis)\ (21\%\ trans) + CH_3CH_2CH{=}CH_2 \; (67\%)$$

$$C_6H_5CH(CH_3)-CH(N(CH_3)_2{\rightarrow}O)CH_3 \xrightarrow{110\text{–}115°C} C_6H_5C(CH_3){=}CHCH_3 \; (93\%\ \text{of product}) + 7\%\ \text{unconjugated}$$

threo ⟶ 94% *trans*, 0·1% *cis*

erythro ⟶ 4% *trans*, 89% *cis*

With alicyclic compounds some restrictions are imposed by the conformation of the leaving groups and the necessity to form the cyclic intermediate. Thus, the cyclohexyl acetate (7) in which the leaving group is axial, in contrast to its isomer (8) does not form a double bond in the direction of the ethoxycarbonyl group, even though it would be conjugated, because the necessary cyclic transition state is not sterically possible from this conformation. With the xanthate (9), however, in which the leaving group is equatorial, the six-membered cyclic transition state can lead equally well to abstraction of a hydrogen atom from either β-carbon, and an equimolecular mixture of two products results (2.17).

H, $CO_2C_2H_5$, H, $OCOCH_3$ (7) $\xrightarrow{435°C}$ $CO_2C_2H_5$

H, $CO_2C_2H_5$, $OCOCH_3$, H (8) $\xrightarrow{435°C}$ $CO_2C_2H_5$ (2.17)

CH_3, H, $OCSCH_3$ (=S), H (9) $\xrightarrow{\Delta}$ CH_3 + CH_3

(51% of product) (49% of product)

Pyrolysis of alicyclic amine oxides does not necessarily lead to product mixtures similar in composition to those obtained from the corresponding acetates and xanthates (DePuy and King, 1960). A notable difference is found in the pyrolysis of 1-methylcyclohexyl derivatives. The acetates and xanthates afford mixtures containing 1-methylcyclohexene and methylenecyclohexane in a ratio of 3:1, whereas pyrolysis of the oxide of 1-dimethylamino-1-methylcyclohexane gives methylenecyclohexane almost exclusively. The reason for this is thought to be that in the oxide the coplanar five-membered cyclic transition state only allows abstraction of hydrogen from the methyl substituent, whereas with the more flexible six-membered transition state of the ester pyrolysis, hydrogen abstraction is also possible from the ring. With larger more flexible rings, where the requisite planar transition state is more easily attained, the cycloalkene is the preponderant product from amine oxides as well.

As a general rule formation of compounds with a double bond exocyclic to a ring is not favoured in pyrolysis of esters, presumably because of their relative instability with respect to other isomers. For instance, 1-cyclohexylethyl acetate on pyrolysis gives mainly vinylcyclohexane.

Sulphoxides with a β-hydrogen atom readily undergo *syn* elimination on pyrolysis to form olefins. These reactions also take place by way of a concerted cyclic pathway (cf. p. 94). They are highly stereoselective, the *erythro* sulphoxide (10), for example, leading predominantly to *trans*-methylstilbene, while the corresponding *threo* isomer gives mainly *cis*-methylstilbene (Kingsbury and Cram, 1960; see also Trost and Leung, 1975) (2.18).

$\bar{O}$ $\overset{+}{S}C_6H_5$ CH_3 C_6H_5 H H C_6H_5 $\equiv$ $\bar{O}$ H $\overset{+}{S}C_6H_5$ C_6H_5 CH_3 C_6H_5 H $\xrightarrow[\text{dioxan}]{80°C}$ C_6H_5 H CH_3 C_6H_5 + C_6H_5SOH (2.18)

(10)

Since sulphoxides are readily obtained by oxidation of sulphides, the reaction provides another useful method for making carbon–carbon double bonds. Equation (2.19) shows how it can be used to prepare terminal olefins from primary alcohols containing one fewer carbon atom, through reaction of the derived bromides or toluene-*p*-sulphonates

$$C_{15}H_{31}CH_2OH \longrightarrow C_{15}H_{31}CH_2OTos$$

$$\downarrow \text{Na}^+\ \bar{C}H_2SOCH_3 \text{ in } CH_3SOCH_3 \qquad (2.19)$$

$$\underset{(80\%)}{C_{15}H_{31}CH{=}CH_2} \xleftarrow[CH_3SOCH_3]{\text{reflux in}} C_{15}H_{31}CH_2CH_2SOCH_3 \quad (85\%)$$

$$\text{Tos} = H_3C\text{-}C_6H_4\text{-}SO_2$$

with the anion of dimethyl sulphoxide. Pyrolysis of sulphoxides is also a feature of a useful new method for introducing unsaturation at the α-position of aldehydes, ketones and esters. Reaction of the corresponding enolates with dimethyl or diphenyl disulphide affords an α-methylthio derivative (as 11) which on oxidation and decomposition in boiling toluene is converted into the αβ-unsaturated carbonyl compound. Yields are usually good. The queen substance of honey bees was synthesised from methyl 9-oxodecanoate as shown in (2.20) or, better, by initial protection of the ketone function as the ketal (Trost and Salzmann, 1975).

O

CO_2CH_3

↓ 1 equiv. $LiN(iso\text{-}C_3H_7)_2$

O O

OCH_3

↓ CH_3SSCH_3

O

CO_2CH_3

SCH_3

(11)

↓ (1) oxidation, $NaIO_4$ (2) 120°C

O

CO_2CH_3 (2.20)

The *trans* isomer usually predominates in reactions leading to disubstituted ethylenes, but tri- and tetra-substituted compounds may be obtained as mixtures of isomers unless, of course the *threo* and *erythro* sulphoxides (as 10) are separated before pyrolysis. For the preparation of αβ-unsaturated carbonyl compounds, this procedure has a number of advantages over the usual sequence of bromination–dehydrobromination. It takes place under comparatively mild conditions, it is specific, since only enolate anions react at the first stage, and it can be carried out in the presence of other functional groups, such as olefinic double bonds, which might be affected by bromination.

Even better results are obtained by using selenoxides. Alkyl phenyl selenoxides with a β-hydrogen atom undergo *syn* elimination to form olefins under milder conditions than the sulphoxides – at room temperature or below – and this mild reaction has been exploited for the preparation of a variety of different kinds of unsaturated compounds. The selenoxides are readily obtained from the corresponding selenides by oxidation with hydrogen peroxide or other reagents, and they generally decompose under the reaction conditions to give the olefin directly. The selenides themselves can be prepared by a number of routes (see examples below) (Sharpless, Young and Lauer, 1973). Elimination following alkylation of the selenides or selenoxides provides a route to alkenes and conjugated dienes (2.21) and, in conjunction with the addition of organocuprates (p. 66) forms a step in a useful method for β-alkylation of $\alpha\beta$-unsaturated ketones (Reich, Renga and Reich, 1974; Reich and Shah, 1975).

$$C_6H_5CH_2Br \xrightarrow[C_2H_5OH]{C_6H_5SeNa} C_6H_5CH_2SeC_6H_5 \xrightarrow[(2)\ C_2H_5Br]{(1)\ LiN(iso\text{-}C_3H_7)_2,\ THF,\ -78^\circ C} C_6H_5CH(C_2H_5)SeC_6H_5$$

$$\xrightarrow{H_2O_2,\ THF} \left[C_6H_5CH(\overset{+}{Se}(C_6H_5)\text{–}\overset{-}{O})\text{–}CH(CH_3)\text{–}H \right] \ (12) \longrightarrow \underset{(80\%)}{C_6H_5CH{=}CHCH_3}$$

$$C_6H_5\text{–}\overset{+}{Se}(\text{–}O^-)\text{–}CH_2CH_2CH_2CH_2C_6H_5 \xrightarrow[(2)\ (CH_3)_2C{=}CHCH_2Br]{(1)\ LiN(iso\text{-}C_3H_7)_2,\ THF,\ 78^\circ C} \left[C_6H_5\overset{+}{Se}(\text{–}O^-)\text{–}CH(CH_2CH_2CH_2C_6H_5)\text{–}CH(H)\text{–}CH{=}C(CH_3)_2 \right]$$

$$\xrightarrow{\text{reflux } CH_2Cl_2} \underset{(64\%)}{C_6H_5CH_2CH_2CH_2CH{=}CH\text{–}CH{=}C(CH_3)_2} \qquad (2.21)$$

Disubstituted alkenes are generally obtained largely as the *trans* isomers, following *syn* elimination from a cyclic transition state (12) in which the two substituent groups are staggered. In some cases, particularly in reactions leading to the conversion of primary alkyl groups into terminal olefins, better yields are obtained by using *o*-nitro- or *p*-chloro-phenyl selenides rather than the phenyl selenides themselves (Sharpless and Young, 1975).

The reaction forms a step in a good method for converting epoxides into allylic alcohols (compare p. 337), and for introducing unsaturation at the α-position of carbonyl compounds. In the latter conversion it is superior to the sulphoxide elimination described on p. 100, because of the

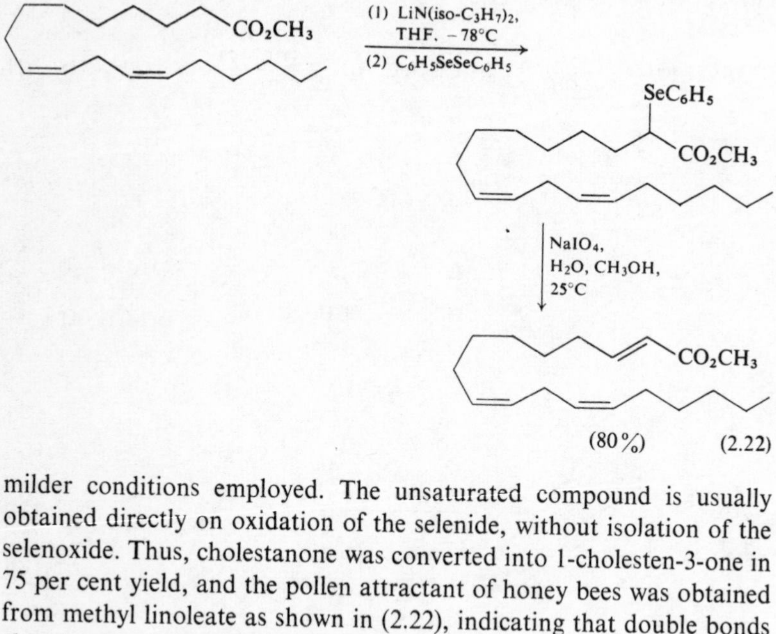

CO_2CH_3 —(1) $LiN(iso\text{-}C_3H_7)_2$, THF, −78°C; (2) $C_6H_5SeSeC_6H_5$→ (SeC_6H_5) CO_2CH_3

↓ $NaIO_4$, H_2O, CH_3OH, 25°C

CO_2CH_3

(80%) (2.22)

milder conditions employed. The unsaturated compound is usually obtained directly on oxidation of the selenide, without isolation of the selenoxide. Thus, cholestanone was converted into 1-cholesten-3-one in 75 per cent yield, and the pollen attractant of honey bees was obtained from methyl linoleate as shown in (2.22), indicating that double bonds elsewhere in the molecule do not interfere with the reaction (Sharpless, Lauer and Teranishi, 1973; Reich, Reich and Renga, 1973).

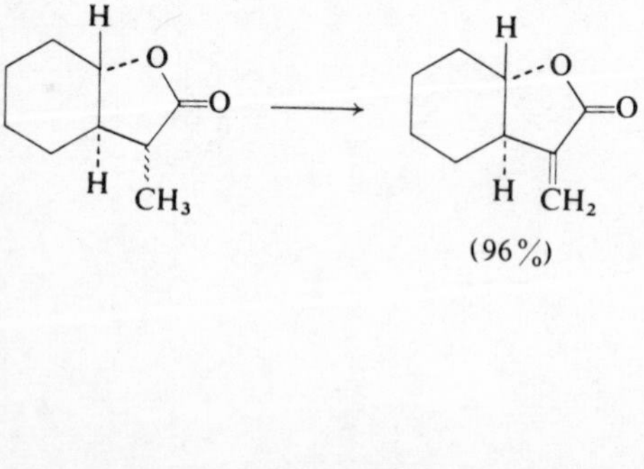

H, O, =O, H, CH_3 ⟶ H, O, =O, H, CH_2 (2.23)

(96%)

The reaction has also been used to obtain the α-methylene lactone structural unit found in cytotoxic sesquiterpenes (2.23) (Grieco and Miyashita, 1974) and in a useful procedure for converting γ-butyrolactones into substituted furans (Grieco, Pogonowski and Burke, 1974).

2.3. Sulphoxide–sulphenate rearrangement Synthesis of allyl alcohols

Allylic sulphoxides, on gentle warming, are partly converted in a reversible reaction into rearranged allyl sulphenate esters (13). The reactions are [2,3]-sigmatropic rearrangements and take place through a five-membered cyclic transition state. The equilibrium is usually much in favour of the sulphoxide, but if the mixture is treated with a 'thiophile' (trimethyl phosphite and diethylamine are often used) the sulphenate is irreversibly converted into the allylic alcohol and, even although the sulphenate is present in low equilibrium concentration with the sulphoxide, its removal by reaction with the thiophile results in conversion of the sulphoxide into the rearranged allylic alcohol in high yield. Combined with alkylation of the sulphoxides the reaction provides a versatile synthesis of di- and tri-substituted allylic alcohols (Evans and Andrews, 1974). Equally good results have been obtained using selenoxides (Reich, 1975).

H_2C O^- / HC C S^+ R / H_2 ⇌ H_2C------O / HC C SR / H_2 ⇌ H_2C——O / HC C SR / H_2

(13) (2.24)

↓ 'thiophile'
methanol, 25°C

$CH_2{=}CHCH_2OH$

As indicated in (2.25), the rearrangement step is highly stereoselective, leading, in the acyclic series, mainly to the *trans* allylic alcohol. The factors controlling this are not clearly understood.

In this series of reactions the anions derived from the sulphoxides, such as (14), are equivalent to the vinyl anions (15) and the sequence provides, in fact, a method for effecting β-alkylation of allylic alcohols. This is exemplified in the second example in (2.25) in which the sulphoxide is obtained by rearrangement of the sulphenate ester prepared from the unsubstituted allyl alcohol.

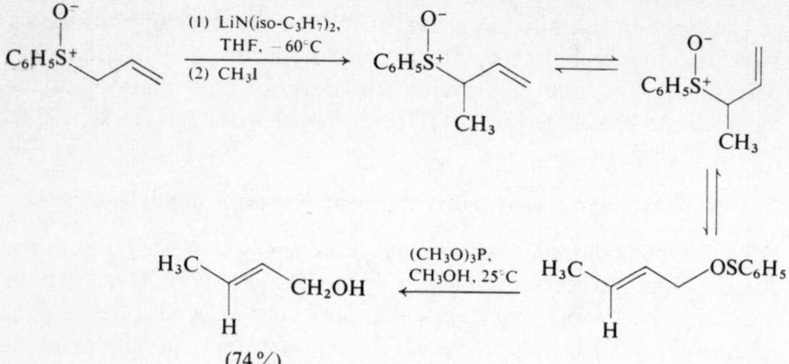

O^- | $C_6H_5S^+$ —(1) $LiN(iso\text{-}C_3H_7)_2$, THF, −60°C; (2) CH_3I→ O^- | $C_6H_5S^+$, CH_3 ⇌ O^- | $C_6H_5S^+$, CH_3

⇅

H_3C, CH_2OH, H (74%) ←$(CH_3O)_3P$, CH_3OH, 25°C— H_3C, OSC_6H_5, H

(2.25)

OH —(1) C_4H_9Li; (2) C_6H_5SCl→ OSC_6H_5 → $\overset{+}{S}C_6H_5$, O^-

↓ (1) $LiN(iso\text{-}C_3H_7)_2$, THF, −60°C; (2) CH_3I

OH, CH_3 ←$(CH_3O)_3P$, CH_3OH— CH_3, $\overset{+}{S}C_6H_5$, ^-O

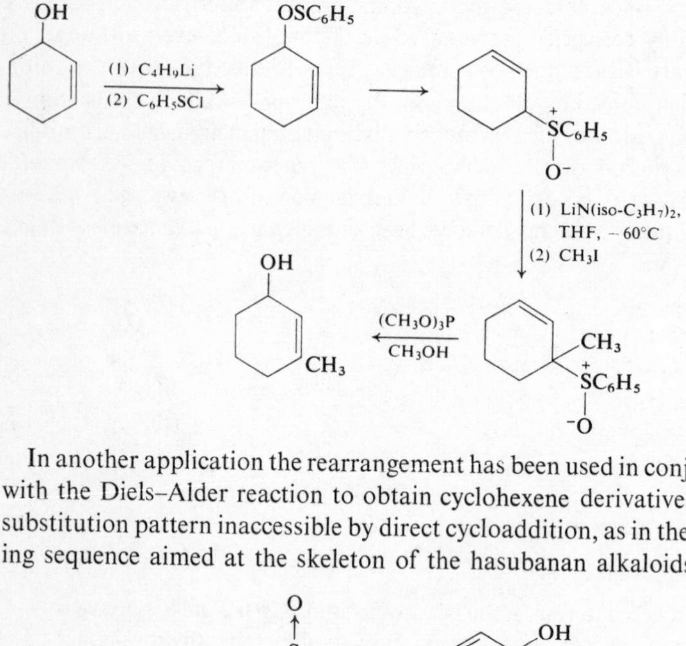

In another application the rearrangement has been used in conjunction with the Diels–Alder reaction to obtain cyclohexene derivatives with a substitution pattern inaccessible by direct cycloaddition, as in the following sequence aimed at the skeleton of the hasubanan alkaloids (2.27).

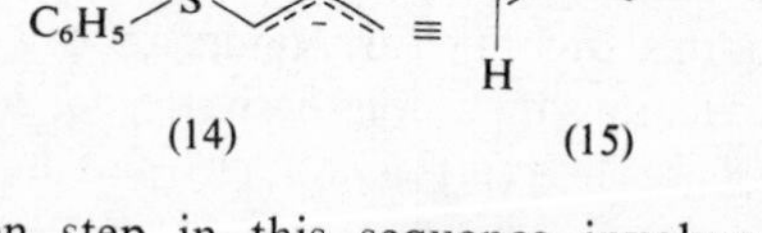

O ↑ C_6H_5–S ≡ OH, H (2.26)

(14) (15)

The cyclo-addition step in this sequence involves an electron-rich dienophile and an electron-deficient diene. The more usual reaction with an electron-deficient dienophile, such as maleic anhydride, and an electron-rich diene, can be achieved by using the butadienyl sulphide

$C_6H_5S{\rightarrow}O$ + N CH_3 → NCH_3 $C_6H_5S{\rightarrow}O$

warm in methanol with a 'thiophile'

HO NCH_3 (2.27)

instead of the sulphoxide. Oxidation to the sulphoxide is then effected after the addition step.

A disadvantage of the reaction sequence described above is that alkylation of the allyl sulphoxide anions frequently gives some unwanted γ-alkylation in addition to α-alkylation, particularly when there is already a substituent on the α-carbon atom. This difficulty can be overcome and yields increased generally by using, instead of the phenyl sulphoxide, the α-pyridyl or *N*-methylimidazoyl sulphides. In these, alkylation is directed largely into the α-position, possibly by chelation of the lithium cation by the nitrogen atom of the heterocycle.

The sulphoxide–sulphenate interconversion has been employed in a synthesis of $\alpha\beta$-unsaturated aldehydes from carbonyl compounds and vinyl halides, by a route involving two different [2,3]-sigmatropic rearrangements. This is exemplified in (2.28) by the preparation of 2-octenal from hexanal and vinylmagnesium bromide.

Typically, an allylic alcohol generated by the addition of a vinyl organometallic compound to a carbonyl compound is treated with benzenesulphenyl chloride to produce the allylic sulphoxide via the usual sigmatropic rearrangement. Conversion into the anion with lithium diisopropylamide and reaction with diphenyl disulphide gives a sulphenylation product which rearranges with loss of diphenyl disulphide to yield the hydroxy enol thioether directly. Hydrolysis liberates the

$C_5H_{11}CHO$ + $CH_2{=}CHMgBr$ ⟶ $C_5H_{11}CH(OMgBr)CH{=}CH_2$ $\xrightarrow{C_6H_5SCl}$ [$C_5H_{11}CH(OSC_6H_5)CH{=}CH_2$] ⟶ $C_5H_{11}CH{=}CHCH_2\overset{+}{S}(O^-)C_6H_5$ $\xrightarrow[(2)\ C_6H_5SSC_6H_5]{(1)\ LiN(iso\text{-}C_3H_7)_2,\ THF,\ -78°C}$ [$C_5H_{11}CH{=}CHCH(SC_6H_5)\overset{+}{S}(O^-)C_6H_5$] ⇌ [$C_5H_{11}CH(O{-}S{-}C_6H_5)CH{=}CHSC_6H_5$ + $^-SC_6H_5$] ⟶ $C_5H_{11}CH(OH)CH{=}CHSC_6H_5$ $\xrightarrow[CH_3CN]{H_2O,\ HgCl_2}$ $C_5H_{11}CH{=}CHCHO$ (80% overall) (2.28)

$\alpha\beta$-unsaturated carbonyl compound (Trost and Stanton, 1975). The sequence provides a method for the selective formation of one product of a 'mixed aldol' condensation, the vinyl bromide acting as the 'active methylene' component. Pinacolone and propenylmagnesium bromide, for example, give the ketone (16).

$(CH_3)_3CCOCH_3 + CH_3CH{=}CHMgBr \longrightarrow (CH_3)_3CC(CH_3){=}CHCOCH_3$ (16) (48%) (2.29)

2.4. The Wittig and related reactions

The reaction occurring between a phosphorane, or phosphonium ylid, and an aldehyde or ketone to form a phosphine oxide and an olefin, has become known as the Wittig reaction, after the German chemist Georg

Wittig, who first showed the value of this procedure in the synthesis of olefins (Trippett, 1963; Maercker, 1965; Johnson, 1966) (2.30). The reaction is easy to carry out and proceeds under mild conditions to give,

$$R_3P{=}CR^1R^2 + \begin{matrix}R^3\\R^4\end{matrix}\!\!>\!C{=}O \longrightarrow R_3P{=}O + \begin{matrix}R^1\\R^2\end{matrix}\!\!>\!C{=}C\!<\!\!\begin{matrix}R^3\\R^4\end{matrix} \quad (2.30)$$

in general, high yields of the olefinic compound. A particularly valuable feature of the Wittig procedure is that, in contrast to the elimination and pyrolytic reactions discussed above, it gives rise to olefins in which the position of the newly formed double bond is unambiguous, as illustrated in the examples (2.31).

$$(C_6H_5)_3P{=}CH_2 + C_7H_{15}COCH_3 \xrightarrow{\text{ether}} C_7H_{15}\overset{CH_3}{\overset{|}{C}}{=}CH_2$$

$$(C_6H_5)_3P{=}CHCH_3 + CH_3COCO_2C_2H_5 \xrightarrow[\text{reflux}]{\text{ether}} CH_3CH{=}\overset{CH_3}{\overset{|}{C}}CO_2C_2H_5 \quad (2.31)$$

Phosphoranes are resonance-stabilised structures in which there is some overlap between the carbon p orbital and one of the d orbitals of phosphorus as shown in (2.32). Reaction with a carbonyl compound

$$\underset{\text{phosphorane form}}{\begin{matrix}R^1\\R^2\end{matrix}\!\!>\!C{=}PR_3} \longleftrightarrow \underset{\text{ylid form}}{\begin{matrix}R^1\\R^2\end{matrix}\!\!>\!\bar{C}{-}\overset{+}{P}R_3} \quad (2.32)$$

takes place by attack of the carbanionoid carbon of the ylid form on the electrophilic carbon of the carbonyl group with the formation of a betaine which collapses to the products by way of a four-membered cyclic transition state (2.33), the driving force being provided by formation of the very strong phosphorus–oxygen bonds. Depending on the

$$R_3\overset{+}{P}{-}\bar{C}\!<\!\!\begin{matrix}R^1\\R^2\end{matrix} + O{=}C\!<\!\!\begin{matrix}R^3\\R^4\end{matrix} \rightleftharpoons \begin{matrix}R_3\overset{+}{P}{-}C(R^1)(R^2)\\ |\\ {}^{-}O{-}C(R^3)(R^4)\end{matrix}$$

$$\downarrow \quad (2.33)$$

$$R_3P{=}O + \begin{matrix}R^1\\R^2\end{matrix}\!\!>\!C{=}C\!<\!\!\begin{matrix}R^3\\R^4\end{matrix} \longleftarrow \begin{matrix}R_3P{-}C(R^1)(R^2)\\ |\quad\quad |\\ O{-}C(R^3)(R^4)\end{matrix}$$

reactants, either the first or the second step may be rate-determining. It has never been observed that the last step is the slowest, and it is uncertain whether the four-membered ring compound is a true intermediate or a transition state. Evidence for the formation of a betaine in the first stage of the reaction is provided by the isolation of compounds of this type in certain cases.

The reactivity of the phosphorane depends on the nature of the groups R, R^1 and R^2. In practice, R is nearly always phenyl. Alkylidene trialkylphosphoranes, in which the formal positive charge on the phosphorus is lessened by the inductive effect of the alkyl groups, are more reactive than alkylidenetriphenylphosphoranes in the initial addition to a carbonyl group to form a betaine, but by the same token decomposition of the betaine becomes more difficult, and alkylidene trialkylphosphoranes are superior to the triphenyl compounds only in certain special cases. In the alkylidene part of the phosphorane, if R^1 or R^2 is an electron-withdrawing group (e.g. CO or CO_2R) the negative charge in the ylid becomes delocalised into R^1 or R^2 and the nucleophilic character, and reactivity towards carbonyl groups, is decreased. Reagents of this type are much more stable and less reactive than those in which R^1 and R^2 are alkyl groups, and with them the rate-determining step in reactions with carbonyl groups is the initial addition to form the betaine. The more electrophilic the carbonyl group the more readily does the reaction proceed. The preparatively important carbalkoxymethylene-triphenylphosphoranes, for example, react fairly readily with aldehydes, but give only poor yields in reaction with the less reactive carbonyl group of ketones (2.34).

$$(C_6H_5)_3P{=}CHCO_2C_2H_5 \xrightarrow{\text{crotonaldehyde, benzene, reflux}} CH_3CH{=}CHCH{=}CHCO_2C_2H_5 \quad (80\%)$$

$$(C_6H_5)_3P{=}CHCO_2C_2H_5 \xrightarrow{\text{cyclohexanone, benzene, reflux}} C_6H_{10}{=}CHCO_2C_2H_5 \quad (25\%) \qquad (2.34)$$

In the majority of Wittig reagents R^1 and R^2 are alkyl groups which have little effect on the carbanionoid character of the molecule. These reagents are markedly nucleophilic and react readily with carbonyl and other polar groups. Addition of the ylid to carbonyl groups takes place rapidly and decomposition of the betaine now becomes the rate-determining step of the reaction. Since the polarity of the carbonyl group is of little consequence when the second step is rate determining, aldehydes and ketones usually react equally well with these reagents.

Alkylidenephosphoranes can be prepared by a number of methods, but in practice they are usually obtained by action of base on alkyltriphenylphosphonium salts, which are themselves readily available from an alkyl halide and triphenylphosphine (2.35). The phosphonium salt

$$(C_6H_5)_3P + R^1R^2CHX \longrightarrow (C_6H_5)_3\overset{+}{P}{-}CHR^1R^2 \quad X^- \quad \underset{\text{base}}{\rightleftharpoons} \quad (C_6H_5)_3P{=}CR^1R^2 + HB \qquad (2.35)$$

can usually be isolated and crystallised but the phosphorane is generally prepared in solution and used without isolation. Formation of the phosphorane is reversible, and the strength of base necessary and the reaction conditions depend entirely on the nature of the ylid. A common procedure is to add the stoichiometric amount of an ethereal solution of phenyl- or butyl-lithium to a solution or suspension of the phosphonium salt in ether, benzene or tetrahydrofuran, followed, after an appropriate interval, by the carbonyl compound. Lithium and sodium alkoxides, in solution in the corresponding alcohol or in dimethylformamide, are also very commonly used. This procedure is simpler than that using organolithium compounds, and has the added advantage that it can be used to prepare phosphoranes from compounds containing functional groups, such as the ethoxycarbonyl group, which would react with organolithium compounds. By this method, also, good yields of condensation products can be obtained from unstable ylids by generating the ylid in presence of the carbonyl compound with which it is to react. The mechanism of the reaction in alcoholic solvents may not be entirely the same as that originally suggested for the Wittig reaction, and in some cases complex products may be formed by migration of a group from phosphorus to the adjacent carbon atom (Allen, Heatley, Hutley and Mellors, 1974).

Reactions involving non-stabilised ylids must be conducted under anhydrous conditions and in an inert atmosphere, because these ylids react both with oxygen and with water. Water effects hydrolysis, with formation of a phosphine oxide and a hydrocarbon; the most electronegative group is cleaved. Benzylidenetriphenylphosphorane, for example, reacts with water to give triphenylphosphine oxide and toluene.

With oxygen, reaction leads in the first place to triphenylphosphine oxide and a carbonyl compound which undergoes a Wittig reaction with unoxidised ylid to form a symmetrical olefin. This is in fact quite a useful route to symmetrical olefins. The reaction is conveniently effected by passing oxygen through a solution of the phosphorane or by reaction of lithium ethoxide with the phosphonium periodate in ethanol (2.36) (Bestmann, Armsen and Wagner, 1969).

$$(C_6H_5)_3P{=}CHC_6H_5 \xrightarrow[\text{or } NaIO_4]{O_2} (C_6H_5)_3PO + C_6H_5CHO \xrightarrow{(C_6H_5)_3P{=}CHC_6H_5} C_6H_5CH{=}CHC_6H_5 \ (55\%) \qquad (2.36)$$

A very useful alternative method for the preparation of resonance-stabilised phosphoranes for use in the Wittig reaction proceeds from phosphonate esters, themselves readily available from alkyl halides and triethyl phosphite via an Arbuzov rearrangement. Reaction with a suitable base gives the corresponding anions (as 17) which are strongly nucleophilic and react readily with the carbonyl groups of aldehydes and ketones to form an olefin and a water-soluble phosphate ester (Boutagy and Thomas, 1974). Thus, the anion (17) from ethyl bromoacetate and

$$(C_2H_5O)_3P + BrCH_2CO_2C_2H_5 \longrightarrow \left[(C_2H_5O)_2\overset{+}{P}(OC_2H_5)\text{—}CH_2CO_2C_2H_5 \ \ Br^-\right] \longrightarrow (C_2H_5O)_2\overset{O}{\overset{\|}{P}}CH_2CO_2C_2H_5$$

$$(C_2H_5O)_2\overset{O}{\overset{\|}{P}}CH_2CO_2C_2H_5 \xrightarrow[\text{dimethoxy-ethane}]{NaH} \underset{(17)}{(C_2H_5O)_2\overset{O}{\overset{\|}{P}}\bar{C}HCO_2C_2H_5} \xrightarrow{\text{cyclohexanone}} \underset{(70\%)}{C_6H_{10}{=}CHCO_2C_2H_5} + (C_2H_5O)_2\overset{O}{\overset{\|}{P}}\text{—}O^- \qquad (2.37)$$

triethyl phosphite reacts rapidly with cyclohexanone at room temperature to give ethyl cyclohexylideneacetate in 70 per cent yield (2.37) compared with a 25 per cent yield obtained in the reaction with the triphenylphosphorane. In general, the phosphonate ester reaction is the method of choice for the interaction of resonance-stabilised phosphoranes with aldehydes and ketones, and this procedure has been widely used for the preparation of $\alpha\beta$-unsaturated esters. It is unsuitable for the preparation of olefins from non-stabilised reagents; in these cases only the conjugate acid of the betaine is isolated and no olefin is produced. Another procedure which can be used to prepare β-keto phosphonates, and thence $\alpha\beta$-unsaturated ketones, from esters is shown in equation (2.38) (Dauben, Beasley, Broadhurst, Muller, Peppard, Pesnelle and Suter, 1975).

THPO ... $CO_2C_2H_5$

$LiCH_2P(O)(OCH_3)_2$, THF, $-70°C$

THPO ... $P(O)(OCH_3)_2$

(1) NaH
(2) $(CH_3)_2CH$–CHO
(3) H_3O^+

HO ... + $(CH_3O)_2P(O)$—O^- (2.38)

(78%)

(THP = tetrahydropyranyl)

Because of the mild conditions under which it proceeds, its versatility and the unambiguous position of the double bond formed, the Wittig reaction is an almost ideal olefin synthesis. Its only disadvantage is that as originally described it is not subject to steric control, and where the structure of the olefin allows it a mixture of *cis* and *trans* isomers is usually produced. In the reaction of resonance-stabilised ylids with aldehydes, the *trans* olefin generally predominates. Non-stabilised ylids, on the other hand, usually give more of the *cis* olefin. Benzylidenetriphenylphosphorane and benzaldehyde, for example, give a product containing 25 per cent *cis*- and 75 per cent *trans*-stilbene, and even higher proportions of *trans* products have been obtained in some other reactions. This preference for the *trans* isomer is a result of the fact that, with stabilised ylids, formation of the intermediate betaines is reversible, allowing interconversion to the more stable *threo* form which collapses to the *trans* olefin. With the non-stabilised ylid, ethylidenetriphenyl-

$$R_3\overset{+}{P}-\bar{C}HR^1 + R^2CHO \rightleftharpoons \left[\begin{array}{l} R_3\overset{+}{P}-C(H)(R^1) \\ \quad | \\ {}^{-}O-C(H)(R^2) \end{array}\right]_{erythro} \longrightarrow \begin{array}{c} R^2 \quad\; R^1 \\ C{=}C \\ H \quad\;\; H \end{array}$$

$$R_3\overset{+}{P}-\bar{C}HR^1 + R^2CHO \rightleftharpoons \left[\begin{array}{l} R_3\overset{+}{P}-C(H)(R^1) \\ \quad | \\ {}^{-}O-C(R^2)(H) \end{array}\right]_{threo} \longrightarrow \begin{array}{c} R^2 \quad\; H \\ C{=}C \\ H \quad\;\; R^1 \end{array} \qquad (2.39)$$

phosphorane, however, benzaldehyde gives a mixture of *cis*- and *trans*-β-methylstyrene containing more of the *cis* isomer (87 per cent). Here formation of the betaine is irreversible and conversion into the olefin proceeds mainly from the kinetically favoured *erythro* betaine. Recent work has shown that the steric course of the reactions can be substantially altered by varying the reaction conditions, and by proper choice of experimental conditions a high degree of steric control can be achieved, at least for reactions leading to disubstituted olefins. Thus, reaction of stabilised ylids with aldehydes in the presence of protic solvents or lithium salts gives increased amounts of *cis* olefins. With non-stabilised ylids, salt-free conditions and non-polar solvents give high selectivity for *cis* olefin. These variations in the ratio of *cis* and *trans* products can be ascribed to the influence of the solvents and additives on the relative stabilities and rates of decomposition of the *threo* and *erythro* betaines (Schlosser, 1970; Reucroft and Sammes, 1971).

Almost pure *trans* olefins can be obtained by equilibration of the intermediate betaines by converting them into betaine ylids by the action of phenyllithium. This is exemplified in the synthesis of *trans*-3-octene, which is represented by Schlosser and Christmann (1966) as shown in (2.40). Interconversion of the stereoisomeric betaine ylids (18) is very

$$\left[(C_6H_5)_3\overset{+}{P}-\underset{\underset{Li}{|}}{CH}-CH_3\right]X^- \xrightarrow{C_5H_{11}CHO} \left[(C_6H_5)_3\overset{+}{P}-\overset{\overset{CH_3}{|}}{CH}-\underset{\underset{OLi}{|}}{\overset{\overset{C_5H_{11}}{|}}{CH}}\right]X^-$$

erythro and *threo*

$$\downarrow C_6H_5Li,\ THF,\ -30\,^\circ C \qquad (2.40)$$

$$(C_6H_5)_3\overset{+}{P},\ OLi,\ Li,\ CH_3,\ H,\ C_5H_{11}\ X^- \ (18a) \rightleftharpoons (C_6H_5)_3\overset{+}{P},\ OLi,\ CH_3,\ Li,\ H,\ C_5H_{11}\ X^- \ (18b)$$

(18*a*) (18*b*)

$$\downarrow (1)\ 1\ equiv.\ HCl,\ ether;\ (2)\ t\text{-}C_4H_9OH,\ t\text{-}C_4H_9OK$$

$$\underset{H}{\overset{CH_3}{}}\!\!>C=C<\!\!\underset{C_5H_{11}}{\overset{H}{}}$$

(70%; 99% *trans*)

rapid and the equilibrium strongly favours the form (18*b*) which affords the *threo*-betaine and thence the *trans* olefin on protonation.

The value of the Wittig reaction is clearly shown by the fact that it has already been used in the synthesis of very many olefins, including a considerable number of natural products (Maercker, 1965). It is a versatile synthesis and can be used for the preparation of mono-, di-, tri-, and tetra-substituted ethylenes, and also of cyclic compounds. The carbonyl components may contain a wide variety of other functional groups such as hydroxyl, ether, ester, halogen and terminal acetylene, which do not interfere with the reaction. In compounds that contain both ester and carbonyl groups the latter react preferentially as long as the Wittig reagent is not present in excess. The mild conditions of the reaction make it an ideal method for the synthesis of sensitive olefins such as carotenoids and other polyunsaturated compounds, as in (2.42).

One especially useful application of the Wittig reaction is in the formation of exocyclic double bonds. The Wittig reaction is the method of choice for converting a cyclic ketone into an exocyclic olefin. The Grignard method gives the endocyclic isomer almost exclusively. Thus, cyclohexanone and methylenetriphenylphosphorane give methylenecyclohexane, and the same reaction has been used to prepare a variety of methylene steroids. Inhoffen's synthesis of vitamins D_2 and D_3 involved the Wittig reaction in three of the steps of which the first is shown in (2.41).

$$+ (C_6H_5)_3P{=}CHCH{=}CH_2 \longrightarrow \quad (2.41)$$

With bifunctional carbonyl compounds and bisphosphoranes cyclic compounds can be obtained, as in the synthesis of 1,2,5,6-dibenzocyclo-octatetraene (2.42).

$$OHCCH{=}CH(C{\equiv}C)_2CH{=}CHCHO + 2(C_6H_5)_3P{=}CH(CH{=}CH)_2CH_3$$

$$\downarrow \quad (2.42)$$

$$CH_3(CH{=}CH)_4(C{\equiv}C)_2(CH{=}CH)_4CH_3$$

$$2C_6H_5CH{=}CHCHO + (C_6H_5)_3P{=}CHCH{=}CHCH{=}P(C_6H_5)_3$$

$$\downarrow$$

$$C_6H_5(CH{=}CH)_5C_6H_5$$

$$o\text{-}C_6H_4(CH_2\overset{+}{P}(C_6H_5)_3)_2 + o\text{-}C_6H_4(CHO)_2 \xrightarrow[\text{dimethylformamide}]{C_2H_5OLi,\ C_2H_5OH} \text{dibenzocyclooctatetraene } (18\%)$$

$$\text{cyclohexanone} + (C_6H_5)_3P{=}CHOCH_3 \longrightarrow \text{cyclohexylidene}{=}CHOCH_3 \xrightarrow{H_3O^+} \text{cyclohexane-CHO (H, CHO)} \quad (2.43)$$

A valuable group of Wittig reagents is derived from α-haloethers. They react with aldehydes and ketones to form vinyl ethers which on acid hydrolysis are converted into aldehydes containing one more carbon atom. Thus cyclohexanone is converted into formylcyclohexane (2.43).

Although there are many recorded examples of the reaction of allylic phosphoranes with $\alpha\beta$-unsaturated aldehydes and ketones to give the expected trienes, in some cases reaction takes a different course and 1,3-cyclohexadienes are formed (Büchi and Wüest, 1971; Bohlmann and Zdero, 1973; Padwa and Brodsky, 1974). Reaction is thought to proceed by conjugate addition of the γ-carbon atom of the ylid to the enone, followed by proton transfer and intramolecular Wittig condensation, as depicted in (2.44) for the reaction of mesityl oxide with allylidenetriphenylphosphorane. It is not completely clear what factors promote

$\overset{+}{P}(C_6H_5)_3$ $\overset{+}{P}(C_6H_5)_3$ $P(C_6H_5)_3$

(2.44)

$+ (C_6H_5)_3P{=}O$

(50%)

formation of cyclohexadienes in preference to the normal Wittig condensation to give a triene. The reaction provides a useful route to some derivatives of cyclohexadiene, and it has been used in a remarkably straighforward synthesis of compounds with strained bridgehead double bonds (Dauben and Ipaktschi, 1973).

The Wittig reaction between allylic phosphoranes and non-conjugated carbonyl compounds can be used to make conjugated dienes, but there are difficulties; here also the ylid may react at either the α- or the γ-position, and mixtures of all four geometrical isomers of the diene may be formed. An alternative approach which leads to a single geometrical isomer of a diene has been described recently (Davidson and Warren,

1976) and is illustrated in (2.45) for the synthesis of *trans*- and *cis*-2-(1-cyclohexenyl)-2-butene. The key intermediates are allylphosphine

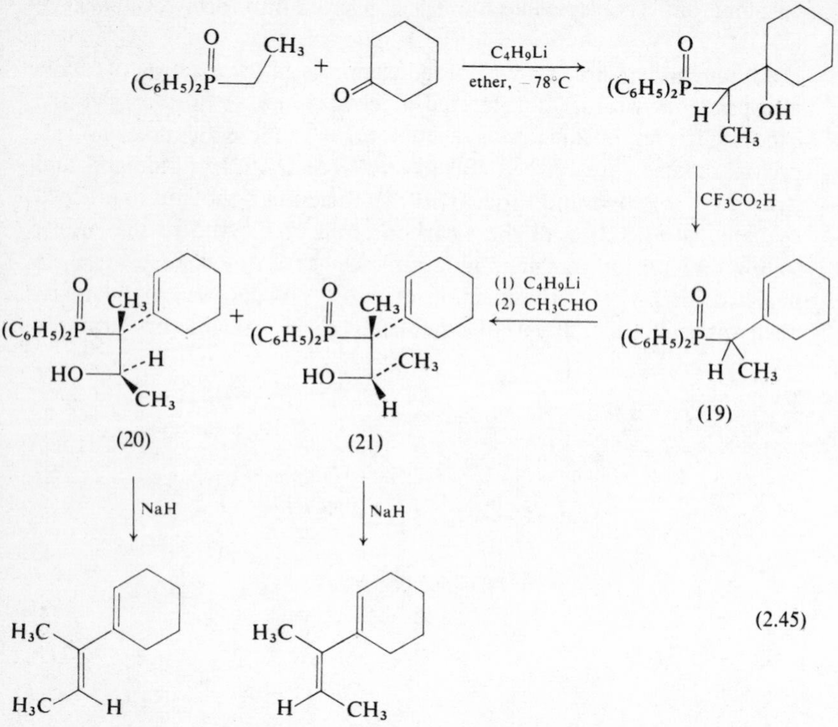

$(C_6H_5)_2P(O)CH_2CH_3$ + cyclohexanone $\xrightarrow[\text{ether, } -78^\circ C]{C_4H_9Li}$ $(C_6H_5)_2P(O)CH(CH_3)C(OH)(C_6H_{10})$

$\xrightarrow{CF_3CO_2H}$ (19)

(19) $\xrightarrow{(1)\ C_4H_9Li;\ (2)\ CH_3CHO}$ (20) + (21)

(20) $\xrightarrow{NaH}$ olefin; (21) $\xrightarrow{NaH}$ olefin

(2.45)

oxides such as (19) which are themselves obtained by dehydration of the alcohols formed by reaction of an alkyldiphenylphosphine oxide with butyl-lithium and a carbonyl compound. The lithio derivative of (19) reacts with acetaldehyde to give a mixture of the diastereomeric alcohols (20) and (21). Separation of these and reaction with sodium hydride gives the two olefins stereospecifically, with elimination of diphenylphosphinic acid $(C_6H_5)_2PO_2H$.

An alternative to the Wittig route to substituted olefins has been described by Corey and Kwiatkowsky (1968). It is based on readily available phosphonbis-*N*,*N*-dialkylamides of type (22) and has a number of advantages over the Wittig procedure. In particular, it can be used for the stereoselective synthesis of *cis* and *trans* isomers. By reaction with butyl-lithium the phosphonamides form lithio derivatives which react readily with aldehydes and ketones to give β-hydroxyphosphonamides. On heating in benzene solution these hydroxyphosphonamides decompose to form olefins in high yield. This method offers wide scope for the synthesis of olefins, for not only can a range of aldehydes and ketones be used, but also a variety of phosphonamides, prepared directly from the appropriate alkyl halide or elaborated from (22) as in the example shown (2.46). Furthermore, diastereomeric β-hydroxyphosphonamides decompose stereospecifically, each giving a single olefin. Separate decomposition of pure diastereomers obtained by fractional crystallisation or other means thus offers a route to the *cis* and *trans* forms of a particular olefin. The reaction involves preferential *syn* elimination, probably by way of a four-membered cyclic transition state similar to that invoked in the Wittig reaction (Corey and Cane, 1969).

$$CH_3Cl + PCl_3 \xrightarrow[(2)\ (CH_3)_2NH]{(1)\ AlCl_3,\ H_2O} CH_3P(=O)[N(CH_3)_2]_2 \quad (22)$$

$$(22) \xrightarrow{C_4H_9Li,\ THF,\ C_6H_{14},\ -80°C} LiCH_2PO[N(CH_3)_2]_2$$

$$LiCH_2PO[N(CH_3)_2]_2 \xrightarrow[(2)\ H_2O]{(1)\ (C_6H_5)_2CO} (C_6H_5)_2C(OH)CH_2PO[N(CH_3)_2]_2 \xrightarrow{C_6H_6,\ reflux} (C_6H_5)_2C{=}CH_2 + [(CH_3)_2N]_2POOH \quad (95\%)$$

$$LiCH_2PO[N(CH_3)_2]_2 \xrightarrow{iso\text{-}C_3H_7I} iso\text{-}C_3H_7CH_2PO[N(CH_3)_2]_2 \xrightarrow{(1)\ C_4H_9Li;\ (2)\ (C_6H_5)_2CO;\ (3)\ H_2O} (C_6H_5)_2C(OH){-}CH(C_3H_7\text{-}iso)PO[N(CH_3)_2]_2$$

$$\xrightarrow{C_6H_6,\ reflux} (C_6H_5)_2C{=}CHC_3H_7\text{-}iso \quad (95\%) \qquad (2.46)$$

There is a silicon version of the Wittig reaction (known as the Peterson reaction) which is discussed more fully on p. 318. It entails the elimination of $(CH_3)_3SiOH$ from a β-hydroxyalkyltrimethylsilane, and has the practical advantage over the Wittig reaction that the by-product is volatile and therefore much easier to remove from the reaction product than triphenylphosphine oxide. Further, the steric course of the reaction seems to be more easily controlled, at least for the generation of disubstituted ethylenes, and both the *cis* and *trans* forms of an olefin can be separately obtained from a single isomer of the hydroxysilane, depending on how the elimination is effected (2.47).

$(CH_3)_3Si(C_3H_7)CH{-}Li + CH_3CHO \longrightarrow (CH_3)_3Si(H)(C_3H_7)C{-}C(H)(OH)(CH_3)$ (*threo* isomer separated)

acid → cis-$C_3H_7CH{=}CHCH_3$ (H, H on same side)

base → trans-$C_3H_7CH{=}CHCH_3$

(2.47)

More recently sulphur ylids have been finding increasing application in synthesis (Trost and Melvin, 1975). Sulphur ylids are formally zwitterions in which a carbanion is stabilised by interaction with an adjacent

$$>\overset{+}{S}-\overset{-}{C}< \quad \longleftrightarrow \quad >S{=}C< \qquad (2.48)$$

sulphonium centre. They are usually prepared by proton abstraction from a sulphonium salt with a suitable base (2.49) or, less commonly, by reaction of a carbene with a sulphide. They do not behave in just the same way as the phosphorus ylids discussed above. In particular, in their reactions with aldehydes and ketones they form epoxides and not olefins. Two of the most widely used reagents are dimethylsulphonium methylide (23) and dimethyloxosulphonium methylide (24), and the reaction of the former with benzophenone to form 2,2-diphenylethylene oxide is shown in (2.49). Reaction begins in the same way as with the phosphorus ylids by attack of the nucleophilic carbon of the ylid on the electrophilic carbon of the carbonyl group, but since sulphur does not have such a high

$$(CH_3)_2\overset{+}{S}{-}CH_3\ \bar{I} \xrightarrow[\text{dimethyl sulphoxide}]{\text{NaH}} (CH_3)_2\overset{+}{S}{-}\bar{C}H_2 \quad (23) \qquad (CH_3)_2\overset{+}{S}(=O){-}\bar{C}H_2 \quad (24)$$

$$(CH_3)_2\overset{+}{S}{-}\bar{C}H_2 + O{=}C(C_6H_5)_2 \longrightarrow \left[(CH_3)_2\overset{+}{S}{-}CH_2{-}C(C_6H_5)_2{-}O^-\right] (25) \longrightarrow \text{epoxide } CH_2{-}O{-}C(C_6H_5)_2 \tag{2.49}$$

affinity for oxygen as phosphorus does, reaction subsequently takes a different course and nucleophilic attack on carbon by the oxy anion leads to formation of the epoxide with displacement of dimethyl sulphide. Dimethyloxosulphonium methylide reacts in the same way with non-conjugated aldehydes and ketones to form epoxides, and since it is more stable than the sulphonium methylide its use might be preferred. The two reagents differ, however, in their reactions with cyclohexanones; in most cases the sulphonium ylid forms an epoxide with a new axial carbon–carbon bond, whereas the oxosulphonium methylide gives an epoxide with an equatorial carbon–carbon bond. This has been ascribed to the fact that addition of the sulphonium methylide to the carbonyl group to form the intermediate zwitterion (as 25) is irreversible, whereas addition of the oxosulphonium methylide is reversible, allowing accumulation of the thermodynamically more stable zwitterion (Johnson, 1973).

Dimethylsulphonium methylide and dimethyloxosulphonium methylide also differ in their reactions with $\alpha\beta$-unsaturated carbonyl compounds. The sulphonium methylide reacts at the carbonyl group to form, again, an epoxide, but with the oxosulphonium methylide a cyclopropane derivative is obtained by Michael addition to the olefinic double bond. The difference is again due to the fact that the kinetically favoured reaction of the sulphonium ylid with carbonyl groups to give a betaine (as 25) is irreversible, whereas the corresponding reaction with the oxosulphonium ylid is reversible, leading to preferential formation of the thermodynamically more stable product from the Michael addition (Johnson, Schroeck and Shanklin, 1973).

Reaction of $\alpha\beta$-unsaturated carbonyl compounds with oxosulphonium alkylids is now one of the most widely used methods for making three-membered rings, along with carbene addition and the Simmons–Smith reaction (see p. 77). Thus, dimethyloxosulphonium methylide adds to

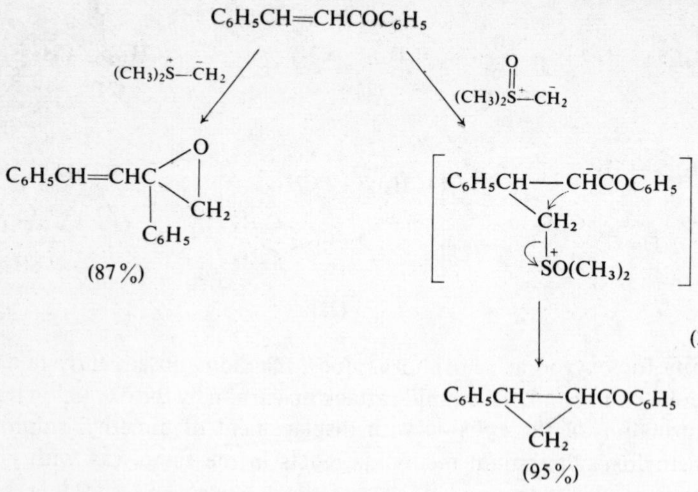

$$C_6H_5CH{=}CHCOC_6H_5$$

$(CH_3)_2\overset{+}{S}{-}\overset{-}{C}H_2$ → $C_6H_5CH{=}CHC(C_6H_5)(-O-CH_2-)$ (epoxide) (87%)

$(CH_3)_2\overset{+}{S}(=O){-}\overset{-}{C}H_2$ → $[C_6H_5CH{-}\overset{-}{C}HCOC_6H_5$, $CH_2\overset{+}{S}O(CH_3)_2]$ → $C_6H_5CH{-}CHCOC_6H_5$ (cyclopropane, ring CH_2) (95%)

(2.50)

the $\alpha\beta$-unsaturated ketone (26), giving the highly hindered cyclopropane derivative (27) in 80 per cent yield. In a more unusual example, (±)-methyl *trans*-chrysanthemate was obtained by reaction of the unsaturated ester (28) with diphenylsulphonium isopropylide (a sulphonium ylid) (2.51).

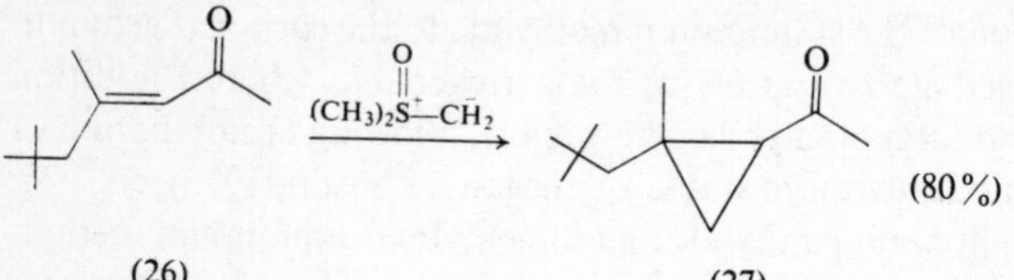

(26) $\xrightarrow{(CH_3)_2\overset{+}{S}(=O){-}\overset{-}{C}H_2}$ (27) (80%)

(2.51)

$$(CH_3)_2C{=}CHCH{=}CHCO_2CH_3 \text{ (28, } trans\text{)} + (C_6H_5)_2\overset{+}{S}{-}\overset{-}{C}(CH_3)_2 \longrightarrow (CH_3)_2C{=}CH{-}\underset{}{C}H{-}CH{-}CO_2CH_3 \text{ (ring } C(CH_3)_2\text{)}$$

(73%)

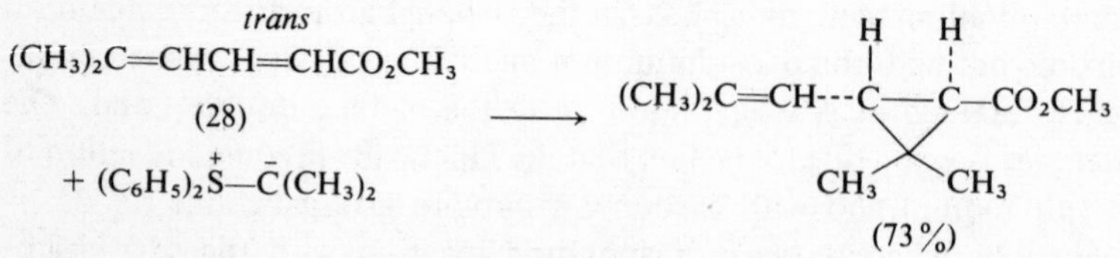

The formation of cyclopropane derivatives from $\alpha\beta$-unsaturated carbonyl compounds and oxosulphonium ylids has hitherto been restricted to reactions involving the parent dimethyloxosulphonium methylid, because of the difficulty of making higher alkylids in this series. This limitation has now been removed by the introduction of a new class of sulphur ylid, the dimethylamino-oxosulphonium ylids (29), formed by deprotonation of *N*,*N*-dialkyl salts of sulphoximines (Johnson,

$$C_6H_5\overset{O}{\overset{\|}{S}}{}^{+}(N(CH_3)_2)-CHR^1R^2\ BF_4^- \xrightarrow[\text{dimethyl sulphoxide}]{NaH} C_6H_5\overset{O}{\overset{\|}{S}}{}^{+}(N(CH_3)_2)-\bar{C}R^1R^2 \quad \textbf{(29)} \tag{2.52}$$

1973). These ylids behave very much like dimethyloxosulphonium methylid, forming epoxides by reaction with carbonyl compounds and cyclopropane derivatives with $\alpha\beta$-unsaturated carbonyl compounds, but they have the advantage that higher members of the series can be made. The isopropylide (29, $R^1 = R^2 = CH_3$), for example, may be obtained ultimately from phenyl isopropyl sulphide or by alkylation of the ethylidene ylid (29, $R^1 = H$, $R^2 = CH_3$) with methyl iodide and regeneration of the ylid with base. Alternatively, epoxides, cyclopropane derivatives and aziridines can be obtained in high yield by reaction of the appropriate unsaturated precursor with the anions derived from *N-p*-tosylsulphoximines (as 30). The sulphoximines are readily available from the corresponding sulphoxides and the anions form another class of nucleophilic alkylidene transfer agents, but they have the advantage over the ylids described above that the by-products of the reactions are water-soluble salts and not neutral molecules. 1,2-Diphenylaziridine, for example, is obtained from benzylideneaniline as shown in (2.53). The anion of the dimethylsulphoximine derivative (31), which is commercially available or is easily prepared from dimethyl sulphoxide and chloramine T, is a

$$\underset{(30)}{C_6H_5\overset{O}{\overset{\|}{S}}(=NTos)CH_2^-Na^+} \xrightarrow{C_6H_5CH{=}NC_6H_5} C_6H_5\overset{O}{\overset{\|}{S}}(=NTos)-CH_2-CH(\bar{N}C_6H_5)C_6H_5 \longrightarrow \underset{(86\%)}{\text{1,2-diphenylaziridine } (C_6H_5N\langle CH_2CHC_6H_5\rangle)} + C_6H_5\overset{O}{\overset{\|}{S}}\bar{N}Tos\ Na^+ \tag{2.53}$$

$$\underset{(31)}{CH_3\overset{O}{\overset{\|}{S}}(=NTos)CH_3}$$

Tos = H_3C–C_6H_4–SO_2

very convenient reagent for the preparation of epoxides from aldehydes and ketones.

Another valuable sulphonium ylid is diphenylsulphoniumcyclopropylid (32), which is usually generated *in situ* by action of potassium hydroxide on cyclopropyldiphenylsulphonium fluoroborate (Bogdanowicz and Trost, 1974; Trost, 1974). In contrast to dimethylsulphonium

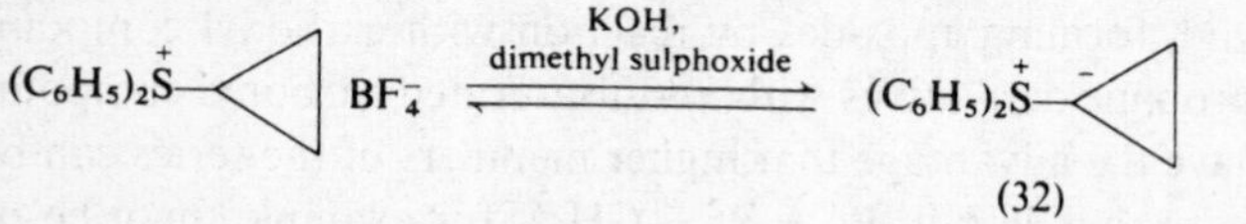

$$(C_6H_5)_2\overset{+}{S}\text{—}C_3H_5 \; BF_4^- \xrightarrow[\text{dimethyl sulphoxide}]{\text{KOH,}} (C_6H_5)_2\overset{+}{S}\text{—}\overset{-}{C}_3H_4 \quad (32) \qquad (2.54)$$

methylid it attacks $\alpha\beta$-unsaturated ketones at the olefinic double bond to form spiropentanes in excellent yield. With aldehydes and ketones, oxaspiropentanes are formed. This reaction is of value in synthesis because of the wide variety of useful transformations which the oxaspiropentanes undergo (Trost, 1974). For example, on treatment with acid or, better, lithium perchlorate or fluoroborate in refluxing benzene, they rearrange to cyclobutanones which are themselves valuable synthetic intermediates (2.55) (see p. 41), and with lithium diethylamide followed

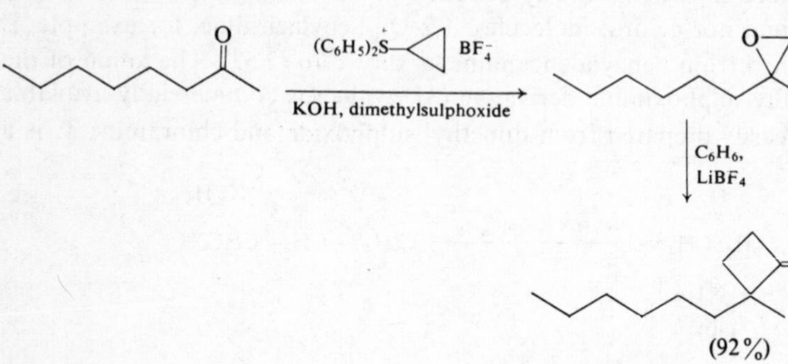

(2.55)

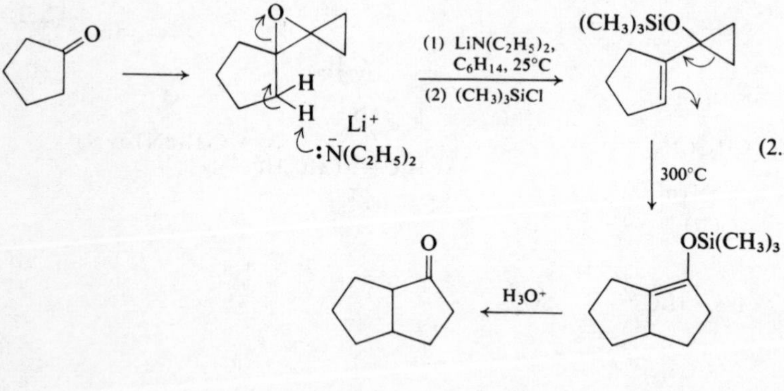

$(C_6H_5)_2\overset{+}{S}$—cyclopropyl BF_4^-; KOH, dimethylsulphoxide

C_6H_6, $LiBF_4$

(92%)

(2.56)

(1) $LiN(C_2H_5)_2$, C_6H_{14}, 25°C

(2) $(CH_3)_3SiCl$

Li^+ $:\bar{N}(C_2H_5)_2$

$(CH_3)_3SiO$

300°C

$OSi(CH_3)_3$

H_3O^+

by quenching with chlorotrimethylsilane they give the trimethylsilyl ethers of 1-alkenylcyclopropanols (cf. p. 373). As vinyl cyclopropanes these rearrange on pyrolysis to trimethylsilyloxycyclopentenes, and the sequence provides a versatile method for synthesising cyclopentanone derivatives, as exemplified in (2.56).

Sulphur ylids undergo a number of rearrangements, some of which have been exploited in synthesis for the formation of carbon–carbon bonds (see pp. 39, 158). Carbon–carbon double bonds are obtained by Stevens rearrangement of the ylid followed by Hofmann elimination of a derived sulphonium salt (2.57). The method has been used very successfully to prepare highly strained unsaturated cyclophanes, as illustrated

$$-CH_2-\overset{+}{S}(CH_3)-\bar{C}H- \longrightarrow -CH_2-CH(SCH_3)- \longrightarrow -CH_2-CH(\overset{+}{S}(CH_3)_2)- \xrightarrow{\text{base}} -CH{=}CH- \qquad (2.57)$$

$(CH_3O)_2\overset{+}{C}HBF_4^-$; CH_3-S^+ ; $^+S-CH_3$; $t\text{-}C_4H_9OK$, THF ; SCH_3 ; CH_3S ; $(CH_3O)_2\overset{+}{C}HBF_4^-$; $\overset{+}{S}(CH_3)_2$; $(CH_3)_2\overset{+}{S}$; $t\text{-}C_4H_9OK$, THF ; H H

(2.58)

in (2.58) for the synthesis of *anti*-[2,2]metacyclophan-1,9-diene (Mitchell and Boekelheide, 1974). This general strategy of bringing two carbon atoms into close proximity through sulphide linkages, followed by direct linking of the carbon atoms with extrusion of sulphur, has been used with advantage in other syntheses, notably in the synthesis of vitamin B12 (Eschenmoser, 1970). In one situation the problem was to combine the lactam (33) and the eneamide (34) to give the bicyclic compound (35) and this was beautifully achieved by the 'sulphide contraction' route shown in (2.59). The intermediate (36), on heating with the thiophile triphenylphosphine formed the required product directly, presumably by way of the episulphide (37).

NC, N, H, O (33) + H_2C, N, H, O (34) → NC, HN, N, O (35) (2.59)

NC, N, H, O —P_2S_5→ NC, N, H, S —CH_2, N, H, O; $(C_6H_5COO)_2$, HCl→ NC, N, S, H, N, O (36)

heat, $(C_6H_5)_3P$ or $(C_2H_5O)_3P$

NC, N, H, S, N, O (37) → NC, N, HN, O (50% overall)

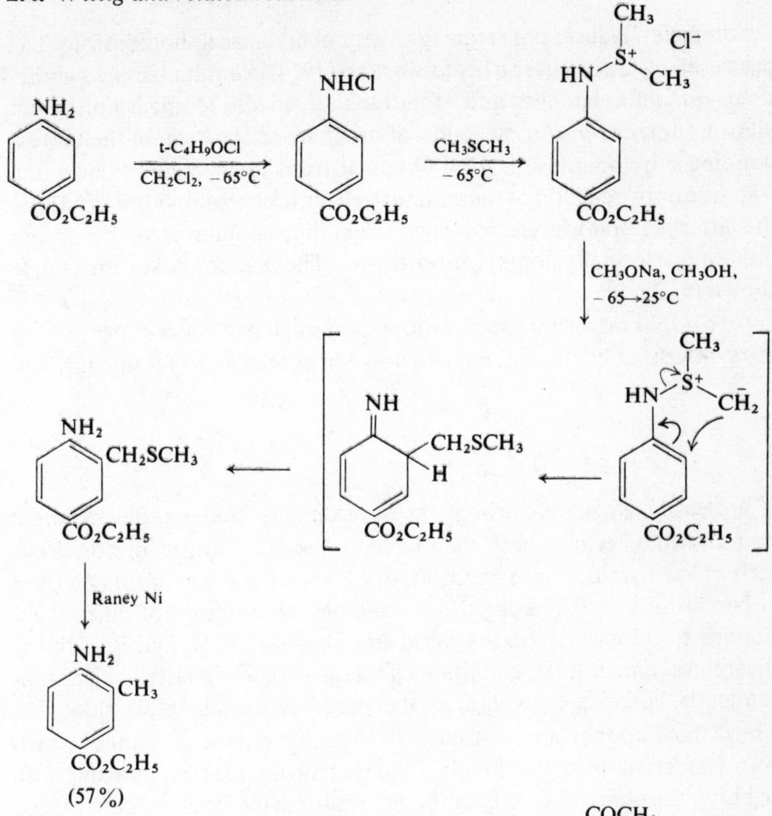

NH_2, $CO_2C_2H_5$ $\xrightarrow[CH_2Cl_2,\ -65°C]{t\text{-}C_4H_9OCl}$ NHCl, $CO_2C_2H_5$ $\xrightarrow[-65°C]{CH_3SCH_3}$ $HN-S^+(CH_3)_2$ Cl^-, $CO_2C_2H_5$

CH_3ONa, CH_3OH, $-65 \rightarrow 25°C$

[$HN-S^+(CH_3)\bar{C}H_2$, $CO_2C_2H_5$ ⟶ NH, CH_2SCH_3, H, $CO_2C_2H_5$] ⟶ NH_2, CH_2SCH_3, $CO_2C_2H_5$

Raney Ni

NH_2, CH_3, $CO_2C_2H_5$
(57%)

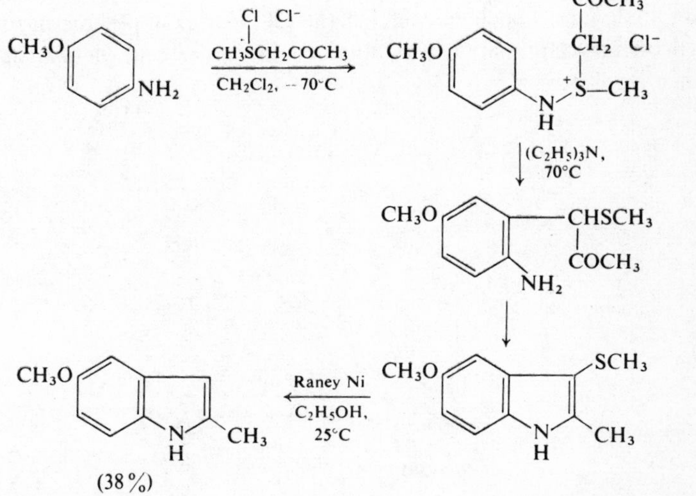

CH_3O, NH_2 $\xrightarrow[CH_2Cl_2,\ -70°C]{CH_3\overset{Cl}{S}CH_2COCH_3\ Cl^-}$ CH_3O, $\underset{H}{N}-\overset{+}{S}(CH_3)-CH_2-COCH_3$ Cl^-

$(C_2H_5)_3N$, 70°C

CH_3O, $CHSCH_3$, $COCH_3$, NH_2

$$(2.60)$$

CH_3O, SCH_3, $\underset{H}{N}$, CH_3 $\xrightarrow[C_2H_5OH,\ 25°C]{Raney\ Ni}$ CH_3O, $\underset{H}{N}$, CH_3
(38%)

Sommelet–Hauser type rearrangement of arylazasulphonium ylids has been used by Gassman and his co-workers (1974) in a valuable new synthesis of *ortho*-alkylanilines and of indole derivatives. The azasulphonium ylids are formed *in situ* by action of an appropriate base on the corresponding sulphonium salt, itself obtained from the *N*-chloraniline and a sulphide or by reaction of the aniline itself with a halosulphonium halide; the latter method is preferred when the aniline contains strong electron-donating, especially alkoxy, substituents. The reaction takes the course shown in (2.60).

A two-fold extrusion process forms the key step in a recent method for the synthesis of highly hindered olefins. The principle here is that if X and

$$R^1R^2C(X)(Y)CR^3R^4 \longrightarrow R^1R^2C{=}CR^3R^4 + X + Y$$

Y are easily extrudable groups then even highly hindered olefins might be accessible because both the olefinic bonds are formed intramolecularly. Good results have been obtained with 1,3,4-thiadiazolines (X = —N=N—, Y = S), prepared, for example, by addition of diazo compounds to thiones (Barton, Guziec and Shahak, 1974). Pyrolysis of the diazolines under mild conditions first gives episulphides which subsequently collapse to olefins in the presence of a suitable thiophile. Thus, thiocamphor and di-t-butyldiazomethane gave an adduct which was converted into the highly hindered olefin (38) on heating with tributylphosphine (2.61). Even better results have been obtained using selenium analogues. Fenchylidenefenchane (39), for example, the most hindered olefin yet prepared, was obtained in 24 per cent yield from

+ N_2C — THF, reflux → ; $(C_4H_9)_3P$, 120°C

(2.61)

(39) (38)

selenofenchone (Back, Barton, Britten-Kelly and Guziec, 1975). Tetra-t-butylethylene, however, still defies isolation. The tetra-isopropyl compound has been obtained by reductive coupling of di-isopropyl ketone with an active titanium powder prepared by reduction of titanium trichloride with potassium (McMurry and Fleming, 1976).

A more flexible route which allows the preparation of tetrasubstituted olefins from two different ketones is exemplified (2.62) by the synthesis of 4-t-butylisopropylidenecyclohexane. The key step is an elimination from a β-hydroxyselenide formed from a carbonyl compound and a selenium-stabilised carbanion, itself produced by cleavage of a selenoacetal with butyl-lithium (Reich and Chow, 1975; Rémion, Dumont and Krief, 1976).

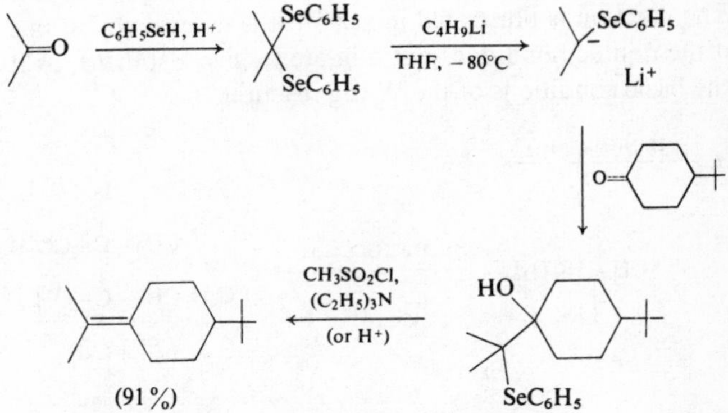

(2.62)

The well-known conversion of α-halosulphones into olefins on treatment with base (the Ramberg–Bäcklund reaction) is believed to proceed as shown in (2.63) with final extrusion of sulphur dioxide from an episulphone (Paquette, 1968; Bordwell, 1970).

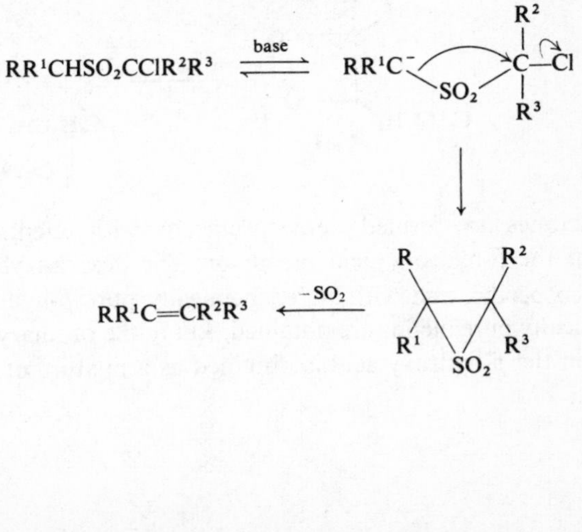

(2.63)

2.5. Decarboxylation of β-lactones

The decomposition of β-lactones to olefins and carbon dioxide at moderate temperature has been known for many years but has not been much used in synthesis because of the inaccessibility of β-lactones. It has recently been found, however, that β-lactones can be made from β-hydroxy acids by action of benzenesulphonyl chloride in pyridine. β-Lactones decompose readily at 140–160°C to give the corresponding olefin in virtually quantitative yield. The β-hydroxy acids themselves can be made by the Reformatsky reaction or, better, by condensation of α-metalated carboxylic salts with aldehydes and ketones. The whole sequence makes a useful alternative to the Wittig reaction particularly for the synthesis of tri- and tetra-substituted olefins (Adam, Baeza and Liu, 1972). The reaction is illustrated in (2.64); it is noteworthy in this example that the double bond does not migrate to give a stilbene, as it does under the basic conditions of the Wittig reaction.

$$CH_3CH_2CO_2H \xrightarrow[THF]{2LiN(iso\text{-}C_3H_7)_2} CH_3\underset{\underset{Li}{|}}{C}HCO_2Li \xrightarrow{C_6H_5CH_2COC_6H_5} \begin{array}{c} H \\ | \\ CH_3-C-CO_2H \\ | \\ C_6H_5CH_2-C-OH \\ | \\ C_6H_5 \end{array}$$

$$\xrightarrow[\text{pyridine, } 0°C]{C_6H_5SO_2Cl} \beta\text{-lactone (CH}_3\text{, H; C}_6H_5CH_2\text{, C}_6H_5\text{; CO–O ring)} \xrightarrow{140°C} \begin{array}{ccc} CH_3 & & H \\ & C{=}C & \\ C_6H_5CH_2 & & C_6H_5 \end{array} \quad (2.64)$$

(> 99% *trans*)

The β-lactones are formed stereospecifically with retention of the geometry of the β-hydroxy acid precursor. The decarboxylation step also is stereospecific, and with stereochemically pure β-hydroxy acids stereochemically pure olefins are obtained. But in the ordinary course of events, when the β-hydroxy acid is obtained as a mixture of *threo* and *erythro* isomers reaction affords a mixture of the *cis* and *trans* olefins where this is structurally possible. Nevertheless this is a good mild method for the regiospecific synthesis of olefins.

2.6. Stereoselective synthesis of tri- and tetra-substituted ethylenes

cis- and *trans*-1,2-Disubstituted ethylenes are readily available by reduction of the corresponding acetylenes or by other means (see p. 111), but isomerically pure tri- and tetra-substituted olefins have been more difficult to obtain. Recently a number of stereoselective syntheses of tri- and tetra-substituted ethylenes have been developed in the course of work aimed at the synthesis of biologically and biosynthetically important polyisoprenoids. The first of these methods due to Cornforth, Cornforth and Mathew (1959) was subsequently applied by them in the synthesis of all-*trans*-squalene. The critical step in this method is the reaction of a Grignard reagent with an α-chloro-aldehyde or -ketone. The most reactive conformation of an α-chloroaldehyde or α-chloroketone is that in which the C=O and C—Cl dipoles are anti-parallel. It was found that addition of a Grignard reagent to this conformation is highly stereoselective and takes place mainly from the side of the carbonyl group which is less sterically hindered by the groups R^1 and R^2 on the α-carbon atom, leading predominantly to the chlorohydrin in which the incoming group R^4 and the larger of the groups R^1 and R^2 are *anti* to each other (2.65). The resulting chlorohydrin is converted, by a series of stereo-

$$R^3\text{—}\overset{\overset{O\,\delta^-}{\|}}{C}{}^{\delta+}\text{—}\overset{\delta+}{C}(R^1)(R^2)\text{—Cl}^{\delta-} \xrightarrow{R^4MgBr} R^4\text{—}C(OH)(R^3)\text{—}C(R^1)(R^2)\text{—Cl} \xrightarrow[\text{steps}]{\text{several}} (R^4)(R^3)C{=}C(R^1)(R^2) \qquad (2.65)$$

selective conversions, into the olefin in which three of the groups on the double bond are derived from the chlorocarbonyl compound, and the other from the Grignard reagent. The method is exemplified in (2.66) for the synthesis of *trans*-3-methyl-2-pentene. A similar series of reactions beginning with the addition of methylmagnesium iodide to 2-chloropentan-3-one gave *cis*-3-methyl-2-pentene.

An entirely different procedure which makes possible the stereospecific conversion of a propargylic alcohol into a 2- or 3-alkylated allylic alcohol has been described by Corey, Katzenellenbogen and Posner (1967) (2.67). This procedure is based on a remarkable, specific, conversion of propargylic alcohols into β- or γ-iodoallylic alcohols by reduction with a modified lithium aluminium hydride reagent and subsequent reaction of the crude reduction product with iodine. If the reduction is effected in presence of sodium methoxide, the final product is exclusively the γ-iodoallylic alcohol, whereas reduction with lithium aluminium hydride and aluminium chloride gives finally the β-iodoallylic

$$CH_3C(=O)\text{–}CHCl(CH_3) \xrightarrow[\text{ether, }-70°C]{C_2H_5MgBr} C_2H_5C(OH)(CH_3)\text{–}CHCl(CH_3)\ (90\%)\ (85\%\text{ stereoselective}) \tag{2.66}$$

$$\xrightarrow{NaOH} \text{epoxide } C_2H_5(CH_3)C\text{–O–}CH(CH_3)\ (91\%) \xrightarrow[CH_3CO_2Na,\ CH_3CO_2H]{NaI} C_2H_5C(OH)(CH_3)\text{–}CHI(CH_3)$$

$$\xrightarrow[\text{pyridine}]{SnCl_2,\ POCl_3} (C_2H_5)(CH_3)C{=}CH(CH_3)\ (81\%)$$

$$R\text{–}C{\equiv}C\text{–}CH_2OH \xrightarrow[(2)\ I_2,\ -78°C]{(1)\ LiAlH_4,\ CH_3ONa,\ THF} R(I)C{=}CH(CH_2OH) \xrightarrow{Li(CH_3)_2Cu} R(CH_3)C{=}CH(CH_2OH) \tag{2.67}$$

$$R\text{–}C{\equiv}C\text{–}CH_2OH \xrightarrow[(2)\ I_2,\ -78°C]{(1)\ LiAlH_4,\ AlCl_3,\ THF} RHC{=}C(I)CH_2OH \xrightarrow{Li(CH_3)_2Cu} RHC{=}C(CH_3)CH_2OH$$

alcohol. Reaction of the resulting iodo compounds with lithium organocuprates (p. 64) then affords the corresponding substituted allylic alcohols in which the substituents originally present in the acetylene are *trans* to each other. This method is applicable to a variety of synthetic problems in which the stereospecific introduction of trisubstituted olefinic linkages is involved. For example, it has been used by Corey, Katzenellenbogen, Gilman, Roman and Erickson (1968) in their synthesis of juvenile hormone, the relevant steps of which are shown in (2.68).

$(CH_2)_2C{\equiv}CCH_2OH$

(1) $LiAlH_4$, CH_3ONa, THF,
(2) I_2
(3) $Li(CH_3)_2Cu$

(2.68)

$(CH_2)_2$ CH_2OH

(54%; 97% *trans*)

Another synthesis leads to $\beta\beta$-dialkylacrylic esters from $\alpha\beta$-acetylenic esters (Corey and Katzenellenbogen, 1969; Siddall, Biskup and Fried, 1969). Lithium dialkylcuprates (p. 64) add rapidly to $\alpha\beta$-acetylenic esters to give the stereoisomeric acrylic esters (40) and (41) in high yield.

$$R^1{-}C{\equiv}C{-}CO_2CH_3 \xrightarrow[(2)\ H_3O^+]{(1)\ LiR^2_2Cu}$$

$$\underset{(40)}{(R^1)(R^2)C{=}C(H)(CO_2CH_3)} + \underset{(41)}{(R^1)(R^2)C{=}C(H)(CO_2CH_3)} \qquad (2.69)$$

The stereochemistry of the products is highly dependent on the reaction temperature and the nature of the solvent, and high yields of the *cis* addition compound (40) are obtained by conducting the reaction at −78°C in tetrahydrofuran solution. In contrast to the procedure with propargyl alcohols described above, this reaction yields an olefin in which the substituents in the acetylenic precursor are *cis* to each other in the olefinic product.

Alkenylcuprates appear to be particularly effective in this reaction giving 2,4-dienoic esters with high stereoselectivity; the stereochemistry of the alkenyl groups is almost completely preserved in the reactions (Näf and Degen, 1971) (2.70). Even better results are obtained using

$$\text{(cis-1-bromopropene)} \xrightarrow[\text{(2) } Cu_2I_2]{\text{(1) Li, ether}}$$

$$(\text{cis-propenyl})_2CuLi \xrightarrow[\text{(2) } CH_3OH,\ -78°C]{\text{(1) } CH{\equiv}CCO_2C_2H_5,\ \text{ether},\ -10°C} \text{(2,4-hexadienoate)}\ CO_2C_2H_5$$

(77%; 95% *trans, cis*)

$$HC{\equiv}CCO_2CH_3 \xrightarrow[\text{(2) } CH_3OH,\ -78°C]{\text{(1) } CH_2{=}CHCu,\ THF,\ -78°C} \text{(2,4-pentadienoate)}\ CO_2CH_3 \qquad (2.70)$$

(85%)

alkenylcopper reagents prepared from one equivalent of an alkenyl-lithium and one equivalent of a cuprous salt (Corey, Kim, Chen and Takeda, 1972). In the simplest case, methyl *trans*-2,4-pentadienoate was obtained in 85 per cent yield from vinylcopper and methyl propynoate (2.70). With allylcopper reagents, 2,5-dienoic acids are readily obtained.

Other useful methods for the stereospecific synthesis of olefins from acetylenes proceed from the derived alkenylalanes and alkenylboranes. Alkenylalanes are readily prepared by hydroalumination of acetylenes with, for example, diisobutylaluminium hydride. The reaction takes place by *cis* addition, giving *cis* alkenylalanes. Reaction of these alkenyl alanes with halogens proceeds with retention of configuration to give the corresponding alkenyl halides, the halogen atoms of which are replaceable by alkyl groups by reaction with an organocopper reagent. Thus iodination of the alane from the reaction of 1-hexyne with diiso-butylaluminium hydride produces the isomerically pure *trans*-1-iodo-1-hexene, while 3-hexyne gives *trans*-3-iodo-3-hexene (2.71). As a comple-ment to the above reaction with terminal acetylenes *cis*-1-halogeno-alkenes can be obtained by addition of dicyclohexylborane (p. 287) to 1-bromo- and 1-iodo-acetylenes followed by protonolysis of the product with acetic acid. Under these conditions 1-iodohexyne gave *cis*-1-iodo-1-hexene in 95 per cent yield (Zweifel and Arzoumanian, 1967).

Another excellent route to *cis*- and *trans*-1-halogenoalkenes from terminal acetylenes uses the *trans*-alkenylboronic acids which are easily obtained by reaction of terminal acetylenes with catecholborane (p. 276) followed by hydrolysis. Reaction of the boronic acids with iodine and

$$C_4H_9C{\equiv}CH \xrightarrow[C_7H_{16}]{(iso\text{-}C_4H_9)_2AlH} \underset{C_4H_9}{\overset{H}{>}}C{=}C\underset{H}{\overset{Al(iso\text{-}C_4H_9)_2}{<}}$$

$$\xrightarrow{(1)\ I_2,\ THF,\ -50°C;\ (2)\ H_3O^+} \underset{C_4H_9}{\overset{H}{>}}C{=}C\underset{H}{\overset{I}{<}} \quad (74\%)$$

(2.71)

$$C_2H_5C{\equiv}CC_2H_5 \xrightarrow[C_7H_{16}]{(iso\text{-}C_4H_9)_2AlH} \underset{H}{\overset{C_2H_5}{>}}C{=}C\underset{Al(iso\text{-}C_4H_9)_2}{\overset{C_2H_5}{<}}$$

$$\xrightarrow{(1)\ I_2,\ THF,\ -50°C;\ (2)\ H_3O^+} \underset{H}{\overset{C_2H_5}{>}}C{=}C\underset{I}{\overset{C_2H_5}{<}} \quad (72\%)$$

$$C_4H_9C{\equiv}CI \xrightarrow[25°C]{(C_6H_{11})_2BH} \underset{H}{\overset{C_4H_9}{>}}C{=}C\underset{B(C_6H_{11})_2}{\overset{I}{<}}$$

$$\xrightarrow{CH_3CO_2H} \underset{H}{\overset{C_4H_9}{>}}C{=}C\underset{H}{\overset{I}{<}} \quad (95\%;\ 99\%\ cis)$$

sodium hydroxide results in replacement of the boronic acid group by iodine with retention of configuration to give the *trans*-1-iodoalkene. But with bromine, followed by base, substitution is accompanied by inversion and the *cis*-1-bromoalkene is formed. In each case the reaction is highly stereoselective. Thus, 1-octyne was converted into *cis*-1-bromo-1-octene in 85 per cent yield with 99 per cent stereoselectivity, and 1-hexyne gave 89 per cent *trans*-1-iodo-1-hexene (Brown, Hamaoka and Ravindran,

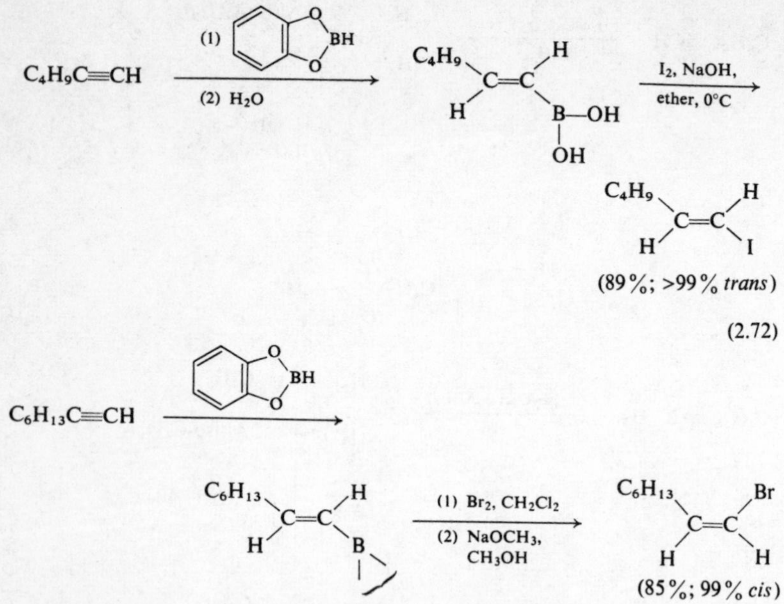

$$C_4H_9C{\equiv}CH \xrightarrow[(2)\ H_2O]{(1)\ \text{catecholborane (BH)}} C_4H_9CH{=}CHB(OH)_2 \xrightarrow[\text{ether, }0°C]{I_2,\ NaOH,} C_4H_9CH{=}CHI$$

(89%; >99% *trans*)

(2.72)

$$C_6H_{13}C{\equiv}CH \xrightarrow{\text{catecholborane (BH)}} C_6H_{13}CH{=}CH\text{–B} \xrightarrow[(2)\ NaOCH_3,\ CH_3OH]{(1)\ Br_2,\ CH_2Cl_2} C_6H_{13}CH{=}CHBr$$

(85%; 99% *cis*)

1973) (2.72). The inversion of configuration in the bromination reaction can be accounted for by the usual *trans* addition of bromine to the double bond followed by base-induced *trans* elimination of boron and bromine, but the course of the iodination reaction is not clear (2.73).

Reaction of alkenylalanes with methyl-lithium affords the corresponding aluminates which are converted into $\alpha\beta$-unsaturated carboxylic acids in excellent yields on reaction with carbon dioxide. Similarly, para-

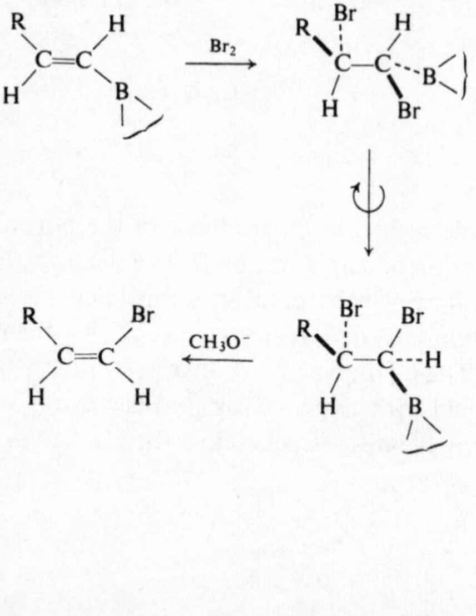

$$RCH{=}CH\text{–B} \xrightarrow{Br_2} RCHBr\text{–}CHBr\text{–B} \xrightarrow{\text{rotation}} RCHBr\text{–}CHBr\text{–B} \xrightarrow{CH_3O^-} RCH{=}CHBr$$

(2.73)

$$CH_3C{\equiv}CCH_3 \xrightarrow{(iso\text{-}C_4H_9)_2AlH} (E)\text{-}H_3C(H)C{=}C(CH_3)Al(iso\text{-}C_4H_9)_2 \xrightarrow[\text{ether, } -30^{\circ}C]{CH_3Li,} \left[H_3C(H)C{=}C(CH_3)\bar{A}l(iso\text{-}C_4H_9)_2(CH_3)\right]Li^+ \xrightarrow{CO_2,\ H_3O^+} H_3C(H)C{=}C(CH_3)CO_2H \quad (76\%)$$

$$CH_3C{\equiv}CCH_3 \xrightarrow{Li[(iso\text{-}C_4H_9)_2AlHCH_3]} \left[H(H_3C)C{=}C(CH_3)\bar{A}l(iso\text{-}C_4H_9)_2(CH_3)\right]Li^+ \xrightarrow{CO_2,\ H_3O^+} H(H_3C)C{=}C(CH_3)CO_2H \quad (72\%)$$

$$\left[H(H_3C)C{=}C(CH_3)\bar{A}l(iso\text{-}C_4H_9)_2(CH_3)\right]Li^+ \xrightarrow[(2)\ H_3O^+]{(1)\ \text{paraformaldehyde}} H(H_3C)C{=}C(CH_3)CH_2OH \quad (68\%)$$

(2.74)

formaldehyde and other aldehydes afford allylic alcohols (Zweifel and Steele, 1967*a*) (2.74).

In contrast to the reaction with diisobutylaluminium hydride discussed above, hydroalumination of disubstituted acetylenes with lithium hydridodiisobutylmethylaluminate, obtained from diisobutylaluminium hydride and methyl-lithium, results in *trans* addition to the triple bond, thus opening the way to a convenient synthesis of isomeric series of olefins from a disubstituted acetylene. Carbonation gives $\alpha\beta$-unsaturated acids, reaction with paraformaldehyde or acetaldehyde affords allylic alcohols and iodine gives alkenyl iodides, all isomeric with the products obtained in the reaction sequences using diisobutylaluminium hydride discussed above (Zweifel and Steele, 1967*b*). Thus the isomeric α-methylcrotonic acids are conveniently obtained from 2-butyne as illustrated in (2.74). In all these reactions the isobutyl groups of the hydroaluminating agent are converted into isobutane in the hydrolysis step, and do not interfere with the isolation of the products.

Alkenylboranes, obtained by hydroboration of acetylenes (p. 281) do not behave in the same way as alkenylalanes. Iodination does not afford the corresponding iodide, but results in stereospecific transfer of one alkyl group from boron to the adjacent carbon, providing another stereospecific synthesis of substituted ethylenes. Thus, addition of iodine and sodium hydroxide to the alkenylborane obtained by hydroboration of 1-hexyne with dicyclohexylborane, affords *cis*-1-cyclohexyl-1-hexene in 75 per cent yield. The reaction is thought to take the course indicated in (2.75) (Zweifel, Arzoumanian and Whitney, 1967). 3-Hexyne is similarly converted into *cis*-3-cyclohexyl-3-hexene in 85 per cent yield. The necessity for *anti* elimination of the iodine and the boron group ensures the stereoselectivity of the olefin-forming step.

This procedure has been used in a synthesis of conjugated *cis*,*trans*-dienes from acetylenes as described on p. 292, and in a novel method for the stereoselective introduction of olefinic side chains into cyclic systems (Zweifel, Fisher, Snow and Whitney, 1971; see also Corey and Ravindranathan, 1972). Thus *cis*-1-(*trans*-2-methylcyclohexyl)-hexene (42) was obtained from bis-(*trans*-2-methylcyclohexyl)borane and 1-hexyne in 85 per cent yield as shown in (2.76), the 2-methylcyclohexyl group migrating from boron to carbon with complete retention of configuration. Since the borane itself is obtained from 1-methylcyclohexene, the sequence results, in effect, in the reductive coupling of an olefinic and an acetylenic unit. The corresponding *trans*-alkene can be obtained from the same borane by modification of the reaction sequence.

$$C_4H_9C\equiv CH + (C_6H_{11})_2BH \longrightarrow (H)(C_4H_9)C{=}C(H)(B(C_6H_{11})_2)$$

$$\xrightarrow{I_2,\ NaOH} \left[\text{iodonium intermediate: } C_6H_{11},\ I^-,\ B{-}C_6H_{11},\ H,\ C{-}C^{+},\ C_4H_9,\ I,\ H \right] \quad (2.75)$$

$$\longrightarrow C_4H_9(H)(I)C{-}C(H)(C_6H_{11})B(C_6H_{11})I \equiv C_4H_9(H)(I)C{-}C(C_6H_{11})(H){\cdots}B(I)(C_6H_{11})$$

$$\longrightarrow (H)(C_4H_9)C{=}C(H)(C_6H_{11})$$

(99% *cis*)

A disadvantage of this procedure is that only one of the carbon ligands attached to boron is used in the coupling reaction; the other is effectively wasted. One solution to this general problem has been to employ 'mixed' dialkylboranes where one of the boron-bound carbon ligands (e.g. thexyl) shows little tendency to migrate (cf. p. 295), but it is now becoming clear that the hierarchy of migratory aptitudes is not as well defined as was once thought, and that in some cases the thexyl group does migrate during the alkenyl–borane iodination reaction. A highly promising way round this difficulty has been found recently by Evans, Thomas and Walker (1976) by using oxygen ligands as the non-migratory groups. Reaction of boronic esters with 1-alkenyl-lithium reagents, and treatment of the resultant boron 'ate' complexes with iodine under appropriate conditions gave the substituted olefin in good yield and with high stereoselectivity. The sequence is illustrated in the second example in (2.76).

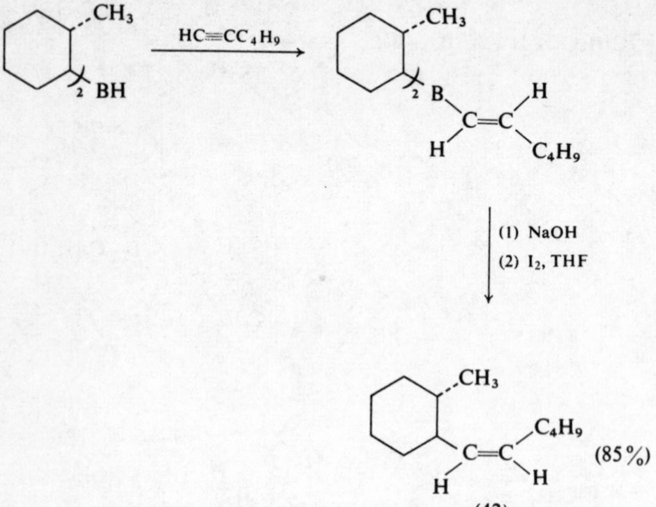

$$\text{(2-methylcyclohexyl)}_2\text{BH} \xrightarrow{HC\equiv CC_4H_9} \text{(2-methylcyclohexyl)}_2\text{B}-CH=CH-C_4H_9$$

(1) NaOH
(2) I_2, THF

(42) (85%)

$$\text{(trans) } LiCH=CHCH(C_5H_{11})OR, \text{ THF, } 0°C$$

$B(OCH_3)_2$ $\bar{B}(OCH_3)_2$ Li^+

C_2H_5 RO C_5H_{11}

I_2, $NaOCH_3$, CH_3OH (2.76)

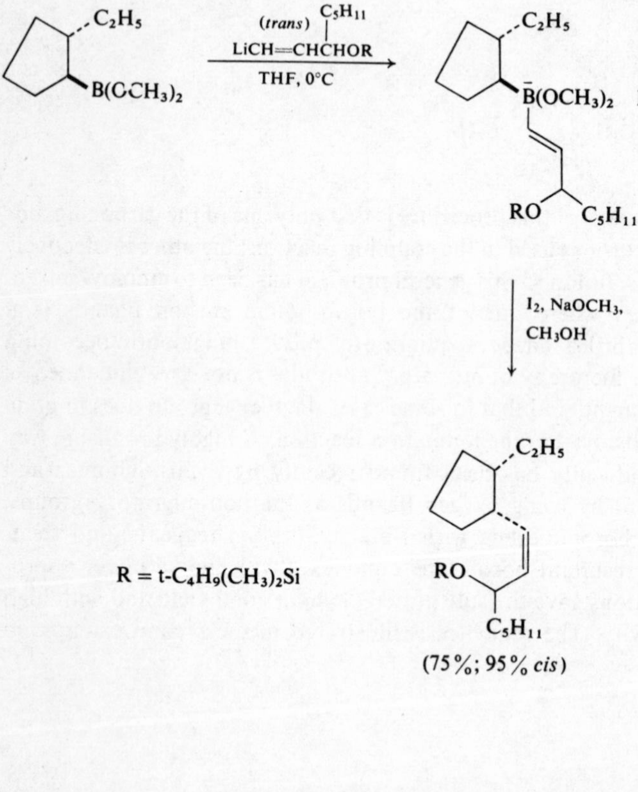

$R = t\text{-}C_4H_9(CH_3)_2Si$

(75%; 95% *cis*)

A method for obtaining trisubstituted olefins of the type (48; R = alkyl) is based on work by Julia, Julia and Tchen (1961), who earlier developed a novel synthesis of homoallylic bromides by rearrangement of cyclopropylcarbinol systems. Thus, the secondary cyclopropylcarbinol (43, $R^1 = CH_3$, $R^2 = H$) on treatment with 48 per cent hydrobromic acid gave the *trans* compound (44, $R^1 = CH_3$, $R^2 = H$) in 80 per cent yield with 90–95 per cent stereoselectivity (2.77). With tertiary

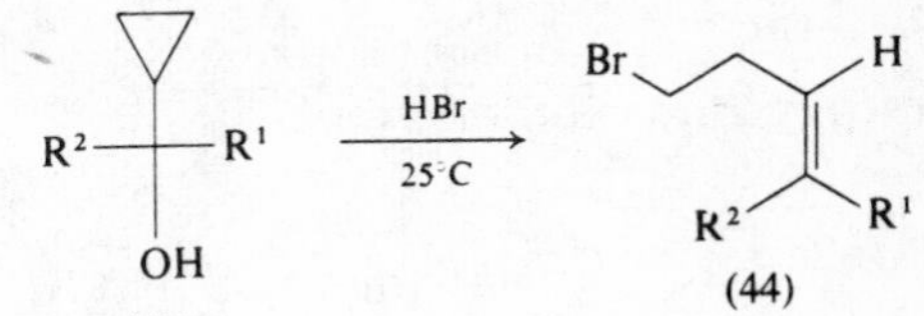

(2.77)

carbinols, unfortunately, the reaction was not so stereoselective; the carbinol (43, $R^1 = C_4H_9$, $R^2 = CH_3$) gave a mixture of the *trans* bromide (44, $R^1 = C_4H_9$, $R^2 = CH_3$) and its *cis* isomer in a ratio of 3 to 1. If $R^2 = H$, then clearly the transition state (45) which leads to the *trans*-olefin (44) will be preferred to the alternative (46). But if R^1 and R^2 are both substituent groups then unless they differ greatly in steric bulk there will be no great preference for one transition state over the other.

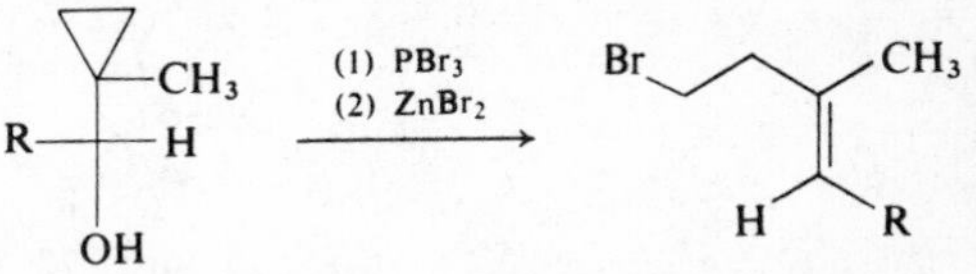

With methylcyclopropylcarbinols of the type (47), however, treatment with phosphorus tribromide followed by anhydrous zinc bromide in ether affords trisubstituted ethylenes (48) in good yield and with a high degree of stereoselectivity (Brady, Ilton and Johnson, 1968; Johnson, Tsung-Tee Li, Faulkener and Campbell, 1968). Thus the alcohol (49) was smoothly converted into the diene (50) in 85–90 per cent yield (2.78). A useful feature of this procedure is that, under the mild conditions of the reaction, acid-catalysed migration of double bonds does not take place. In the reaction with (49) there was no isomerisation to give an isopropylidene derivative. The reason for the high stereoselectivity can be seen by considering the transition states for the two modes of reaction as seen in the Newman projection formulae (51) and (52). Because of non-

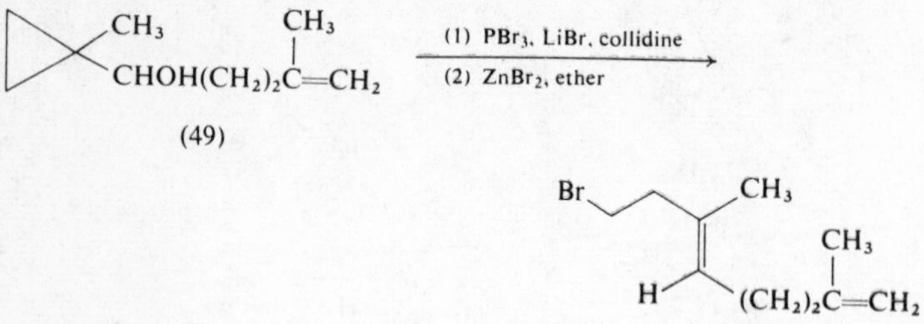

$$\text{(49)}\ \xrightarrow[\text{(2) } ZnBr_2\text{, ether}]{\text{(1) } PBr_3\text{, LiBr, collidine}}\ \text{(50) (97\% } trans\text{)} \qquad (2.78)$$

(49) (50) (97% *trans*) (2.78)

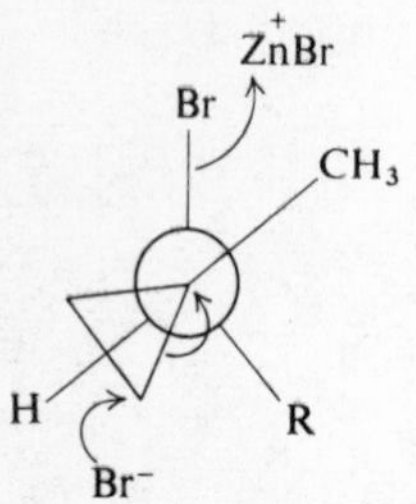

(51) 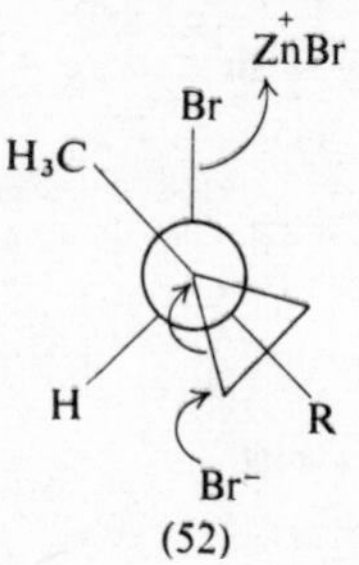 (52)

bonded interactions between the cyclopropyl ring and the substituent R in (52) the conformation (51) is preferred and the usual anti-parallel electronic reorganisation leads to the olefin in which R and CH_3 are *cis* to each other.

A more recent modification of the reaction leads stereoselectively to functionalised allylic (not homoallylic) alcohols (Nakamura, Yamamoto and Nozaki, 1973).

Another promising stereoselective synthesis of trisubstituted olefins starting from allylic alcohols depends critically on the stereoselective epoxidation of allylic alcohols described on p. 364 (Tanaka, Yamamato, Nozaki, Sharpless, Michaelson and Cutting, 1974). Thus, for the synthesis of *cis*-6-methyl-5-undecene, the allylic alcohol (53) was oxidised with t-butyl hydroperoxide and a vanadium catalyst, giving selectively the *erythro*-epoxide (54) which was converted into the vicinal diol (55) by reaction with lithium dibutylcuprate. Stereospecific deoxygenation of the diol was accomplished by reaction with *N,N*-dimethylformamide dimethyl acetal and decomposition of the resulting dioxolane derivative by heating with acetic anhydride and gave the *cis* olefin with very high stereochemical purity (2.79) (cf. p. 149). Furthermore, the same diol (55)

C_4H_9, HO, CH_3 (53) $\xrightarrow[VO(acac)_2]{t\text{-}C_4H_9OOH}$ (54) (96% *erythro*) $\xrightarrow[ether,\ -26°C]{Li(C_4H_9)_2Cu}$ (55) $\xrightarrow{(CH_3)_2NCH(OCH_3)_2,\ 25°C}$ dioxolane (C_4H_9, C_5H_{11}, H, CH_3, H, $N(CH_3)_2$) $\xrightarrow[130°C]{(CH_3CO)_2O}$ C_4H_9, C_5H_{11}, H, CH_3 olefin (2.79)

was used to prepare *trans*-6-methyl-5-undecene by conversion into the epoxide and deoxygenation by Cornforth's procedure (see p. 129), showing the versatility and stereochemical control possible in this approach.

2.7. Fragmentation reactions

A number of other methods for forming carbon–carbon double bonds have less general application than those discussed above, but are of value in particular circumstances. One of these makes use of the fragmentation of the monotoluene-*p*-sulphonates or methanesulphonates of suitable cyclic 1,3-diols on treatment with base (Grob and Schiess, 1967; Clayton, Henbest and Smith, 1957) (2.80). A feature of these reactions is that when the C–X bond and the $C_{(a)}$–$C_{(b)}$ bond have the *trans* anti-parallel

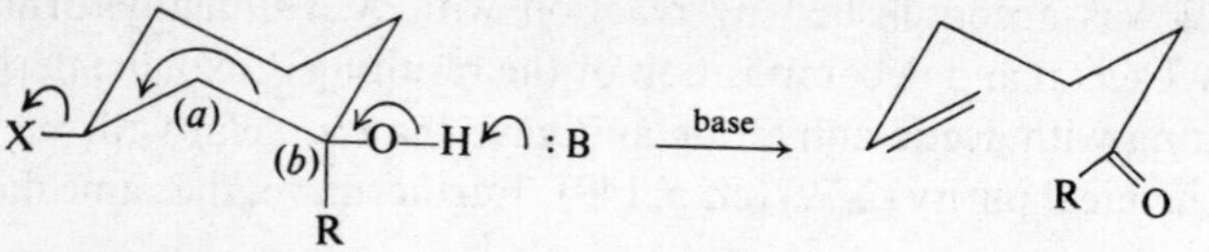

(2.80)

X = leaving group, e.g.: $—OSO_2C_6H_4CH_3$-*p*, $—OSO_2CH_3$

arrangement the reaction proceeds very readily by a concerted pathway to give an olefin the stereochemistry of which is governed solely by the relative orientation of groups in the cyclic precursor. For example, the decalin derivative (56) in which the tosyloxy group and the adjacent angular hydrogen atom are *cis* affords *trans*-5-cyclodecenone in high

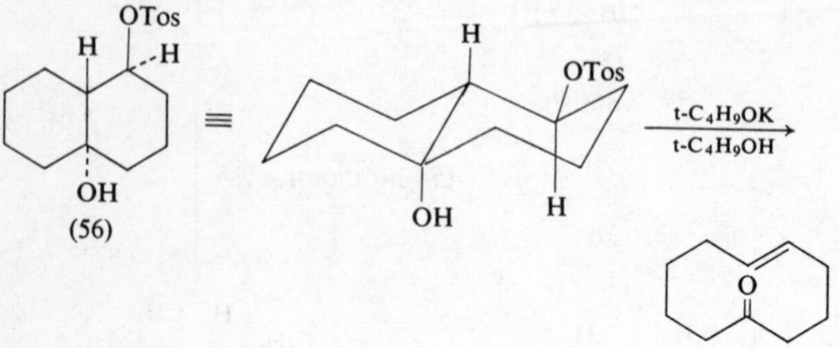

(56)

t-C_4H_9OK / t-C_4H_9OH

(> 90%)

(2.81)

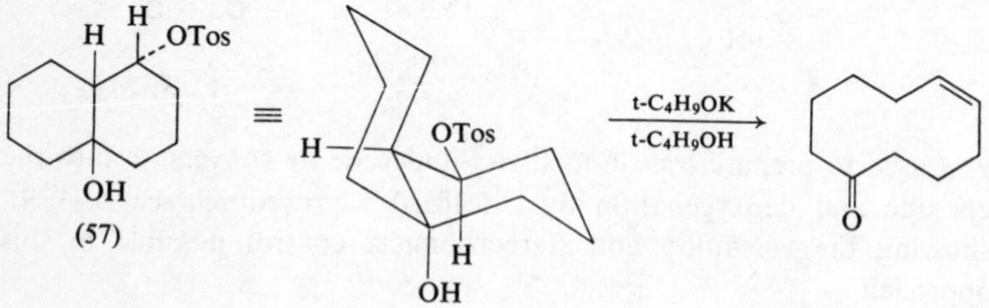

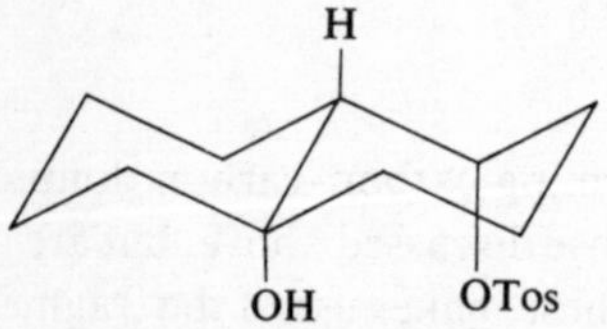

(57)

t-C_4H_9OK / t-C_4H_9OH

(58)

Tos = H_3C–C_6H_4–SO_2

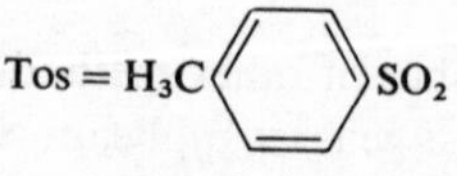

yield, whereas the isomer (57) in which the tosyloxy substituent and the hydrogen atom are *trans* affords the *cis*-cyclodecenone (i.e. in each case the relative orientation of the hydrogen atoms in the precursor is retained in the olefin). In these derivatives a *trans* anti-parallel arrangement of the breaking bonds is easily attained, but this is not so in the isomer (58) and this compound on treatment with base gives a mixture of products containing only a very small amount of the *trans*-cyclodecenone (2.81). This reaction has been used to prepare a variety of *cis*- and *trans*-cyclodecenone and cyclononenone derivatives, notably in the course of the synthesis of caryophyllene (Corey, Mitra and Uda, 1964). Marshall (1969, 1971) has extended the reaction to the preparation of cyclodecadienes by fragmentation of appropriately substituted decalylboranes, themselves easily available by hydroboration of the appropriate olefin (2.82) (see p. 274). Fragmentation reactions may also be used to

OSO_2CH_3, H_3C, CH_3, CH_3 $\xrightarrow{B_2H_6}$

CH_3, OSO_2CH_3, H, CH_3, CH_3, H, B, OH^- $\xrightarrow{OH^-}$ CH_3, CH_3, CH_3 (2.82)

prepare acyclic olefins from cyclic precursors. Control of olefin geometry is thereby transposed to control of relative stereochemistry in cyclic systems. The ketone (60), for example, an intermediate in a synthesis of juvenile hormone, was obtained stereospecifically from the bicyclic compound (59) using two successive fragmentation steps (Zurflüh, Wall, Siddall and Edwards, 1968) (2.83). The geometry of the intermediates (59) and (61) is such as to allow easy fragmentation at each stage.

Simpler acyclic 1,5-dienes have been obtained by fragmentation of cyclohexylboronic acid mesylates (as 62), which can themselves be

(59) NaH, THF → (quantitative) several steps → (61) NaH, THF → (60) (80%; 100% *cis,trans*) (2.83)

synthesised, ultimately, from anisole derivatives (2.84) (Marshall and Babler, 1970).

(1) B_2H_6, THF (2) H_2O → (62) ≡ NaOH → (2.84)

2.8. Oxidative decarboxylation of carboxylic acids

Another useful method for generating olefinic double bonds is by oxidative decarboxylation of vicinal dicarboxylic acids (2.85) (Sheldon and Kochi, 1972). This transformation can be brought about in a number

$$CO_2H,\ CO_2H \longrightarrow \; + 2CO_2 + 2H^+ + 2e \qquad (2.85)$$

of ways. The familiar procedure with lead tetra-acetate in boiling benzene gives poor and variable yields and is not applicable to bicyclic diacids with nearby double bonds (Criegee, 1965). It has been found by Cimarusti and Wolinsky (1968), however, that much improved yields are obtained when reaction is effected in presence of oxygen (2.86). An even better procedure is by electrolysis of the acid in pyridine solution in

$$CO_2H,\ CO_2H \xrightarrow[\text{pyridine, 67°C}]{Pb(OCOCH_3)_4,\, O_2} \quad (76\%) \qquad (2.86)$$

presence of trimethylamine. This method gives good yields and since reaction proceeds under mild conditions this is an attractive procedure for preparing highly strained unsaturated small and bridged ring compounds. Dewar benzene, for example, is best prepared by this route from bicyclo[2,2,0]hex-2-en-5,6-dicarboxylic acid (Radlick, Klem, Spurlock, Sims, van Tamelen and Whitesides, 1968) (2.87). Another useful method which proceeds smoothly under mild conditions is the thermal or photolytic decomposition of di-t-butyl peresters, which are readily obtained from the diacid chlorides and t-butyl hydroperoxide. This photolytic process can be used for the synthesis of thermally labile alkenes (Cain, Vukov and Masamune, 1969). The vicinal dicarboxylic

$$CO_2H,\ CO_2H \xrightarrow[\text{pyridine, } (CH_3)_3N]{\text{electrolysis}} \quad (35\%)$$

$$(2.87)$$

$$CO_2OC_4H_9\text{-t},\ CO_2OC_4H_9\text{-t} \xrightarrow[C_6H_6,\ 25°C]{h\nu} \quad (34\%)$$

acids required as starting materials in these reactions are readily available by Diels–Alder or photosensitised addition of maleic anhydride to dienes or olefins.

Related to these reactions is the oxidative decarboxylation of monocarboxylic acids with lead tetra-acetate. Under ordinary conditions the course of the reaction depends on the structure of the acid and poor yields, and mixtures of products are often obtained. In the presence of catalytic amounts of cupric acetate, however, preparatively useful yields of alkenes are formed from primary and secondary carboxylic acids,

$$CH_3(CH_2)_7CO_2H + Pb(OAc)_4 \xrightarrow[C_6H_6,\ h\nu,\ 30°C]{Cu(OCOCH_3)_2} CH_3(CH_2)_5CH{=}CH_2 \quad \text{quantitative}$$

$$\text{cyclobutanecarboxylic acid} \xrightarrow[C_6H_6,\ \text{reflux}]{Pb(OCOCH_3)_4,\ Cu(OCOCH_3)_2} \text{cyclobutene} \quad (77\%) \qquad (2.88)$$

either photolytically or thermally, in boiling benzene (Bacha and Kochi, 1968; Beckwith, Cross and Gream, 1974). Thus, nonanoic acid is converted into 1-octene in quantitative yield and cyclobutanecarboxylic acid gives cyclobutene in 77 per cent yield (2.88).

The following radical chain mechanism is proposed.

$$\begin{aligned} RCO_2Pb^{III} &\longrightarrow R^{\bullet} + CO_2 + Pb^{II} \\ R^{\bullet} + Cu^{II} &\longrightarrow \text{olefin} + H^+ + Cu^{I} \\ Cu^{I} + RCO_2Pb^{IV} &\longrightarrow Cu^{II} + RCO_2Pb^{III}, \text{etc.} \end{aligned} \qquad (2.89)$$

Birch and Slobbe (1976) have used this reaction in a useful sequence leading to alkylation–decarboxylation of aromatic acids. The carbanion obtained from the acid with lithium and ammonia (p. 449) is alkylated directly by reaction with an alkyl halide, and the resulting dihydroaromatic carboxylic acid is decarboxylated with lead tetra-acetate, regenerating the aromatic nucleus in which, however, the carboxyl group is now replaced by an alkyl group. Rosefuran, for example, was obtained from 3-methyl-2-furoic acid as shown in (2.90).

A different mode of oxidative decarboxylation, resulting in the conversion of a carboxylic acid, not into an olefin, but into a ketone, with the loss of one carbon atom, can be effected as illustrated in (2.91). Sulphenylation of the dianion of the acid with dimethyl disulphide and reaction of the sulphenylated acid with *N*-chlorosuccinimide in ethanol

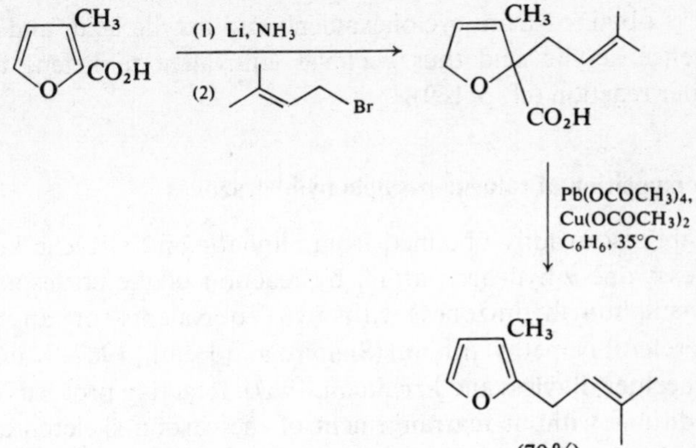

CH_3 CO_2H O

(1) Li, NH_3
(2) Br

CH_3 O CO_2H

$Pb(OCOCH_3)_4$,
$Cu(OCOCH_3)_2$
C_6H_6, 35°C

(2.90)

CH_3 O

(70%)

R H R^1 CO_2H

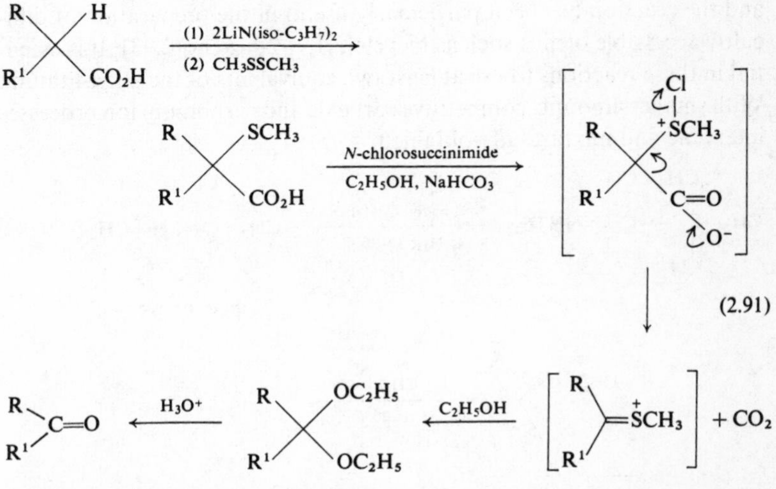

(1) $2LiN(iso\text{-}C_3H_7)_2$
(2) CH_3SSCH_3

R SCH_3 R^1 CO_2H

N-chlorosuccinimide
C_2H_5OH, $NaHCO_3$

[R Cl $^+SCH_3$ R^1 C=O O^-]

(2.91)

R C=O R^1 $\xleftarrow{H_3O^+}$ R OC_2H_5 R^1 OC_2H_5 $\xleftarrow{C_2H_5OH}$ [R R^1 =$\overset{+}{S}CH_3$] + CO_2

in presence of sodium hydrogen carbonate results in rapid loss of carbon dioxide to give a diacetal, which is readily hydrolysed to the ketone with dilute acid (Trost and Tamaru, 1975). The bicyclo-octenecarboxylic acid (63), for example, is readily converted into the ketone (64) at 0°C.

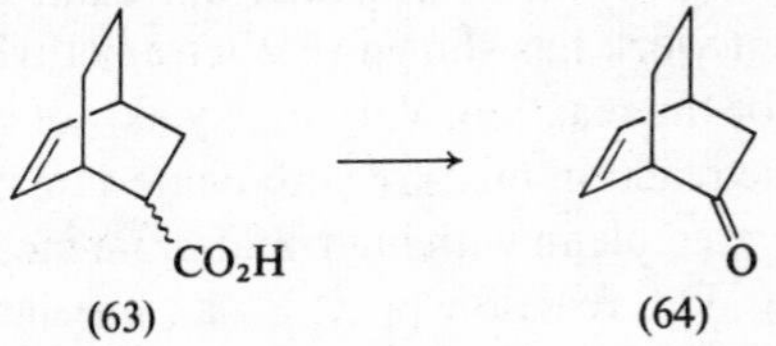

CO_2H ⟶ O

(2.92)

(63) (64)

The acid is obtained from cyclohexadiene and acrylic acid, and using this sequence acrylic acid thus becomes equivalent to ketene in the Diels–Alder reaction (cf. p. 169).

2.9. Decomposition of toluene-*p*-sulphonylhydrazones

Alkenes are also readily obtained from aliphatic and alicyclic ketones with at least one α-hydrogen atom, by reaction of the corresponding toluene-*p*-sulphonylhydrazones with two equivalents of an alkyl-lithium, preferably methyl lithium (Shapiro and Heath, 1967; Kaufman, Cook, Schechter, Bayless and Friedman, 1967). Reaction proceeds under mild conditions without rearrangement of the carbon skeleton and in general leads to the less substituted double bond, where there is a choice. $\alpha\beta$-Unsaturated tosylhydrazones are likewise converted into dienes, and the reaction has been particularly useful in the preparation of difficultly accessible olefins such as bicyclo[2,1,1]-2-hexene (2.93). It is essential in these reactions to use at least two equivalents of the alkyl-lithium. With smaller amounts competitive carbenic and carbonium ion processes intervene and mixtures are obtained.

$$(CH_3)_3C{-}C(CH_3){=}NNHTos \xrightarrow[C_6H_{14},\ 70\text{–}90°C]{2\ \text{equiv.}\ C_4H_9Li} CH_3{-}C(CH_3)_2{-}CH{=}CH_2 \quad \text{quantitative} \qquad (2.93)$$

TosNHN= (bicyclo[2,1,1]hexan-2-one tosylhydrazone) $\xrightarrow[25°C]{CH_3Li,\ ether}$ bicyclo[2,1,1]-2-hexene (25%)

Tos = H_3C–C_6H_4–SO_2

Reaction appears to proceed by way of a carbanion intermediate (2.94) which extracts a proton from the solvent (ether) to give the product. Incorporation of deuterium from deuterium oxide is low under these conditions. Recent work has shown that tetramethylethylenediamine is a better solvent for the reaction. Very high yields of olefins are obtained and quenching the reaction mixture with deuterium oxide now gives the specifically deuterated olefin with high deuterium incorporation (Stemke and Bond, 1975). The reaction provides an excellent method for the

$$
\begin{array}{c}
\overset{Li^+}{-C{=}N{-}\bar{N}{-}Tos} \\ | \\ -C{-}H \quad R^- \\ |
\end{array}
\longrightarrow
\begin{array}{c}
-C{-}N{=}\bar{N} \quad Li^+ \\ \| \\ -C \\ |
\end{array}
\longrightarrow
\begin{array}{c}
-C^- \; Li^+ \\ \| \\ -C \\ |
\end{array}
+ N_2
\xrightarrow[\text{or solvent}]{H_2O}
\begin{array}{c}
-CH \\ \| \\ -C \\ |
\end{array}
\qquad (2.94)
$$

preparation of specifically deuterated olefins. 2-Methylcyclohexanone, for example, was converted into 2-deuterio-3-methylcyclohexene in 98 per cent yield with 80 per cent incorporation of deuterium, and octan-2-one gave entirely 2-deuterio-1-octene, confirming the specific formation of the less substituted olefin in the reaction (but see Dauben, *et al.*, 1976). Aromatic tosylhydrazones with no α-hydrogen atom form nucleophilic substitution products on reaction with alkyl-lithium, instead of olefins. Fluorenone tosylhydrazone, for example, with methyl-lithium, gives 9-methylfluorene. Variable amounts of substitution products have also been obtained in the reaction of some alicyclic derivatives (see Herz and Gonzalez, 1969).

2.10. Stereospecific synthesis from 1,2-diols

Several approaches to the regiospecific and stereospecific generation of double bonds from 1,2-diols have been devised, all proceeding by decomposition of an intermediate of the type (65). One of the best methods uses the cyclic thionocarbonates (67) which are converted into olefins by heating with triethyl phosphite (Corey, Carey and Winter, 1965; Paquette, Itoh and Farnham, 1975) or with the zerovalent nickel complex bis(1,5-cyclo-octadiene)nickel(0) (Semmelhack and Stauffer, 1973), either by a concerted process or by way of an intermediate carbene (66). The reactions proceed with complete stereospecificity by a *syn* elimination pathway allowing the stereospecific synthesis of strained cyclo-olefins and, together with the *trans*-perhydroxylation reaction (p. 361) providing a general and unambiguous method for the interconversion of *cis* and *trans* olefins. Thus, *meso*-1,4-diphenylbutan-2,3-diol was converted into *cis*-1,4-diphenyl-2-butene in 96 per cent yield,

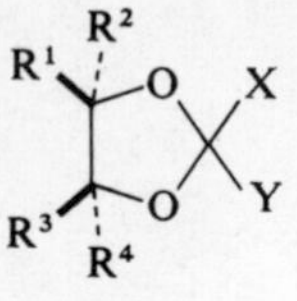

(65) 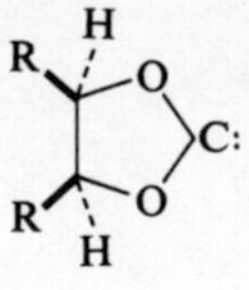 (66)

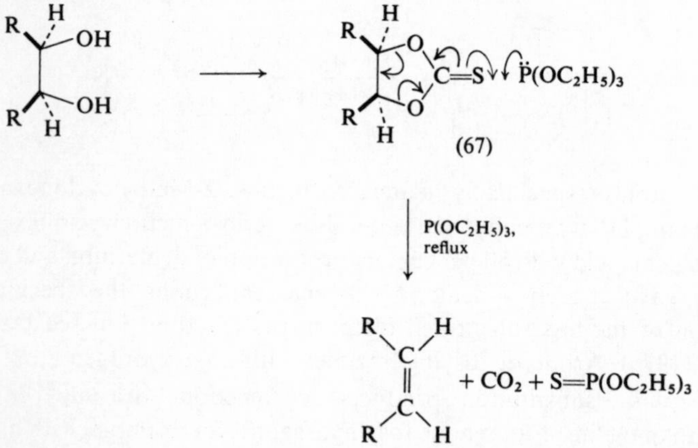

(67)

$P(OC_2H_5)_3$, reflux (2.95)

$+ CO_2 + S{=}P(OC_2H_5)_3$

while the *dl*-compound gave the *trans* olefin. *cis*-Cyclo-octene was converted into *trans*-cyclo-octene as shown in (2.96). An alternative procedure which avoids the disadvantages of Corey's method (prolonged reaction at high temperature and relative inaccessibility of reagents) was found by Hines, Peagram, Thomas and Whitham (1973). The benzylidene

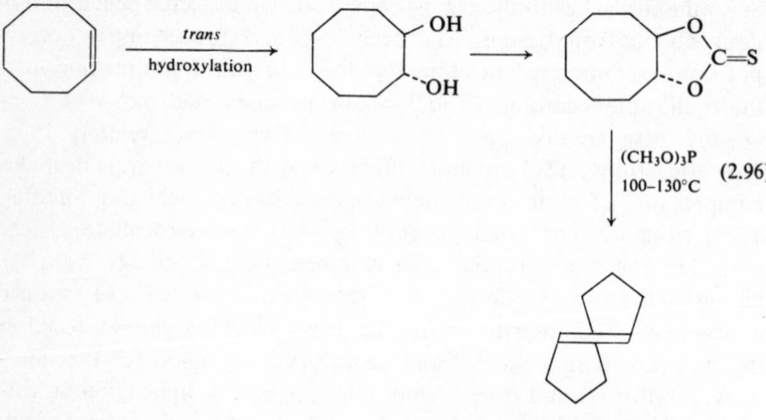

trans hydroxylation

$(CH_3O)_3P$ 100–130°C (2.96)

(75%; 99% *trans*)

derivative of *trans*-cyclo-octan-1,2-diol on treatment with butyl-lithium in hexane at 20°C and then with water gave *trans*-cyclo-octene in 75 per cent yield. But this method, although valuable, is of limited application and is said to be suitable for only lightly substituted alkenes. Other good stereospecific routes proceed through acetals formed by heating the diol with ethyl orthoformate (Hiyama and Nozaki, 1973) or *N,N*-dimethylformamide dimethyl acetal (Eastwood, Harrington, Josan and Pura, 1970), but they suffer from the disadvantage of the high temperatures required. 1,2-Diols have also been converted into olefins by reaction with active titanium metal (see p. 495) or with tungsten salts, K_2WX_6, but neither method is stereospecific.

2.11. Claisen rearrangement of allyl vinyl ethers

The Claisen rearrangement of allyl vinyl ethers provides an excellent stereoselective route to $\gamma\delta$-unsaturated carbonyl compounds (aldehydes, ketones and esters) from allyl alcohols, and has formed an important step in the synthesis of a number of natural products (Rhoads, 1963; Rhoads and Raulins, 1975). The reaction is a [3,3]-sigmatropic rearrangement, and takes place by a concerted mechanism through a cyclic six-membered transition state (2.97). Much of its value in synthesis stems from the fact

(68) (2.97)

R^1 = H, alkyl, OR, NR_2, $OSi(CH_3)_3$

that it is highly stereoselective, particularly when $R^1 \neq H$, leading predominantly then to the *trans* configuration of the newly formed double bond. A chair conformation is preferred for the cyclic transition state, and the high stereoselectivity favouring the *trans* olefin is attributed to non-bonded interaction between the substituents R^1 and R^3 in the transition state for the *cis* olefin (Faulkner and Petersen, 1973).

The allyl vinyl ethers used in the reaction are prepared from allyl alcohols by ether exchange. Thus reaction of an allyl alcohol with ethyl vinyl ether and pyrolysis of the resulting vinyl ether affords a $\gamma\delta$-unsaturated aldehyde as in the example (2.98).

Using appropriately substituted vinyl ethers, $\gamma\delta$-unsaturated ketones are obtained. These reactions, where $R^1 \neq H$ in (68) are highly stereo-

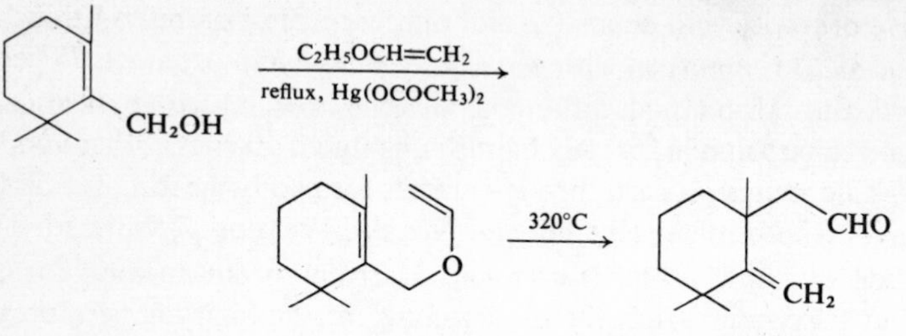

$C_2H_5OCH{=}CH_2$, reflux, $Hg(OCOCH_3)_2$; CH₂OH → 320°C → CHO, CH₂ (85%) (2.98)

selective. Thus, reaction of the allylic alcohol (69) with 2-methoxy-3-methyl-1,3-butadiene gave the *trans* $\gamma\delta$-unsaturated ketone (70) directly without isolation of the intermediate allyl vinyl ether. None of the *cis* isomer was detected. Reduction of the carbonyl group with sodium borohydride and repetition of the process gave the ketone (71), showing how this procedure can be used to build up polyisoprenoid chains of the type commonly found in natural products (Faulkner and Petersen, 1973; Johnson, Brocksom, Loew, Rich, Werthemann, Arnold, Tsung-tee Li and Faulkner, 1970) (2.99).

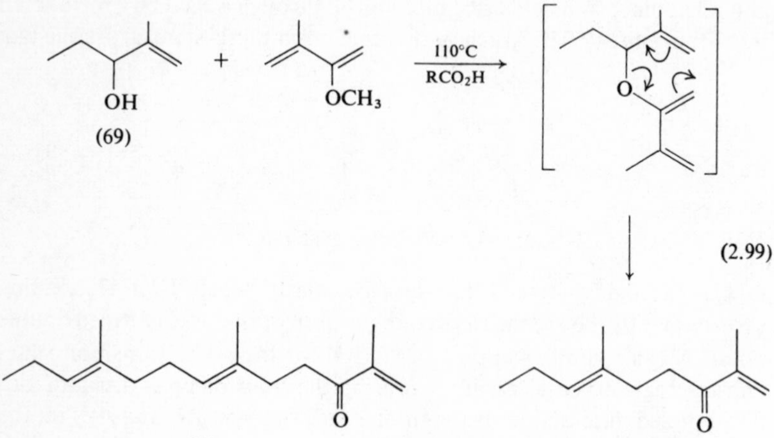

OH (69) + OCH₃ → 110°C, RCO_2H → [O] → (2.99)

(71) (70) (67%; 100% *trans*)

$\gamma\delta$-Unsaturated esters are obtained by heating the allylic alcohol with ethyl or methyl orthoacetate in presence of a weak acid (propionic acid is often used), as in the following synthesis of the pheromone of the male Queen butterfly (72) (2.100) (Miles, Loew, Johnson, Kluge and

Meinwald, 1972). A mixed orthoester is first formed and loses methanol to form a ketene acetal which rearranges to the $\gamma\delta$-unsaturated ester. This orthoester reaction, like the methoxyisoprene method described above, lends itself readily to the synthesis of structures with successive isoprene units containing *trans* double bonds and joined in the head-to-tail manner found in many natural products. All-*trans*-squalene was synthesised by both routes, with 95 per cent stereochemical purity (Johnson, Werthemann, Bartlett, Brocksom, Tsung-tee Li, Faulkner and Petersen, 1970).

$\gamma\delta$-Unsaturated acids are also conveniently obtained from carboxylic esters of allylic alcohols by Claisen rearrangement of the lithium enolates or, better, the derived trimethylsilyl or t-butyldimethylsilyl enol ethers.

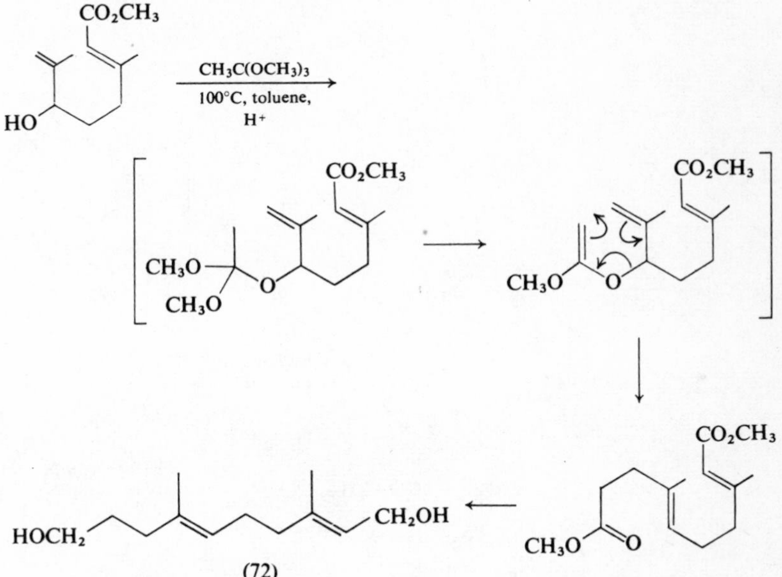

CO_2CH_3

HO

$CH_3C(OCH_3)_3$

100°C, toluene, H^+

CO_2CH_3

CH_3O

CH_3O

O

CO_2CH_3

CH_3O

O

CO_2CH_3

$HOCH_2$

CH_2OH

(72)

CH_3O

O

(73%; 95% *trans*)

(2.100)

(continued on p. 154)

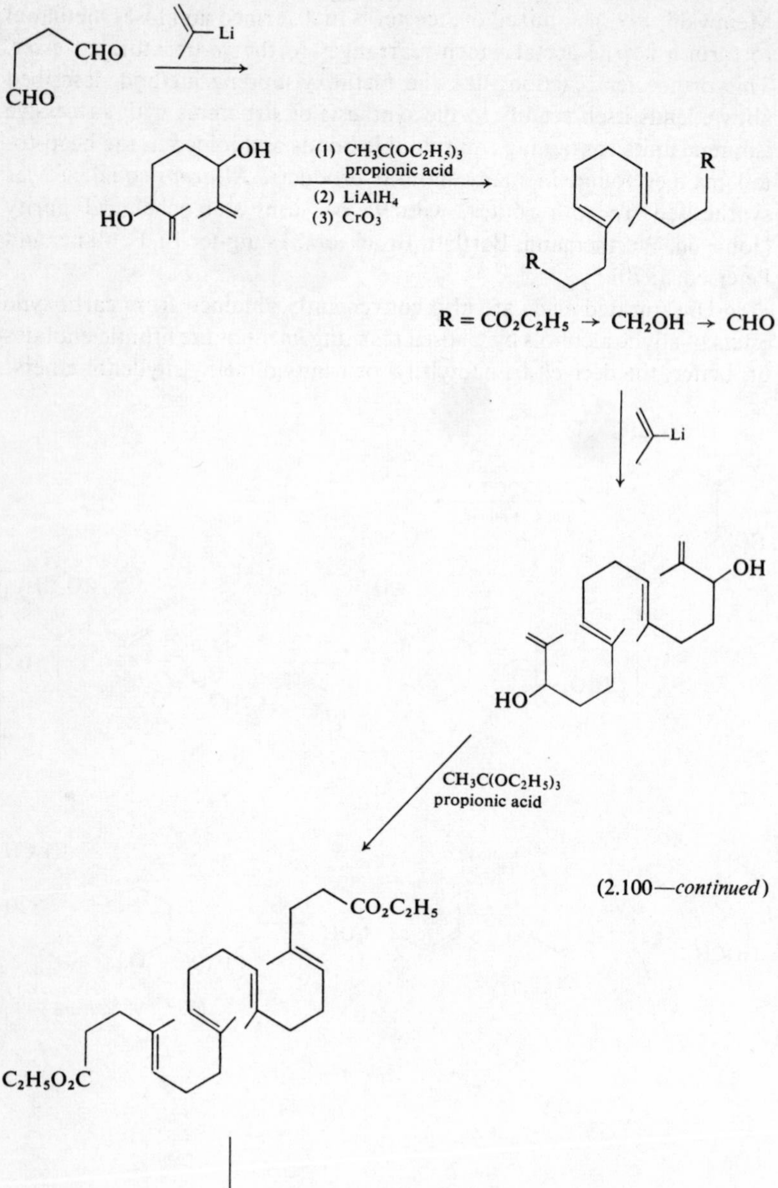

CHO
CHO
Li

OH
HO
(1) $CH_3C(OC_2H_5)_3$
propionic acid
(2) $LiAlH_4$
(3) CrO_3

R
R

$R = CO_2C_2H_5 \rightarrow CH_2OH \rightarrow CHO$

Li

OH
HO

$CH_3C(OC_2H_5)_3$
propionic acid

(2.100—*continued*)

$CO_2C_2H_5$
$C_2H_5O_2C$

all-*trans*-squalene

High yields of the unsaturated acids are obtained under remarkably mild conditions, and the preference for the formation of *trans* double bonds is again observed (Ireland, Mueller and Willard, 1976; Katzenellenbogen and Christy, 1974). Thus, the allylic acetate (73) is converted into the $\gamma\delta$-unsaturated acid (74) in 80 per cent yield and with almost complete *trans* selectivity.

CH_3 C_6H_{13} O O (73)

(1) $LiN(iso\text{-}C_3H_7)_2$, THF, −78°C
(2) $(CH_3)_3SiCl$

CH_3 C_6H_{13} O $OSi(CH_3)_3$

warm, 70°C, then acidified CH_3CO_2H

(2.101)

CH_3 C_6H_{13} CO_2H

(74)

(80%; > 98% *trans*)

Usefully, the stereochemistry about the newly formed *single* bond depends on the solvent used at the enolisation stage. Thus, *trans*-crotyl propionate by enolisation in tetrahydrofuran solution, followed by reaction with t-butyldimethylsilyl chloride and rearrangement, gives mainly the *erythro* acid (75), but if the solvent contains hexamethylphosphoramide the main product in the *threo* acid (76). This has been exploited in a synthesis of the irregular monoterpene methyl santolinate (77) (Boyd, Epstein & Fráter, 1976).

The Claisen rearrangement of allyl vinyl ethers may be regarded as a particular case of the well-known Cope [3,3]-sigmatropic rearrangement of 1,5-hexadienes (Rhoads and Raulins, 1975). The so-called oxy-Cope rearrangement of 1,5-hexadienes with hydroxy substituents at C-3 and C-4 provides a useful method for the preparation of $\delta\epsilon$-unsaturated aldehydes and of 1,6-dicarbonyl compounds (2.103). It has recently been found that very large increases in rate in these rearrangements can be

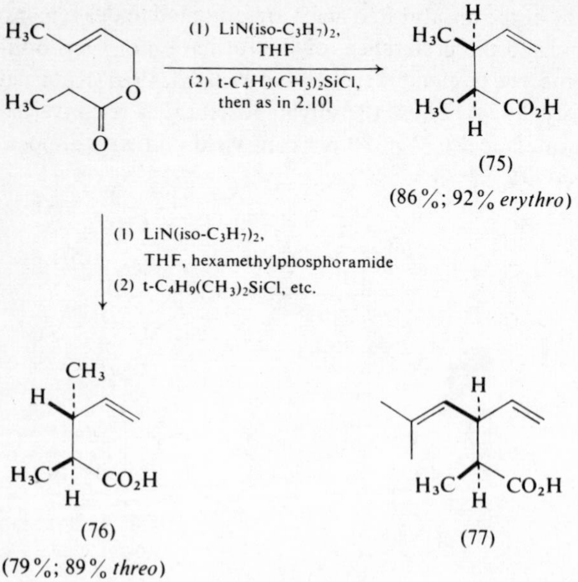

H_3C … O … H_3C … O

(1) $LiN(iso\text{-}C_3H_7)_2$, THF

(2) $t\text{-}C_4H_9(CH_3)_2SiCl$, then as in 2.101

H, H_3C, H_3C, CO_2H, H

(75)

(86%; 92% *erythro*)

(1) $LiN(iso\text{-}C_3H_7)_2$, THF, hexamethylphosphoramide

(2) $t\text{-}C_4H_9(CH_3)_2SiCl$, etc.

(2.102)

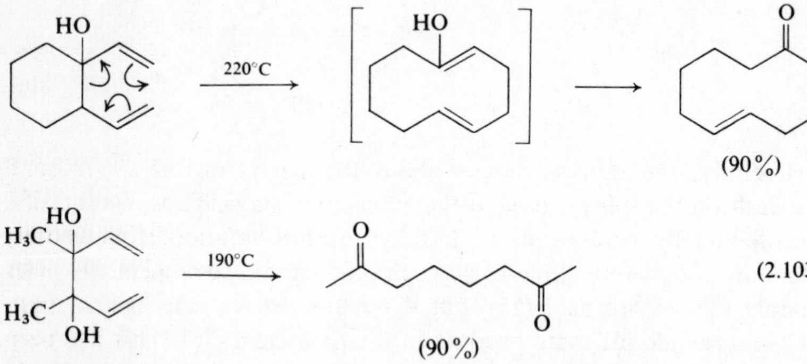

CH_3, H, H_3C, CO_2H, H

(76)

(79%; 89% *threo*)

H, H_3C, CO_2H, H

(77)

HO … 220°C … [HO] … O

(90%)

HO, H_3C, H_3C, OH … 190°C … O, O

(2.103)

(90%)

obtained by using the potassium alkoxides, in which the metal ion is free of the oxyanion, rather than the hydroxy compounds themselves (Evans and Golob, 1975). With some cyclic compounds, however, (vinylcycloalkenols) reaction takes a different course. A [1,3]-sigmatropic rearrangement supervenes, and the sequence provides a method for ring expansion by two carbon atoms. Much better yields are obtained by using the trimethylsilyl ethers rather than the hydroxy compounds themselves (Thies, 1972; Thies and Billigmeier, 1974).

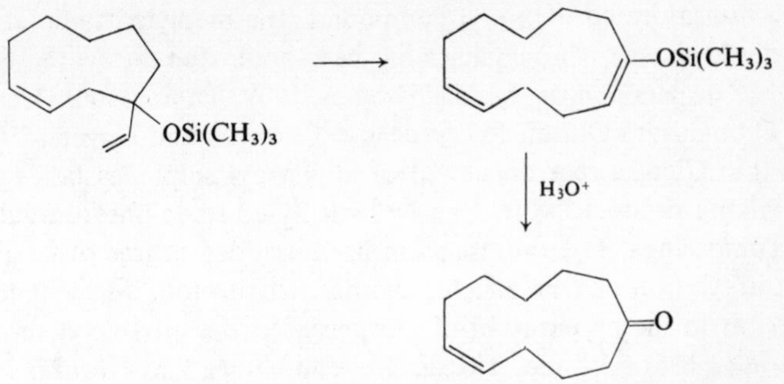

$OSi(CH_3)_3$ → $OSi(CH_3)_3$

H_3O^+ (2.104)

O

A number of synthetic schemes in the terpene field have used sequential Claisen and Cope rearrangements to build up polyisoprenoid chains. Thus, in one variation, chain lengthening of an allylic alcohol by one isoprene unit is achieved as shown in (2.105) for the conversion of geraniol into methyl farnesate, using the silyloxy version of the Claisen rearrangement (Fráter, 1975). The overall reaction results in γ-allylation

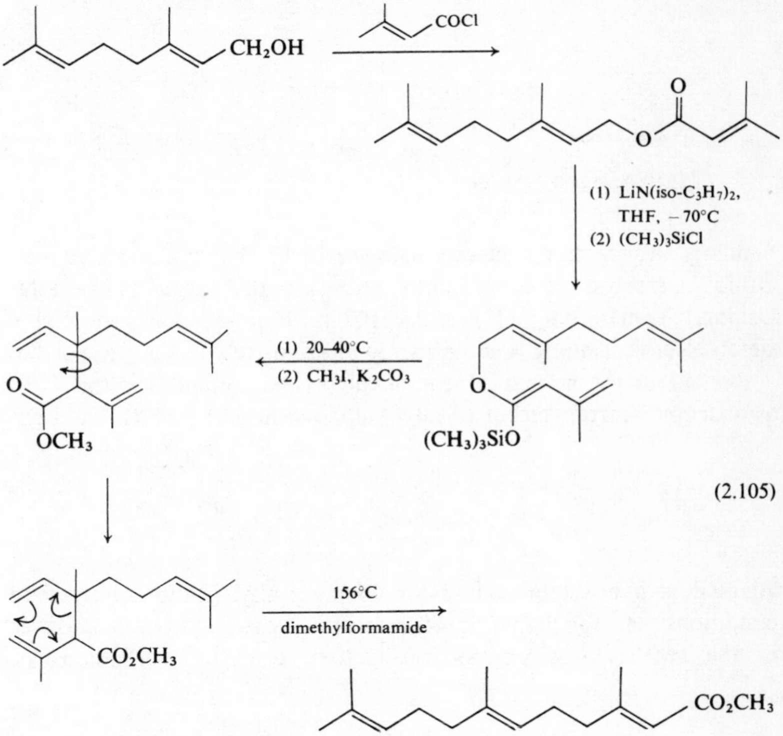

CH_2OH → COCl

O

O

(1) $LiN(iso\text{-}C_3H_7)_2$, THF, $-70°C$
(2) $(CH_3)_3SiCl$

(1) 20–40°C
(2) CH_3I, K_2CO_3

O

OCH_3

O

$(CH_3)_3SiO$

(2.105)

CO_2CH_3

156°C
dimethylformamide

CO_2CH_3

of an $\alpha\beta$-unsaturated carbonyl compound (the methylcrotonic acid in the example above). The sequence has been applied in the synthesis of a number of terpenoid compounds (Thomas, 1969; Thomas and Ozainne, 1969; Thomas and Ohloff, 1970; see also Cookson and Rogers, 1973).

The thio Claisen rearrangement of allyl vinyl sulphides takes place easily, giving products which can be hydrolysed to $\gamma\delta$-unsaturated carbonyl compounds. The thio reaction has an added degree of flexibility in that alkylation can be effected at the carbon atom adjacent to the sulphur, as in the preparation of 4-tridecenal from allyl vinyl sulphide shown in (2.106) (Oshima, Takahashi, Yamamoto and Nozaki, 1973).

(1) s-C_4H_9Li, THF, $-78°C$
(2) $C_8H_{17}Br$

C_8H_{17} S

aq. dimethoxyethane, $CaCO_3$, reflux

C_8H_{17} ... CHO ⟵ [C_8H_{17} ... C(=S)—H]

(57%; 100% *trans*)

(2.106)

Similarly, in a reaction sequence analogous to the silyloxy Claisen reaction dithio esters are converted into $\gamma\delta$-unsaturated acids (Takahashi, Oshima, Yamamoto and Nozaki, 1973). Both reactions are highly stereoselective, leading predominantly or exclusively to the *trans* olefin.

Related to the rearrangement of allyl vinyl sulphides is the [2,3]-sigmatropic rearrangement of allyl sulphonium ylids, which has been

R–$\overset{+}{S}$ ⟷ R–S ⟶ RS (2.107)

adapted in a novel procedure for coupling allyl groups under mild conditions, and for preparing $\alpha\beta$-unsaturated carbonyl compounds, as in the synthesis of γ-cyclocitral (2.108). α-Allylthiocarbanions or

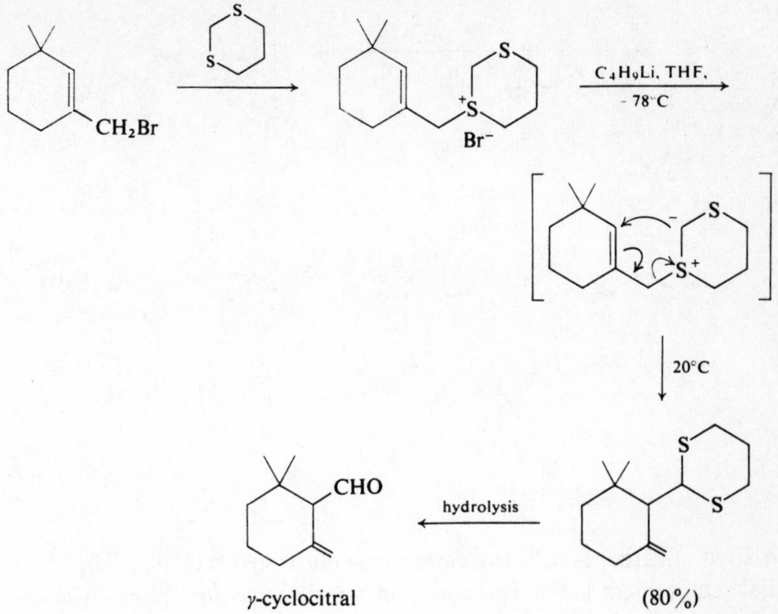

S

S

CH_2Br

S

$\overset{+}{S}$

Br^-

C_4H_9Li, THF, -78°C

S

S^+

20°C

S

S

CHO

hydrolysis

γ-cyclocitral

(80%)

(2.108)

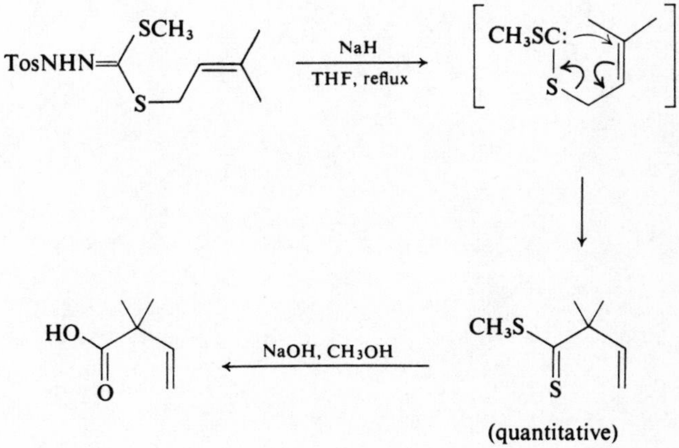

SCH_3

TosNHN=

S

NaH

THF, reflux

CH_3SC:

S

HO

O

NaOH, CH_3OH

CH_3S

S

(quantitative)

(2.109)

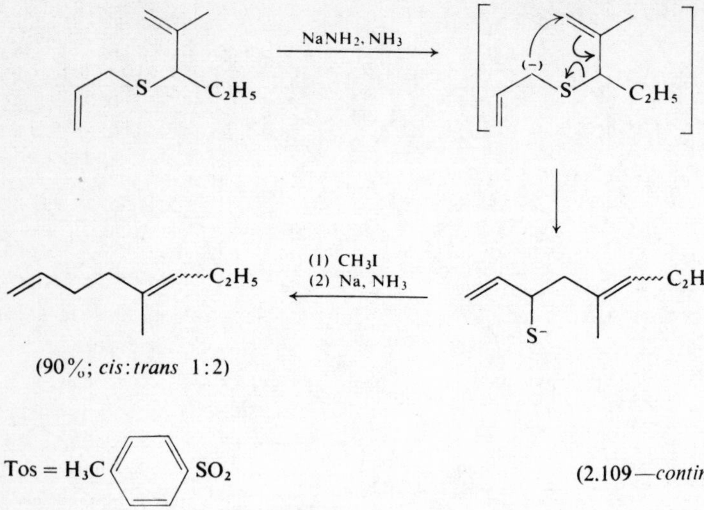

NaNH$_2$, NH$_3$

S

C_2H_5

(–)

S

C_2H_5

C_2H_5

(1) CH_3I
(2) Na, NH_3

C_2H_5

S^-

(90%; *cis*:*trans* 1:2)

Tos = H_3C–C_6H_4–SO_2

(2.109—*continued*)

α-allylthiocarbenes will also rearrange as illustrated in (2.109). The essential requirement is the availability of six electrons for the electrocyclic reaction (Kreiser and Wurziger, 1975; Baldwin and Walker, 1972).

3 The Diels–Alder reaction

3.1. General

The Diels–Alder reaction, one of the most useful synthetic reactions in organic chemistry, is one of a general class of cyclo-addition reactions (Huisgen, Grashey and Sauer, 1964; Huisgen, 1968; Schmidt, 1973). In it a 1,3-diene reacts with an olefinic or acetylenic dienophile to form an adduct with a six-membered hydroaromatic ring (3.1). In the reaction two new σ-bonds are formed at the expense of two π-bonds in the starting materials (Onischenko, 1964; Sauer, 1966, 1967).

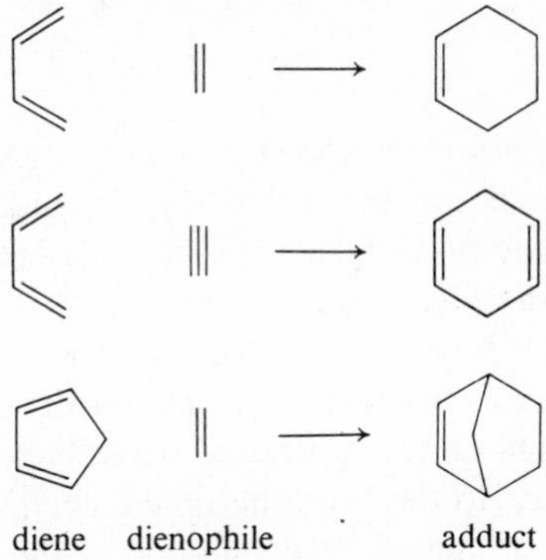

(3.1) (*a*)

(*b*)

(*c*)

diene dienophile adduct

In general the reaction takes place easily, simply by mixing the components at room temperature or by gentle warming in a suitable solvent, although in some cases with unreactive dienes or dienophiles more vigorous conditions may be necessary. The Diels–Alder reaction is reversible, and many adducts dissociate into their components at quite low temperatures, particularly those formed from cyclic dienes such as cyclopentadiene, fulvene or furan. In these cases heating is disadvantageous and better yields are obtained by using an excess of one of the components, or a solvent from which the adduct separates readily. Many Diels–Alder reactions are accelerated by Lewis acid catalysts (see p. 198).

The usefulness of the Diels–Alder reaction in synthesis arises from its versatility and from its remarkable stereoselectivity. By varying the nature of the diene and the dienophile many different types of structures can be built up. In the majority of cases all six atoms involved in forming the new ring are carbon atoms, but ring closure may also take place at atoms other than carbon, giving rise to heterocyclic compounds. It is very frequently found, moreover, that although reaction could conceivably give rise to a number of structurally- or stereo-isomeric products, one isomer is formed exclusively or at least in preponderant amount.

Many dienes can exist in a *cisoid* and a *transoid* conformation, and it is only the *cisoid* form which can undergo addition. If the diene does not have or cannot adopt a *cisoid* conformation no reaction occurs.

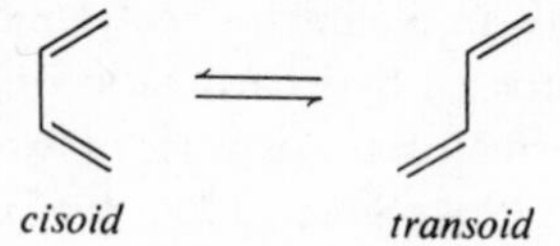

cisoid ⇌ *transoid* (3.2)

3.2. The dienophile

Ethylenic and acetylenic dienophiles; quinones. Many different kinds of dienophile can take part in the Diels–Alder reaction. They may be derivatives of ethylene or acetylene (the majority of cases) or reagents in which one or both of the reacting atoms is a heteroatom. All dienophiles do not undergo the reaction with equal ease; the reactivity depends greatly on the structure. In general, the greater the number of electron-attracting substituents on the double or triple bond the more reactive is the dienophile, due to the lowering of the energy of the lowest unoccupied molecular orbital of the dienophile by the substituents (see p. 199). Thus, ethylene reacts only slowly with butadiene at 20°C and 91×10^5 N m^{-2} pressure (3.1*a*), whereas maleic anhydride affords a quantitative yield of adduct in boiling benzene or, more slowly, at room temperature (3.3). Tetracyanoethylene, with four electron-attracting substituents, reacts extremely rapidly even at 0°C. Similarly, acetylene reacts with electron-rich dienes only under severe conditions, but pro-

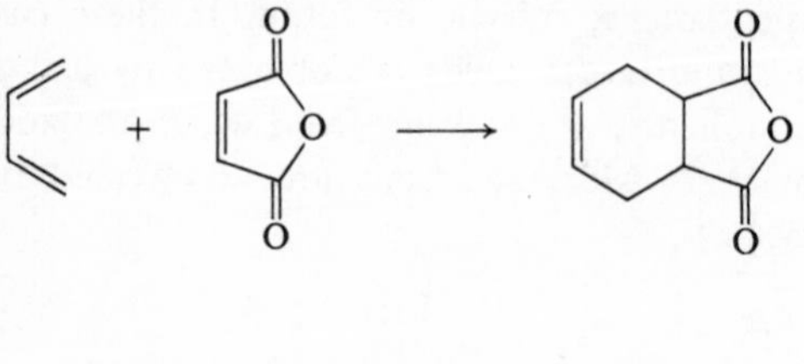

(3.3)

TABLE 3.1 *Reaction of dienophiles with cyclopentadiene and 9,10-dimethylanthracene*

Dienophile	Cyclopentadiene $10^5 k_1$/l mol^{-1} s^{-1}	9,10-Dimethylanthracene $10^5 k_1$/l mol^{-1} s^{-1}
Tetracyanoethylene	*c.* 43 000 000	*c.* 1 300 000 000
Tricyanoethylene	*c.* 480 000	590 000
1,1-Dicyanoethylene	45 500	12 700
Acrylonitrile	1·04	0·089
Dimethyl fumarate	74	215
Dimethyl acetylene dicarboxylate	31	140

piolic acid, phenylpropiolic acid and acetylene dicarboxylic acid react readily, and have been frequently used as dienophiles in the Diels–Alder reaction (Holmes, 1948; Fuks and Viehe, 1969). Table 3.1 gives some values for the rates of addition of a number of dienophiles to cyclopentadiene and 9,10-dimethylanthracene in dioxan at 20°C.

It should be noted, however, that there are a number of Diels–Alder reactions in which the above generalisation does not hold and in which reaction takes place between electron-*rich* dienophiles and electron-*poor* dienes such as *o*-quinones, thiophen-1,1-dioxides and perchlorocyclopentadiene (see Sauer, 1967). The essential feature is that the two components should have complementary electronic character. The vast majority of Diels–Alder reactions involve an electron-rich diene and an electron-deficient dienophile.

The most commonly encountered activating substituents for the 'normal' Diels–Alder reaction are —CO—, —COOR, —C≡N and —NO_2, and dienophiles containing one or more of these groups in conjugation with the double or triple bond react readily with dienes (Kloetzel, 1948; Holmes, 1948).

$\alpha\beta$-Unsaturated carbonyl compounds are very reactive dienophiles and are probably the most widely used dienophiles in synthesis. Typical examples are acrolein, acrylic acid and its esters, maleic acid and its anhydride and acetylenedicarboxylic acid. Thus, acrolein reacts rapidly with butadiene in benzene solution at 0°C to give tetrahydrobenzaldehyde in quantitative yield (3.4), and acetylenedicarboxylic acid and butadiene give 3,6-dihydrophthalic acid.

Substituents exert a pronounced steric effect on the reactivity of dienophiles. Comparative experiments show that the yields of adducts

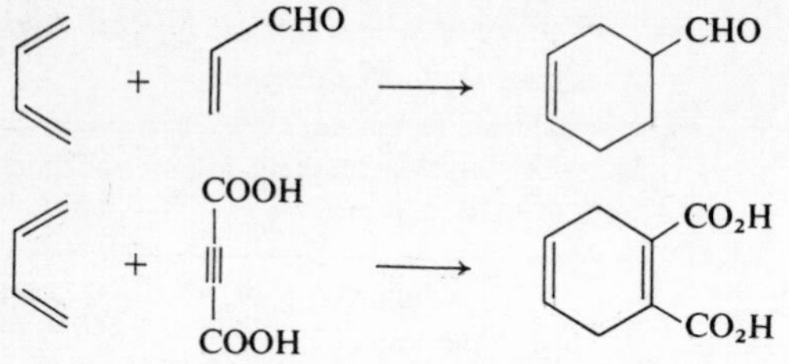

CHO + → CHO

COOH + COOH → CO_2H CO_2H (3.4)

obtained in the condensation of butadiene and 2,3-dimethylbutadiene with acrylic dienophiles decrease with the introduction of substituents into the α-position of the dienophile molecule, and αβ-unsaturated ketones with two alkyl substituents in the β-position react very slowly. Similarly, the reactions of butadiene and 2,3-dimethylbutadiene with citraconic anhydride (methylmaleic anhydride) require more drastic conditions than the reactions with maleic anhydride.

It is of interest that esters of maleic and citraconic acid condense with dienes far less readily than their *trans* isomers. Diene condensations with *cis* dienophiles occur readily only in the case of anhydrides.

Another important group of dienophiles of the αβ-unsaturated carbonyl class are quinones (Butz and Rytina, 1949). *p*-Benzoquinone itself reacts readily with butadiene at room temperature to form a high yield of the mono-adduct, tetrahydronaphthaquinone (3.5). Under more

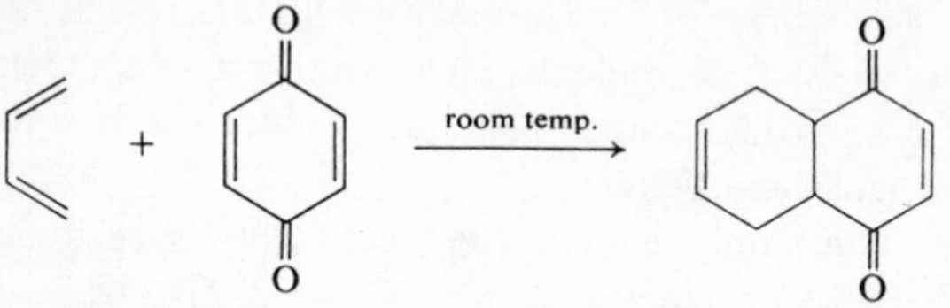

+ O O —room temp.→ O O (3.5)

vigorous conditions a bis-adduct is obtained which can be converted into anthraquinone by oxidation of an alkaline solution with atmospheric oxygen. 1,4-Naphthaquinone behaves similarly, readily furnishing adducts with acyclic dienes which can be oxidised to anthraquinones (3.6). If the reaction is conducted in nitrobenzene solution the anthra-

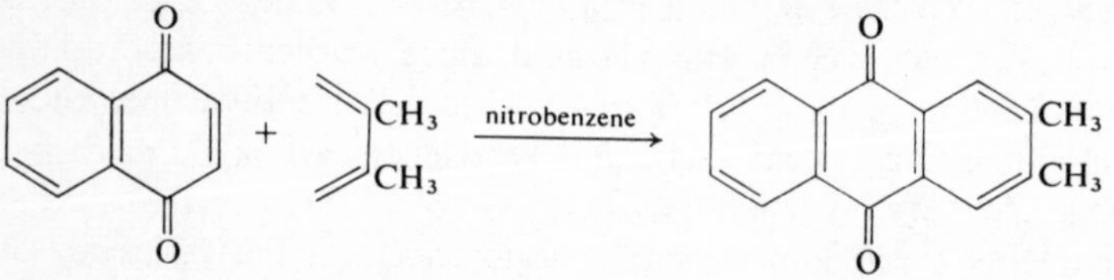

O O + CH_3 CH_3 —nitrobenzene→ O O CH_3 CH_3 (3.6)

(1)

quinone, e.g. (1), is often obtained directly by dehydrogenation of the initial adduct by the solvent.

As with other dienophiles, substitution on the double bond leads to a weakening of the dienophilic properties of quinones. It is found in general that monosubstituted *p*-benzoquinones add dienes at the unsubstituted double bond only.

A double bond of a conjugated diene can sometimes itself act as a dienophile, being activated by the neighbouring double bond. This is seen in the dimerisation of dienes, for example of cyclopentadiene and of isoprene (3.7). Self-condensation of simple conjugated dienes sometimes has an adverse effect on the yields obtained in Diels–Alder reactions, particularly in the case of reactions which proceed slowly or require vigorous conditions.

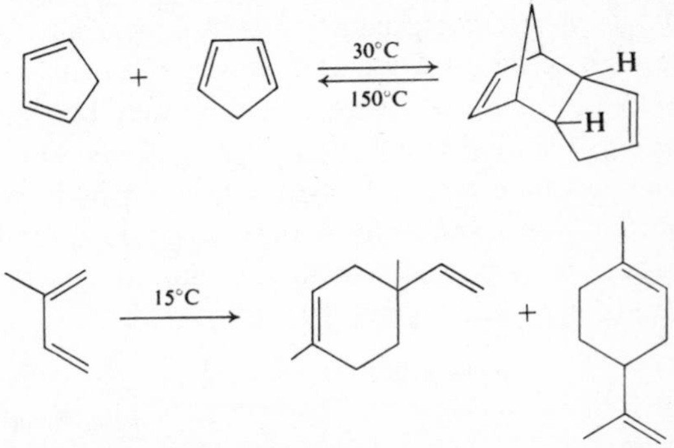

(3.7)

In contrast to the reactive dienophiles discussed above, in which the double or triple bond is activated by conjugation with unsaturated electron-attracting groups, ethylenic compounds such as allyl alcohol and its esters, allyl halides, vinyl compounds and styrene are relatively unreactive, although they can frequently be induced to react with dienes under forcing conditions. Thus vinyl acetate and butadiene give only a 6 per cent yield of 4-acetoxycyclohexene even at 180°C for 12 h although with the more reactive diene, cyclopentadiene, dihydronorbornyl acetate is formed in almost quantitative yield under the same conditions.

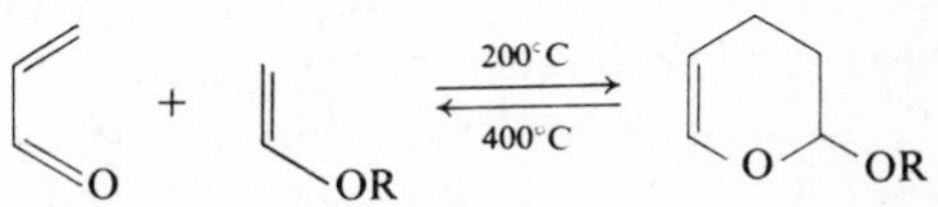

(3.8)

Vinyl ethers and esters have been widely used in the synthesis of dihydropyrans and chromans by reaction with $\alpha\beta$-unsaturated carbonyl compounds. 2-Alkoxydihydropyrans are obtained in good yields at temperatures between 150 and 200°C (Colonge and Descotes, 1967) (3.8). They are useful intermediates for the preparation of glutaraldehydes.

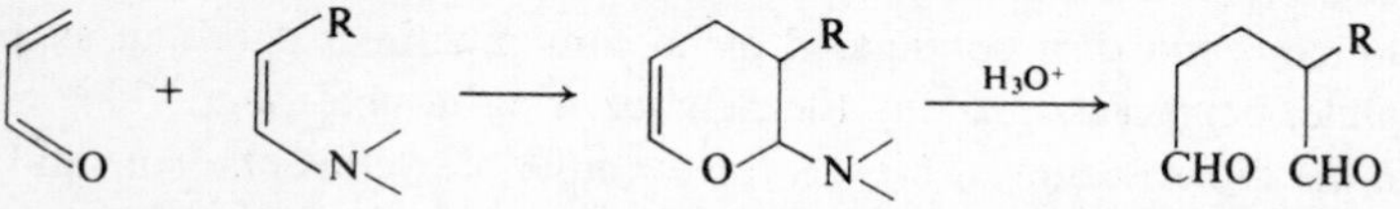

H_3O^+

CHO CHO

(3.9)

Enamines and vinyl carbamates also react readily as dienophiles with $\alpha\beta$-unsaturated carbonyl compounds. The products are again easily hydrolysed in acid solution to glutaraldehydes (3.9). It is not certain that these cyclo-additions are concerted; they may be stepwise ionic reactions (see e.g. Fleming and Kargar, 1967). Diels–Alder reactions with cyclic enamines are useful in the synthesis of polycyclic compounds. Thus, an important step in the synthesis of the alkaloid (±)-minovine was achieved by addition of a cyclic enamine to an indolylacrylic ester as shown in (3.10) (Ziegler and Spitzer, 1970).

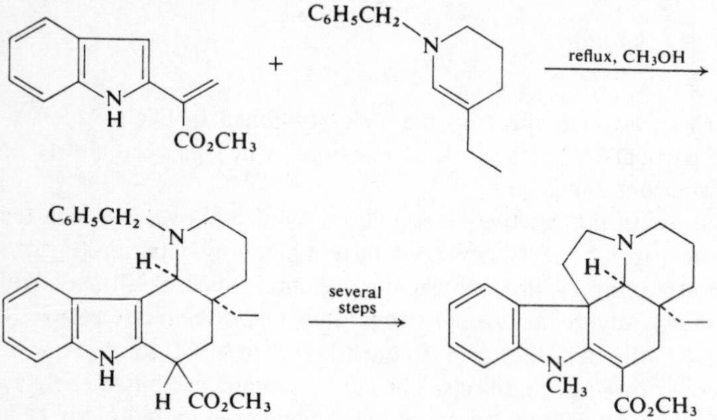

$C_6H_5CH_2$

reflux, CH_3OH

CO_2CH_3

$C_6H_5CH_2$

several steps

H CO_2CH_3

(60%)

CH_3

CO_2CH_3

(±)-Minovine

(3.10)

Cycloalkenes and cycloalkynes. A number of cyclic olefins and acetylenes with pronounced angular strain are reactive dienophiles. The driving force for these reactions is thought to be the reduction in angular strain associated with the transition state for the addition. Thus, cyclopropene reacts rapidly and stereospecifically with cyclopentadiene at 0°C to form the *endo* adduct (2) in 96 per cent yield, and butadiene forms norcarene (3) in 37 per cent yield. Many cyclopropene derivatives behave similarly (Deem, 1972).

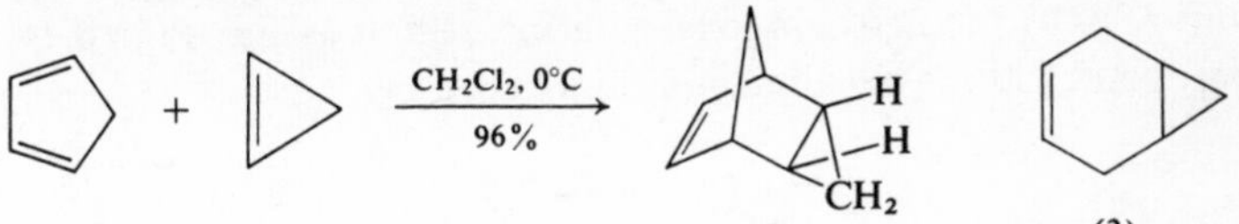

CH_2Cl_2, 0°C
96%
H
H
CH_2
(2)
(3)
(3.11)

In cyclobutene derivatives as well, the angular strain has a reaction-promoting effect. 3,3,4,4-Tetrafluorocyclobutene reacts readily with a number of dienes (3.12) and so do unsaturated four-membered cyclic sulphones. The existence of the transient species benzocyclobutadiene (4)

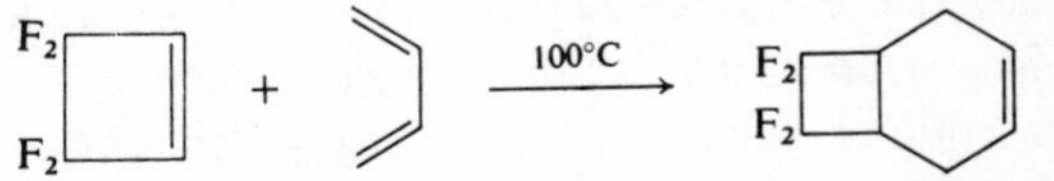

F_2
F_2
100°C
F_2
F_2
(3.12)

has been shown by trapping experiments with the reactive dienophile diphenylisobenzofuran (3.13).

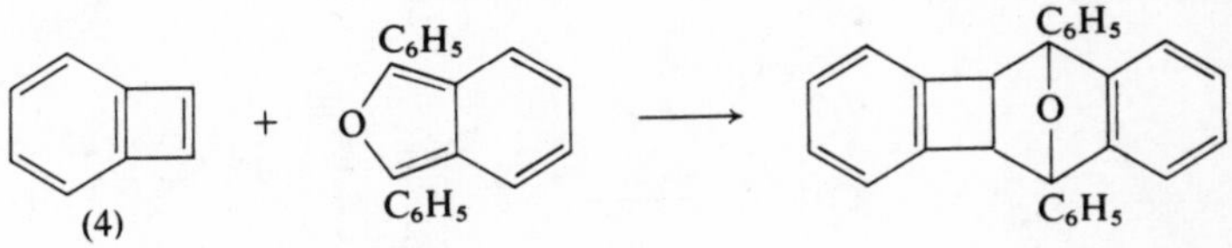

C_6H_5
O
C_6H_5
(4)
C_6H_5
O
C_6H_5
(3.13)

Some cyclic acetylenic compounds are also powerful dienophiles. Because of its linear structure a carbon–carbon triple bond can only be incorporated without strain into a ring with nine or more members. The increasing strain with decreasing ring size in the sequence cyclo-octyne to cyclopentyne is shown in an increasing tendency to take part in 1,4-cyclo-addition reactions. Cyclo-octyne has been prepared as a stable liquid with pronounced dienophilic properties. It reacts readily with diphenylisobenzofuran to give an adduct in 91 per cent yield (3.14).

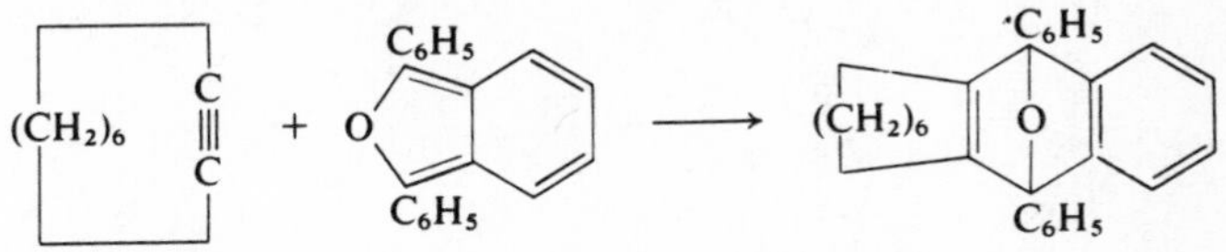

$(CH_2)_6$
C
C
C_6H_5
O
C_6H_5
C_6H_5
$(CH_2)_6$
O
C_6H_5
(3.14)

The lower cycloalkynes have not been isolated but their existence has been shown by trapping them with diphenylisobenzofuran (Wittig, 1962).

Arynes, such as dehydrobenzene, also readily undergo Diels–Alder addition reactions. Cyclopentadiene, cyclohexadiene and even benzene and naphthalene add to the highly reactive species C_6H_4 (Wittig, 1962; Hoffmann, 1967) (3.15). Analogous addition reactions are shown by dehydroaromatics in the pyridine and thiophen series.

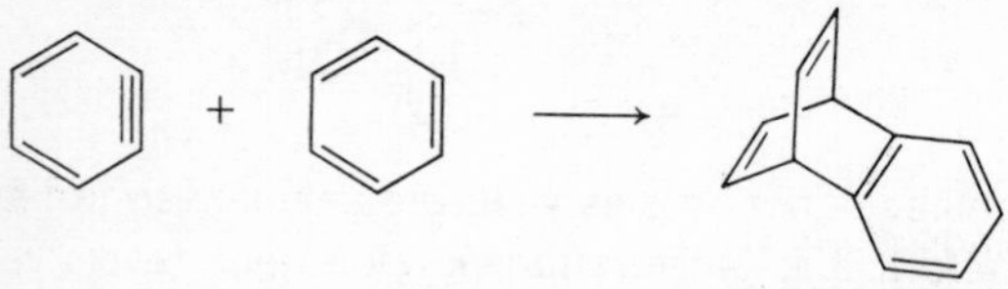

+ other products (3.15)

Ketenes and allenes. It is evident that, under the appropriate conditions, most olefinic and acetylenic compounds can function as dienophiles. A notable exception is ketene. The C=C linkage in the system C=C=O does indeed react with dienes but the additions are not diene syntheses; the products are four-membered ring compounds formed by 1,2-addition. For example, cyclopentadiene and dimethylketene form the cyclobutanone derivative (5) (3.16).

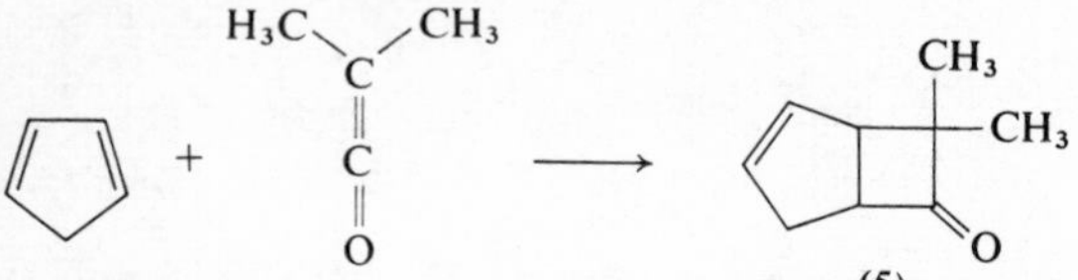

H₃C CH₃ C C O CH₃ CH₃ O (5) (3.16)

These are now regarded as concerted $\pi 2s + \pi 2a$ cyclo-additions, with the ketene acting as the antarafacial component (Woodward and

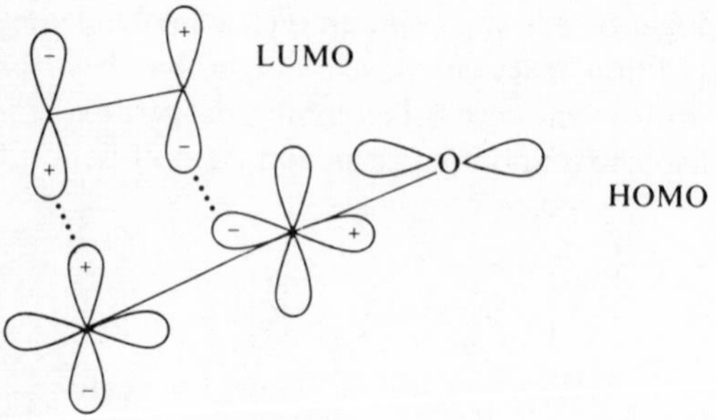

LUMO HOMO

LUMO = lowest unoccupied molecular orbital of the olefin;
HOMO = highest occupied molecular orbital of the ketene.

Hoffmann, 1970). However, indirect methods have been developed to achieve the conversion corresponding to Diels–Alder addition of ketene to 1,3-dienes. Most of these involve addition of a suitably chosen acrylic

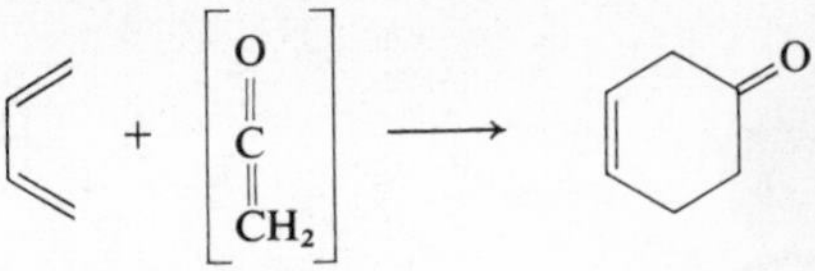

(3.17)

acid derivative to the diene followed by conversion of the initial adduct into the required ketone. A number of reagents have been used; the best so far appear to be 2-chloroacrylonitrile (Evans, Scott and Truesdale, 1972) and 2-chloroacryloyl chloride (Corey, Ravindranathan and Terashima, 1971), and acrylic acid itself has also been recommended (Trost and Tamaru, 1975). Addition of the nitrile to dienes is strongly catalysed by cupric chloride, so that it can be used with sensitive or unreactive dienes. Another advantage of the nitrile is that addition to unsymmetrical dienes is highly regioselective. Conversion of the initial adducts into the desired ketones is easily effected by hydrolysis.

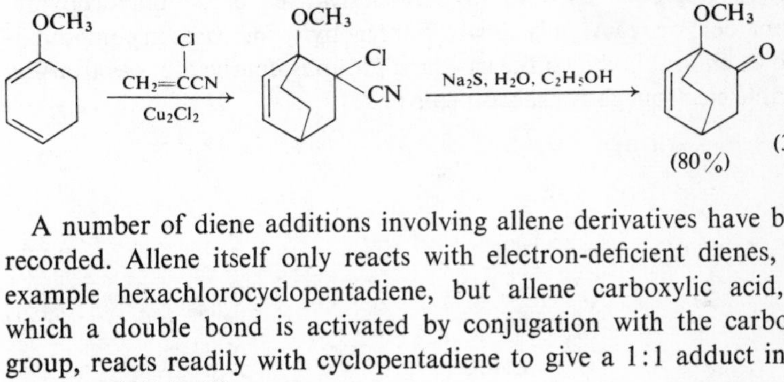

OCH_3 — $CH_2{=}C(Cl)CN$, Cu_2Cl_2 → OCH_3, Cl, CN — Na_2S, H_2O, C_2H_5OH → OCH_3, O (80%) (3.18)

A number of diene additions involving allene derivatives have been recorded. Allene itself only reacts with electron-deficient dienes, for example hexachlorocyclopentadiene, but allene carboxylic acid, in which a double bond is activated by conjugation with the carboxyl group, reacts readily with cyclopentadiene to give a 1:1 adduct in 84 per cent yield (3.19).

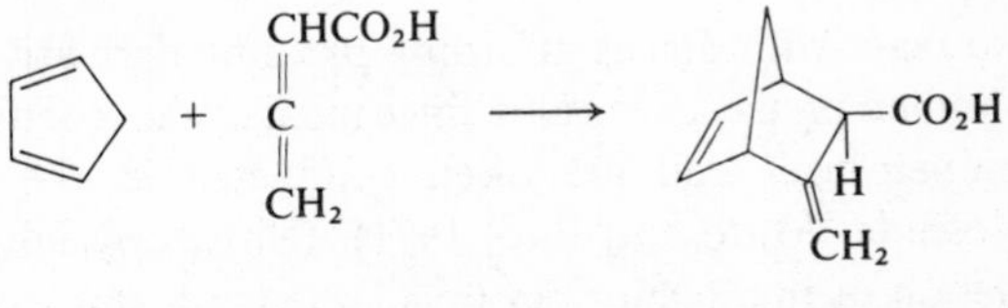

$CHCO_2H{=}C{=}CH_2$ → CO_2H, H, CH_2 (3.19)

An 'allene equivalent' is vinyltriphenylphosphonium bromide, which is reported to react with a number of dienes to form cyclic phosphonium salts which can be converted into methylene compounds by the usual Wittig procedure (Ruden and Bonjouklian, 1974).

$$\text{CH}_2{=}\text{CH}\overset{+}{\text{P}}(\text{C}_6\text{H}_5)_3\ \overset{-}{\text{Br}} + \text{2,3-dimethylbutadiene} \xrightarrow[150^\circ 3]{\text{CH}_3\text{CN}} \text{3,4-dimethylcyclohex-3-enyl-}\overset{+}{\text{P}}(\text{C}_6\text{H}_5)_3\ \overset{-}{\text{Br}} \xrightarrow{\text{LiN(iso-C}_3\text{H}_7)_2} \text{ylide }{=}\text{P}(\text{C}_6\text{H}_5)_3 \xrightarrow{\text{CH}_2\text{O}} \text{3,4-dimethyl-1-methylenecyclohex-3-ene} \quad (3.20)$$

Heterodienophiles. One or both of the carbon atoms of the dienophile multiple bond may frequently be replaced by a heteroatom without loss of activity. But in general Diels–Alder reactions involving heterodienophiles have not been so widely employed in synthesis as those using olefines and acetylenes (Needleman and Chang Kuo, 1962).

Carbonyl groups in aldehydes and ketones add to dienes, and the reaction has been used to prepare derivatives of 5,6-dihydropyran. Formaldehyde reacts only slowly, but reactivity increases in compounds such as chloral in which the carbonyl group is deprived of electrons by suitable electronegative substituents (3.21).

$$\text{1-ethoxybutadiene (OC}_2\text{H}_5\text{)} + \text{O}{=}\text{CH}_2 \xrightarrow{180^\circ\text{C}} \text{2-ethoxy-5,6-dihydropyran (OC}_2\text{H}_5\text{)}$$

$$\text{isoprene (H}_3\text{C)} + \text{O}{=}\text{CHCCl}_3 \xrightarrow{150^\circ\text{C}} \text{4-methyl-6-trichloromethyl-5,6-dihydropyran (H}_3\text{C, CCl}_3\text{)} \quad (3.21)$$

Nitriles also react with dienes affording pyridine derivatives, but the method is of limited preparative value since most of the reactions require very high temperatures and are often conducted in the gas phase (Butsugan, Yoshida, Muto and Bito, 1971). Dihydropyridines are obtained by reaction of iminochlorides (available from amides and phosphorus oxychloride) with aliphatic dienes. For example 2,3-dimethylbutadiene and acetamide in presence of phosphorus oxychloride readily afford 3,4,6-trimethyl-1,2-dihydropyridine (3.22). With styrenes, derivatives of 3,4-dihydroisoquinoline are obtained.

$$H_3C\!-\!C(=CH_2)\!-\!C(=CH_2)\!-\!CH_3 \;+\; HN{=}C(CH_3)Cl \longrightarrow$$

$$\text{(tetrahydropyridine: } H_3C, H_3C, NH, CH_3, Cl) \longrightarrow \text{(dihydropyridine: } H_3C, H_3C, NH, CH_3) \qquad (3.22)$$

Nitroso compounds also react with dienes to form oxazine derivatives (Hamer and Ahmad, 1967). Aromatic nitroso compounds are the most reactive. Thus with butadiene nitrosobenzene gives a high yield of *N*-phenyl-3,6-dihydro-oxazine (3.23) at 0°C, and cyclopentadiene forms a

$$C_6H_5NO \;+\; \text{butadiene} \xrightarrow[0^\circ C]{C_2H_5OH} C_6H_5\!-\!N\text{(dihydro-oxazine, O)} \quad (95\,\%) \qquad (3.23)$$

related unstable adduct at –40°C. Electron-releasing substituents in the *ortho* and *para* positions of the nitrosobenzene lower its activity considerably.

With aliphatic nitroso compounds only derivatives with an electron-withdrawing group on the α-carbon atom undergo the reaction. Thus, α-chloro or α-cyano nitroso compounds react readily with butadienes.

Nitrosocarbonyl compounds, formed transiently from hydroxamic acids by oxidation with periodate, react readily with 1,3-dienes to give *N*-acyl or *N*-aroyldihydro-oxazines (Kirby and Sweeny, 1973).

$$R\!-\!C(=O)NHOH \xrightarrow{IO_4^-} \left[R\!-\!C(=O)\!-\!N{=}O\right] \xrightarrow{\text{diene}} \text{(dihydro-oxazine: O, NCOR)} \qquad (3.24)$$

Some azo compounds with electron-attracting groups attached to the nitrogen atoms are very reactive dienophiles (Gillis, 1967), and react with a variety of dienes leading to derivatives of tetrahydropyridazine. The most commonly used reagent of this type is ethyl azodicarboxylate and the reaction of this dienophile with cyclopentadiene formed one of the earliest-known examples of Diels–Alder addition. Most azodicarboxylic esters have the *trans* configuration; recent work has shown that the *cis* compounds are even more reactive, partly owing to the decrease in steric hindrance. This contrasts with the reactivity of isomeric pairs of olefinic dienophiles (see p. 164).

The Diels–Alder reaction of azo dienophiles is of interest because it provides a convenient route to the pyridazine ring system. Thus, 2,3-dimethylbutadiene and ethyl azodicarboxylate react readily at room temperature giving an almost quantitative yield of the tetrahydropyridazine, which is readily converted into the cyclic hydrazine by the steps shown (3.25). Ethyl azodicarboxylate forms adducts with a wide variety of other dienophiles, including anthracene and furan, both of which react readily.

H_3C, H_3C + $NCO_2C_2H_5$ ‖ $NCO_2C_2H_5$ ⟶ H_3C, H_3C / $NCO_2C_2H_5$ | $NCO_2C_2H_5$

H_2, Pt (3.25)

H_3C, H_3C / NH | NH ⟵ KOH, H_2O — H_3C, H_3C / $NCO_2C_2H_5$ | $NCO_2C_2H_5$

A useful scheme involving a heterodienophile which results in the formal addition of carbon monoxide to 1,3-dienes has been described by Corey and Walinsky (1972). Carbon monoxide cannot be added directly

+ [CO] ⟶ =O (3.26)

to dienes, but the readily available 1,3-dithienium ion (6) reacts easily with butadiene and some of its derivatives to form $(4+2)\pi$ adducts of the type (7). Reaction of these with butyl-lithium brings about rearrangement to a vinylcyclopropane derivative which rearranges readily on pyrolysis to the trimethylenedithiocyclopentene; hydrolysis gives the cyclopentenone (3.27). All the reactions proceed in high yield and the sequence provides a potentially valuable route to cyclopentenone derivatives.

Oxygen as a dienophile. The reaction of oxygen with 1,3-dienes to form endoperoxides is of outstanding interest (Gollnick and Schenck, 1967). Generally addition of oxygen to dienes is effected under the influence of light, either directly or in presence of a photosensitiser, but the reaction can also be effected with sodium hypochlorite and hydrogen peroxide

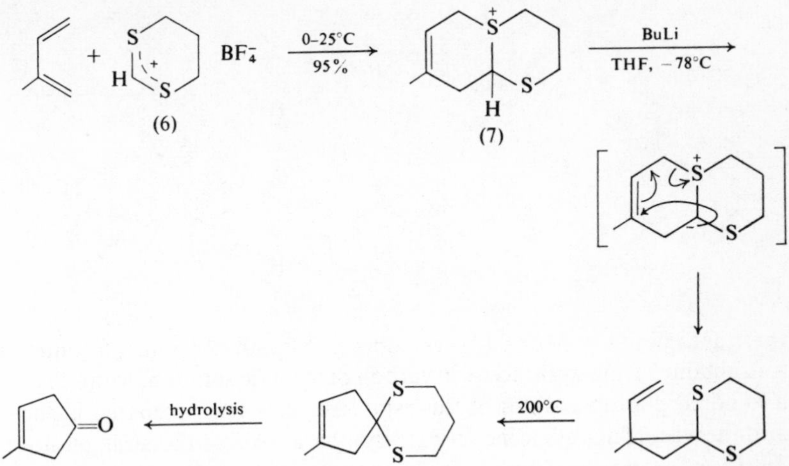

0–25°C 95%

BuLi THF, −78°C

(6) (7)

hydrolysis

200°C

(3.27)

and by other means, and it is generally believed that singlet oxygen is the reactive species in each case (Corey and Taylor, 1964; Foote and Wexler, 1964; Foote, 1968; Gollnick, 1968; Kearns, 1971). The reaction bears

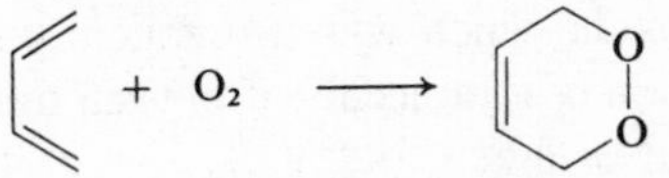

$$+ \ O_2 \longrightarrow \qquad (3.28)$$

many similarities to the conventional thermal Diels–Alder reaction. It appears to be concerted, as suggested by the fact that addition of oxygen to 1,1′-bicyclohexenyl gave only the *cis*-endoperoxide and it has been

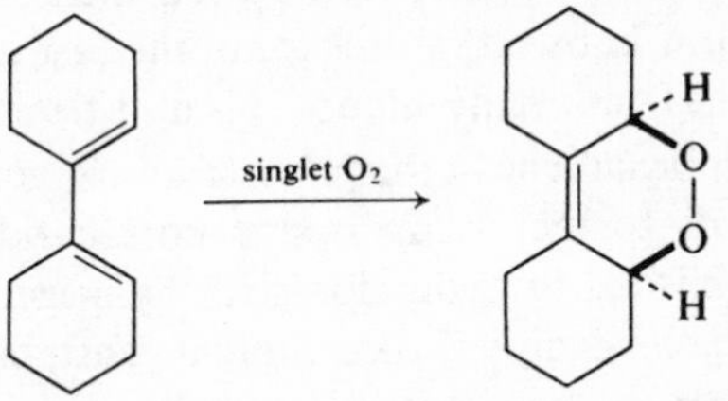

singlet O_2

(3.29)

suggested that the reaction proceeds through a six-membered cyclic transition state as does the Diels–Alder reaction (Kearns, 1971).

The light-induced 1,4-addition of oxygen to dienes was discovered independently by Clar and by Windaus. Clar found in 1930 that when a solution of the linear pentacyclic aromatic hydrocarbon pentacene (8) in benzene was irradiated with ultraviolet light in presence of oxygen,

$h\nu$, O_2 / C_6H_6

(8) → (9) (3.30)

the transannular peroxide (9) was obtained (3.30). A similar product was obtained from anthracene in carbon disulphide solution. More than a hundred photoperoxides of this type have now been prepared in the anthracene and naphthacene series (Schönberg, 1968). In general, photosensitisers are not used in these reactions, but the rate of formation of the oxide is often strongly dependent on the solvent.

The photosensitised addition of oxygen to conjugated alicyclic dienes was discovered by Windaus in 1928 in the course of his classical studies of the conversion of ergosterol into vitamin D. Irradiation of a solution of ergosterol in alcohol in presence of oxygen and a sensitiser led to the formation of a peroxide which was subsequently shown to have been formed by 1,4-addition of a molecule of oxygen to the conjugated diene system (Windaus and Brunken, 1928).

Numerous other endoperoxides of the same type have since been obtained by sensitised photo-oxidation of other steroidal dienes with *cisoid* 1,3-diene systems. For many years it was believed that photosensitised addition of oxygen to conjugated dienes was restricted to steroids, but it is now known that this is not the case and endoperoxides have been obtained from many mono-, bi- and tri-cyclic dienes. Thus, irradiation of cyclohexadiene in the presence of oxygen with chlorophyll as a sensitiser leads to the endoperoxide norascaridole, the structure of which was established by reduction with hydrogen and platinum to *cis*-1,4-dihydroxycyclohexane (3.31). Similar unstable peroxides have been obtained from cyclopentadiene and from cycloheptadiene; reduction of the product in each case led to the *cis*-1,4-diol. Open-chain

$h\nu$, O_2 / sensitiser; H_2, Pt (3.31)

dienes also (e.g. isoprene) are readily converted into cyclic peroxides by irradiation in presence of oxygen and a sensitiser (Kondo and Matsumoto, 1972).

Photosensitised oxidation of 1,3-dienes has been used with conspicuous success as a key step in the synthesis of a number of natural products. The longest known, and so far the only endoperoxide found in nature, is the terpene peroxide ascaridole (10) the principal component of chenopodium oil (3.32). Its structure was elucidated by hydrogenation to *p*-menthane-*cis*-1,4-diol. The compound was readily synthesised by Schenck and Ziegler by photosensitised oxidation of α-terpinene. The

$h\nu$, O_2, alcohol; chlorophyll (3.32)

(10)

product from chenopodium oil is optically inactive and it has been suggested that, in the plant, ascaridole is formed by chlorophyll-sensitised photo-oxidation of α-terpinene rather than by an enzymic process which would be expected to lead to optically active ascaridole. In the synthesis of the growth inhibitory substance abscissic acid (12) the unsaturated ketol grouping was introduced via photosensitised addition of oxygen to the conjugated diene (11) (Cornforth, Milborrow and Ryback, 1965) (3.33).

CH_3 CO_2H $h\nu$, O_2, eosin; benzene–ethanol

H_3C CH_3 CH_3

(11)

CH_3 CO_2H dil. NaOH; 100°C O CH_3 CO_2H OH (3.33)

H_3C CH_3 CH_3 H_3C CH_3 CH_3

(12)

An efficient alternative route to *endo*-peroxides from some 1,3-dienes by reaction with *triplet* oxygen in presence of triphenylmethyl cation or Lewis acids has recently been reported. Thus α-terpinene is converted into ascaridole in high yield with oxygen and trityl cation in methylene chloride at −78°C. The precise mechanism of these reactions is still uncertain. Light is not always required, and it appears that two pathways are available, but it is believed that in each case the catalyst provides a mechanism for overcoming the spin barrier (Barton, Haynes, Leclerc, Magnus and Menzies, 1975).

3.3. The diene

It has already been pointed out (p. 162) that the diene component must have or must be able to adopt the *cisoid* conformation before it can take part in Diels–Alder reactions with dienophiles. The majority of dienes which satisfy this condition undergo the reaction more or less easily depending on their structure.

Acyclic dienes. Acyclic conjugated dienes react readily often forming the adducts in almost quantitative yield. For example, butadiene itself reacts quantitatively with maleic anhydride in benzene at 100°C in 5 h, or more slowly at room temperature, to form *cis*-1,2,3,6-tetrahydrophthalic anhydride.

Examples have already been given (p. 169) of 'dienophile equivalents' which have proved valuable in some synthetic transformations. A useful 'equivalent' for butadiene is the crystalline 2,5-dihydrothiophen sulphone (sulpholene), the cyclic adduct of butadiene and sulphur dioxide, which regenerates butadiene *in situ* on heating. Thus, with diethyl fumarate at 105–110°C, diethyl *trans*-tetrahydrophthalate is obtained in 70 per cent yield (Sample and Hatch, 1970). By use of this reagent an excess of butadiene, often employed to force the usual reaction, with consequent formation of polymeric byproducts, can be avoided.

Substituents in the butadiene molecule influence the rate of cycloaddition both through their electronic nature and by a steric effect on the conformational equilibrium. Alder found that the rate of the reaction is often increased by electron-donating substituents (e.g.—NMe_2, —OMe, —Me) in the diene as well as by electron-attracting substituents in the dienophile. Bulky substituents which discourage the diene from adopting the *cisoid* conformation hinder the reaction. Thus, whereas 2-methyl-, 2,3-dimethyl- and 2-t-butylbutadiene react normally with maleic anhydride the 2,3-diphenyl compound is less reactive and 2,3-di-t-butylbutadiene is completely unreactive. Apparently the molecule of the 2,3-di-t-butyl compound is prevented from attaining the necessary planar *cisoid* conformation by the steric effect of the two bulky t-butyl substituents. In contrast, 1,3-di-t-butylbutadiene, in which the substituents do not interfere with each other even in the *cisoid* form, reacts readily with maleic anhydride.

cis Alkyl or aryl substituents in the 1-position of the diene reduce its reactivity by sterically hindering formation of the *cisoid* conformation through non-bonded interaction with a hydrogen atom at position 4. Accordingly, a *trans* substituted 1,3-butadiene reacts with dienophiles much more readily than the *cis* isomer. Thus, *cis*-piperylene gave only a 4 per cent yield of adduct when heated with maleic anhydride at 100°C for 8 h, whereas the *trans* isomer formed an adduct in almost quantitative yield in benzene at 0°C, (3.34).

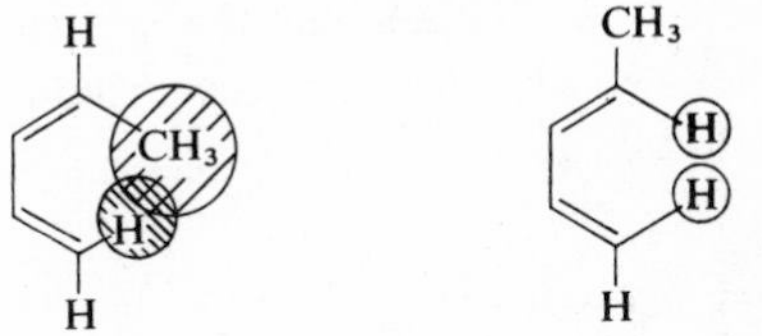

(3.34)

Similarly, *trans,trans*-1,4-dimethylbutadiene reacts readily with many dienophiles, but the *cis,trans* isomer yields an adduct only when the components are heated in benzene at 150°C. *cis* Substituents in both the 1- and 4- positions prevent reaction.

1,1-Disubstituted butadienes also react with difficulty, and with such compounds addition may be preceded by isomerisation of the diene to a more reactive species. Thus, in the reaction of 1,1-dimethylbutadiene with acrylonitrile the diene first isomerises to 1,3-dimethylbutadiene which then reacts in the normal way.

This difference in reactivity towards maleic anhydride of *cis*- and *trans*-1-substituted dienes has been exploited as a method for the determination of the stereochemistry of diene systems in a number of naturally occurring dienes and polyenes (see p. 181; also Alder and Schumacher, 1953).

Very many Diels–Alder reactions with alkyl- and aryl-substituted butadienes have been effected (cf. Onischenko, 1964). Derivatives of hydroxybutadiene also react with dienophiles, providing a route to synthetically useful intermediates. Thus, 2-alkoxybutadienes react easily with maleic anhydride and other dienophiles to form adducts which, as enol ethers, are readily hydrolysed to cyclohexanone derivatives (3.35). Again, the reaction of *trans,trans*-1,4-diacetoxybutadiene with methyl acrylate formed the first step in a stereospecific total synthesis of the

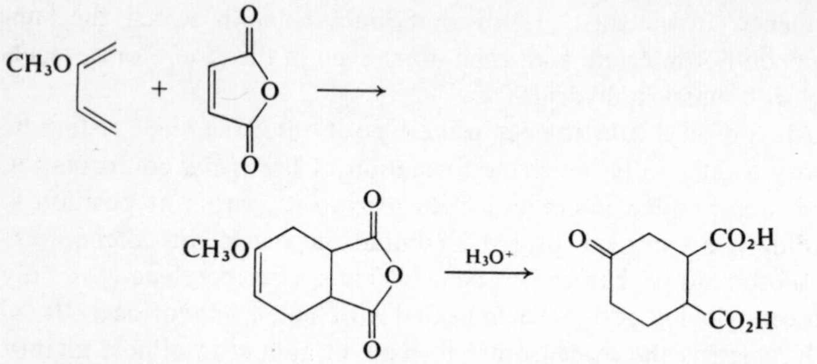

$$CH_3O\text{-butadiene} + \text{maleic anhydride} \longrightarrow$$

$$\xrightarrow{H_3O^+} \qquad (3.35)$$

CO₂H

important biosynthetic intermediate shikimic acid (13) (Smissman, Suh, Oxman and Daniels, 1962) (3.36). Another route to this compound begins with the Diels–Alder reaction between butadiene and propiolic acid (Grewe and Hinrich, 1964).

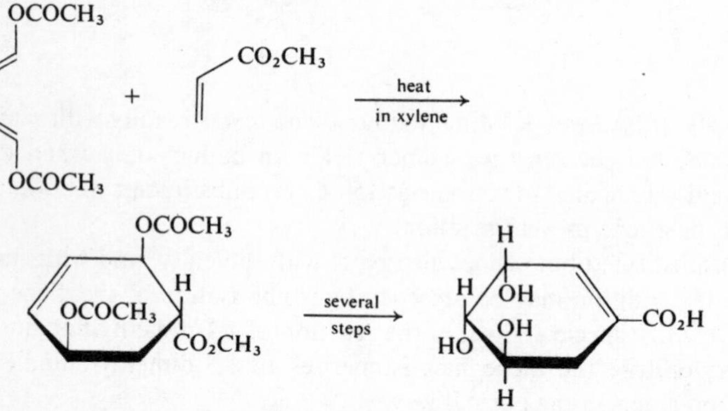

$$\xrightarrow[\text{in xylene}]{\text{heat}}$$

$$\xrightarrow[\text{steps}]{\text{several}} \qquad (3.36)$$

(13)

Reaction of diacetoxybutadiene with olefinic and acetylenic dienophiles provides a useful synthetic route to benzene derivatives, for the initial adducts readily eliminate acetic acid to give the aromatic compound. In many cases it is unnecessary to isolate the intermediate adduct and the benzene derivative is obtained directly by heating the components together at 100–120°C (Hill and Carlson, 1965) (3.37). 1,3-Dialkoxybutadienes are difficult to obtain and have been little studied, but Danishefsky and Kitahara (1974, 1975) have demonstrated the usefulness of 1-methoxy-3-trimethylsilyloxybutadiene. This compound is readily available from methoxyvinyl methyl ketone and reacts with a variety of dienophiles. The initial adducts yield cyclohexenones on acid hydrolysis; the reaction with cyclohexenones is the key step in a

$OCOCH_3$ X Y $OCOCH_3$ X Y $OCOCH_3$ 120°C X Y $OCOCH_3$

CO_2CH_3 C C CO_2CH_3

CO_2CH_3 CO_2CH_3 $OCOCH_3$

(3.37)

OCH_3 O CH_3 $(CH_3)_3SiCl$ $(C_2H_5)_3N$ OCH_3 $(CH_3)_3SiO$ CHO reflux, C_6H_6

OCH_3 CHO CH_3 $(CH_3)_3SiO$ H_3O^+ CHO CH_3 O (77%)

OCH_3 $(CH_3)_3SiO$ + $CO_2C_2H_5$ (1) heat (2) H_3O^+ $CO_2C_2H_5$ O H (53%) (3.38) good route to *cis*-octalones (3.38). 2-Trimethylsilyloxybutadiene has also been used to make derivatives of cyclohexanone (Jung and McCombs, 1976). 1-Trimethylsilylbutadiene also reacts readily with dienophiles, giving products which can be converted into cyclic allylic alcohols (see p. 328).

CH_3O C_6H_5S + → CH_3O C_6H_5S (75%)

(14) CO_2CH_3 + CH_3 —150°C→ [(15) CH_3, O, CO, H, CO_2CH_3] → (16) CH_3, H, CO_2CH_3 + CO_2 (3.39)

Two other derivatives of butadiene which have some promise in synthesis are 2-methoxy-3-(phenylthio)butadiene and the pyrone (14). The former compound reacts with dienophiles to give cyclic products containing masked β-phenylthio ketone groups, which themselves undergo a number of synthetically useful transformations (see p. 39). The pyrone (14) reacts with dienophiles with loss of carbon dioxide to form cyclic dienes. Thus, reaction with 4-methyl-3-cyclohexenone affords the *cis*-hexahydronaphthalene derivative (16) by way of the intermediate (15) (Corey and Watt, 1973). For another example of the use of a pyrone as dienophile see p. 182.

Conjugated polyenes react normally with dienophiles by 1,4-addition to form six-membered rings. The position of attack on the polyene chain is governed by the same principles as those operating in the case of butadienes. Thus, *trans,trans*-1,3,5-heptatriene reacts readily with maleic anhydride in ether at room temperature to give a 90 per cent yield of a mixture of the two possible adducts (3.40). In contrast, the isomeric

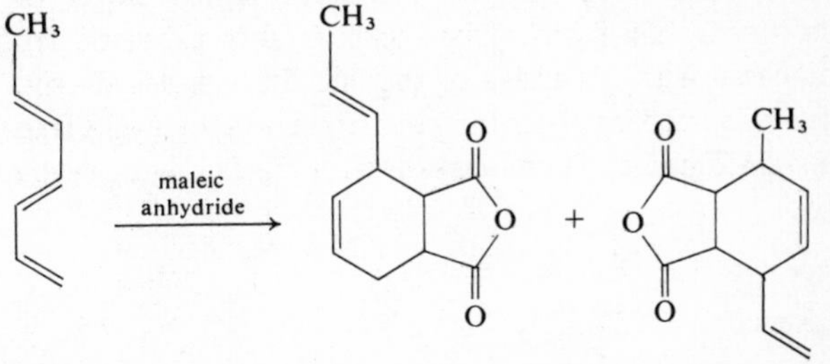

CH_3 —maleic anhydride→ CH_3 ... + ... CH_3 (3.40)

cis,trans-heptatriene yields only one adduct by reaction at the *trans*-substituted butadiene grouping.

If the *cisoid* conformations are not hindered a tetraene can add two molecules of maleic anhydride. Thus the naturally occurring β-parinaric acid (17) has been shown to be the all-*trans* compound by the formation of a bis-adduct with maleic anhydride.

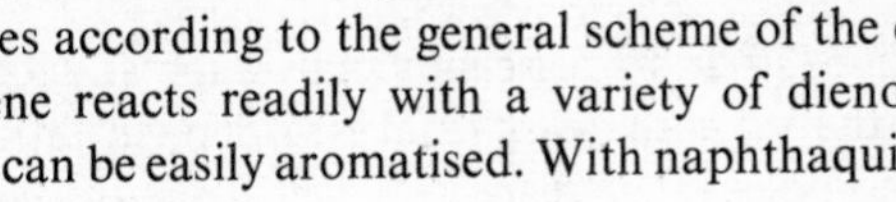

C_2H_5—CH=CH—CH=CH—CH=CH—CH=CH—$(CH_2)_7CO_2H$

(17)

Trienes with an allene arrangement of the double bonds condense with dienophiles according to the general scheme of the diene synthesis. 1,2,4-Pentatriene reacts readily with a variety of dienophiles forming adducts which can be easily aromatised. With naphthaquinone an adduct

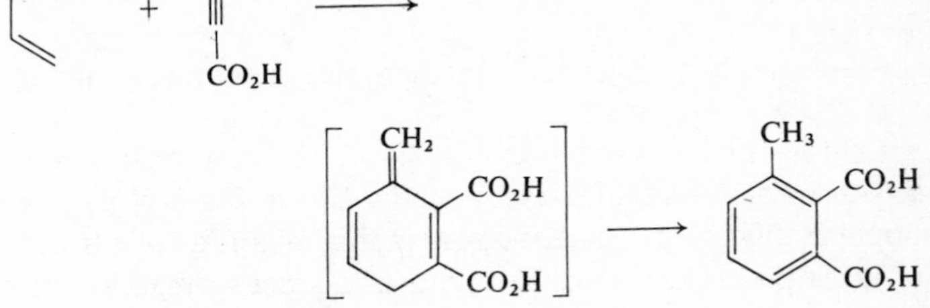

+ HO_2C—C≡C—CO_2H —90°C→ [CH_2, CO_2H, CO_2H] → CH_3, CO_2H, CO_2H (3.41)

is obtained which is converted into 1-methylanthraquinone with palladium in boiling ethanol, and acetylenedicarboxylic acid gives 3-methylphthalic acid directly at 90°C (3.41).

The formally analogous vinyl ketenes are unsuitable for use as dienes because of their inaccessibility, their instability and their tendency to form (2 + 2) adducts. Corey and Kozikowsky (1975) have found, however, that the hydroxypyrone (18) is a useful equivalent of vinylketene itself. The crystalline hydroxypyrone is readily available from mucic acid and reacts with a variety of olefinic dienophiles to give adducts which are converted easily into dihydrophenols or cyclohexenones by loss of carbon dioxide. Thus reaction with methyl methacrylate affords

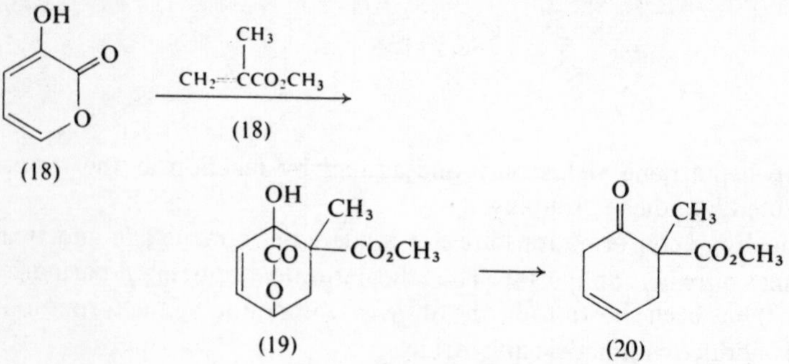

$$(18) \xrightarrow[(18)]{CH_2{=}C(CH_3)CO_2CH_3} (19) \longrightarrow (20) \qquad (3.42)$$

(18) (19) (20)

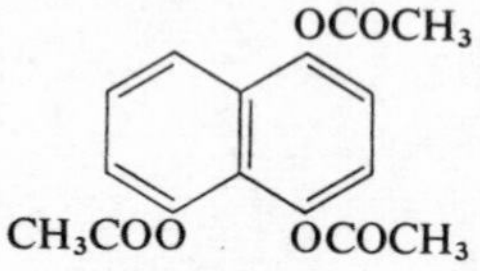

(21)

the cyclohexenone (20) in 85 per cent yield via the adduct (19), and with benzoquinone the naphthalene (21) is obtained after acetylation. Acetylenic dienophiles also react to give aromatic compounds; for example, dimethyl 3-hydroxyphthalate is obtained from dimethyl acetylenedicarboxylate.

Eneynes and dienynes. Eneynes and dienynes, with the groupings $-\overset{|}{C}{=}\overset{|}{C}-C{\equiv}C-$ and $-\overset{|}{C}{=}\overset{|}{C}-C{\equiv}C-\overset{|}{C}{=}\overset{|}{C}-$, react readily with dienophiles to form 1,4-addition compounds. The reaction of eneynes with dienophiles differs from that of dienes in that migration of a hydrogen atom takes place during the reaction. It has been suggested that addition is preceded by tautomerism of the eneyne to a zwitterionic diene, as illustrated (3.43) for the dimerisation of 3-methyl-3-penten-1-yne (22) to (23). Many examples of diene condensations involving enynes

CH_3 H_3C CH C C CH (22) ⇌ CH_3 H_3C C^- C CH CH^+ + CH C C H_3C $CHCH_3$ ⟶

CH_3 H_3C CH_3 C $CHCH_3$ (23) (3.43)

have been recorded. With acetylenic dienophiles aromatic compounds are formed directly, as in the example cited above.

Heterodienes. Heterodienes, in which one or more of the atoms of the conjugated diene is a heteroatom, are not so numerous as hetero-dienophiles (see p. 170) and only the 1,4-addition reactions of $\alpha\beta$-unsaturated carbonyl compounds have been used synthetically to any extent so far (Colonge and Descotes, 1967). Diene condensations of heterodienes containing nitrogen atoms in the diene system have been much less studied (see Needleman and Chang Kuo, 1962).

$\alpha\beta$-Unsaturated carbonyl compounds react most readily as dienes with electron-rich dienophiles such as enol ethers and enamines (see p. 166). With less reactive dienophiles dimerisation of the $\alpha\beta$-unsaturated carbonyl compound is a competing reaction. For example, acrolein is converted into a dimer, 2-formyl-2,3-dihydropyran, in 40 per cent yield by heating in benzene solution, one molecule acting as the diene component and the other as dienophile. The reaction with *N*-vinyl compounds, such as enamines and *N*-vinyl carbamates, and with enol ethers, proceeds readily to yield derivatives of dihydropyran which can be converted into synthetically useful 1,5-dicarbonyl compounds (3.44).

In the nitrogen series 1-dimethylamino-2-azabutadienes react with electrophilic acetylenic and olefinic dienophiles with loss of dimethyl-

$$\ce{H3C-CO-CH=CH2} + \ce{C3H7-CH=CH-NC5H10} \xrightarrow[80\%]{20^\circ C}$$

$$\text{(dihydropyran: } \ce{H3C}, \ce{C3H7}, \ce{NC5H10}) \xrightarrow{HCl} \ce{CH3CO-CH2-CH2-CH(C3H7)-CHO} \qquad (3.44)$$

amine, to give pyridines or dihydropyridines (Demoulin, Gorissen, Hesbain-Frisque and Ghosez, 1975).

$$\ce{H3C-C(=CH2)-N=CH-N(CH3)2} + \ce{CH3O2C-C#C-CO2CH3} \xrightarrow[20^\circ C]{CH_3CN} \text{(pyridine: } \ce{H3C}, \ce{CO2CH3}, \ce{CO2CH3}) \; (58\%) \qquad (3.45)$$

1,2-Dimethylenecycloalkanes. In 1,2-dimethylenecycloalkanes the *cisoid* conformation of the double bonds necessary for 1,4-addition is fixed and dienes of this type are excellent reagents for dienophiles often forming the adducts in almost quantitative yield. This and the ready availability of the compounds make 1,4-additions to dimethylene-cycloalkanes a convenient route to polycyclic compounds.

Thus, 1,2-dimethylenecyclohexane reacts exothermally with maleic anhydride, and with benzoquinone a bis-adduct (24) is obtained which can be readily converted into the polycyclic aromatic hydrocarbon pentacene (8, p. 174) (3.46). A variety of alkyl derivatives of this and other linearly condensed polycyclic aromatic compounds has been conveniently obtained by this general route.

$$\text{1,2-dimethylenecyclohexane } (\ce{CH2}, \ce{CH2}) + \text{benzoquinone} \xrightarrow[77\%]{100^\circ C,\ 2\ h} \text{(24)} \qquad (3.46)$$

(24)

Vinylcycloalkenes and vinylarenes. Many vinylcycloalkenes react readily with dienophiles. In contrast, dienes of the types (3.47) in

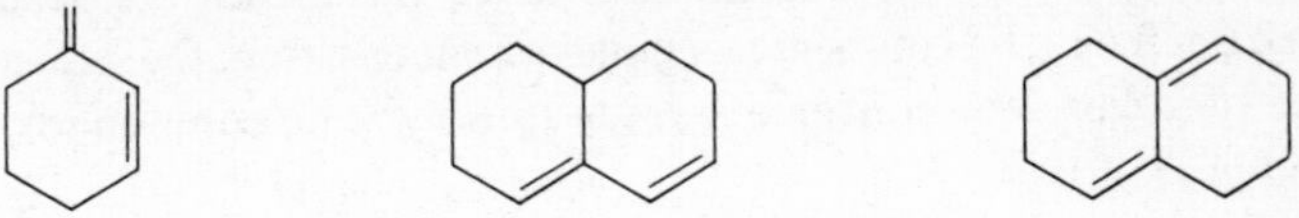

(3.47)

which the double bonds are constrained in the *transoid* conformation do not react. 1-Vinylcyclohexene itself reacts exothermally with maleic anhydride to form a high yield of the adduct (25), (3.48). 1-Vinyl-3,4-

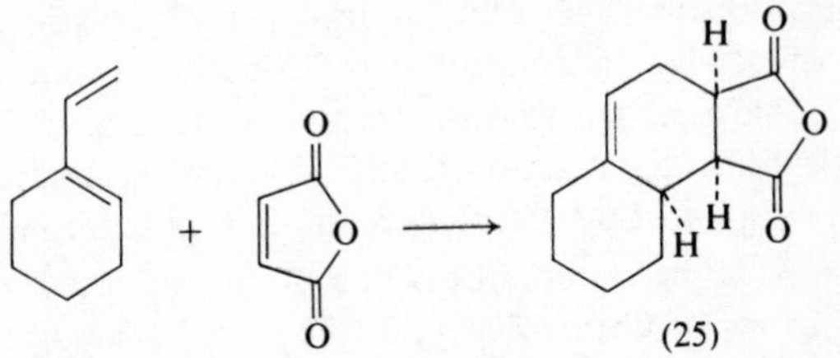

(3.48)

(25)

dihydronaphthalene similarly forms a chrysene derivative with benzoquinone. Reactions of this type using cyclopentenones and cyclohexenones as dienophiles have been extensively used to build up tetracyclic structures related to steroid systems (cf. Onischenko, 1964).

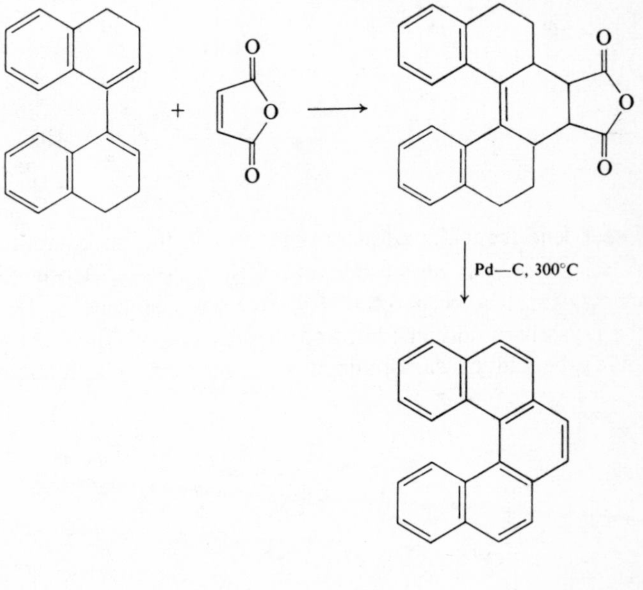

Pd—C, 300°C

(3.49)

1,1′-Dicyclohexenyls and related compounds also react readily with dienophiles provided the molecule is not prevented by steric factors from assuming a planar *cisoid* conformation. Since the dienes are readily available from the cyclic ketones via a pinacol condensation, the sequence constitutes another very convenient route to polycyclic compounds, as illustrated in (3.49).

Many vinyl aromatic compounds react normally with dienophiles, even though the initial reaction results in partial loss of aromatic conjugation.

Aromatic hydrocarbons. A number of types of polycyclic aromatic hydrocarbons react with dienophiles by 1,4-addition, but the reaction is particularly characteristic of anthracene and the higher linear acenes (Badger, 1954). Since the reaction results in loss of aromatic resonance, it is not surprising to find that benzene does not normally undergo thermal Diels–Alder reaction, although a bis adduct has been obtained under the influence of ultraviolet light (see p. 221). Interestingly, however, reaction of benzene with hexafluoro-2-butyne affords a small yield of *o*-bis(trifluoromethyl)benzene (3.50). This substance is thought to arise by initial 1,4-addition of the dienophile to benzene, followed by

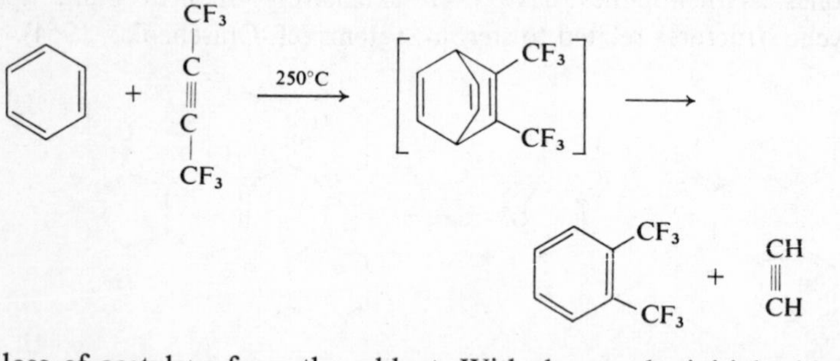

CF_3–C≡C–CF_3 250°C CF_3 CF_3

CF_3 CF_3 + CH≡CH (3.50)

loss of acetylene from the adduct. With durene the initial adduct can be isolated (Krespan, McKusick and Cairns, 1961). More recently dicyanoacetylene has been found to react with benzene at 180°C to form the crystalline adduct (26), and in presence of $AlCl_3$ as catalyst (see p. 198) the adduct was obtained in 63 per cent yield at room tem-

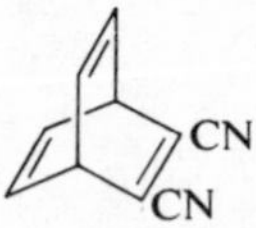

CN CN

(26)

perature (Ciganek, 1967). *p*-Xylene gave a mixture of two isomeric adducts and with hexamethylbenzene the activating effect of methyl groups is shown by the fact that the adduct was isolated in 83 per cent yield from an uncatalysed reaction at 130°C.

Naphthalene reacts very slowly with maleic anhydride but the polymethylnaphthalenes are more reactive, and in presence of an excess of maleic anhydride in boiling benzene almost quantitative yields of 1,4-adducts have been obtained.

Anthracene reacts with an equimolecular amount of maleic anhydride in boiling xylene to form the 9,10-addition product (27) in quantitative yield. Many anthracene derivatives and anthracene benzologues react

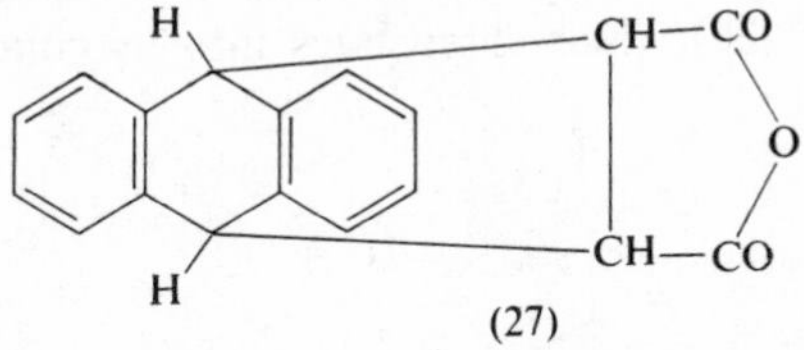

(27)

similarly, although the ease of the reaction varies with the structure of the hydrocarbon (cf. Badger, 1954). Naphthacene, pentacene and the higher acenes react with dienophiles even more readily than anthracene. These addition reactions are reversible, and in many cases the hydrocarbon can be recovered by sublimation or distillation of the adduct.

Cyclic dienes. Cyclopentadiene, in which the double bonds are constrained in a planar *cisoid* conformation, reacts easily with a variety of dienophiles. 1,3-Cyclohexadiene is also reactive, but with increase in the size of the ring the reactivity rapidly decreases because the double bonds can no longer adopt the necessary coplanar configuration because of non-bonded interaction of methylene groups in the planar molecule. *cis,cis*- and *cis,trans*-1,3-Cyclo-octadienes form only copolymers when treated with maleic anhydride and *cis,cis*- and *cis,trans*-1,3-cyclodecadienes similarly do not form adducts with maleic anhydride. Dienes with fourteen- and fifteen-membered rings again react with dienophiles but only under relatively severe conditions.

Cyclopentadiene is a very reactive diene and reacts easily with acetylenic and olefinic dienophiles to form bridged compounds of the bicyclo[2,2,1]-heptane series such as (28) and (29) formed by reaction of maleic anhydride and tetrolic acid.

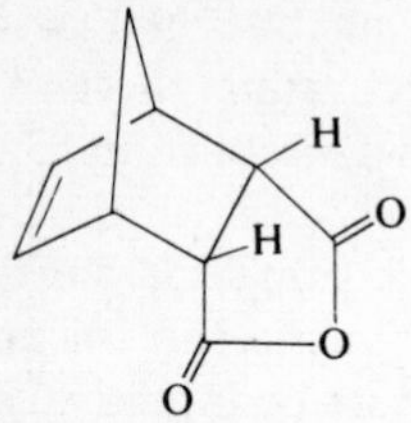

(28)

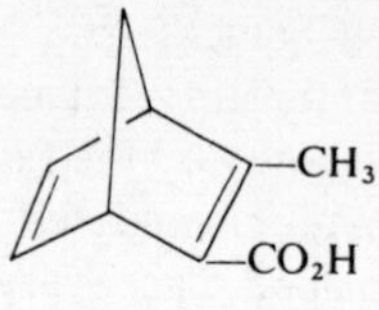

(29)

Less reactive dienophiles such as allyl alcohol, vinyl chloride and vinyl acetate react with cyclopentadiene at elevated temperatures, and even ethylene and propylene form adducts under forcing conditions.

Cyclopentadiene itself can act both as a diene and a dienophile, and readily forms a dimer (30) which dissociates into its components on

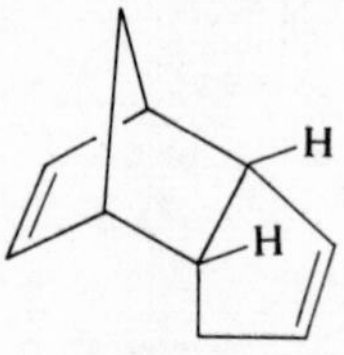

(30)

moderate heating. At higher temperatures the dimer itself can act as a dienophile and trimers, tetramers, etc. are formed which do not readily dissociate.

The reaction of cyclopentadiene with mono- and *cis*-di-substituted ethylenes could apparently give rise to two stereochemically distinct products, the *endo*- and the *exo*-bicyclo[2,2,1]heptene derivatives (3.51).

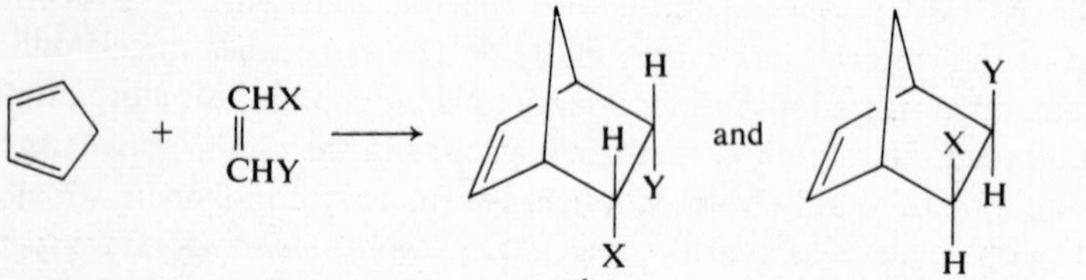

$$\text{cyclopentadiene} + \text{CHX}{=}\text{CHY} \longrightarrow \textit{endo} \text{ and } \textit{exo} \quad (3.51)$$

endo *exo*

It is found in practice, however, that the *endo* isomer always predominates, except under conditions where isomerisation of the original adduct occurs (see p. 204).

Compounds of the bicyclo[2,2,1]heptane type are widely distributed in nature among the bicyclic terpenes, and the Diels–Alder reaction provides a convenient method for their synthesis. Thus, cyclopentadiene and vinylacetate react smoothly when warmed together to form the acetate (31) which is easily transformed into norcamphor (32) and related compounds (3.52).

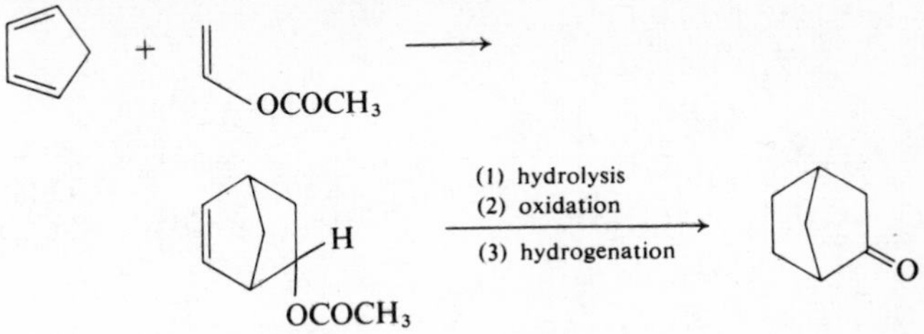

+ $OCOCH_3$ ⟶

(1) hydrolysis
(2) oxidation
(3) hydrogenation

H, $OCOCH_3$ (31) ⟶ O (32) (3.52)

1,3-Cyclohexadienes react with ethylenic dienophiles to form derivatives of bicyclo[2,2,2]octene. In general, the additions proceed more slowly than the corresponding reactions with cyclopentadiene.

With acetylenic dienophiles derivatives of bicyclo[2,2,2]octadiene are formed initially, but these often undergo a retro Diels–Alder reaction (see p. 195) with elimination of an ethylene and formation of a benzene derivative, as illustrated in (3.53).

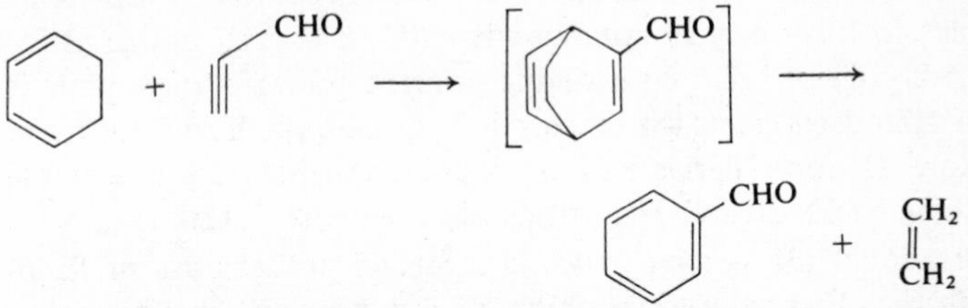

+ CHO ⟶ [CHO] ⟶

CHO + CH_2 ‖ CH_2 (3.53)

Cyclopentadienones and ortho-*quinones*. Derivatives of cyclopentadienone and of *o*-quinones also form adducts with ethylenic and acetylenic dienophiles. Many cyclopentadienones can act both as dienes and dienophiles and this is particularly true of cyclopentadienone itself which dimerises spontaneously. The adducts obtained from cyclopentadienone lose carbon monoxide easily on heating, with formation of dihydrobenzene or benzene derivatives. Thus, tetraphenylcyclopentadienone and diphenylacetylene form hexaphenylbenzene when heated together at 250°C (3.54).

o-Quinones can also react both as dienes and dienophiles. *o*-Benzoquinone spontaneously forms the stable crystalline dimer (33) in acetone solution at room temperature; 4,5-dimethylbenzoquinone behaves

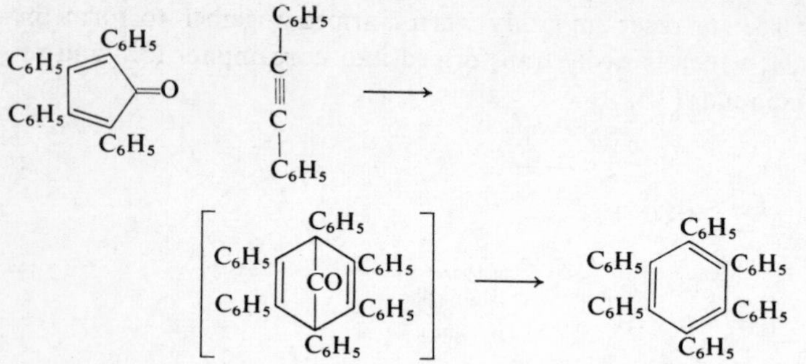

$$(3.54)$$

similarly. But it appears that in general their capacity for acting as dienes predominates (Ansell *et al.*, 1971).

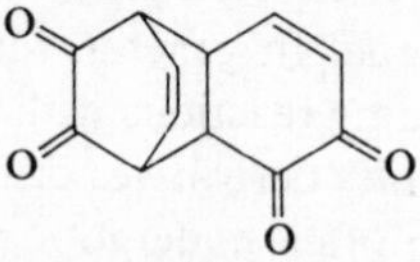

(33)

Furans. Many furan derivatives react with ethylenic and acetylenic dienophiles to form bicyclo compounds with an oxygen bridge (3.55) (Alder, 1948) (cf. p. 167). With simple furans a powerful dienophile is needed. Furan itself reacts with maleic anhydride or maleimide at room temperature to form derivatives of oxabicycloheptene, for example (34), in high yield. Acetylenedicarboxylic ester also reacts readily to form the adduct (35) which may react further with excess of furan (3.56). On the other hand, no reaction takes place with dimethylmaleic anhydride or with tetrolic acid.

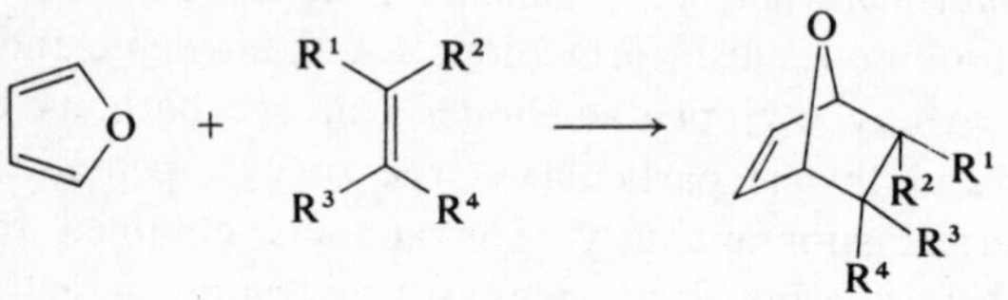

$$(3.55)$$

(34)

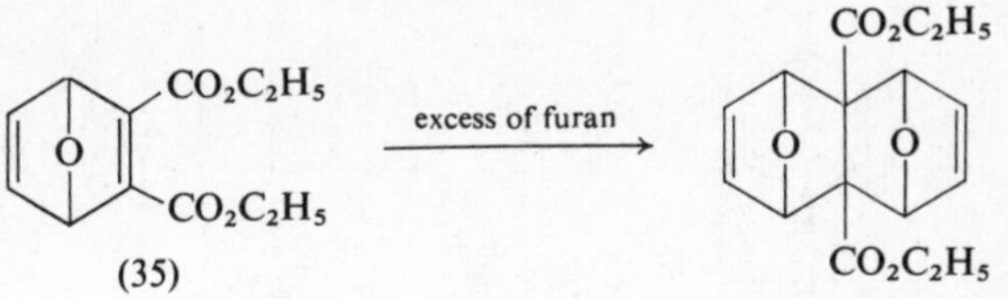

(35) $\xrightarrow{\text{excess of furan}}$ (3.56)

Most of the adducts obtained from furan derivatives are thermally labile and readily dissociate into their components on warming. For this reason, of course, heat cannot be used to speed up their rate of formation. But this limitation is not serious for addition generally proceeds rapidly at room temperature and by using a solvent from which the adduct crystallises out, high yields can be obtained.

The adduct from furan and maleic anhydride has been shown to have the *exo* structure (36) apparently violating the rule (p. 204) that the

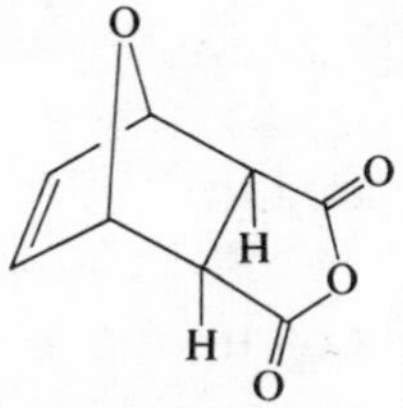

(36)

endo isomer predominates. The reason for this is found in the related observation that the normal *endo* adduct formed from maleimide and furan at 20°C dissociates at temperatures only slightly above room temperature and more rapidly on warming, allowing conversion of the *endo* adduct formed in the kinetically controlled reaction into the thermodynamically more stable *exo* isomer. With the maleic anhydride adduct, equilibration takes place below room temperature so that the *endo* adduct formed under kinetic control is not observed.

Pyrrole and its derivatives are unsuitable as dienes in the Diels–Alder reaction because the susceptibility of the nucleus to electrophilic substitution leads to side reactions. Normal adducts at the 2- and 5-positions have been obtained from some *N*-substituted pyrroles and acetylenic dienophiles under vigorous conditions. However (Barlow, Haszeldine and Hubbard, 1969), the very reactive dienophile (37) reacts with pyrrole itself in ether to form the 1:1 adduct (38) in 58 per cent yield (3.57).

It has long been thought that thiophen and its simple derivatives, unlike furans, do not undergo (4 + 2) cyclo-addition reactions, but it

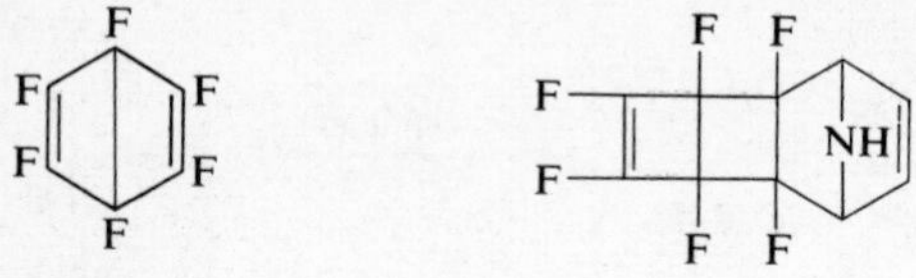

(37) (38) (3.57)

has now been found that they react with acetylenic dienophiles under vigorous conditions to form benzene derivatives by extrusion of sulphur from the initially formed Diels–Alder adducts (Kuhn and Gollnick, 1972).

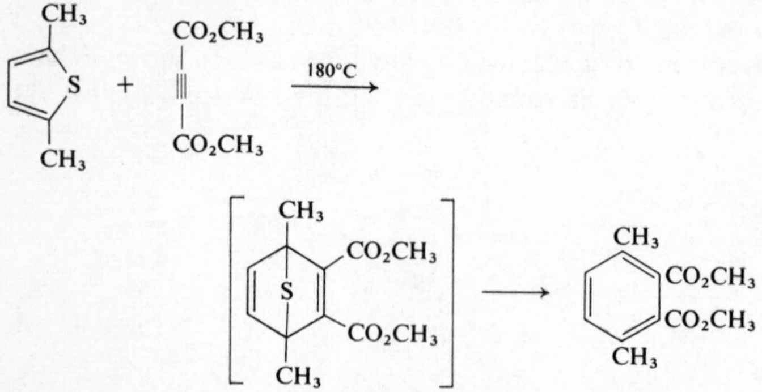

(3.58)

The 1,1-dioxide behaves both as a diene and dienophile; all attempts to prepare it have led to the sulphone (39) through loss of sulphur dioxide from the intermediate dimer.

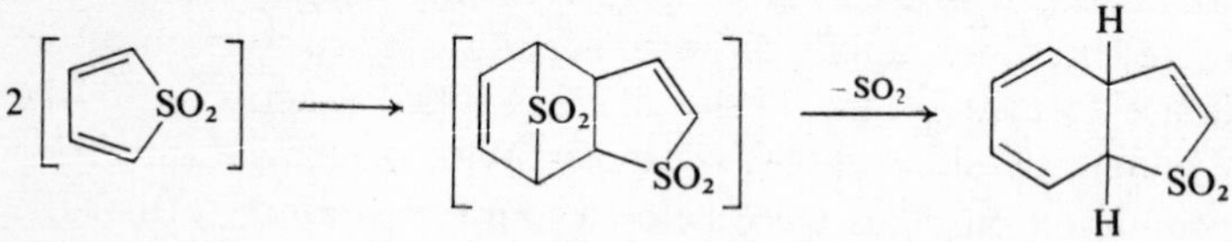

(39) (3.59)

3.4. Intramolecular Diels–Alder reactions

The usefulness of intermolecular Diels–Alder reactions in synthesis is evident from the examples which have already been given. In the past the intramolecular reaction, in which the diene and dienophile components form part of the same molecule, has been employed only rarely but it is now becoming apparent that this procedure offers a very convenient and highly stereoselective route to some complex ring systems. Because of the favourable entropy factor the intramolecular reactions often proceed in higher yield and under milder conditions than comparable intermolecular reactions. Thus the triene (40) affords the *cis*-tetrahydro-indanone (41) in 70 per cent yield, and patchouli alcohol, an important material in the perfumery industry, was synthesised neatly by a stereo-selective route the key step of which involved a base-catalysed intra-molecular Diels–Alder reaction (Näf and Ohloff, 1974).

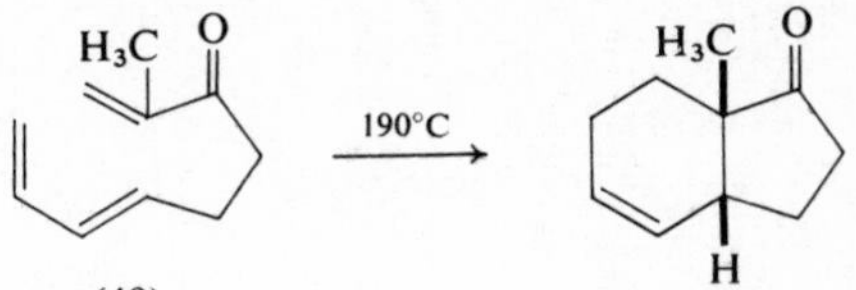

190°C

(40) (41)

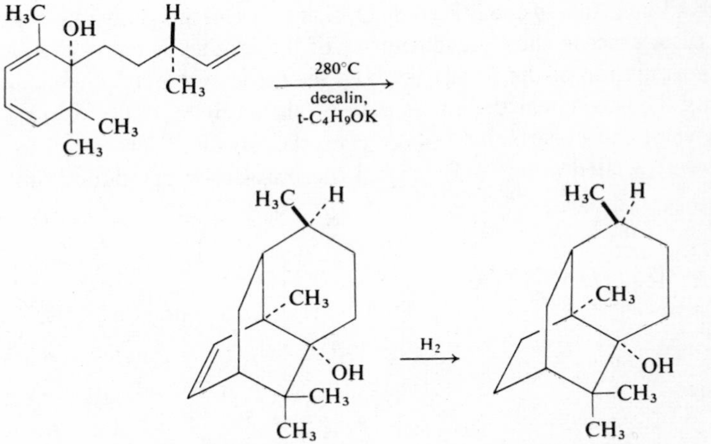

280°C
decalin,
t-C_4H_9OK

H_2

patchouli alcohol

(3.60)

With unsymmetrical dienes and dienophiles reaction may lead to a mixture of two structural isomers formed by addition of the dienophile to the diene in the two alternative ways. Generally, however, one mode of addition is highly favoured by geometric factors and reaction leads exclusively or predominantly to a single regioisomer. Some of the factors which influence the direction of intramolecular cyclo-additions have been discussed by Krantz and Lin (1973; see also House and Cronin, 1965).

A valuable feature of the intramolecular reactions, as of intermolecular Diels–Alder reactions, is their high stereoselectivity. Thus, the piperidine derivative (42) cyclises spontaneously to the tetracyclic product (43) in which four new chiral centres are formed stereo-specifically in one step (Geschwend, 1973). Generally *endo* products are obtained, as in the

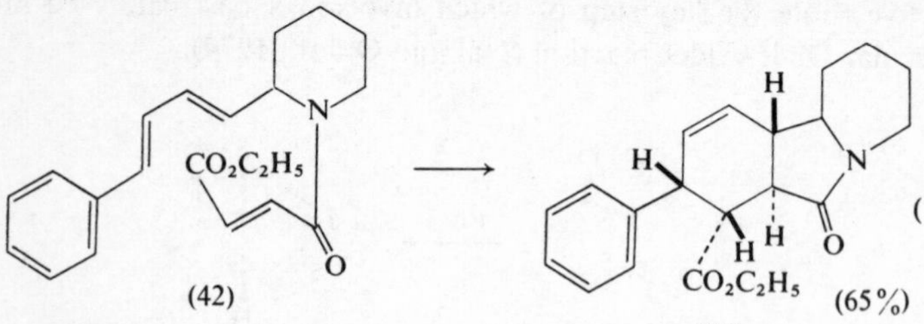

(42) → (43) (65%) (3.61)

above example, but in certain cases non-bonding interaction of substituent groups in the transition state may favour the formation of the *exo* adduct. In a systematic study Oppolzer (1974) has shown how, in some cases at least, the stereochemistry of the product is controlled by the conformation of the bridge between the diene and dienophile components. Thus, whereas the amide (45), produced by thermolysis of the benzocyclobutene derivative (44), gave the *cis*-fused product (46), the closely related amine (47) formed the *trans*-fused compound (48).

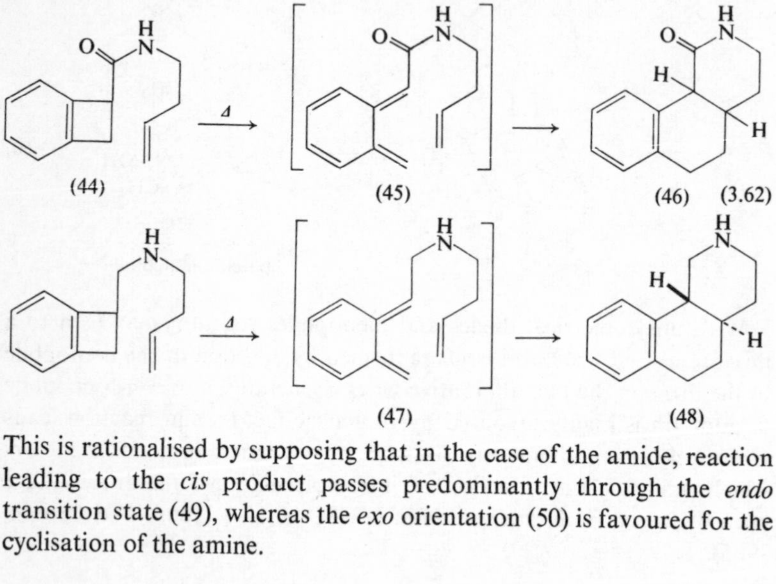

(44) → (45) → (46) (3.62)

(47) (48)

This is rationalised by supposing that in the case of the amide, reaction leading to the *cis* product passes predominantly through the *endo* transition state (49), whereas the *exo* orientation (50) is favoured for the cyclisation of the amine.

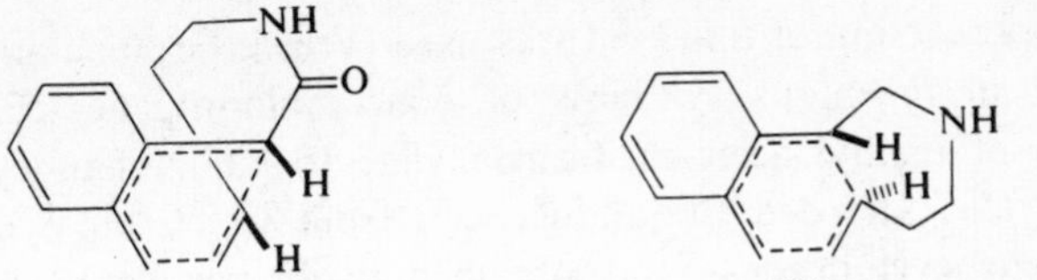

(49) (50) (3.63)

3.5. The retro Diels–Alder reaction

Diels–Alder reactions are reversible, and on heating many adducts dissociate into their components, sometimes under quite mild conditions. This can be made use of as in, for example, the separation of anthracene derivatives from mixtures with other hydrocarbons through their adducts with maleic anhydride, and in the preparation of pure D vitamins from the mixtures obtained by irradiation of the provitamins (cf. Alder, 1948).

More interesting are reactions in which the original adduct is modified chemically and subsequently dissociated to yield a new diene or dienophile (Kwart and King, 1968). Thus, catalytic hydrogenation of the adduct obtained from 4-vinylcyclohexene and anthracene followed by thermal dissociation affords vinylcyclohexane, whereas direct hydrogenation of vinylcyclohexene itself results in reduction of both double bonds. In this case adduct formation has been used to protect one of the double bonds of the vinylcyclohexene.

A similar technique is used to obtain 22,33-dihydroergosterol from ergosterol by protection of the cyclic 1,3-diene system through the adduct with maleic anhydride, followed by hydrogenation of the side-chain double bond and regeneration of the diene system.

It is not always the bonds formed in the original diene addition which are broken in the retro reaction. Examples have already been noted for the adducts formed from cyclohexadiene and acetylenic dienophiles

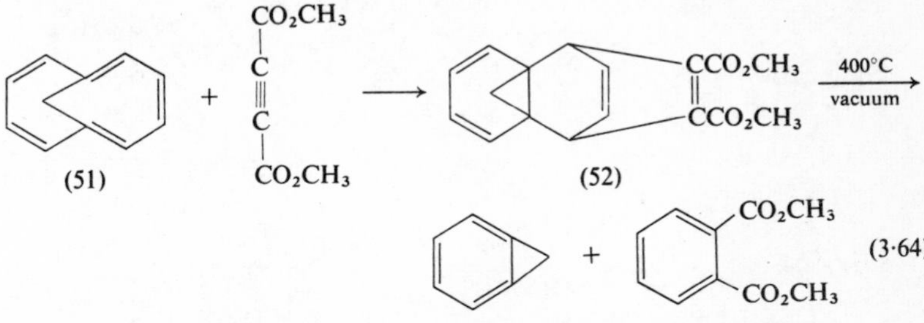

(51) (52) (53) (3·64)

(p. 189). A reaction of this kind was used (Vogel, Grimme and Korte, 1963) in an ingenious synthesis of benzocyclopropene (53) (3.64). Addition of methyl acetylenedicarboxylate to 1,6-methanocyclodecapentaene (51) afforded the adduct (52) from which on pyrolysis at 400°C benzocyclopropene was obtained in 45 per cent yield, with elimination of dimethyl phthalate. In another example furan 3,4-dicarboxylic acid was readily prepared by decomposition of the hydrogenated adduct from furan and acetylenedicarboxylic ester (3.65).

CO_2CH_3 / CO_2CH_3 $\xrightarrow{H_2, Pd}$ CO_2CH_3 / CO_2CH_3 $\xrightarrow{heat}$ $CH_2{=}CH_2$ + furan-3,4-(CO_2CH_3)$_2$ (3.65)

The use of cyclo-octatetraene as a source of four carbon atoms in sequences terminating in a reverse Diels–Alder reaction is exemplified in the preparation of *trans,trans*-1,4-diacetoxybutadiene, a valuable diene in Diels–Alder syntheses (cf. p. 208) (Carlson and Hill, 1970).

In the heterocyclic series, retro Diels–Alder cleavage of 5,6-dihydro-1,2-oxazines forms the key step in procedures developed by Eschenmoser

$\xrightarrow{Hg(OCOCH_3)_2}$ $OCOCH_3$ / $OCOCH_3$ $\xrightarrow[\text{warm } C_6H_6]{CCO_2CH_3 \equiv CCO_2CH_3}$ [$OCOCH_3$, CH_3COO, CO_2CH_3, CO_2CH_3] $\longrightarrow$ (3.66)

CH_3COO–CH=CH–CH=CH–$OCOCH_3$ (49%) + $C_6H_4(CO_2CH_3)_2$

for oxidative cleavage of carbon–carbon double bonds and for the conversion of acetylenes into $\alpha\beta$-unsaturated carbonyl compounds, exemplified in (3.67). The oxazines are themselves obtained by reaction of α-chloronitrones with the olefin or acetylene in presence of silver ion, and are smoothly converted into the fragmentation products on treatment with base (Gygax, Das Gupta and Eschenmoser, 1972; Schatzmiller and Eschenmoser, 1973).

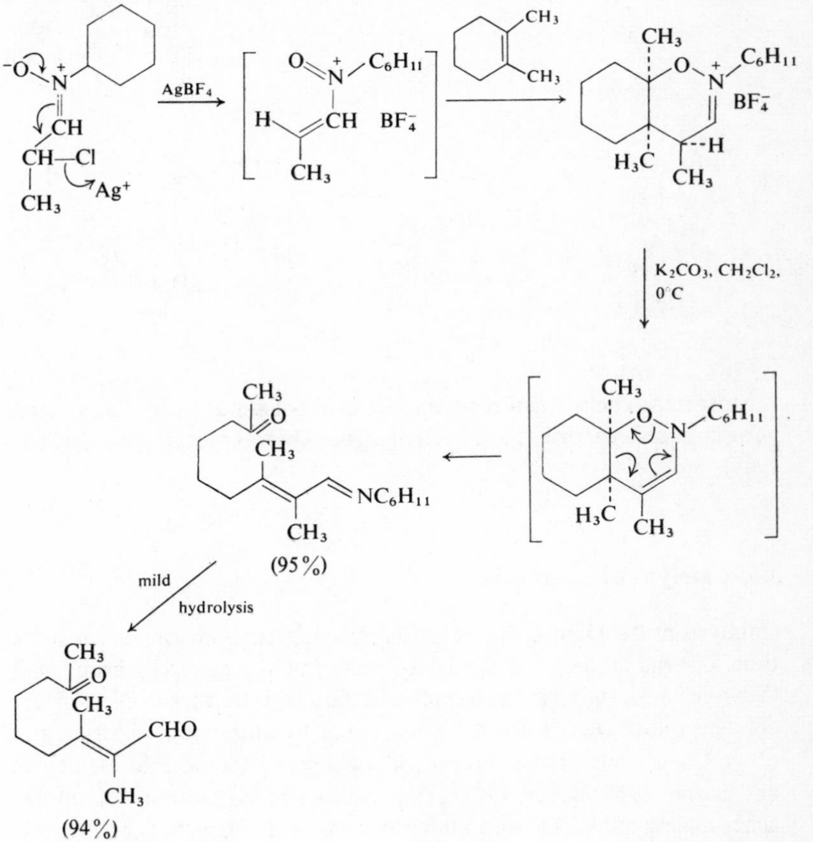

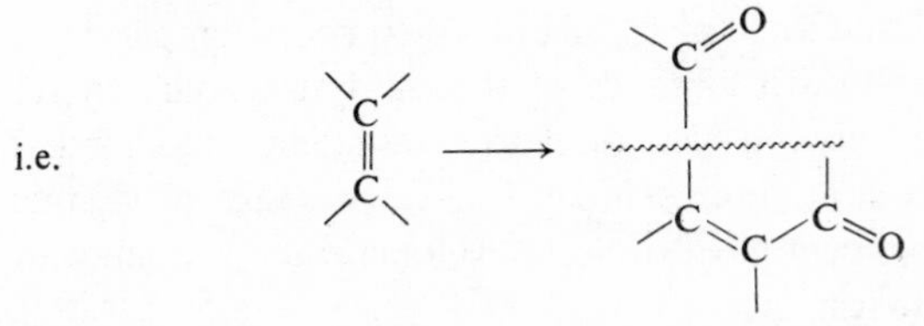

$AgBF_4$

K_2CO_3, CH_2Cl_2, 0°C

(95%)

mild hydrolysis

(94%)

i.e.

(3.67)

(continued on next page)

C_5H_{11}–C≡C–H + $\bar{O}$–$\overset{+}{N}$(C_6H_{11})=CH–CHCl–CH_3 $\xrightarrow[SO_2]{AgBF_4}$ C_5H_{11}, O, $\overset{+}{N}$, C_6H_{11}, H, H, H, CH_3

$\downarrow$ OH^- (3.67 continued)

C_5H_{11}, O, H, H, CH_3 (70%) + HNC_6H_{11}–CHO $\leftarrow$ [C_5H_{11}, O, N, C_6H_{11}, OH, H, H, H, CH_3]

Most retro Diels–Alder reactions are brought about by heat, but some photoinduced reactions have been observed (Nozake, Kato and Noyori, 1969).

3.6. Catalysis by Lewis acids

Catalysis of the Diels–Alder reaction by acids has been known for some time, but the influence of catalysts on the rate has generally been small (Wassermann, 1965). It has been found, however, that some Diels–Alder condensations are accelerated remarkably by aluminium chloride and other Lewis acids such as boron trifluoride and stannic chloride (Yates and Eaton, 1960; Sauer, 1967). Thus equimolecular amounts of anthracene, maleic anhydride and aluminium chloride in methylene chloride solution gave a quantitative yield of adduct in 90 s at room temperature. It is estimated that reaction in absence of the catalyst would require 4800 h for 95 per cent completion. Similarly, butadiene and methyl vinyl ketone react in 1 h at room temperature in presence of stannic chloride to give a 75 per cent yield of acetylcyclohexene. In absence of a catalyst no reaction takes place.

Pure *cis* addition is observed in reactions carried out in presence of Lewis acids just as in the uncatalysed reaction (see p. 203), that is the relative orientation of substituents in the diene and dienophile is preserved in the adduct. But the ratio of structurally isomeric or stereoisomeric 1:1 adducts may be different in the catalysed reactions and generally greatly increased selectivity is observed. Thus on addition of methyl vinyl ketone to isoprene two structural isomers are formed, of which the predominant one is that with the greatest separation between the two groups (3.68). The proportion of this product is even greater in the catalysed reaction (Lutz and Bailey, 1964). Many other examples of this effect have been recorded (see e.g. Inukai and Kojima, 1970, 1971; Kreiser, Haumesser and Thomas, 1974).

COCH₃ COCH₃ COCH₃ (3.68)

no catalyst, toluene, 120°C; ratio 71:29

$SnCl_4.5H_2O$, benzene, <25°C; ratio 93:7

Similarly, in the addition of acrylic acid to cyclopentadiene the proportion of *endo* adduct was found to increase noticeably in presence of aluminium chloride etherate (Sauer and Kredel, 1966*a*) (3.69). Other

CH_2=$CHCO_2H$ → H, CO_2H + CO_2H, H (3.69)

		endo	*exo*
0°C, no catalyst;	ratio	84	16
0°C, + 47% $AlCl_3.O(C_2H_5)_2$;	ratio	93	7
−70°C, + 47% $AlCl_3.O(C_2H_5)_2$;	ratio	97	3

Lewis acid catalysts had the same effect (see also Inukai and Kojima, 1966). These effects are ascribed to complex formation between the Lewis acid and the polar groups of the dienophile which brings about changes in the energies and orbital coefficients of the frontier orbitals of the dienophile.

The influence of Lewis acids on the course of Diels–Alder reactions has been rationalised by application of frontier orbital theory (Houk and Strozier, 1973; Anh and Seyden-Penne, 1973). In a normal Diels–Alder reaction, that is one involving an electron-deficient dienophile and an electron-rich diene, the main interaction is that between the highest occupied molecular orbital (HOMO) of the diene and the lowest unoccupied orbital (LUMO) of the dienophile, and the smaller the energy difference between these orbitals and the better the overlap the more readily the reaction occurs (Houk, 1975; Sustmann, 1974). Co-ordination of a Lewis acid with, say, the non-bonding electrons of a carbonyl or cyano group in the dienophile lowers the energies of the frontier orbitals of the dienophile and alters the distribution of the atomic orbital coefficients. The energy difference between the HOMO of the diene and the LUMO of the dienophile is thus reduced and, while the energy difference between the HOMO of the dienophile and the LUMO of the diene is increased at the same time, the former effect predominates resulting in stabilisation of the reaction complex and faster reaction. This is illustrated diagramatically in the figure for the reaction between butadiene and methyl acrylate.

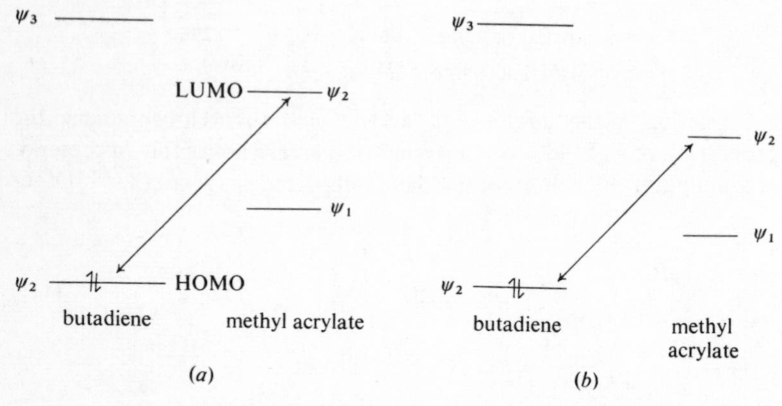

Effect of Lewis acid catalysts on the interaction of the frontier orbitals of butadiene and methyl acrylate (*a*) non-catalysed reaction, (*b*) catalysed reaction.

In agreement with these general ideas, lowering of the HOMO of the diene by conjugation with an electron-attracting substituent results in a slower reaction. In Diels–Alder reactions with inverse electron demand, interaction of the LUMO of the *diene* and the HOMO of the *dienophile* is the controlling factor, and reaction is facilitated by electron release in the dienophile which raises the energy of the frontier orbitals.

The increased selectivity shown in catalysed reactions was formerly ascribed to steric effects in the complexed dienophile, but it now appears that this also can be accounted for by consideration of frontier orbital interactions. The orientation of the compounds in the transition state of a Diels–Alder reaction is governed by the orbital coefficients of the reacting atoms, the atom with the larger coefficient in the dienophile interacts preferentially with that with the larger coefficient in the diene, since this leads to more efficient overlap of orbitals. For the reaction of 2-phenylbutadiene with methyl acrylate, for example, the interaction (*a*) is favoured over (*b*), and the ratio of *para* to *meta* products found is

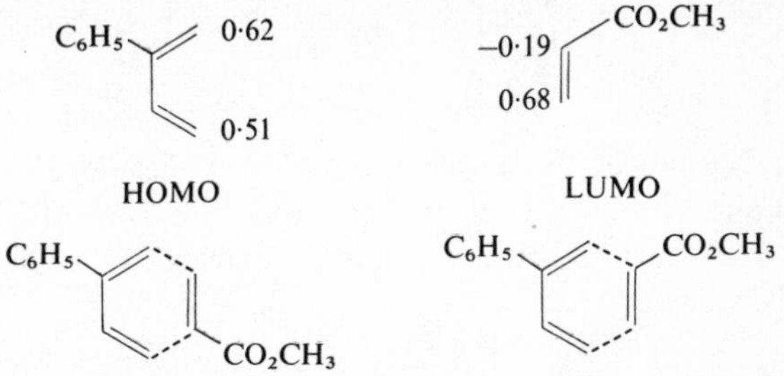

C_6H_5 0·62 0·51 HOMO

CO_2CH_3 −0·19 0·68 LUMO

(3.70)

C_6H_5 CO_2CH_3 (*a*)

C_6H_5 CO_2CH_3 (*b*)

80:20. In the catalysed reaction, interaction of the catalyst with the ester group of the methyl acrylate increases the difference between the coefficients at C-2 and C-3 with the result that the reaction becomes more selective; transition state (*a*) is favoured even more and the *para:meta* ratio rises to 97:3.

Secondary orbital interactions may also play a part in determining the orientation of the adducts obtained in some catalysed reactions (Alston and Ottenbrite, 1975). A striking example is seen in the reaction between 2-methyl-1-phenylbutadiene and β-nitrostyrene. Here the HOMO coefficients at the terminal carbon atoms of the diene have significantly different magnitudes and favour formation of adduct (55), but secondary orbital interactions between the LUMO of the nitrogen atom of the nitro group and the HOMO at C-2 and C-3 of the diene favour isomer (54). Since secondary orbital interactions are weak in the uncatalysed reaction, the terminal interactions dominate and (55) is obtained experimentally. In the catalysed reaction, however, secondary orbital interactions are increased to such an extent that only isomer (54) is obtained.

The increase in *endo:exo* ratio in catalysed reactions is also ascribed to increased secondary orbital interactions. In a typical non-catalysed reaction, e.g. that between cyclopentadiene and acrolein, the preferred formation of the *endo* product is due to secondary interactions involving

C_6H_5 H_3C + NO_2 C_6H_5 ⟶

C_6H_5 H_3C NO_2 C_6H_5 (54) + C_6H_5 H_3C C_6H_5 NO_2 (55) (3.71)

the carbonyl group of the acrolein and C-2 of the diene (see figure) (Hoffmann and Woodward, 1965). This interaction is greatly increased in the catalysed reaction because of a large increase in the coefficient of the carbonyl carbon atom.

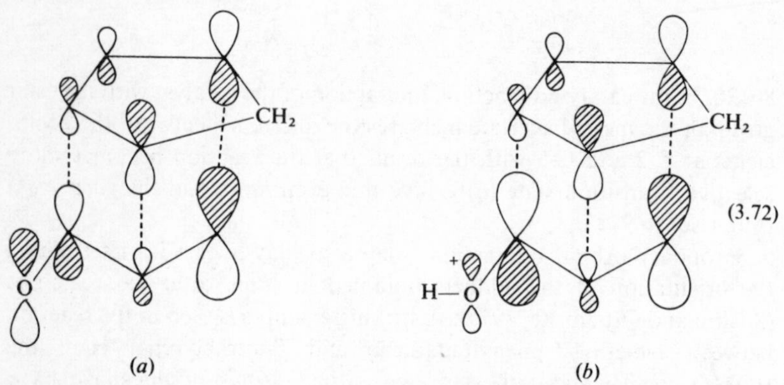

Diene HOMO–dienophile LUMO interactions in *endo* transition states with (*a*) acrolein and (*b*) protonated acrolein.

3.7. Stereochemistry of the Diels–Alder reaction

The great synthetic usefulness of the Diels–Alder reaction depends not only on the fact that it provides easy access to a variety of six-membered ring compounds, but also on its remarkable stereoselectivity. This factor more than any other has contributed to its successful application in the synthesis of a number of complex natural products (cf. Alder and Schumacher, 1953; Martin and Hill, 1961). It should be noted, however, that the high stereoselectivity applies only to the kinetically controlled reaction and may be lost by epimerisation of the product or starting materials, or by easy dissociation of the adduct allowing thermodynamic control of the reaction. These factors are fully discussed by Martin and Hill (1961).

The cis *principle.* The stereochemistry of the adduct obtained in many Diels–Alder reactions can be selected on the basis of two empirical rules formulated by Alder and Stein in 1937. According to the '*cis* principle', which is very widely followed, the relative stereochemistry of substituents in both the dienophile and the diene is retained in the adduct. That is, a dienophile with *trans* substituents will give an adduct in which the *trans* configuration of the substituents is retained, while a *cis* disubstituted dienophile will form an adduct in which the substituents are *cis* to each other. For example, in the reaction of cyclopentadiene with dimethyl maleate the *cis* adducts (56) and (57) are formed while in the reaction with dimethyl fumarate the *trans* configuration of the ester groups is retained in the adduct (58) (3.73).

$CHCO_2CH_3$ ‖ $CHCO_2CH_3$ → (56) + (57)

$HCCO_2CH_3$ ‖ H_3CO_2CCH → (58)

(3.73)

Similarly with the diene component, the relative configuration of the substituents in the 1- and 4- positions is retained in the adduct; a *trans,trans*-1,4-disubstituted diene gives rise to adducts in which the 1-

and 4- substituents are *cis* to each other, and a *cis,trans*-disubstituted diene gives adducts with *trans* substituents (3.74).

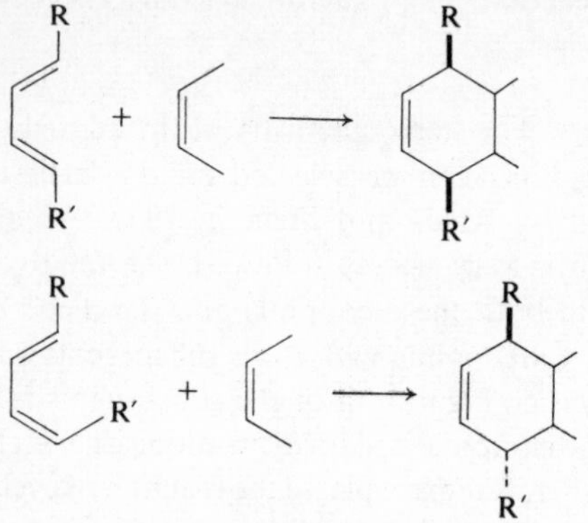

(3.74)

The almost universal application of the *cis* principle provides strong evidence for a mechanism for the Diels–Alder reaction in which both the new bonds between the diene and the dienophile are formed at the same time. But a two-step mechanism is not completely excluded, for the same stereochemical result would obtain if the rate of formation of the second bond in the (diradical or zwitterionic) intermediate (59), (3.75), were faster than the rate of rotation about a carbon–carbon bond (compare Bartlett, 1970, 1971).

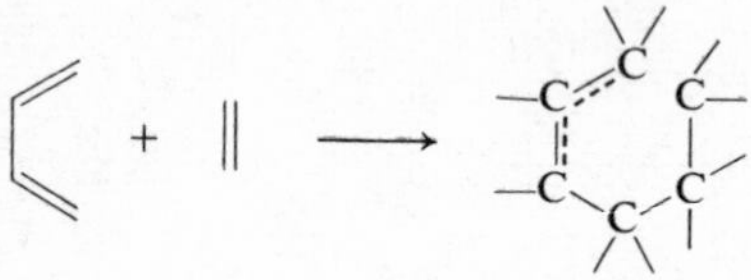

(3.75)

(59)

The endo *addition rule*. In the addition of maleic anhydride to cyclopentadiene, two different products, the *endo* and the *exo*, might conceivably be formed depending on the manner in which the diene and the dienophile are disposed in the transition state. According to Alder's *endo* addition rule, in a diene addition reaction the two components arrange themselves in parallel planes, and the most stable transition state arises from the orientation in which there is 'maximum accumulation of double bonds'. Not only the double bonds which actually take part in the addition are taken into account, but also the π-bonds of the activating groups in the dienophile. The rule appears to be strictly applicable only to the addition of cyclic dienophiles to cyclic dienes, but it is a useful guide in many other additions as well.

Thus, in the addition of maleic anhydride to cyclopentadiene the *endo* product, formed from the orientation with maximum accumulation of double bonds, is produced almost exclusively (3.76). The thermodynamically more stable *exo* compound is formed in yields of less than

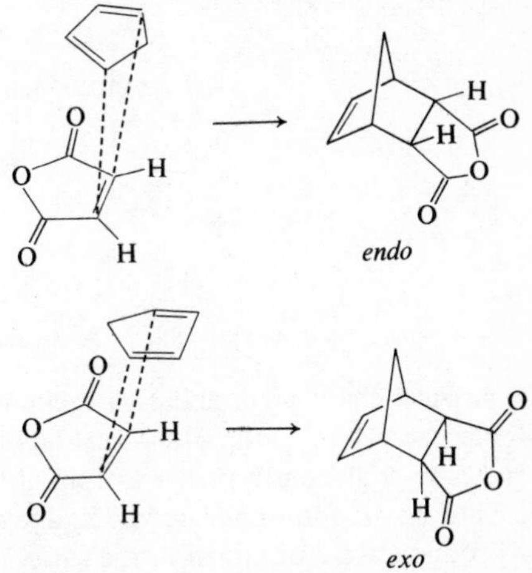

endo

exo

(3.76)

1.5 per cent. From benzoquinone and cyclopentadiene again only the *endo* adduct (60) was isolated, the configuration of the product being shown by its conversion into the 'caged' compound (61) with ultraviolet light (3.77).

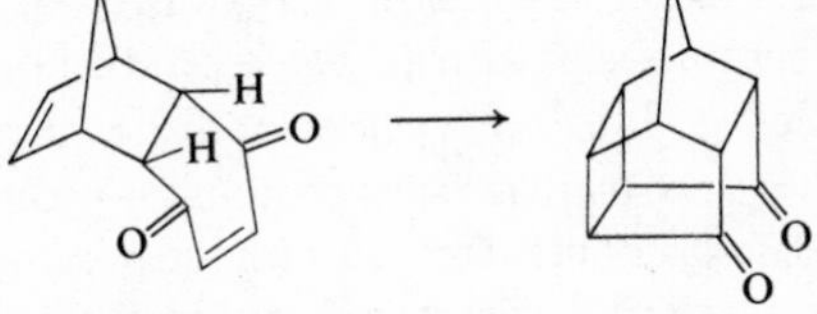

(60) (61)

(3.77)

The products obtained from the cyclic diene furan and maleic anhydride and from diene addition reactions of fulvene do not obey the *endo* rule. The reason is that the initial *endo* adducts easily dissociate at moderate temperatures, allowing conversion of the kinetic *endo* adduct into the thermodynamically more stable *exo* isomer (cf. p. 191). In other cases prolonged reaction times may lead to the formation of some *exo* isomer at the expense of the *endo*.

TABLE 3.2 *Proportion of* endo *and* exo *acids formed in addition of α-substituted acrylic acids to cyclopentadiene*

$CH_2{=}C(X)CO_2H$ + cyclopentadiene ⟶ endo CO_2H + exo CO_2H

(62)

X	*endo* CO_2H	*exo* CO_2H
H	75	25
CH_3	35	65
C_2H_5	—	100
C_6H_5	60	40
Br	30	70

In the addition of open-chain dienophiles to cyclic dienes, the *endo* rule is not always obeyed and the composition of the mixture obtained may depend on the precise structure of the dienophile and on the reaction conditions. Thus, in the addition of acrylic acid to cyclopentadiene the *endo* and *exo* products were obtained in the ratio 75:25 but in the α-substituted acrylic acids (62) the product ratio varied depending on the nature of the group X (Martin and Hill, 1961, p. 550) (see Table 3.2). Equally variable ratios are observed in reactions with β-substituted acrylic acids. With acrylic acid itself, the proportion of *endo* adduct formed was noticeably increased by the presence of Lewis acid catalysts (p. 199) (see also Kobuke, Fueno and Furukawa, 1970).

Solvent and temperature may also affect the product ratio. Thus, in the kinetically controlled addition of cyclopentadiene to methyl acrylate, methyl methacrylate and methyl *trans*-crotonate in different solvents, the proportion of *endo* product increased with the polarity of the solvent, and the product ratio was also slightly affected by the temperature of the reaction. In all cases mixtures of the *endo* and *exo* products were obtained and with methyl methacrylate the *exo* isomer was the predominant product under all experimental conditions. With methyl crotonate the *exo* adduct was predominant in some solvents (e.g. trimethylamine at 30°C) and the *endo* in others (ethanol, acetic acid) (see Fig. 3.1) (Berson, Hamlet and Mueller, 1962).

The adducts obtained from acyclic dienes and cyclic dienophiles are frequently formed in accordance with the *endo* rule. Thus, in the addition

A = methyl acrylate
B = methyl *trans*-crotonate
C = methyl methacrylate

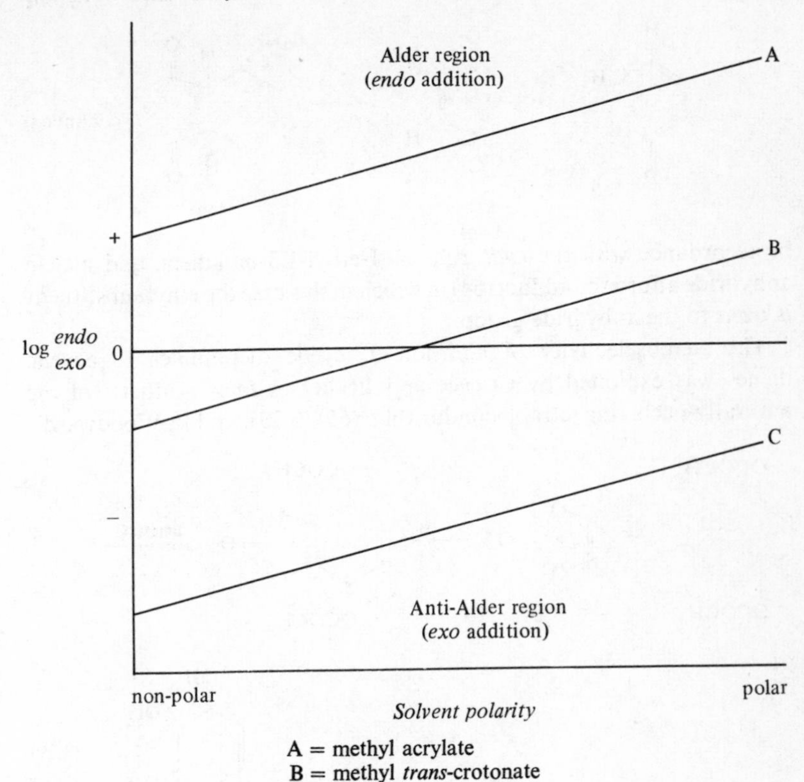

Fig. 3.1. *Endo:exo* product ratio as a function of the dienophile and solvent polarity in addition of methyl acrylate derivatives to cyclopentadiene (schematic). (From Berson, Hamlet and Mueller, 1962.)

of maleic anhydride to *trans,trans*-1,4-diphenylbutadiene the *cis* adduct (63) is formed almost exclusively (3.78) through the orientation of diene and dienophile with 'maximum accumulation of double bonds'. Again

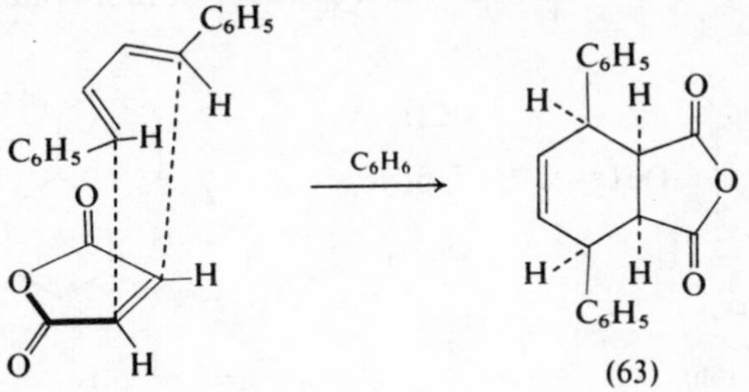

(63)

(3.78)

(continued on next page)

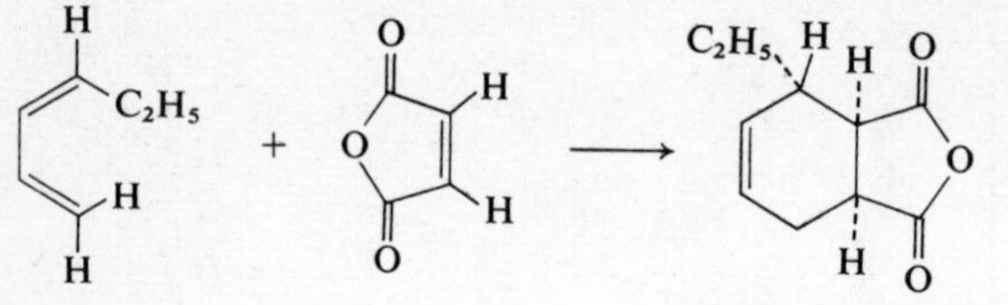

(3.78 continued)

(64)

in accordance with the *endo* rule *cis*-1-ethyl-1,3-butadiene and maleic anhydride afford the adduct (64) in which in this case the ethyl substituent is *trans* to the anhydride group.

The stereoselectivity of addition of cyclic dienophiles to acyclic dienes was exploited by Criegee and Becher in their synthesis of the naturally-occurring tetraol, conduritol D (65) (3.79), and in Woodward's

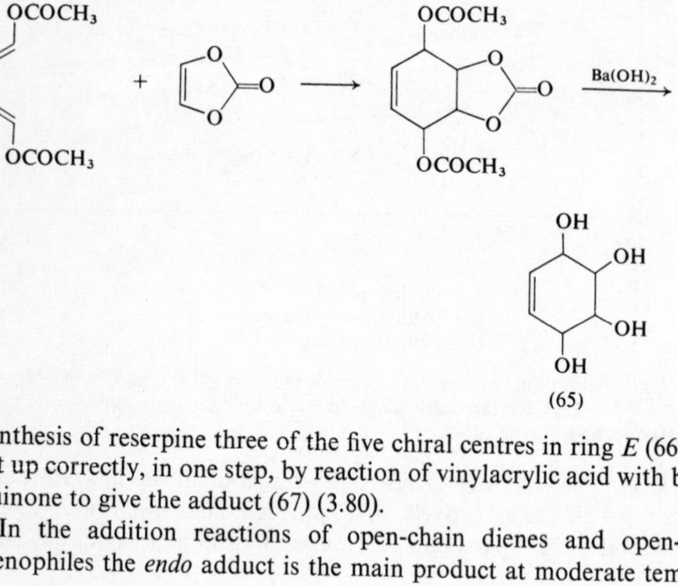

(3.79)

(65)

synthesis of reserpine three of the five chiral centres in ring *E* (66) were set up correctly, in one step, by reaction of vinylacrylic acid with benzoquinone to give the adduct (67) (3.80).

In the addition reactions of open-chain dienes and open-chain dienophiles the *endo* adduct is the main product at moderate tempera-

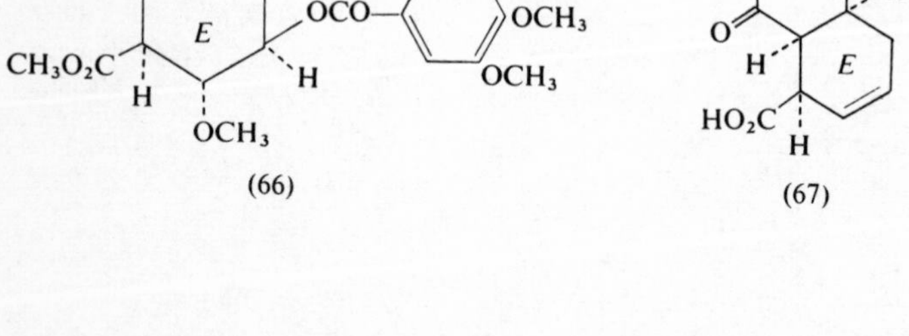

(66) (67) (3.80)

TABLE 3.3 *Effect of temperature on ratio of* cis- *and* trans-*adducts formed in reaction of* trans-*butadiene*-1-*carboxylic acid with acrylic acid* (from Sauer, 1967)

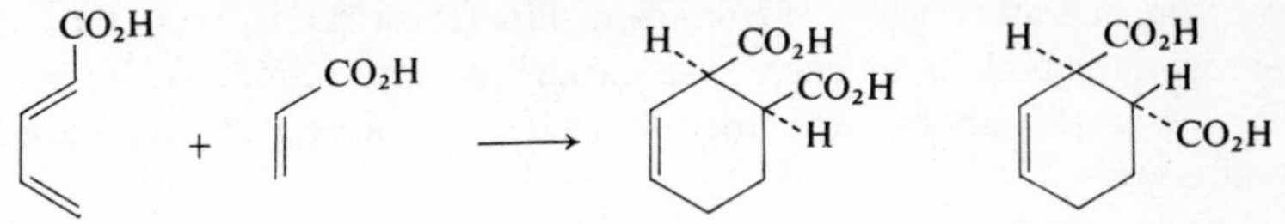

Temperature °C	75	90	100	110	130
Ratio *cis*:*trans*	*cis* only	7:1	4·5:1	2:1	1:1

tures, but in a number of cases it has been found that the proportion of the *exo* isomer increases with rise in temperature. Thus, in the addition of acrylonitrile to *trans*-1-phenylbutadiene at 100°C the *cis* (*endo*) isomer (68) was obtained, whereas at reflux temperature the main product was

(68) (69) (3.81)

the *trans* (*exo*) isomer (69) (3.81). Table 3.3 shows the effect of temperature on the ratio of *cis* and *trans* adducts obtained by reaction of *trans*-butadiene-1-carboxylic acid and acrylic acid.

The factors which determine the steric course of diene additions are still not completely clear. It appears that a number of different forces operate in the transition state and the precise steric composition of the product depends on the balance among these (see Martin and Hill, 1961; Berson, Hamlet and Mueller, 1962; Wassermann, 1965). The predominance of *endo* addition has been ascribed to the tendency of dienophile substituents to be so oriented in the favoured transition state of the reaction that they lie directly above the residual unsaturation of the diene, either for reasons of spatial orbital overlap or for reasons of steric accommodation. Non-bonded charge-transfer type interactions may also play a part in the stabilisation of the transition state in some cases. Alder and Stein originally summarised the steric nature of the transition state by their principle of 'maximum accumulation of double bonds'.

This has been rationalised by Woodward and Hoffmann (1969) as a stabilisation of the *endo* transition state by secondary orbital interaction. Thus, in the reaction of cyclopentadiene with maleic anhydride, secondary orbital interactions, represented by the dashed lines in (3.82), lower the energy of the *endo* transition state (shown) relative to that of the *exo* transition state, where these secondary interactions are absent; hence the *endo* adduct is the one obtained under kinetically controlled conditions.

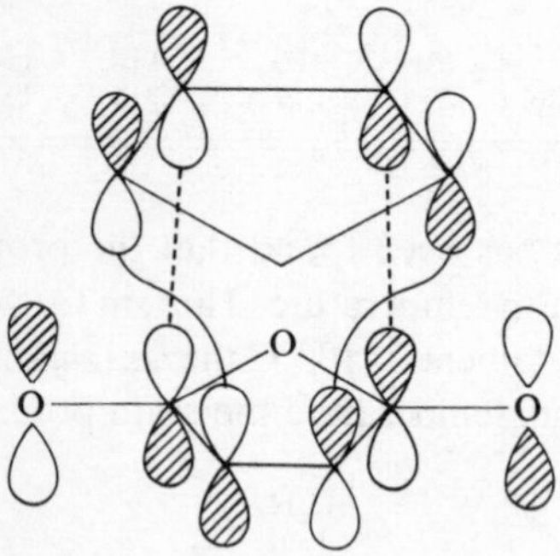

HOMO of cyclopentadiene

LUMO of maleic anhydride

(3.82)

However, this is clearly not the whole story (see pp. 204–209) and it appears that these attractive forces are easily overweighed by steric factors and, in some cases, by changes in the experimental conditions. Thus in the reaction of 1,2-dicyanocyclobutene with cyclopentadiene the only product formed is that with the cyano groups in the *exo* position, in defiance of the *endo* rule, whereas with furan the main product is the expected *endo* derivative. It is suggested that in the former case secondary orbital interactions in the *endo* transition state are outweighed by non-

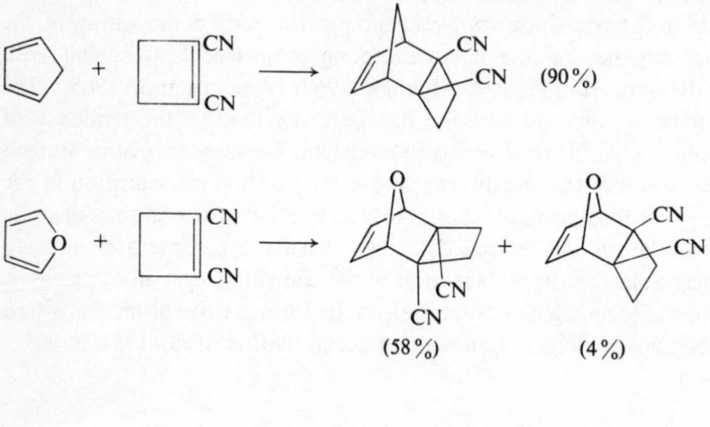

(90%)

(3.83)

(58%) (4%)

bonded repulsive interaction between the methylene group of the diene and the cyclobutane ring, leading to a more favourable *exo* transition state. With furan this steric effect is absent or lessened, and secondary orbital interaction in the *endo* transition state directs the course of the reaction (Belluš, von Bredow, Sauter and Weis, 1973; Belluš, Mez and Rihs, 1974; see also Mellor and Webb, 1974, and Cantello, Mellor and Webb, 1974). Similar factors may account for the fact that reaction of methyl acrylate, methyl methacrylate and methyl crotonate with cyclopentadiene gives only a slight (or no) excess of the *endo* adduct (no solvent was used; cf. p. 207), while with 2,5-dimethyl-3,4-diphenylcyclopentadienone the *endo* compounds are the main products (Houk and Luskus, 1971).

It appears that the transition state which is best stabilised by spatial orbital overlap and by non-bonded interactions and simultaneously least destabilised by unfavourable steric repulsions has the lowest energy and consequently predominates in the kinetically controlled product. For reactions with cyclic dienophiles this is often the *endo* adduct, but with open-chain dienophiles the interplay of the different factors makes it more difficult to predict precisely the steric course of the additions.

Orientation of addition of unsymmetrical components. Addition of an unsymmetrical diene to an unsymmetrical dienophile could take place in two ways to give two structurally isomeric products. It is found in practice, however, that formation of one of the isomers is strongly favoured (Sauer, 1967). Thus, in the addition of acrylic acid derivatives to 1-substituted butadienes the '*ortho*' (1,2)-adduct is favoured, irrespective of the electronic nature of the substituent (see Table 3.4). The orientating forces are relatively weak and with greater steric demand of the substituents the proportion of the '*meta*' (1,3-) isomer approaches the statistical value. The very strong '*ortho*' orientation observed with acrylic acid and butadiene-1-carboxylic acid is lost when the anions are used, presumably because of the Coulomb repulsion of the two charged groups. It has been thought that increase in temperature led to formation of an increasing proportion of '*meta*' adduct, but recent work in which the product distribution was determined by gas–liquid chromatography indicates that this is not the case.

Similarly, in the addition of methyl acrylate to 2-substituted butadienes the '*para*' (1,4-) adduct is formed predominantly, irrespective of the electronic nature of the substituent (see Table 3.5).

TABLE 3.4 *Proportions of structural isomers formed in addition of acrylic acid derivatives to 1-substituted butadienes*

1,2- 1,3-

R^1	R^2	R	T °C	Ratio of products 1,2-	1,3-
$N(C_2H_5)_2$	H	C_2H_5	20	1,2- only	—
CH_3	H	CH_3	20	18	1
CO_2H	H	H	70	1,2- only	—
CO_2Na	H	Na	220	1	1
$C(CH_3)_3$	$C(CH_3)_3$	CH_3	200	0·9	1

Addition of acetylenic dienophiles to 1- and 2- substituted butadienes also results in preferential formation of the '*ortho*' and '*para*' adducts as with olefinic dienophiles.

As would be expected, in 1,3-disubstituted butadienes the directive influence of the substituents is additive. Thus, in the addition of 1,3-dimethylbutadiene and acrylic acid the adduct (70) is formed almost exclusively.

The reason for these orientation effects has long been puzzling, but they have been rationalised recently using frontier orbital theory (Houk, 1973, 1975; Sustmann, 1974).

TABLE 3.5 *Proportions of structural isomers formed in addition of methyl acrylate to 2-substituted butadienes*

1,4- 1,3-

R_1	T °C	Ratio 1,4- to 1,3-
OC_2H_5	160	1,4- only
C_6H_5	150	4·5 to 1
CN	95	1,4- only

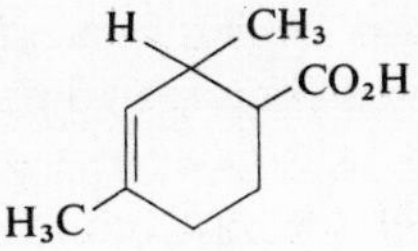

(70)

In a ‘normal’ Diels–Alder reaction (i.e. one involving a conjugated or electron-rich diene and an electron-deficient dienophile), the main interaction in the transition state is between the HOMO of the diene and the LUMO of the dienophile and the orientation of the product obtained from an unsymmetrical diene and an unsymmetrical dienophile is governed largely by the atomic orbital coefficients at the termini of the conjugated systems concerned. The atoms with the larger terminal coefficients on each addend bond preferentially in the transition state, since this leads to better orbital overlap. It turns out that in most cases this leads mainly to the 1,2(‘*ortho*’)-adduct with 1-substituted butadienes and to the 1,4(‘*para*’)-adduct with 2-substituted butadienes. Thus, for butadiene-1-carboxylic acid and acrylic acid (Table 3.4) the frontier orbitals are polarised as shown below, where the size of the circles is roughly proportional to the size of the coefficients. Shaded and empty circles represent lobes of opposite sign, an ‘allowed’ reaction involving overlap of lobes of the same sign. Similarly for reaction of 2-phenylbutadiene and

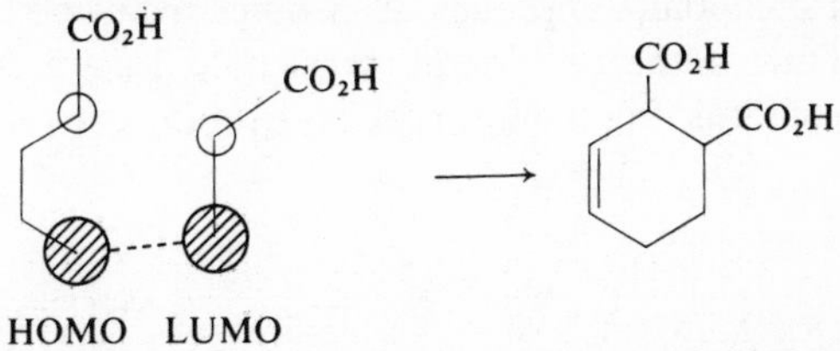

(3.84)

methyl acrylate, preferential formation of the ‘*para*’ adduct is predicted and found (Table 3.5). With 2-methyl-1,3-butadiene, however, where the coefficients of the terminal carbon atoms in the HOMO do not differ

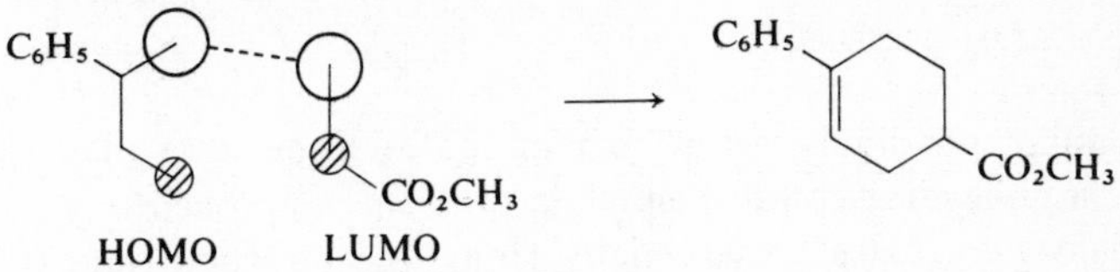

(3.85)

from each other so much as they do in the 2-phenyl compound, reaction with methyl acrylate or acrolein gives larger amounts of the ‘*meta*’(1,3)-adduct.

In certain cases secondary orbital interactions may also play an important part in determining the course of the reaction (Alston, Ottenbrite and Shillady, 1973).

The relative amounts of the structurally isomeric products formed in these reactions are strongly influenced by the presence of Lewis acid catalysts, and when these conditions are applicable very high yields of a single isomer can be obtained. Thus, for the addition of methyl vinyl ketone or acrolein to isoprene the proportion of the '*para*' adduct was increased in presence of stannic chloride, so that it became almost the exclusive product of the reaction (Lutz and Bailey, 1964) (3.86). Similar effects were noted for the addition of isoprene to methyl acrylate (Inukai and Kojima, 1966) (see pp. 198–202).

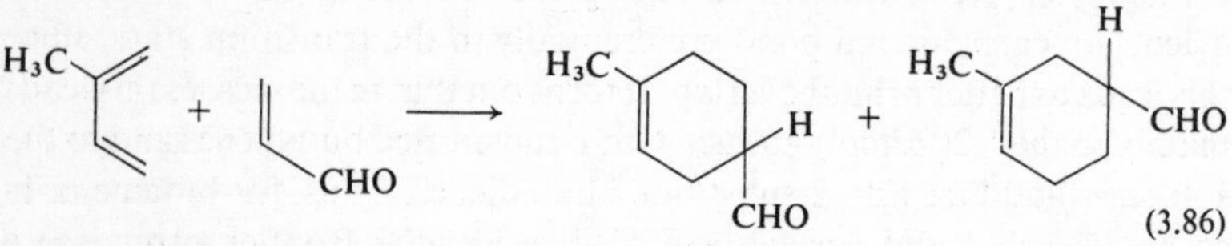

(3.86)

toluene, 120°C, no catalyst;	ratio	59	41
benzene, 25°C, $SnCl_4.5H_2O$;	ratio	96	4

Asymmetric synthesis. In reactions of disymmetric dienes and dienophiles in which the plane of the double bonds is not a plane of symmetry, the diene and dienophile approach each other from the less hindered side of each. Thus, in the addition of maleic anhydride to cyclo-octatetraene, which reacts as if in the form (71), the product is exclusively (72) (3.87).

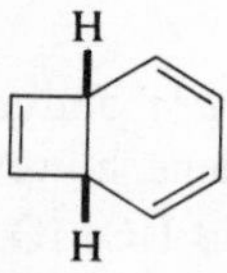

(71) 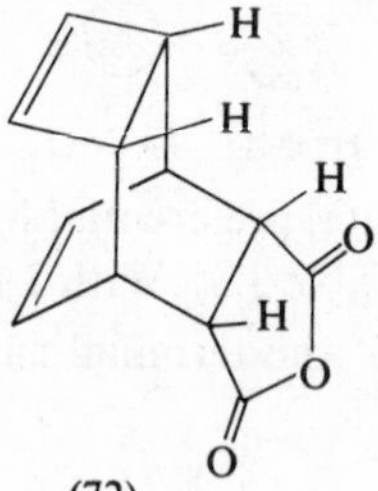 (72) (3.87)

The interesting possibility arises that in the addition of a planar diene or dienophile to a chiral dienophile or diene, approach of the reagents might take place preferentially from one direction, and if reaction gives rise to a new asymmetric centre this might be formed in non-statistical ratio, resulting in an asymmetric synthesis. This possibility has been realised in a number of instances recently. In the normal thermal reaction optical yields are low, but in reactions catalysed by Lewis acid catalysts, particularly when carried out at low temperatures, very high optical yields have been obtained, presumably due to the formation of a more stable transition state in the catalysed reactions because of increased secondary orbital interactions, as well as possible conformational changes in the complexed dienophile (Houk and Strozier, 1973) (cf. p. 201). Furthermore, the absolute stereochemistry of the predominant isomer can often be predicted, and the catalysed sequence offers a convenient new method for the asymmetric synthesis of a variety of optically active compounds of known absolute configuration (Sauer and Kredel, 1966; Farmer and Hamer, 1966).

Thus, reaction of (–)-menthyl fumarate with butadiene or isoprene followed by reduction of the adduct with lithium aluminium hydride with removal of the auxiliary asymmetric (–)-menthol, gave optically active specimens of the compounds (73), (R = H and CH_3) in optical yields of 1–9 per cent (3.88). Again, addition of R-(–)-menthyl acrylate to cyclopentadiene and reduction with lithium aluminium hydride gave the optically active product (74) in 1–9 per cent optical yield, but when the addition was effected at –70°C in presence of boron trifluoride etherate the (+)-isomer of (74) was obtained in nearly 90 per cent optical yield.

R H CH_2OH H CH_2OH H CH_2OH (3.88)

(73) (74)

3.8. Mechanism of the Diels–Alder reaction

The precise mechanism of the Diels–Alder reaction has been the subject of much debate. There is general agreement that the rate-determining step in adduct formation is bimolecular and that the two components approach each other in parallel planes roughly orthogonal to the direction of the new bonds about to be formed. Formation of the two new σ-bonds takes place by overlap of molecular π-orbitals in a direction corresponding to endwise overlap of atomic *p*-orbitals. But until recently there has been uncertainty about the nature of the transition state and in particular, about the timing of the changes in covalency that result in the formation of the new bonds (see Sauer, 1967; Wassermann, 1965).

Two main views have been considered: (*a*) The reaction is a concerted addition in which both of the new single bonds are formed at the same time (3.89). (*b*) The reaction takes place in two steps, the first of which, the formation of a single bond between atoms of the reactants, is rate controlling; the addition is then completed by formation of the second bond in a fast reaction. The intermediate in the second alternative may have either zwitterionic or diradical character.

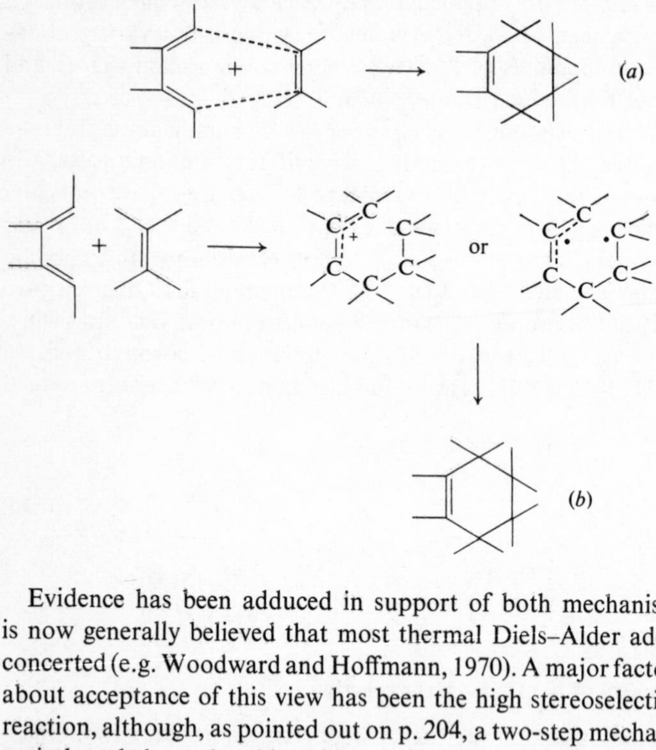

(*a*)

or

(3.89)

(*b*)

Evidence has been adduced in support of both mechanisms, but it is now generally believed that most thermal Diels–Alder additions are concerted (e.g. Woodward and Hoffmann, 1970). A major factor bringing about acceptance of this view has been the high stereoselectivity of the reaction, although, as pointed out on p. 204, a two-step mechanism is not entirely ruled out by this evidence if we assume that rotation about carbon–carbon single bonds in the intermediate might be slow compared with the rate of formation of the second bond. In this connection it is noteworthy that the Diels–Alder reactions of *cis*- and *trans*-1,2-dichloro-ethylene with cyclopentadiene, gave strict *cis* addition: a two-step mechanism would have given the di-radicals (75) and (76) which would be expected to be long lived and to undergo some interconversion (3.90). On the other hand, in the addition of dichlorodifluoroethylene to the isomeric hexadienes to form cyclobutane derivatives, which certainly

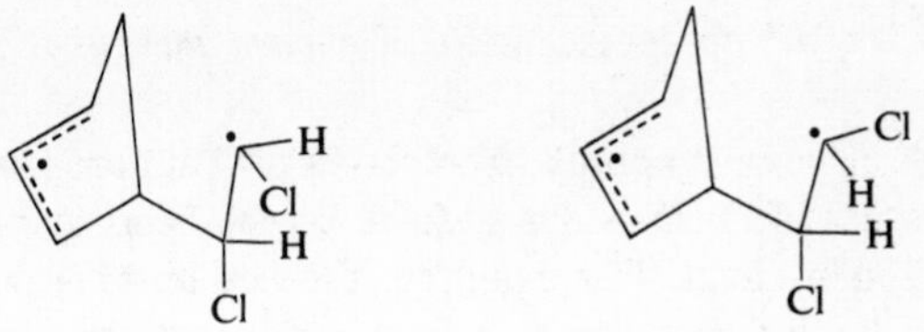

(75) (76) (3.90)

proceeds by a two-step mechanism with a biradical intermediate, the step leading to formation of the ring is not stereoselective, because the rate of rotation about a carbon–carbon single bond in the intermediate is comparable with the rate of formation of the ring (3.91).

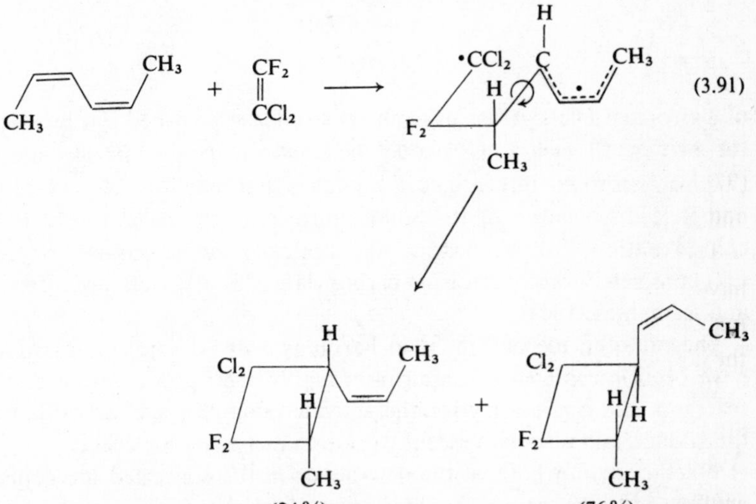

(3.91)

(24%) (76%)

Attempts to show the involvement of free radicals in the Diels–Alder reaction have been negative. No biradical intermediates have been detected, and compounds that catalyse singlet–triplet transitions have no effect on the reaction. Similarly, the kinetic effects of *para* substituents in 1-phenylbutadiene, although large in absolute terms, are considered much too small for a rate-determining transition state corresponding to a zwitterion intermediate, and indicate a synchronous mechanism. Thus, replacement of methoxyl by nitro in the reaction of *para* substituted phenylbutadienes with maleic anhydride results in a decrease in the rate constant by a factor of only 10, whereas the rate of solvolysis of *para* substituted α,α-dimethylbenzyl chlorides to give the ion pairs decreases by a factor of 7×10^9 on replacement of a *para* methoxyl by a nitro substituent.

The two-step mechanism similarly receives no support from studies of the isomerisation of Diels–Alder adducts. Many Diels–Alder adducts can be isomerised by heat. For example, the adduct (77) from cyclopentadiene and methyl methacrylate is isomerised to the isomer (78) on heating in decalin solution (3.92). Strong support for the possibility

CO_2CH_3, CH_3 (77) $\xrightarrow[\text{decalin}]{\text{boiling}}$ CH_3, CO_2CH_3 (78) $\qquad$ $\cdot C{-}CH_3$, CO_2CH_3 (79) $\qquad$ (3.92)

of a two-step Diels–Alder mechanism would be provided if a biradical (or zwitterion) such as (79) could be shown to be an intermediate in this isomerisation. But in fact it is known that reaction takes place by complete dissociation of the adduct into its components followed by recombination, for on heating the optically active *exo*-carboxylate (77) completely racemised *endo*-carboxylate (78) was obtained (Berson and Remanick, 1961).

The one-step mechanism is in harmony with kinetic studies which have been interpreted as requiring a highly special orientation of the reactants and a geometry for the activated complex similar to that of the adduct, and also with recent work on kinetic isotope effects.

Whether or not both of the new bonds in the concerted mechanism are formed to the same extent at the transition state is an open question. A current view is that although they both begin to be formed at the same time the process may take place at different rates in the two bonds so that the transition state is 'lop-sided', with one bond formed to a greater extent than the other (Woodward and Katz, 1959; Doering, Franck-Neumann, Hasselmann and Kaye, 1972). There may be a gradation of mechanisms for different Diels–Alder reactions, extending from a completely concerted four-centre mechanism with a symmetrical transition state at one extreme to something approaching a two-step process at the other. It has been generally assumed, however, that in the reaction between symmetrical dienes and dienophiles the transition state is symmetrical or nearly so, but this view is challenged by calculations which suggest that the prototypical reaction between butadiene and ethylene proceeds via a highly unsymmetrical transition state in which one new δ-bond is almost completely formed while the other has hardly formed at all (Dewar, Griffin and Kirschner, 1974).

It has always been difficult to understand why addition of olefins to conjugated systems to form six-membered rings takes place so easily, while thermal addition to form cyclobutanes is so uncommon, although it is easily effected photochemically. From a consideration of molecular orbital symmetries, Woodward and Hoffmann (1969, 1970) have been able to explain why this is so. Using extended Hückel calculations, they have shown that the hypothetical concerted reaction of two ethylene molecules to form cyclobutane is a highly unfavourable ground state process, because of symmetry restrictions, and requires an activation energy comparable to that of electronic excitation of ethylene, an energy unavailable in ordinary thermal reactions. On the other hand, concerted formation of a six-membered ring by addition of an olefin to a conjugated system involves ground-state molecules only and is a thermally allowed reaction. Woodward and Hoffmann have embodied their results in a set of selection rules, according to which cyclo-addition reactions involving ethylenic bonds and conjugated systems take place easily in a concerted fashion when six-membered rings are being formed, but not when the products are four- or eight-membered rings.

Another way of looking at this is based on the idea that the transition state for the Diels–Alder reaction, which involves six π-electrons in a cyclic array, may be considered as 'aromatic' and thus energetically favoured, while for a $(2+2)\pi$ cyclo-addition the transition state, with only four π-electrons, is anti-aromatic and energetically disfavoured, at least for the thermal reaction (Dewar, 1971; Gilchrist and Storr, 1972).

Not all Diels–Alder reactions proceed with equal ease. Frontier molecular orbital theory has been used to show how the ease of reaction with substituted dienes and dienophiles can be correlated with the effect of the substituents on the energies of the frontier orbitals concerned in the reaction (cf. p. 199) (Houk, 1973; Alston and Ottenbrite, 1974; Sustmann, 1974).

$CF_2{=}CCl_2$ + butadiene ⟶ (F_2, Cl_2 vinylcyclobutane) + (F_2, Cl_2 cyclohexene) (3.93)

(99%) (1%)

Bartlett (1968, 1970) and others have described a number of thermal reactions involving 1,2-addition of olefins to conjugated dienes. 1,1-Dichloro-2,2-difluoroethylene, for example, reacts with butadiene at 79°C to form 99 per cent of the 1,2-addition product (3.93) and α-acetoxyacrylonitrile gives a mixture of products formed by 1,2- and 1,4-addition (3.94).

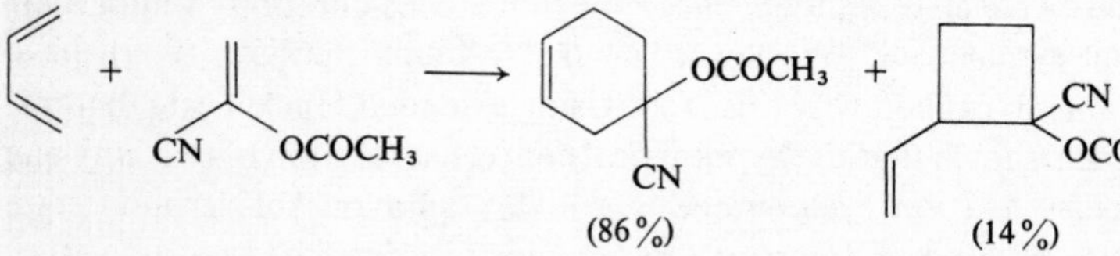

$OCOCH_3$ + CN OCOCH$_3$ → CN $OCOCH_3$ + CN $OCOCH_3$

(86%) (14%) (3.94)

In these reactions it is highly probable that the formation of the four-membered rings takes place by a stepwise process involving bi-radicals, and not by a concerted mechanism.

3.9. Photosensitised Diels–Alder reactions

A number of light-induced Diels–Alder reactions have been described. It is probable that the mechanism of these photoreactions differs from that of thermal Diels–Alder additions and none is known to be concerted. Concerted suprafacial (4 + 2) photocyclo-additions are forbidden by the Woodward–Hoffmann selection rules and a different mechanism is indicated also by the fact that the products of the photo

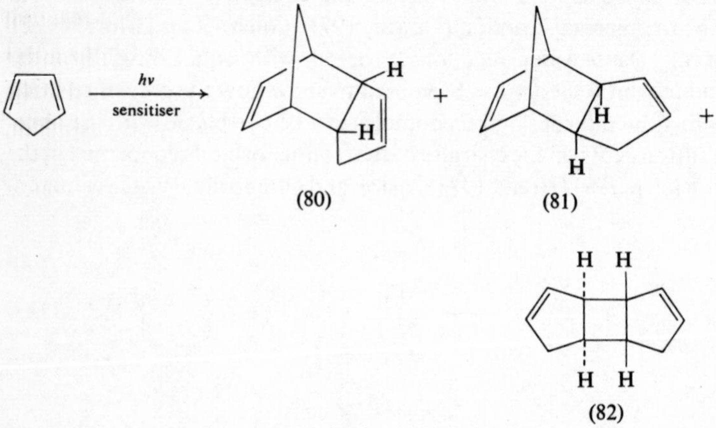

hv sensitiser → H H (80) + H H (81) + H H H H (82) (3.95)

and thermal reactions are not the same. Thus, in the photodimerisation of cyclopentadiene the two dimers (81) and (82) are obtained in equal amount with the normal thermal adduct (80) (3.95), and irradiation of butadiene in presence of ketonic sensitisers affords a mixture of *cis*- and *trans*-divinylcyclobutane along with the normal Diels–Alder adduct vinylcyclohexene (83) (Hammond, Turro and Liu, 1963) (3.96). Cyclohexadiene also affords a mixture of three products on irradiation. None of the thermal *endo* adduct is produced in this case.

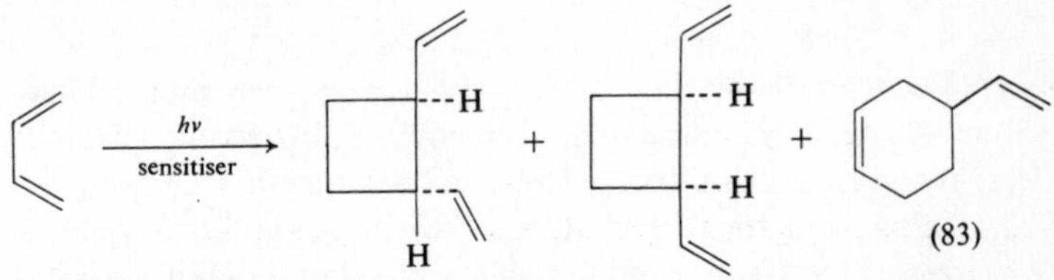

hν, sensitiser (83) (3.96)

Hammond, Turro and Liu (1963) have suggested that these photosensitised Diels–Alder products are formed through diallylic biradicals which themselves arise by addition of triplet diene to another molecule of diene.

The formation of the stable 2:1 adduct (84) by ultraviolet irradiation of a solution of maleic anhydride in benzene, formerly thought to involve a photochemical Diels–Alder reaction, is now known to proceed by way of the photochemical (2 + 2) adduct (85) followed by thermal Diels–Alder addition of maleic anhydride to the cyclohexadiene system (Hartmann, Heine and Schrader, 1974; Bryce-Smith, Deshpande and Gilbert, 1975).

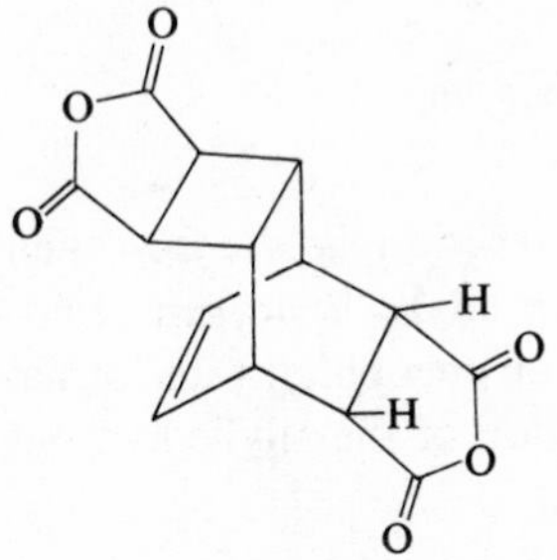

(84) 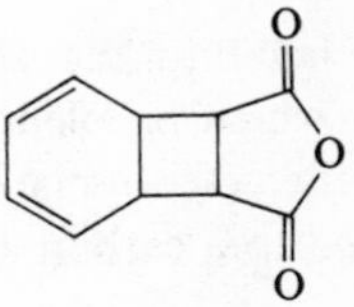 (85) (3.97)

3.10. The homo Diels–Alder reaction

In the homo Diels–Alder reaction of a 1,4-diene with a dienophile, *three* π-bonds are converted into three σ-bonds with formation of two new rings. Most recorded reactions of this type involve additions to

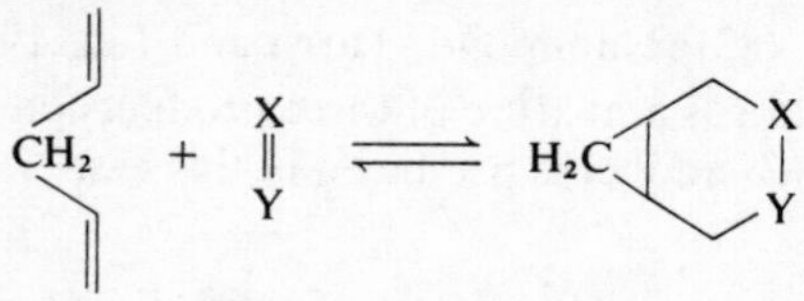

CH_2 + X=Y ⇌ H_2C (X–Y)

bicycloheptadiene, and the dienophiles which have been used include tetracyanoethylene, dicyanoacetylene, 4-phenyl-1,2,4-triazolin-3,5-dione and (less readily) acrylonitrile. Thus, tetracyanoethylene and bicyclo[2,2,1]heptadiene form the adduct (86) in quantitative yield in boiling benzene (3.98) and methyl azodicarboxylate readily gives the

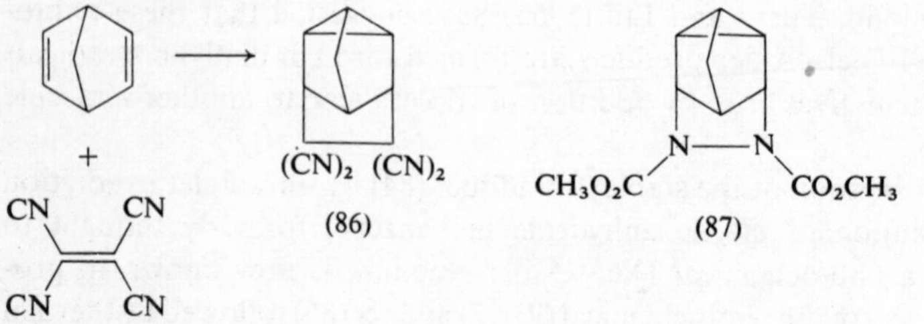

(CN)₂ (CN)₂ (86) CH_3O_2C N—N CO_2CH_3 (87) (3.98)

+ CN CN CN CN

adduct (87). There is evidence (Berson and Olin, 1969) that the homo Diels–Alder reaction, like the Diels–Alder reaction itself, is a concerted process. A concerted thermal addition of this type, since it involves $(4+2)$ π-electrons is 'allowed' by Woodward and Hoffmann's selection rules.

3.11. The 'ene' synthesis

Formally related to the Diels–Alder reaction is the so-called 'ene' synthesis in which an olefin carrying an allylic hydrogen atom reacts thermally with a dienophile (now called an eneophile) with formation of a new σ-bond to a carbon atom, migration of the allylic hydrogen and

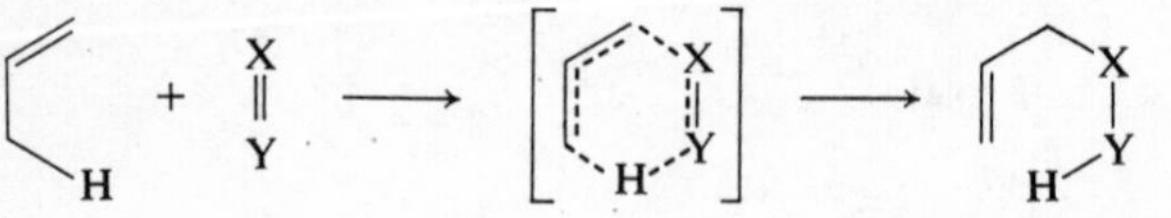

H + X=Y → [H···Y, X] → H–Y, X (3.99)

change in the position of the double bond of the olefin (Alder and von Brachel, 1962; Hoffmann, 1969; Keung and Alper, 1972). In this reaction the two electrons of the allylic C—H σ-bond take the place of two π-electrons of the diene in the ordinary Diels–Alder reaction.

Thus, maleic anhydride and propene react at 200°C to give the product (88), and cyclohexene forms the derivative (89) (3.100). Particularly interesting synthetically is the reaction of allyl alcohol with enophiles; the initial adduct is the enolic form of an aldehyde.

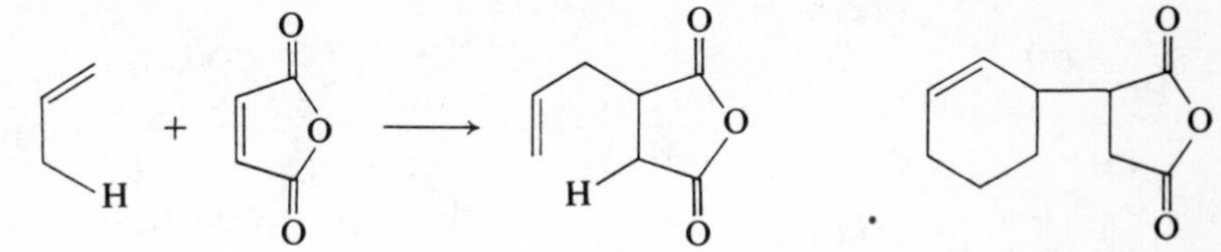

H + O O O ⟶ H O O O (88) O O O (89) (3.100)

H H OH + CO_2CH_3 $\xrightarrow{200°C}$ [HO H H CO_2CH_3] ⟶ CHO CO_2CH_3

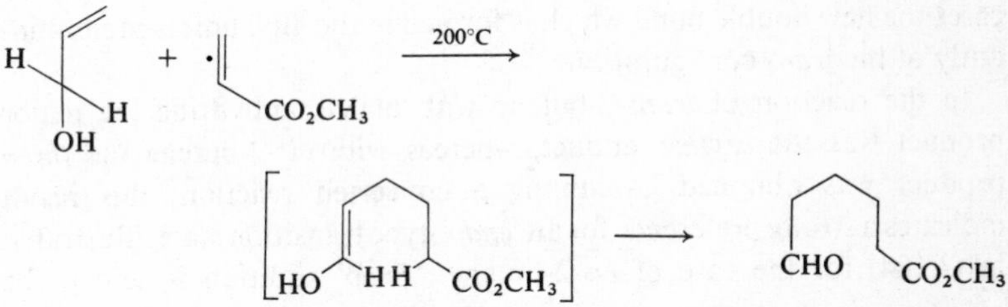

Like the Diels–Alder reaction itself the ene synthesis is reversible. This is shown for example by the decomposition of 1-pentene at 400°C to give ethylene and propylene (3.101). A synthetically useful reaction

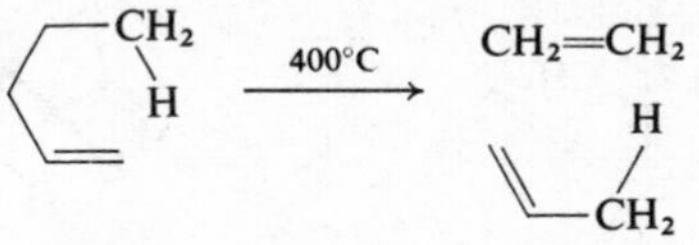

CH_2 H $\xrightarrow{400°C}$ $CH_2{=}CH_2$ H CH_2 (3.101)

of this type is the thermal decomposition of $\beta\gamma$-unsaturated primary alcohols to olefins and formaldehyde (3.102).

The ene synthesis resembles the Diels–Alder reaction also in its stereoselectivity, showing *cis* addition and a preference for the formation of '*endo*' products. *Cis* addition is illustrated by the reaction of 1-heptene with dimethyl acetylenedicarboxylate to form the adduct (90) (3.103). None of the fumaric acid derivative was obtained. Similarly, in the reaction with propiolic ester a *trans* $\alpha\beta$-unsaturated ester was formed.

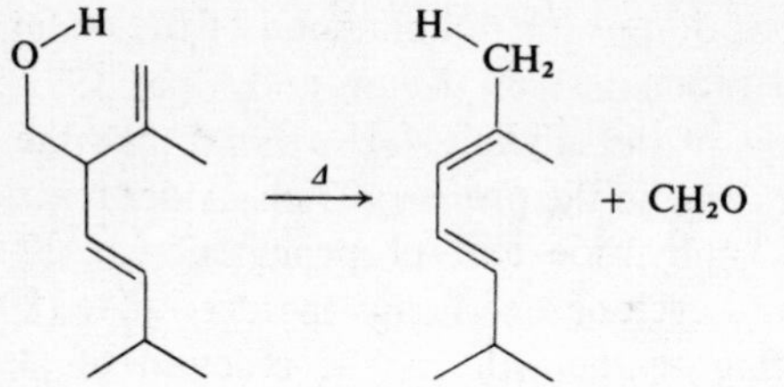

$$\xrightarrow{\Delta} \quad + \; CH_2O \tag{3.102}$$

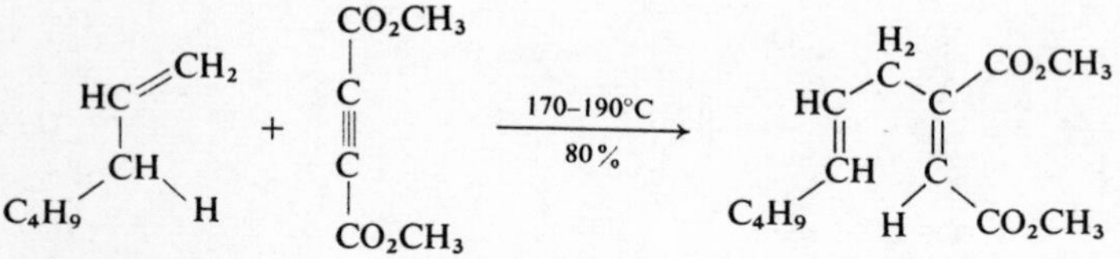

$$\text{(3.103)}$$

(90)

Although there has been no systematic study, it appears that in most cases the new double bond which is formed in the allyl unit is predominantly of the *trans* configuration.

In the reaction of *trans*-2-butene with maleic anhydride the major product was the *erythro* adduct, whereas with *cis*-2-butene the *threo* product was obtained. Assuming a concerted reaction, this result indicates a strong preference for an '*endo*'-type transition state, illustrated in (3.104) for the case of *cis*-2-butene. '*Endo*' addition is said to be

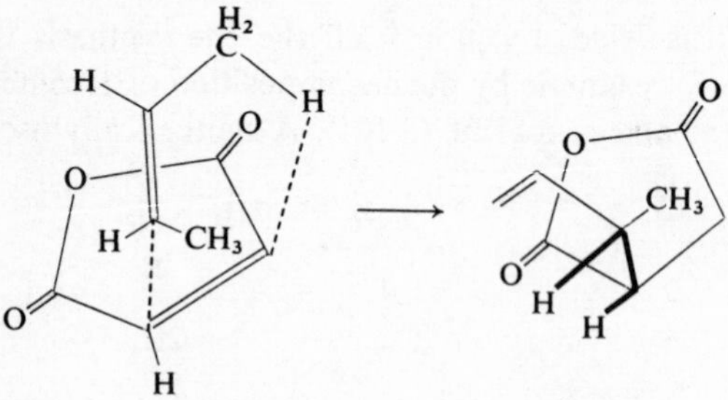

$$\text{(3.104)}$$

endo *threo*

favoured by orbital symmetry relationships, although the preference is not so marked as in the Diels–Alder reaction (Berson, Wall and Perlmutter, 1966).

It is now generally believed that in most cases the reaction proceeds by a concerted mechanism through a cyclic six-membered transition state (Shen-Hong Dai and Dolbier, 1972; Hill, Morgan, Shetty and

Synerholm, 1974; Garsky, Koster and Arnold, 1974), unless this is prohibited by steric factors (cf. Lambert and Napoli, 1973). This view is suggested by the observation that the new C—C and C—H bonds are formed *cis* to each other as in the examples above, and by the fact that optically active products are formed in reactions where the hydrogen atom transferred is initially attached to a chiral centre (Hill and Rabinovitz, 1964). Further support for the concerted mechanism is pro-

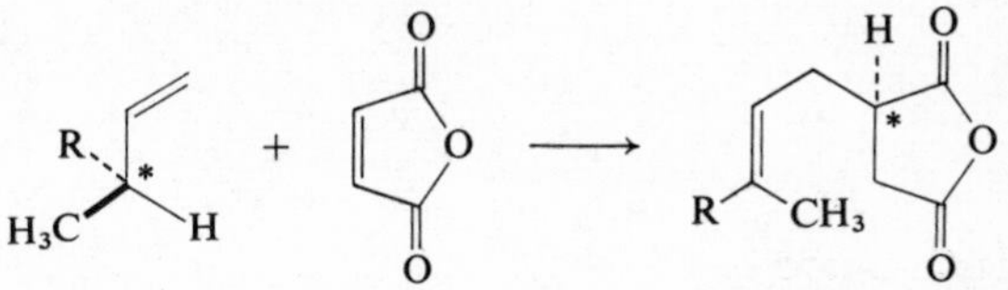

(3.105)

vided by recent work using deuterated α-pinenes which showed that the ene reaction with several eneophiles is highly stereoselective, taking place in each case by transfer of the axial hydrogen atom *trans* to the *gem*-dimethyl bridge (Hill, Morgan, Shetty and Synerholm, 1974; Garsky, Koster and Arnold, 1974).

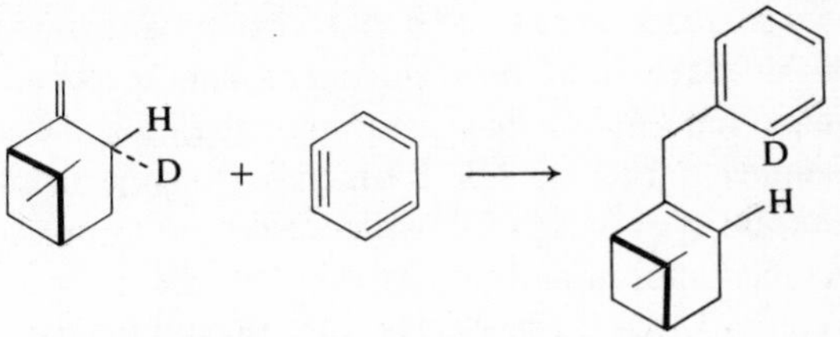

(3.106)

This work also established, in agreement with the earlier conclusion of Berson, Wall and Perlmutter (1966) that the ene reaction takes place preferentially through an *endo* transition state (e.g. 91), at least for the intermolecular reaction.

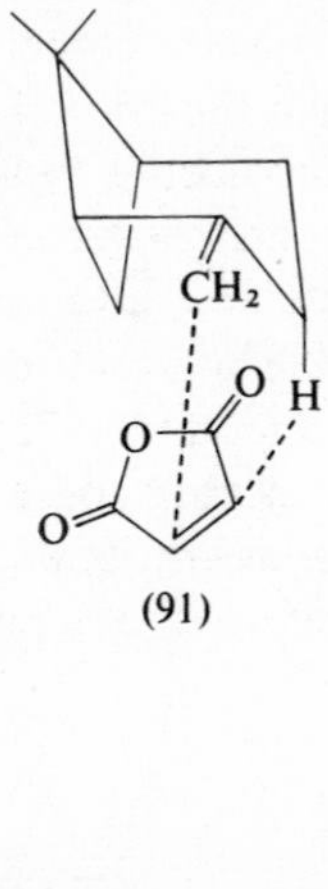

(3.107)

(91)

However, this is not always the case in intramolecular reactions where steric constraints may favour the *exo* transition state. Thus, thermolysis of the *cis*-butenylamide (92) led specifically to the *cis*-pyrrolidine derivative (94) by way of the unstrained *exo* transition state (93) (Oppolzer, Pfenninger and Keller, 1973).

$NCOCH_3$ (92) → $NCOCH_3$ (93) → $NCOCH_3$ H_3C (88%) (94) (3.108)

So far the 'ene' reaction has not been so extensively used in synthesis as the Diels–Alder reaction, but it is evident from recent work that the intramolecular ene reaction has great potential for the synthesis of cyclic compounds, particularly of five-membered rings from 1,6-dienes. Thus, the thermal cyclisation of 5,6-ethylenic ketones provides a useful route to cyclopentanones and cyclopentyl ketones by way of the corresponding enols (3.109) (Leyendecker, Drouin and Conia, 1974; Conia and Le Perchec, 1975). For reactions leading to five-membered rings, a *syn* transition state is preferred, giving a product in which the carbonyl group and the newly formed methyl group are *cis* to each other.

Dienes in which the 'ene' component is part of a ring give rise to bicyclic systems stereospecifically and in high yield. The reaction provides a convenient route to spiro compounds which are not readily accessible by other routes. An illustration is provided by the elegant synthesis of the sesquiterpene alcohol (±)-acorenol shown in (3.110) where the spiro[5,4]decane system was constructed by an ene reaction with the appropriate 1,6-diene and, at a later stage in the synthesis, a shift in the position of a double bond was effected through a retro 'ene' reaction of an allylic acetal (Oppolzer, 1973).

Six-membered rings can also be formed by 'ene' cyclisation of 1,7-dienes, but in general lower yields are obtained than in the reactions with 1,6-dienes and higher temperatures are required.

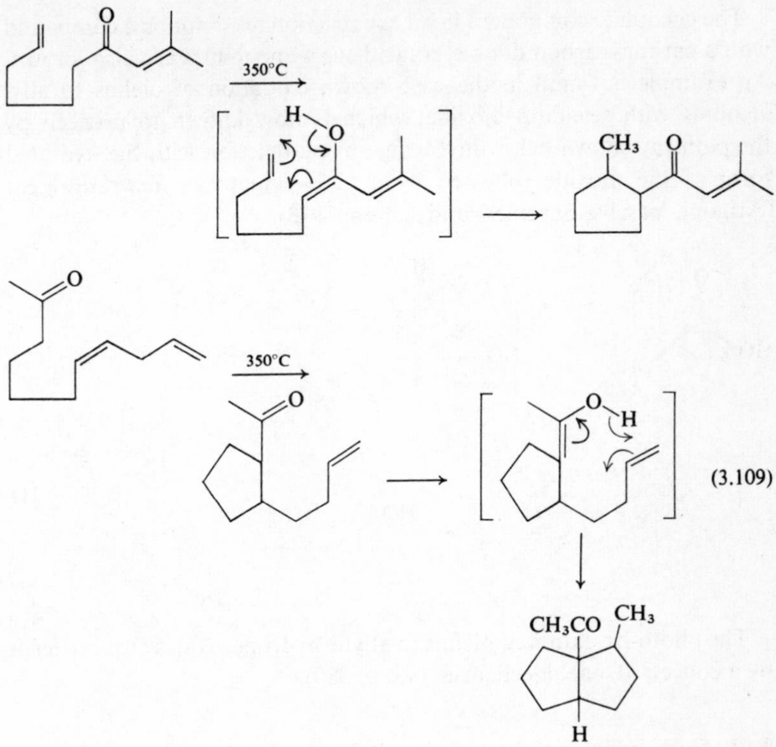

350°C

CH_3 O

350°C

(3.109)

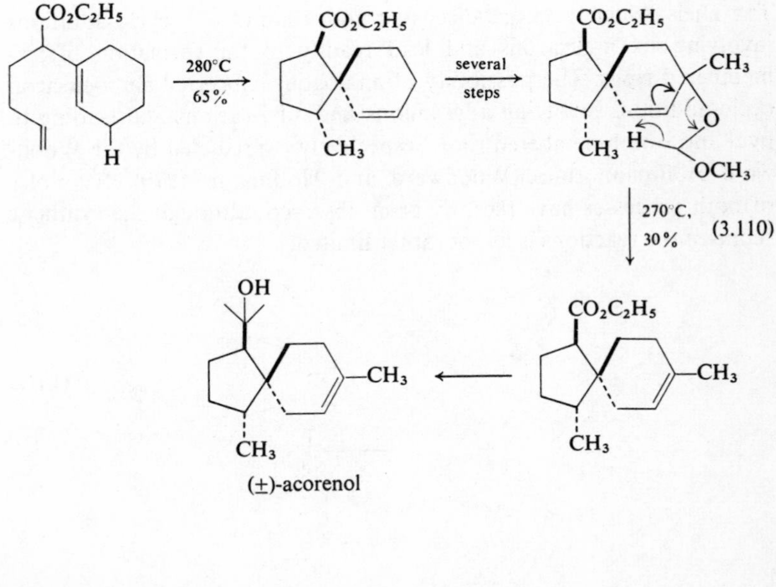

CH_3CO CH_3

H

$CO_2C_2H_5$

280°C
65%

H

$CO_2C_2H_5$

CH_3

several
steps

$CO_2C_2H_5$

CH_3

O

H

CH_3

OCH_3

270°C,
30%

(3.110)

OH

CH_3

CH_3

(±)-acorenol

$CO_2C_2H_5$

CH_3

CH_3

The eneophile component in an ene reaction need not be a compound with a carbon–carbon double bond; hetero-eneophiles are also known. An example is found in the well-known oxidation of olefins to allyl alcohols with selenium dioxide, which is now known to proceed by the pathway shown below involving an ene reaction with the hydrated form of the dioxide followed by a [2,3]-sigmatropic rearrangement (Arigoni, Vasella, Sharpless and Jensen, 1973).

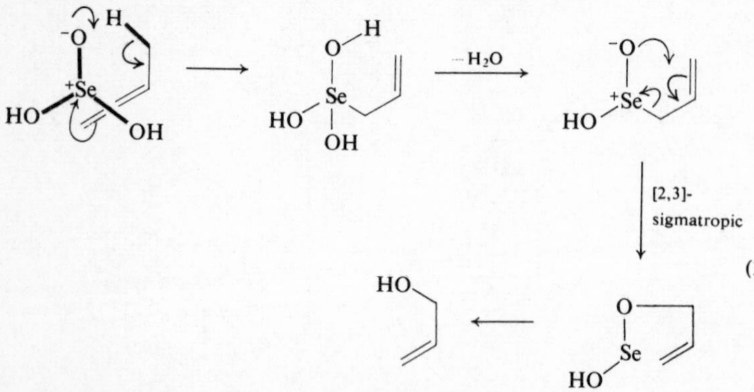

(3.111)

The photo-oxidation of olefins to allylic hydroperoxides also proceeds by a concerted 'ene' mechanism (see p. 380).

3.12. Cyclo-addition reactions with allyl cations and allyl anions

The Diels–Alder reactions discussed above are (4 + 2) cyclo-additions involving six π-electrons and lead readily to the formation of six-membered rings. The possibility of analogous concerted six π-electron cyclo-additions involving allyl anions and allyl cations and leading to five- and seven-membered rings respectively is predicted by the Woodward–Hoffmann rules (Woodward and Hoffmann, 1970). Examples of both processes have recently been observed, although the synthetic scope of the reactions is as yet rather limited.

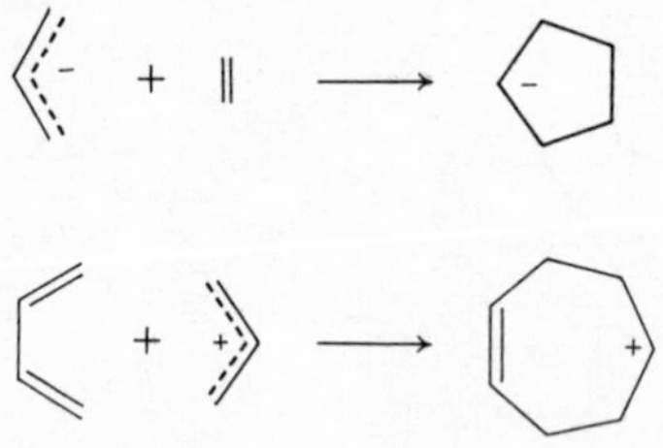

(3.112)

In the anion series best results have been obtained with the 2-aza-allyl anion which adds to a variety of olefins to form pyrrolidine derivatives. Allyl-lithium compounds themselves seem to undergo cyclo-

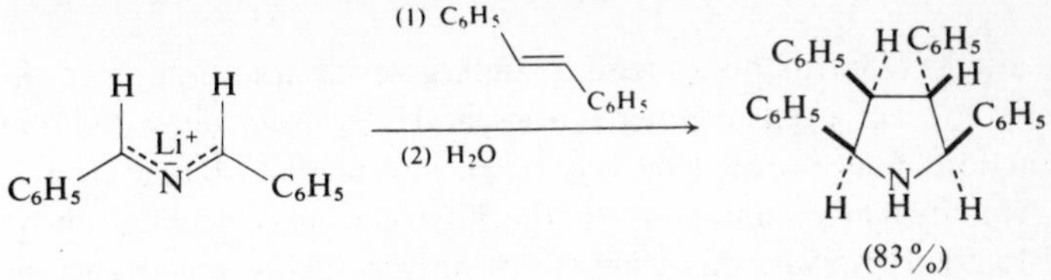

(3.113)

addition to carbon–carbon double bonds only when an electron-attracting group such as C_6H_5 or CN is attached to C-2 of the allyl system (Kauffmann, 1974).

A promising application of the cyclo-addition of the anion derived from a vinylsulphide, which provides a regiospecific and stereospecific method for building a five-membered ring on to an existing ring system ('penta-annelation') has been described by Marino and Mesbergen (1974). Thus the allyl anion (96), readily available from the vinylsulphide (95), reacted smoothly with ethyl acrylate at room temperature to give the adduct (97) in 65 per cent yield.

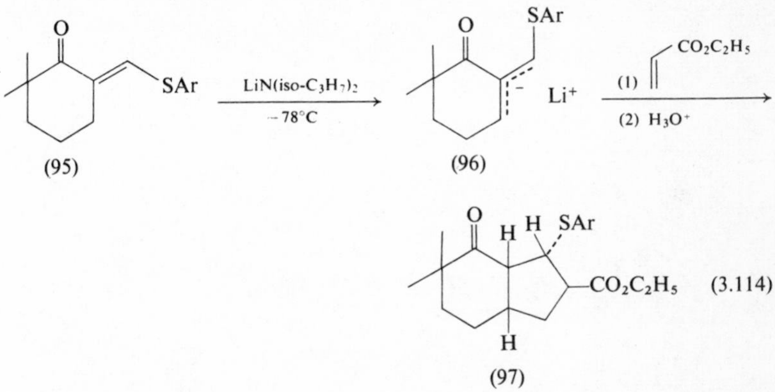

(3.114)

Cyclo-addition of allyl cations to conjugated dienes provides a route to seven-membered carbocycles (Hoffmann, 1973). Several methods are used to generate the allyl cations: one convenient route is from allyl halides and silver trifluoroacetate. Reaction proceeds best with cyclic dienes. Thus cyclohexadiene and methylallyl cation gave 3-methylbicyclo[3,2,2]nona-2,6-diene in 30 per cent yield.

CH_3 + CH_2I $\xrightarrow[\text{iso-}C_5H_{12},\ -78^\circ C]{CF_3CO_2Ag}$ CH_3 (3.115)

Some alkylcyclopropanones readily undergo cyclo-addition reactions with dienes to form seven-membered rings, and it has been suggested that these reactions may proceed by way of an intermediate bidentate 1,3-dipole (98), formally analogous to the allyl cations discussed above (Turro, 1969). However, this view is not universally accepted and the reactions may be of an entirely different type (Hoffmann, 1973).

O + O ⟶ O O

(3.116)

O^- H_3C H + CH_3 H

(98)

4 Reactions at unactivated C—H bonds

Most synthetically useful ionic organic reactions at saturated carbon atoms take place either by displacement of a suitable leaving group, or by replacement of a hydrogen atom at a carbon atom which is 'activated' by the presence of a neighbouring activating group (cf. Ch. 1). Ionic attack at unactivated C—H bonds is uncommon. Free radicals, on the other hand, can often be obtained with enough energy to break unactivated C—H bonds (see Walling, 1957), but intermolecular reactions of this type are of limited value synthetically, because the reagents are unselective and mixtures of products generally result. It has recently been found, however, that in many molecules which meet certain structural and geometrical requirements *intramolecular* free radical attack at unactivated C—H bonds can become quite specific, leading to the introduction of functional groups at the site of specific C—H bonds. A number of reactions of this type have become of synthetic importance.

The key step in these reactions is an intramolecular abstraction of hydrogen from a carbon atom, resulting in the transfer of a hydrogen atom from carbon to the attacking free radical in the same molecule. Because of the geometrical requirements of the transition state for this step, the most frequently observed intramolecular hydrogen transfers of this type are 1,5-shifts, corresponding to specific attack on a hydrogen atom attached to a δ-carbon. The reactions can be generally represented as in (4.1). Homolytic cleavage of the Y—X bond is followed by hydrogen

$$\text{H–C(–C–C–C–Y–X)} \xrightarrow{-X\cdot} \text{H–C(–C–C–C–Y}\cdot) \longrightarrow \cdot\text{C(–C–C–C–YH)} \xrightarrow{+X'\cdot} \text{X'–C(–C–C–C–YH)} \qquad (4.1)$$

transfer from the δ-carbon atom to Y·, and the resulting carbon radical finally reacts with a free radical X'· which may or may not be identical with X·.

A number of important reactions of this type in which Y = O are discussed below. Another useful reaction in which Y = N is the well-known Hofmann–Loeffler–Freytag reaction.

4.1. The Hofmann–Loeffler–Freytag reaction

This reaction provides a convenient and useful method for the synthesis of pyrrolidine derivatives from *N*-halogenated amines (Wolff, 1963) (4.2). The reaction is effected by warming a solution of the halogenated amine

H_2SO_4, heat or $h\nu$; NaOH (4.2)

in strong acid (concentrated sulphuric acid or trifluoroacetic acid are often used), or by irradiation of the acid solution with ultraviolet light. The immediate product of the reaction is the δ-halogenated amine, but this is not generally isolated, and by basification of the reaction mixture it is converted directly into the pyrrolidine. Both *N*-bromo- and *N*-chloro-amines have been used as starting material, but the *N*-chloro-amines are said to give better yields. They are easily obtained from the amines by action of sodium hypochlorite or *N*-chlorosuccinimide.

The first example of this reaction was reported by A. W. Hofmann in 1883. In the course of a study of the reactions of *N*-bromoamides and *N*-bromoamines, he treated *N*-bromoconiine (1) with hot sulphuric acid and obtained, after basification, a tertiary base which was later identified as δ-coneceine (2) (4.3). Later, further examples of the reaction were

(1) H_2SO_4, 140°C; (2) NaOH (4.3)

(1) (2)

reported by Loeffler, including a neat synthesis of the alkaloid nicotine (3) (4.4). Numerous other cyclisations leading to both simple pyrrolidines and to more complex polycyclic structures have since been recorded. Thus, *N*-methyl-*N*-chloro-octylamine affords 1-methyl-2-butylpyrroli-

(4.4)

(3)

$$\underset{\displaystyle Cl}{CH_3\overset{}{N}}CH_2CH_2CH_2CH_2C_4H_9 \xrightarrow[\text{(2) NaOH}]{\text{(1) } H_2SO_4\text{, heat}} \text{[pyrrolidine ring: N–}CH_3\text{, 2-}C_4H_9] \qquad (4.5)$$

dine in high yield (4.5), and the cyclopentyl compound (4) is converted into the bridged pyrrolidine (5) (4.6).

$$\text{(4) cyclopentyl–}CH_2N(C_2H_5)Cl \xrightarrow[\text{(2) NaOH}]{\text{(1) } H_2SO_4\text{, } h\nu} \text{(5) bridged pyrrolidine, N–}C_2H_5 \qquad (4.6)$$

It was long believed that only *N*-halogenated derivatives of secondary amines would undergo the reaction, but it is now known that pyrrolidines lacking an *N*-alkyl substituent can be obtained from *N*-chloro primary amines in strongly acid solution by using ferrous ions as initiators (Schmitz and Murawski, 1966). Thus, *N*-chlorobutylamine is converted into pyrrolidine in 72 per cent yield, and 2-propylpyrrolidine is readily obtained from both *N*-chloro-1-aminoheptane and *N*-chloro-4-aminoheptane (4.7).

$$\text{ClNH–(CH}_2)_6CH_3 \longrightarrow \text{2-propylpyrrolidine (N–H)} \longleftarrow \text{4-heptyl–NHCl} \qquad (4.7)$$

Although the majority of applications of the reaction have employed *N*-haloamines as starting materials, *N*-haloamides have been used occasionally and these also give rise to secondary pyrrolidines by loss of the acyl group. *N*-Butyl-*N*-chloroacetamide, for example, when heated with sulphuric acid and subsequently treated with alkali, is converted into pyrrolidine in 50 per cent yield.

With *N*-halocycloalkylamines cyclisation leads to bridged-ring structures, but in these cases the products are not exclusively pyrrolidine derivatives. Thus, whereas *N*-bromo-*N*-methylcycloheptylamine was converted into tropane (6) in 40 per cent yield when warmed with sulphuric acid, *N*-chloro-4-ethylpiperidine gave a mixture of the azabicycloheptane (7) and quinuclidine (8), and *N*-chloro-*N*-methylcyclooctylamine is reported to yield *N*-methylgranatamine (9) exclusively (4.8). The reasons for the preferential formation of a six-membered

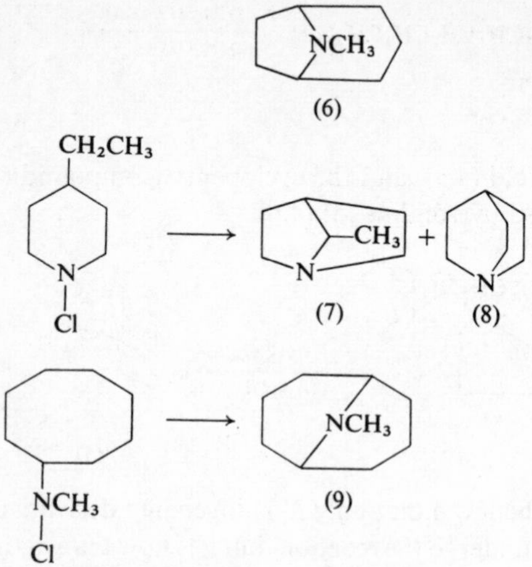

NCH_3

(6)

CH_2CH_3 N Cl ⟶ CH_3 N (7) + N (8) (4.8)

⟶ NCH_3 (9)

NCH_3 Cl

ring in this case are not clearly understood (Wawzonek and Thelen, 1950; Wawzonek, Nelson and Thelen, 1951).

In general, yields obtained in these 'bicyclo' reactions, and the ease of reaction, are noticeably less than with open-chain amines. This is because it may not be so easy in the cycloalkylamines for the molecule to adopt the necessary conformation in the transition state to allow the nitrogen radical to abstract a hydrogen atom from the δ-carbon. An extreme example of this is shown in *N*-chloro-*N*-methylcyclohexylamine in which irradiation in sulphuric acid gave only a very low yield of cyclised product. In this compound the cyclohexane ring has to adopt the unfavourable boat conformation to allow 1,5-intramolecular hydrogen transfer from C-4 to N to take place.

On the other hand, with *N*-chlorocamphidine (10), in which the reacting groups are already suitably disposed in space for ready formation of the six-membered transition state, irradiation in sulphuric acid led to the tertiary amine cyclocamphidine (11) in nearly 70 per cent yield (4.9).

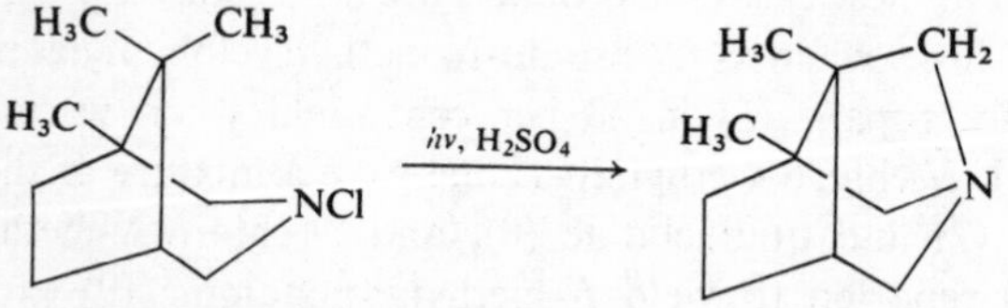

H_3C CH_3 H_3C NCl (10) $\xrightarrow{h\nu,\ H_2SO_4}$ H_3C CH_2 H_3C N (11) (4.9)

The reaction is believed to proceed by a free radical chain process, and the reaction path shown in (4.10) has been proposed (Corey and Hertler, 1960).

$$R{-}CH_2CH_2CH_2CH_2\overset{+}{N}ClHR' \longrightarrow R{-}CH_2CH_2CH_2CH_2\overset{+}{N}{\cdot}HR' + Cl{\cdot}$$

$$R{-}CH_2CH_2CH_2CH_2\overset{+}{N}{\cdot}HR' \longrightarrow R{-}\dot{C}HCH_2CH_2CH_2\overset{+}{N}H_2R' \xrightarrow{R''\overset{+}{N}HClR'} R{-}CHClCH_2CH_2CH_2\overset{+}{N}H_2R' \xrightarrow{\text{base}} \text{R–CH(CH}_2)_3\text{N–R}' \text{ (cyclic)} \qquad (4.10)$$

The reacting species may be either the free chloramine or the chlorammonium ion. This dissociates under the reaction conditions to form the ammonium radical, which then abstracts a suitably situated hydrogen atom on the δ-carbon to give the corresponding carbon radical. This in turn abstracts a chlorine atom from another molecule of chloramine, thus propagating the chain and at the same time forming the δ-chloroamine, from which the cyclic amine is subsequently obtained.

The free radical nature of the reaction is suggested by the fact that it does not proceed in the dark at 25°C, and that it is initiated by heat, light or ferrous ions and inhibited by oxygen. The hydrogen abstraction step must be intramolecular for only thus can the specificity of reaction at the δ-carbon be understood. Strong evidence that the decomposition of *N*-chloroamines in acid involves an intermediate in which the δ-carbon atom is trigonal is provided by the observation that the optically active chloroamine (12) on decomposition in sulphuric acid at 95°C gave a 43 per cent yield of pure 1,2-dimethylpyrrolidine (13) which was optically inactive (4.11). The intermediacy of δ-chloroamines in the reaction has been confirmed by their isolation in a number of cases.

$$CH_3CHD(CH_2)_3N(Cl)CH_3 \; (12) \longrightarrow \text{1,2-dimethylpyrrolidine} \; (13) \qquad (4.11)$$

Where alternative modes of cyclisation are possible it is found that, as in other free radical reactions, secondary hydrogen atoms react more readily than primary. In nearly all cases reaction at the δ-carbon atom with formation of a pyrrolidine is favoured. Thus, in the reaction with *N*-chlorobutylpentylamine, attack by the nitrogen radical on the δ-methyl group would lead subsequently to 1-pentylpyrrolidine, whereas attack on the δ-methylene would result in formation of 1-butyl-2-methylpyrrolidine. Only the latter compound was formed (4.12). Tertiary

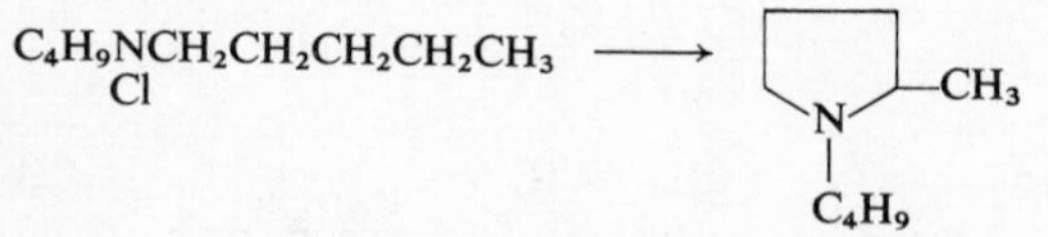

$$C_4H_9N(Cl)CH_2CH_2CH_2CH_2CH_3 \longrightarrow \text{1-}C_4H_9\text{-2-}CH_3\text{-pyrrolidine} \qquad (4.12)$$

hydrogens react very readily, but the resulting tertiary chlorides are rapidly solvolysed under the conditions of the reaction, and no cyclisation products are formed.

An outstanding application of the Hofmann–Loeffler–Freytag reaction is found in the synthesis of the steroidal alkaloid derivative dihydroconessine (14) illustrated in (4.13). In this synthesis the five-membered nitrogen ring is constructed by attack on the unactivated C-18 angular methyl group of the precursor by a suitably placed nitrogen radical at C-20 (Corey and Hertler, 1959). In a similar series of reactions the *N*-chloro derivative (15) was converted into the conanine (16) (4.14).

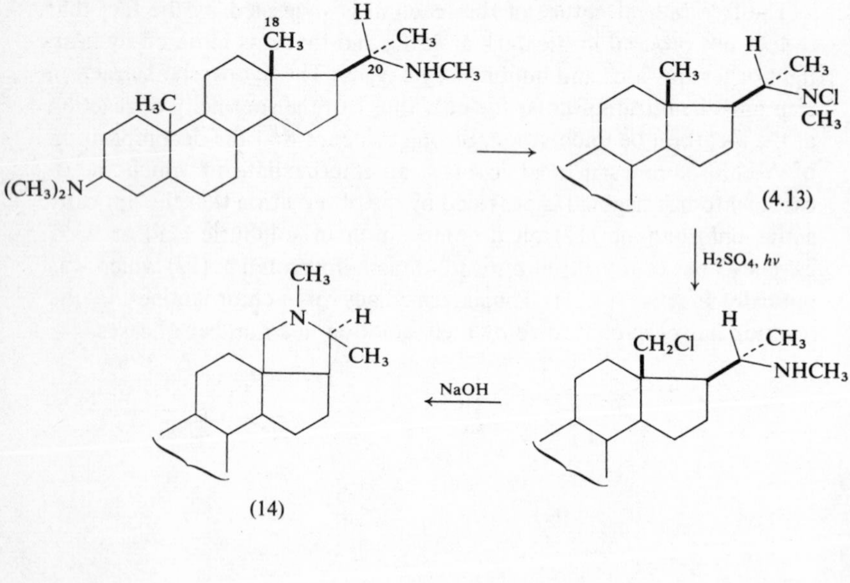

18 CH_3, H, CH_3, 20, $NHCH_3$, H_3C, $(CH_3)_2N$ → CH_3, H, CH_3, NCl, CH_3 (4.13)

H_2SO_4, *hν*

CH_2Cl, H, CH_3, $NHCH_3$ ← NaOH — CH_3, N, H, CH_3

(14)

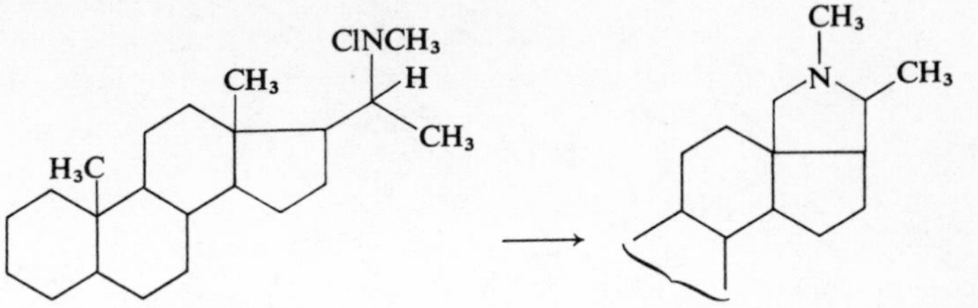

(15) (16) (4.14)

The ease of these reactions is undoubtedly due to the fact that in the rigid steroid framework the β-C-18 angular methyl group and the β-C-20 side chain carrying the nitrogen radical are suitably disposed in space to allow easy formation of the chair-like six-membered transition state necessary for 1,5-hydrogen transfer from the methyl group to nitrogen (4.15).

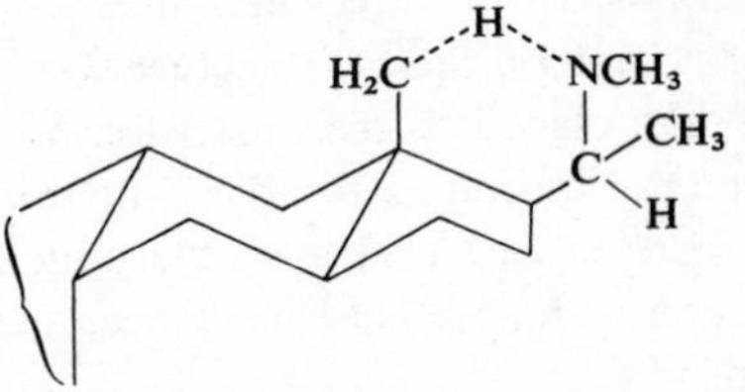

(4.15)

4.2. Cyclisation reactions of nitrenes

Intramolecular reactions leading to the formation of five-membered rings containing nitrogen atoms also take place on thermal decomposition of azides (Abramovitch and Davis, 1964; Smolinsky and Feuer, 1964). Thus, the azide (17) on heating in boiling diphenyl ether is converted into the indoline (18) in 45–50 per cent yield (4.16), and the cyclohexyl derivative shown in similarly converted into hexahydrocarbazole

(4.16)

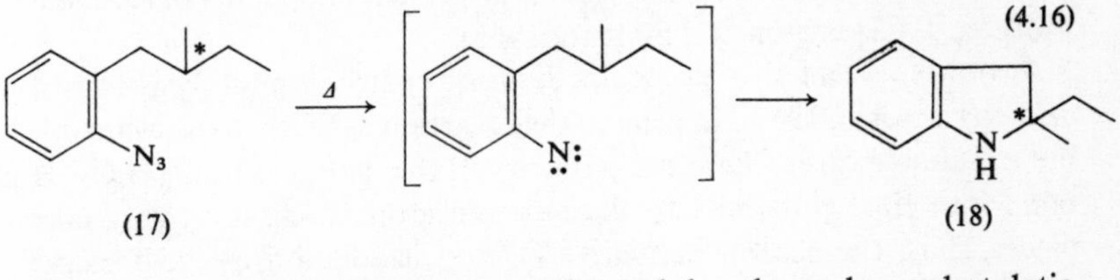

(17) (18)

(4.17). Carbazoles themselves are formed by thermal or photolytic decomposition of 2-azidodiphenyls.

(4.17)

There is much evidence to suggest that these reactions involve the intermediate formation of nitrenes (electron deficient species in which nitrogen has only six electrons in its outer shell; they may have either a singlet R—N̈: or a triplet diradical structure R—N̈·). The cyclisation step is thought to take place by direct insertion of singlet nitrene into the C—H bond, and not through a diradical intermediate, for reaction with the optically active azide (17) gave the optically active indoline (18). A diradical intermediate would have been expected to give an optically inactive product.

It has been reported that photolysis of alkyl azides also gives rise to pyrrolidine derivatives. Thus, 1-butyl azide on irradiation in cyclohexane solution is stated to give pyrrolidine in 22 per cent yield, and 1-octyl azide to form 2-butylpyrrolidine in 35 per cent yield. It has been suggested that in these reactions cyclisation does involve a diradical intermediate formed by 1,5-hydrogen abstraction in the initially formed nitrene. However, the experimental conditions for cyclisation appear to be very critical and have not yet been clearly defined. Subsequent attempts to repeat some of this work have led only to the formation of imines, and the status of the earlier experiments is at present uncertain (Barton and Morgan, 1962; Barton and Starratt, 1965).

Acylnitrenes can also be obtained, most readily by photolysis of acyl azides (Lwowski, 1967). In general they rearrange readily to isocyanates, but in suitable cases where the geometry of the molecule brings a C—H bond into close proximity to the nitrogen atom, cyclisation may take place. Thus, the decalin derivative (19) on irradiation in cyclohexane solution affords the two lactams (20) and (21), as well as the isocyanate (4.18). This reaction has been used to prepare compounds related to the

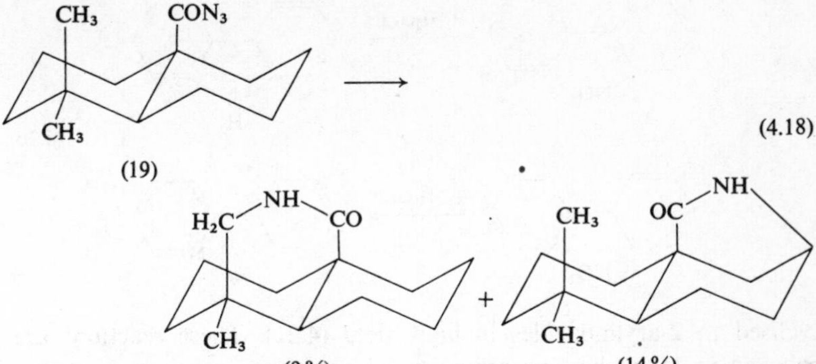

(19) → (20) (9%) + (21) (14%) (4.18)

diterpene alkaloids. Thus, the azide of podocarpic acid methyl ether (22) on photolysis in cyclohexane afforded the lactam (23) by attack on the unactivated angular methyl group (4.19), a synthesis which completed the structural proof of the atisine family of alkaloids (ApSimon and Edwards, 1962). A small amount of γ-lactam was also produced, but the

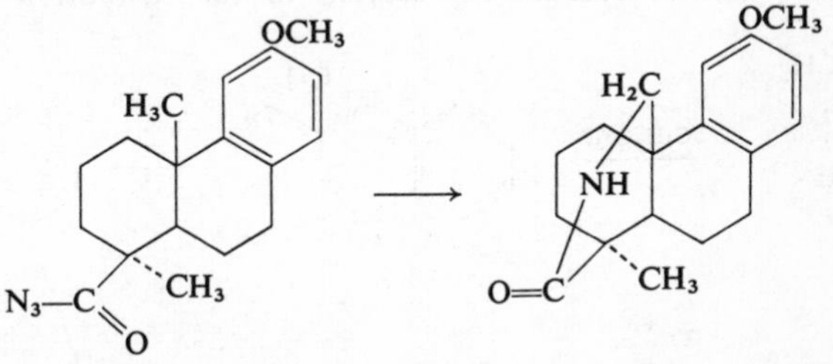

(22) → (23) (4.19)

δ-lactam was the main product. The factors affecting the relative proportions of γ- and δ- lactams produced on photolysis of acyl azides are not clearly understood.

Nitrenes are also thought to be intermediates in the reductive cyclisation of aromatic nitro compounds with triethyl phosphite, a reaction which provides a convenient route to carbazole derivatives and a variety of other aromatic heterocyclic systems (Cadogan, 1968; Cadogan and Mackie, 1974). 2-Nitrobiphenyl, on heating with triethyl phosphite, affords carbazole in high yield, and 2-*o*-nitrophenylpyridine is similarly converted into pyrid[1,2-*b*]indazole (4.20). *o*-Nitro-styrenes or -stilbenes give rise to indoles, and, analogously, *o*-nitrobenzylideneanilines are

$$\text{2-nitrobiphenyl} \xrightarrow{(C_2H_5O)_3P} \text{carbazole} \qquad (4.20)$$

$$\text{2-(2-nitrophenyl)pyridine} \xrightarrow{(C_2H_5O)_3P} \text{pyrido[1,2-b]indazole}$$

cyclised to 2-arylindazoles in high yield (4.21). These reactions are believed to take place by reduction of the nitro group to a nitrene, followed by direct insertion of the singlet nitrene into a C—H bond.

$$\text{o-}O_2N\text{-}C_6H_4\text{-}CH{=}N\text{-}Ar \xrightarrow{(C_2H_5O)_3P} \text{2-aryl-2H-indazole (NAr)} \qquad (4.21)$$

This is supported by the finding that the optically active nitro compound (24) affords the indoline shown which retains optical activity. In nearly every case studied, cyclisation resulted in the formation of a five-

$$\text{(24)} \xrightarrow{(C_2H_5O)_3P} [\text{nitrene} \longrightarrow \text{N-----H transition state}] \longrightarrow \text{2-methyl-2-ethylindoline} \qquad (4.22)$$

membered ring, even when cyclisation to a six-membered ring seemed favourable.

4.3. The Barton reaction and related processes

In the general equation for intramolecular free radical hydrogen transfer reactions (4.1), Y may be an oxygen atom, and a number of synthetically useful reactions of this type, leading to attack on unactivated C—H bonds, have recently been discovered in which Y = O and X = NO, Cl, I or OH.

Photolysis of organic nitrites. It has been known for many years that vapour phase pyrolysis or photolysis of organic nitrites (Y = O, X = NO) affords alkoxyl radicals and nitric oxide. But in many cases the radicals produced are consumed in synthetically useless reactions such as fragmentation, disproportionation and non-selective intermolecular hydrogen abstraction. It has been found, however, that when the structure of the molecule is such as to bring the $-\overset{|}{\underset{|}{C}}-O-NO$ group and a C—H bond into close proximity, or potentially close proximity, the alkoxyl radicals produced by photolysis of the nitrites in solution have sufficient energy to bring about selective *intramolecular* hydrogen abstraction according to the general scheme (4.1), with subsequent capture of nitric oxide by the carbon radical and formation of a nitroso-alcohol, which may be isolated as the dimer or, where the structure permits, rearranged to an oxime (Barton, Beaton, Geller and Pechet, 1961) (4.23). The nitroso (oximino) compounds produced may be

$$-\overset{H}{\underset{|}{\overset{|}{C}}}-(C_2)-\overset{ONO}{\underset{|}{\overset{|}{C}}}- \xrightarrow{h\nu} -\overset{NO}{\underset{|}{\overset{|}{C}}}-(C_2)-\overset{OH}{\underset{|}{\overset{|}{C}}}- \longrightarrow \text{dimer} \qquad (4.23)$$

$$\downarrow\uparrow$$

$$-\overset{NOH}{\overset{\|}{C}}-(C_2)-\overset{OH}{\underset{|}{\overset{|}{C}}}-$$

further transformed into other functional derivatives such as carbonyl compounds, amines or cyano derivatives. The photolytic conversion of organic nitrites into nitroso alcohols has become known as the Barton reaction, after its inventor, and the sequence has been widely used (see Nussbaum and Robinson, 1962), particularly in the synthesis of biologically important steroid derivatives.

The reaction is effected by irradiation under nitrogen of a solution of the nitrite in a suitable non-hydroxylic solvent with light from a high pressure mercury arc lamp. A pyrex filter is usually employed to limit the radiation to wavelengths greater than about 300 nm, thus avoiding deleterious side-reactions induced by more energetic lower wavelength radiation. This is possible because of the multiplicity of weak absorption bands of organic nitrites in the region 320–380 nm. It is the absorption of the light due to these weak bands which brings about the dissociation of the nitrite.

Detailed examination of a range of examples leads to the conclusion that the reaction proceeds in discrete steps by an intramolecular radical mechanism, and is strongly favoured by the possibility of a cyclic six-membered transition state. Mechanistic studies using ^{15}N have shown that photolysis of a nitrite gives, in a reversible step, an alkoxyl radical and nitric oxide which are completely dissociated from each other. The alkoxyl radical rearranges rapidly to a carbon radical which can be captured by deuterium atom transfer or by radical trapping reagents to give 'transfer products'. The normal fate of the carbon radical in the absence of trapping reagents is to react relatively slowly with nitric oxide to give the nitroso alcohol (Akhtar and Pechet, 1964). The sequence is pictured in (4.24).

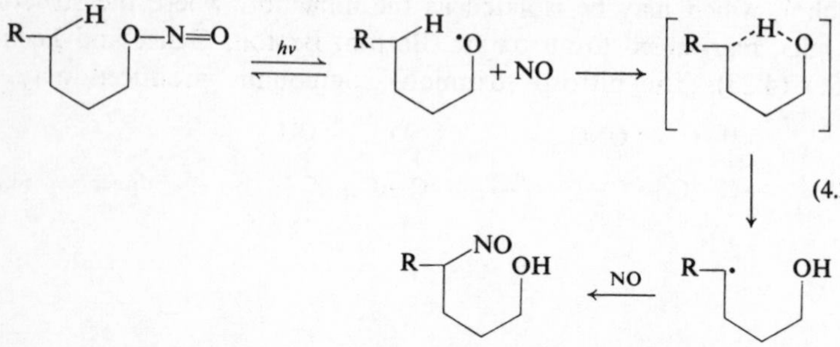

R—CH(H)…O—N=O $\underset{}{\overset{h\nu}{\rightleftharpoons}}$ R—CH(H)…•O + NO ⟶ [R—C…H…O]
↓
R—CH(NO)…OH $\xleftarrow{NO}$ R—C•…OH (4.24)

There is ample evidence to support the view that the hydrogen transfer step takes place through a six-membered cyclic transition state. In practice, reaction occurs almost exclusively by abstraction of a hydrogen atom from the δ-carbon atom. Only a very few examples in which this is not the case have been found. Thus, photolysis of 1-octyl nitrite in benzene solution gave a 45 per cent yield of the dimer of 4-nitroso-1-octanol, by way of the transition state (4.25). No evidence for the formation of any other nitroso-octanol was found. Similarly, 4-phenyl-1-butyl nitrite and 5-phenyl-1-pentyl nitrite are readily converted into nitroso

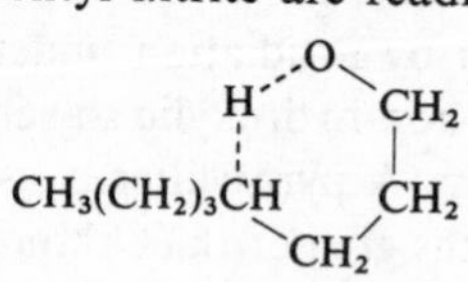

$CH_3(CH_2)_3CH$—H…O—CH_2—CH_2—CH_2 (cyclic) (4.25)

dimers whose structures can be rationalised by assuming a six-membered transition state in the hydrogen transfer step (4.26). In the latter example,

$$C_6H_5(CH_2)_3CH_2ONO \xrightarrow{h\nu} C_6H_5CH(NO)(CH_2)_2CH_2OH$$

$$C_6H_5(CH_2)_4CH_2ONO \longrightarrow C_6H_5(CH_2)_4CH_2O\cdot \qquad (4.26)$$

$$\longrightarrow C_6H_5CH_2CH(NO)(CH_2)_2CH_2OH \quad \text{(via six-membered transition state)}$$

$$\not\longrightarrow \quad \text{(via seven-membered transition state, } C_6H_5\text{—CH—H}\cdots\text{O)}$$

none of the product formed by abstraction of a benzylic hydrogen atom, which would have required a seven-membered transition state, was obtained. In the same way, photolysis of 3-phenylpropyl nitrite did not yield any product corresponding to abstraction of a benzylic hydrogen atom through a five-membered transition state, even although abstraction of the benzylic hydrogen should be highly favourable thermodynamically.

The exact spatial arrangement of the six atomic nuclei forming the transition state is not known with certainty and may not be the same in every case. The chair, boat and semi-chair forms have all been invoked, as well as the form in which the C, H and O atoms are approximately linear (4.27), but reaction in many cases appears to be favoured by the availability of a chair-form transition state (Heusler and Kalvoda, 1964). But even in geometrically favourable cases reaction only takes place provided the distance between the reacting centres is not too great. This may become the deciding factor in rigid molecules. For example, in the radical (25) the distance between the δ-methyl group and the oxygen atom is too large for hydrogen abstraction to take place and the alkoxyl radical undergoes fragmentation instead (4.28). On the other

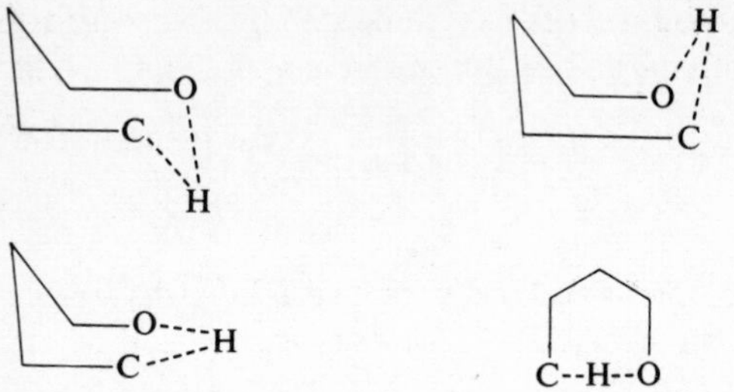

(4.27)

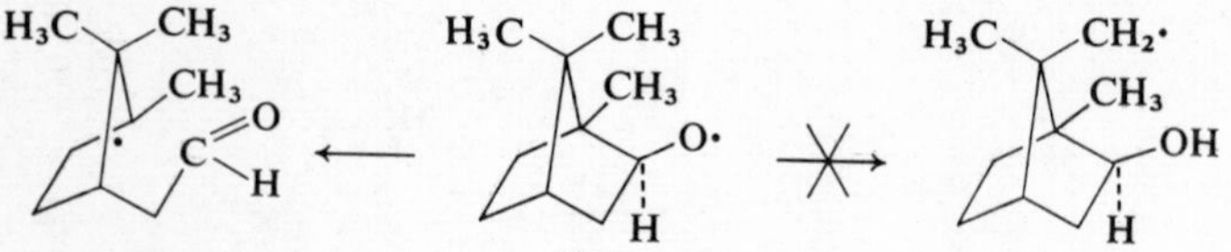

(25)

(4.28)

hand, in the pinane derivative (26), in which the methyl and hydroxyl groups are nearer each other, attack of the oxy radical on the methyl hydrogen atoms takes place readily (4.29), thus providing a convenient method for functionalising the unactivated bridge methyl groups of the pinane structure (Gibson and Erman, 1967; Bosworth and Magnus, 1972). Other similar cases are found in the steroid series. It is estimated

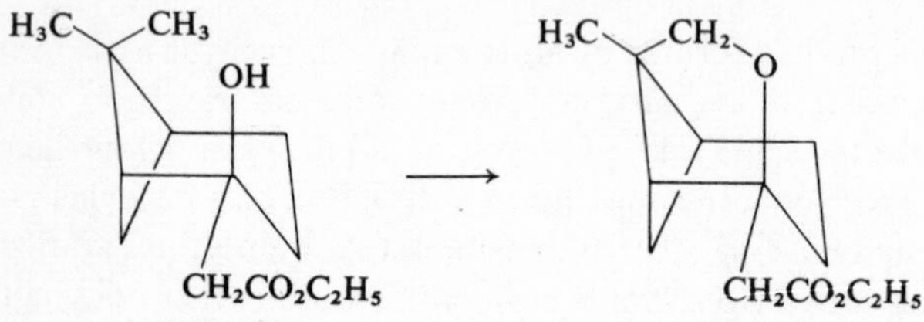

(26)

(4.29)

that the activation energy for the intramolecular abstraction of hydrogen reaches a minimum in fixed systems with an O—C distance of 0.25 to 0.27 nm in the starting material. For distances exceeding 0.28 nm the rate of this reaction falls below that of intermolecular hydrogen abstraction or fragmentation reactions.

Interesting results which further support the proposed reaction pathway for the Barton reaction have been obtained in the photolysis of certain cyclohexyl nitrites. Nitroso dimers or monomers were obtained in only small amounts when cyclohexyl nitrite or *cis-* or *trans-*3-methylcyclohexyl nitrite were irradiated. In these compounds, formation of a six-membered cyclic transition state is unfavourable on conformational grounds. In contrast, rearrangement of both *cis*- and *trans*-2-ethylcyclohexyl nitrites proceeded readily and in good yield, clearly facilitated by the fact that the transition states for the hydrogen transfer steps can adopt stable conformations similar to the *cis*- and *trans*-decalins (4.30).

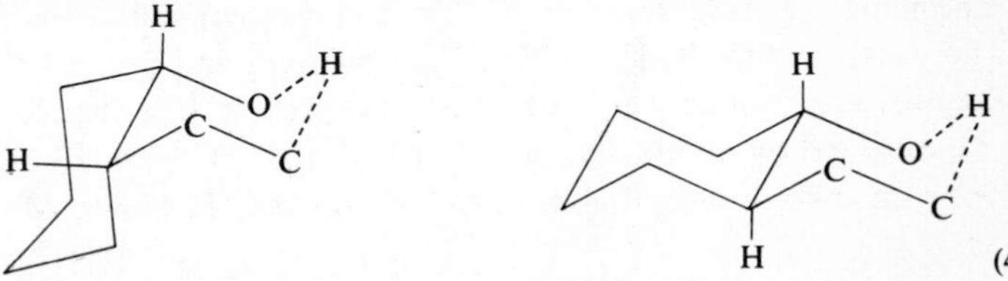

(4.30)

The products obtained from other simple alicyclic nitrites depend on the size of the ring. With cyclobutyl and cyclopentyl nitrites, photolysis results in ring cleavage to give the corresponding linear nitroso aldehydes. With cycloheptyl and cyclo-octyl nitrites, however, 4-nitroso-1-cycloheptanol and 4-nitroso-1-cyclo-octanol are obtained by transannular attack of the oxy radical through a six-membered transition state.

Other factors being equal, it is found that in general, abstraction of a secondary hydrogen atom by an alkoxy radical is much easier than abstraction of a primary hydrogen, although in certain cases attack at the primary hydrogens may be facilitated by favourable geometrical factors. Thus 2-hexyl nitrite afforded a 30 per cent yield of nitroso dimer on photolysis, but with 2-pentyl nitrite, where formation of a six-membered cyclic transition state requires abstraction of a primary hydrogen atom, the yield fell to 6 per cent. Presumably a tertiary hydrogen would react even more easily than a secondary, but apparently no comparative studies have been made (Kabasakalian, Townley and Yudis, 1962).

Intramolecular hydrogen abstraction by alkoxy radicals is always accompanied to a greater or lesser extent by disproportionation, radical decomposition and intermolecular reactions. In the case of oxy radicals derived from tertiary nitrites, as in (27), reaction follows the normal course if tertiary or secondary hydrogen atoms are available for abstraction. But if only primary hydrogens are available the Barton reaction is superseded by alkoxy radical decomposition (4.31). With primary and

$$CH_3{-}C(CH_3)(ONO){-}(CH_2)_2CH_3 \xrightarrow{h\nu} (CH_3)_2C{=}O + ON(CH_2)_2CH_3 \qquad (4.31)$$

(27)

secondary alkoxy radicals, disproportionation to form an alcohol and a carbonyl compound is often favoured over decomposition and always takes place to an appreciable extent, and particularly when the geometric and structural requirements of the Barton reaction are not met.

The most important synthetic applications of the Barton reaction have been in the steroid series, particularly in the functionalisation of the two non-activated C-18 and C-19 angular methyl groups by photolysis of the nitrites of suitably disposed hydroxyl groups. In principle C-18 can be attacked by an alkoxy radical at C-8, C-11, C-15 or C-20, and C-19 by an alkoxy radical at C-2, C-4, C-6 and C-11 (4.32). Most of

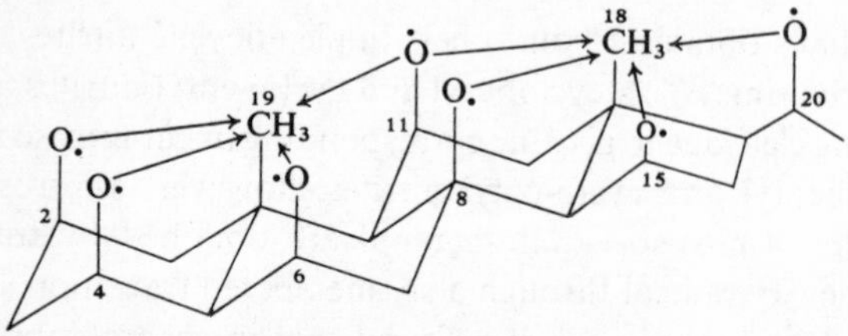

(4.32)

these approaches have been realised in practice, either through the Barton reaction or by one of the related reactions described below (see Akhtar, 1962; Nussbaum and Robinson, 1962; Heusler and Kalvoda, 1964; Kalvoda and Heusler, 1971). The reactions are facilitated by the conformational rigidity of the steroid framework and by the 1,3-diaxial relationship of the interacting groups, which allows easy formation of conformationally favoured six-membered cyclic transition states. Because of this, attack on the primary hydrogen atoms of the methyl groups is much easier than in the aliphatic series, and good yields of nitroso monomers or dimers are often obtained.

Thus the nitrite of 3β-acetoxycholestan-6β-ol (28) in which the methyl and nitrite groups are in the favourable 1,3-diaxial relation, on photolysis in toluene solution gave a 67 per cent yield of the rearranged nitroso dimer which was converted into the oxime (29) by refluxing in propanol solution. The reaction was used by Barton and his co-workers to effect the key step in their elegant synthesis of aldosterone (32), (R = H), (4.34), a biologically important hormone of the adrenal cortex (Barton, Beaton, Geller and Pechet, 1961), by photolysis of the 11β-nitrite (30) in toluene solution. The oxime (31) separated from the solution in 21 per cent yield and on hydrolysis with nitrous acid afforded aldosterone-21-acetate (32), (R = CH_3CO) directly.

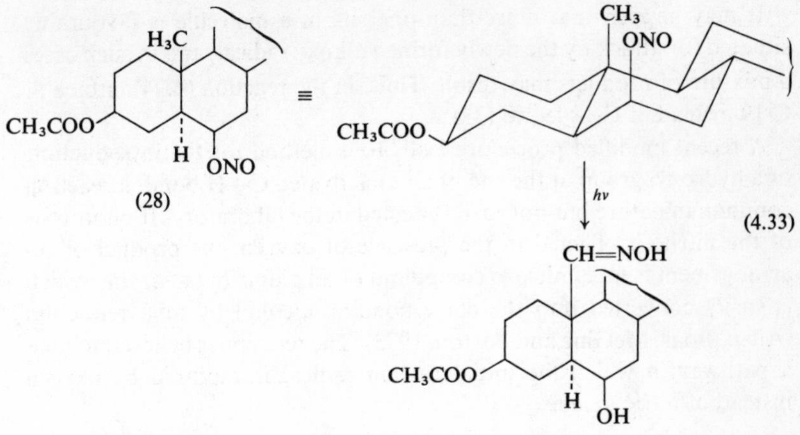

(28)

(4.33)

(29)

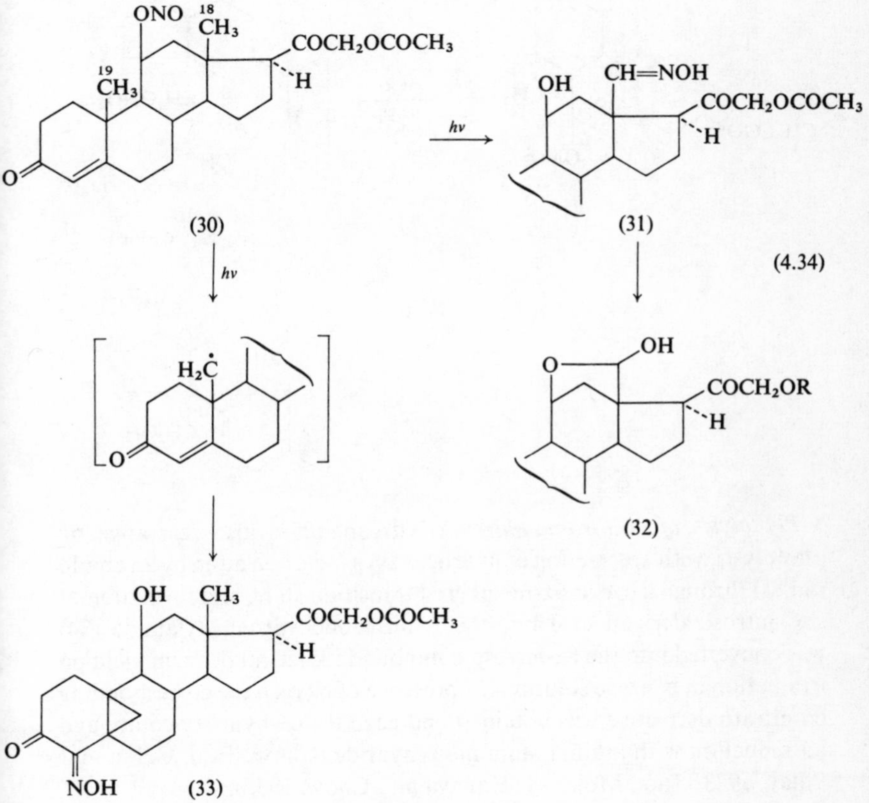

(30)

(31)

(4.34)

(32)

(33)

It may happen that more than one site in a molecule is favourably situated for attack by the newly formed alkoxy radical, and in such cases a mixture of products may result. Thus, in the reaction (4.34), attack at C-19 instead of C-18 led to (33).

A recent modified procedure leads to a method for the introduction of a hydroxyl group at the site of an unactivated C—H bond, a reaction common in nature but not easily effected in the laboratory. If photolysis of the nitrite is effected in the presence of oxygen, the product of re-arrangement is not a nitroso compound or an oxime but a *nitrate*, which is easily converted into the corresponding alcohol by mild reduction (Allen, Boar, McGhie and Barton, 1973). The reaction is believed to take a pathway in which the initial carbon radical is captured by oxygen instead of nitric oxide.

C_8H_{17} — CH_3 — CH_3COO — ONO $\xrightarrow[C_6H_6]{h\nu,\ O_2}$ C_8H_{17} — CH_2ONO_2 — OH

(1) $-H_2O$
(2) Zn, CH_3COOH

C_8H_{17} — CH_2OH

(4.35)

Photolysis of N-nitrosoamides. N-Nitrosoamides also rearrange on photolysis, with abstraction of hydrogen on a δ-carbon atom by an amido radical through a cyclic six-membered transition state, and formation of a C-nitroso derivative. Thus, the *N*-nitrosodehydroabietylamide (34) was converted into the 6α-nitroso compound (35) in 40 per cent yield on irradiation in benzene solution. In presence of oxygen the corresponding 6α-nitrato derivative was obtained and gave the 6α-hydroxy compound on reduction with lithium aluminium hydride (Chow, Tam, Colón and Pillay, 1973; Tam, Mojelsky, Hanaya and Chow, 1975).

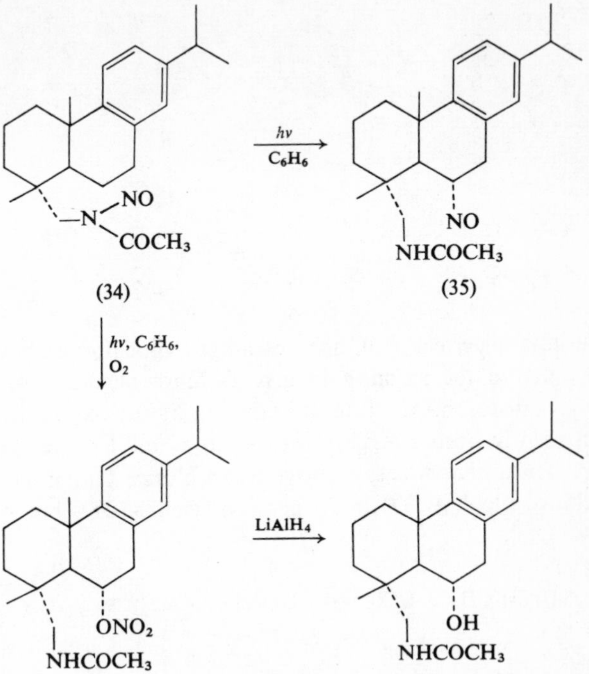

$$\xrightarrow[C_6H_6]{h\nu}$$

(34) (35)

$h\nu$, C_6H_6, O_2

$$\xrightarrow{LiAlH_4} \quad (4{\cdot}36)$$

NO, N, $COCH_3$, NO, $NHCOCH_3$, ONO_2, $NHCOCH_3$, OH, $NHCOCH_3$

Photolysis of hypohalites. The generation of alkoxy radicals which can undergo intramolecular hydrogen abstraction can also be achieved by photolysis of hypochlorites. The initial products of the reaction are 1,4-chlorohydrins, formed again by preferential abstraction of hydrogen attached to the δ-carbon atom. In some cases small amounts of 1,5-chloro alcohols are formed, but in no case has the product of a 1,2-1,3-, 1,4- or 1,7- hydrogen shift been observed. The 1,4-chlorohydrins produced are easily converted into tetrahydrofurans by treatment with alkali and the sequence provides a convenient route to these compounds. The reactions are thought to involve long chains, and, as in the photolysis of nitrites, probably proceed through a six-membered cyclic transition state (Walling and Padwa, 1961, 1963; Jenner, 1962; Akhtar and Barton, 1961, 1964) (4.37).

The hypochlorites are prepared by the action of chlorine dioxide on the corresponding alcohols. They show two sets of ultraviolet absorption bands at 250–260 nm and 300–320 nm and since reaction takes place by irradiation through Pyrex it is the longer wavelength absorption which brings about the photolysis.

$$\text{C(H)}\cdots\text{OCl (ring C–C–C–C)} \xrightarrow{h\nu} \text{C(H)}\cdots\cdot\text{O} \longrightarrow [\text{C}\cdots\text{H}\cdots\text{O}] \longrightarrow \text{C}\cdot\ \ \text{OH} \xrightarrow{\text{ROCl}} \text{C(Cl)}\ \ \text{OH} \xrightarrow{\text{NaOH}} \text{C—O (tetrahydrofuran ring)} \qquad (4.37)$$

In the aliphatic series, a number of alkyl hypochlorites have been shown to undergo the reaction to give δ-chloro alcohols. As in the photolysis of nitrites, abstraction of hydrogen by the oxy radical takes place most readily from a tertiary carbon atom and least readily from a primary. Thus, the dimethylpentyl hypochlorite (36) is converted into the chloro alcohol (37) in 70 per cent yield (4.38), but with the

$$\underset{(36)}{CH_3CH_2CH_2CH_2C(CH_3)_2{-}OCl} \xrightarrow{h\nu} \underset{(37)}{CH_3CHClCH_2CH_2C(CH_3)_2{-}OH} \qquad (4.38)$$

$$\underset{(38)}{(CH_3)_2CHCH_2C(CH_3)_2OCl} \longrightarrow \underset{(39)}{(ClCH_2)(CH_3)CHCH_2C(CH_3)_2{-}OH}$$

isomeric hypochlorite (38) in which a six-membered cyclic transition state requires attack on a primary hydrogen atom, the yield of chloride (39) fell to 29 per cent. In the latter example, no product from attack on the tertiary hydrogen, by 1,4-hydrogen shift, was observed.

With 1-methylcyclo-octyl hypochlorite, transannular abstraction of hydrogen by the oxy radical took place, and a mixture of *cis*- and *trans*-4- and 5-chloro-1-methylcyclo-octanol was obtained. The reactions are most easily effected with tertiary hypochlorites, but with suitable precautions 1,4-chlorohydrins can be obtained from primary and secondary hypochlorites as well.

A competing reaction is β-cleavage of the oxy radical to form, in the case of a tertiary hypochlorite, a ketone and an alkyl chloride (4.39),

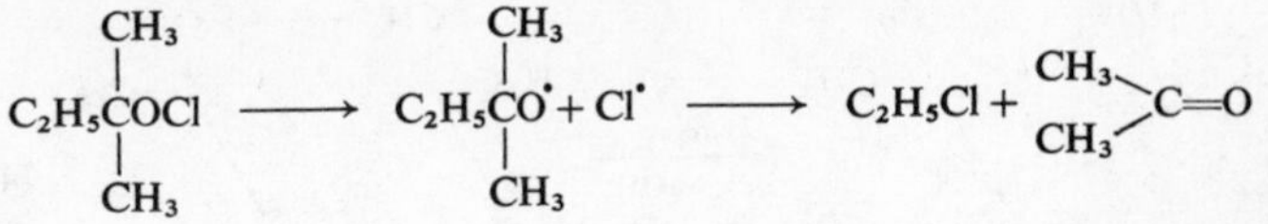

$$\mathrm{C_2H_5C(CH_3)_2OCl} \longrightarrow \mathrm{C_2H_5C(CH_3)_2O^{\bullet}} + \mathrm{Cl^{\bullet}} \longrightarrow \mathrm{C_2H_5Cl} + (\mathrm{CH_3})_2\mathrm{C{=}O} \quad (4.39)$$

The extent of this reaction varies with the structure of the hypochlorite. It becomes the predominant reaction if the carbon chain is too short to permit 1,5-abstraction of hydrogen, or if the geometrical requirements for a six-membered transition state are not met.

A number of reactions in the steroid series involving attack on the non-activated angular methyl groups have been reported. Thus, the hypochlorite (40) on irradiation in benzene solution gave the 1,4-chlorohydrin (41) in 25 per cent yield, smoothly cyclised to the ether (42) with methanolic potassium hydroxide (4.40).

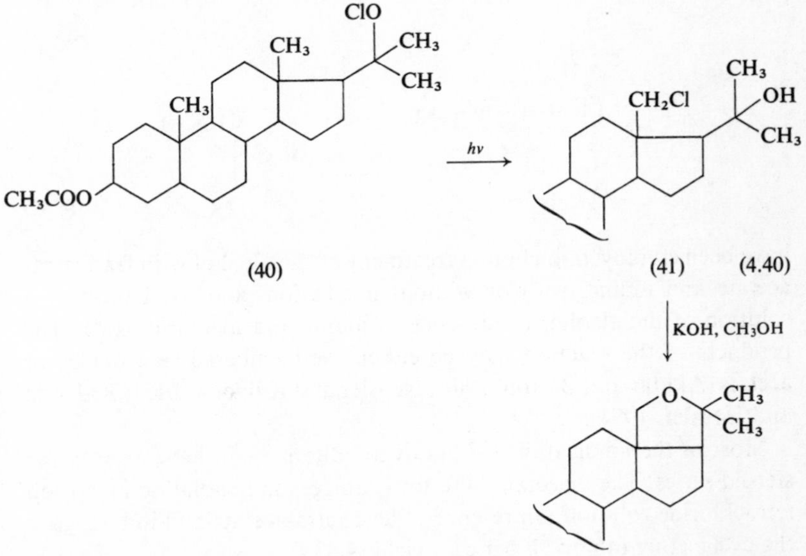

ClO CH₃ CH₃ CH₃ CH₃ CH₃COO hν CH₂Cl CH₃ OH CH₃

(40) (41) (4.40)

KOH, CH₃OH

O CH₃ CH₃

(42)

Hypobromites have also been employed in a limited number of cases (see Brun and Waegell, 1976). Thus, the trimethylcyclohexyl hypobromite (43), on irradiation and treatment with alkali, was converted into the cyclic ether (44) in 60 per cent yield, presumably by way of the 1,4-bromohydrin (4.41). Interestingly, it has been reported that the same change can be effected in 75 per cent yield, in absence of light, when the alcohol is treated with bromine and silver oxide at room temperature. An ionic

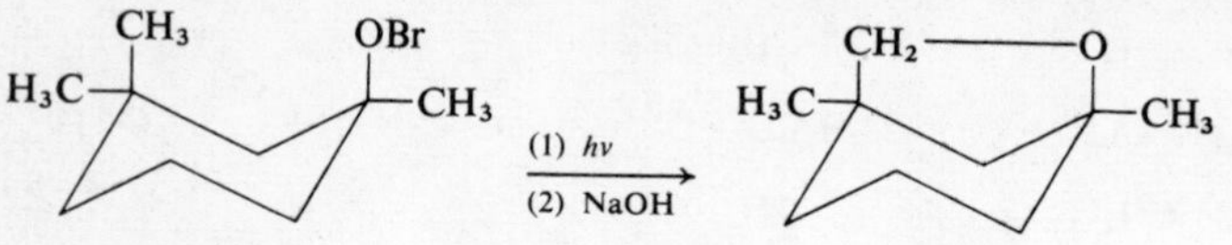

(43) $\xrightarrow[(2)\ NaOH]{(1)\ h\nu}$ (44) (4.41)

mechanism is suggested (Sneen and Matheny, 1964; see also Bosworth and Magnus, 1972) (4.42).

In practice, a more convenient method for generating oxy radicals for intramolecular hydrogen abstractions is by cleavage of hypoiodites prepared *in situ* from the corresponding alcohol. A number of methods

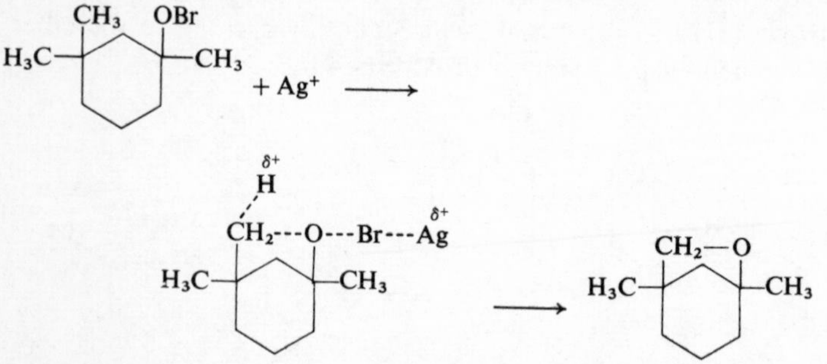

(4.42)

have been employed, including treatment of the alcohol with lead tetra-acetate and iodine, with or without irradiation, and irradiation of a solution of the alcohol in presence of iodine and mercuric oxide. The products of the reaction may be either five-membered ring oxides or acetals (Akhtar and Barton, 1964; Heusler and Kalvoda, 1964; Kalvoda and Heusler, 1971).

Most of the applications of the hypoiodite method have been in the steroid series. The diacetate (45), for example, on irradiation in carbon tetrachloride solution in presence of lead tetra-acetate and iodine, gave the cyclic ether (46) in 90 per cent yield (4.43).

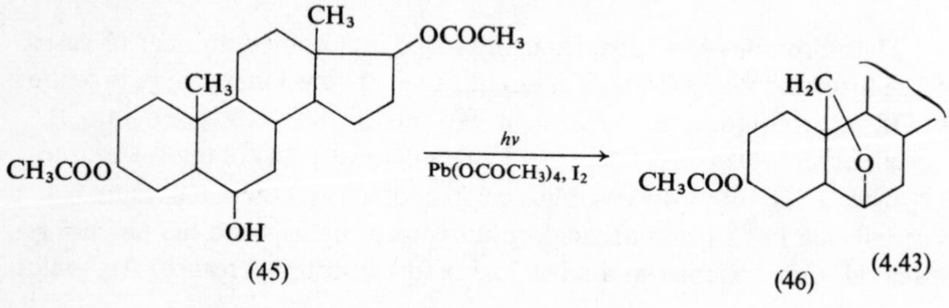

(45) $\xrightarrow[Pb(OCOCH_3)_4,\ I_2]{h\nu}$ (46) (4.43)

Under certain conditions, especially with lead tetra-acetate and iodine, a second substitution of the group being attacked can take place, and hemiacetals are produced. These can be oxidised to lactones (4.44).

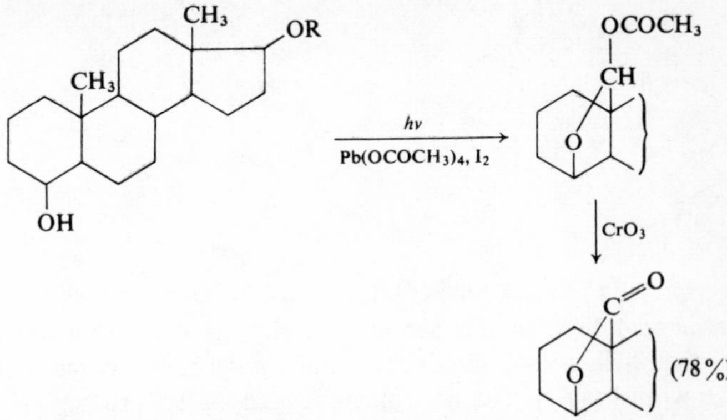

CH_3 OR CH_3 OH

$\xrightarrow[Pb(OCOCH_3)_4,\ I_2]{h\nu}$

$OCOCH_3$ CH O

$\downarrow CrO_3$ (4.44)

C=O O (78 %)

All these reactions are thought to take place by initial formation of the unstable hypoiodite which is converted into the 1,4-iodohydrin in a Barton-type transformation (4.45). In subsequent steps which are influenced by steric factors, the iodohydrin may be converted into the

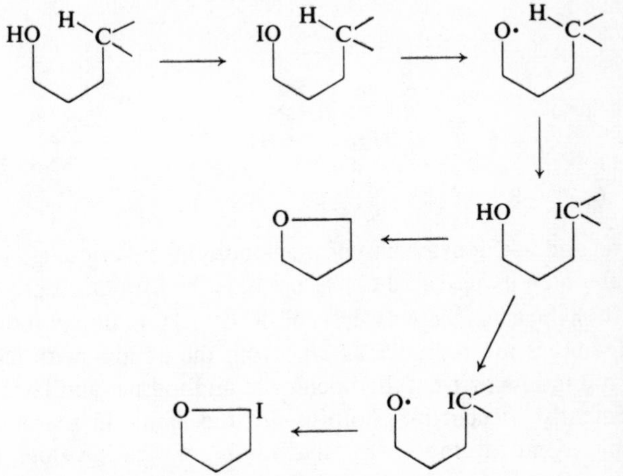

HO H C → IO H C → O· H C

↓

O ← HO IC (4.45)

↙

O I ← O· IC

tetrahydrofuran, or, by further reaction with the reagent, into an iodo derivative from which the hemiacetal is derived.

A related reaction results in the conversion of a cyanohydrin of type (47) (for example a 20-hydroxy-20-cyanosteroid) into the γ-cyanoketone

(48) by action of lead tetra-acetate and iodine, possibly by the pathway shown in (4.47) (Kalvoda, 1970). A new radical reaction which can be used to effect repeated functionalisation at specific sites along an alkyl

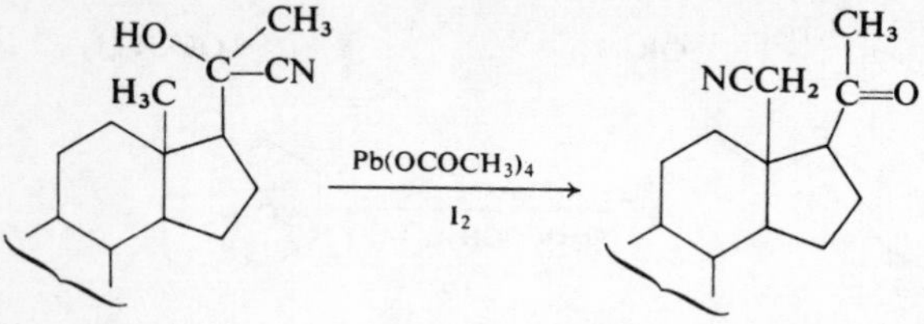

$$\xrightarrow[I_2]{Pb(OCOCH_3)_4} \qquad (4.46)$$

(47) (48)

chain is related to this. A nitrile is treated with a base and then oxygen to give, after acylation, an acyl peroxide. Photolysis of this leads to migration of the nitrile to the fourth carbon along the chain, possibly by the steps shown in (4.47). The new nitrile is ready, after protection of the ketone function, for another such reaction (4.48) (Watt, 1976).

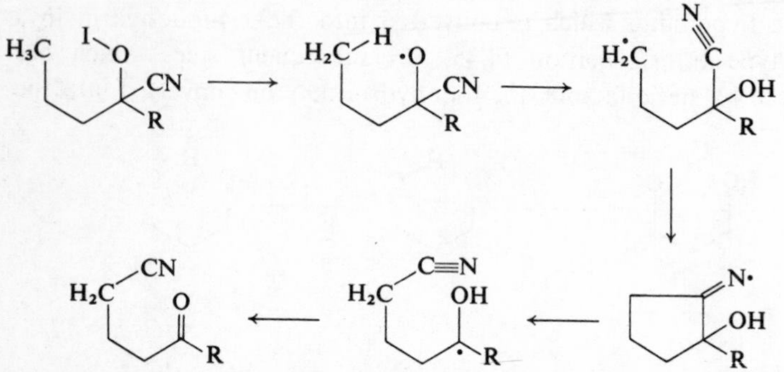

(4.47)

A novel and useful extension of the hypoiodite reaction, again involving attack on an unactivated C—H bond, is the formation of γ-lactones from carboxylic acids by photolysis of the corresponding *N*-iodoamides. The iodoamides are prepared *in situ* from the amides with lead tetra-acetate and iodine or t-butylhypochlorite and iodine, and lactonisation is conveniently effected by photolysis of the amide in presence of the iodinating agent, followed by hydrolysis of the product (Barton, Beckwith and Goosen, 1965). Under these conditions stearamide forms γ-stearolactone together with a smaller amount of the δ-lactone, and 4-phenylbutyramide is converted into phenylbutyrolactone (4.49).

The reaction is thought to proceed by homolysis of the N—I bond and intramolecular hydrogen abstraction by the nitrogen radical through a

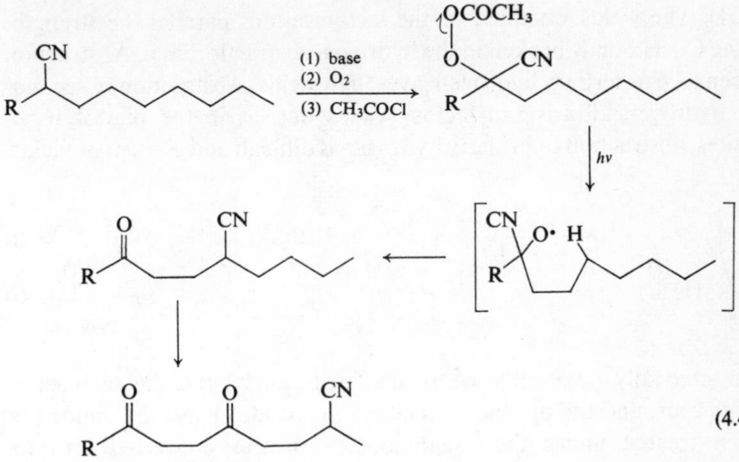

(4.48)

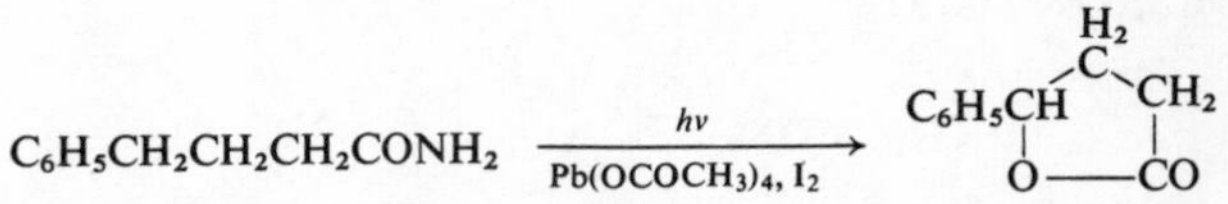

$$C_6H_5CH_2CH_2CH_2CONH_2 \xrightarrow[Pb(OCOCH_3)_4,\ I_2]{h\nu} C_6H_5\overset{}{C}H\text{–}CH_2\text{–}CH_2\text{–}CO\text{–}O \text{ (ring)}$$ (4.49)

six-membered cyclic transition state, followed by formation of the γ-iodoamide. Intramolecular substitution to an imino lactone and hydrolysis then leads to the γ-lactone (4.50). In agreement with the

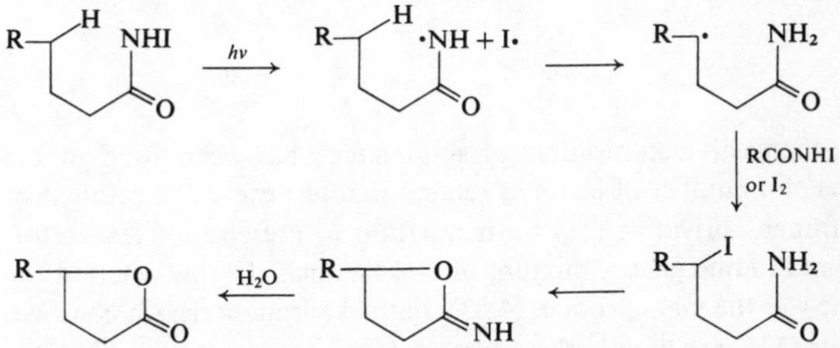

(4.50)

proposed mechanism, the isolation of a derivative of the postulated imino lactone was achieved in the reaction with γ-phenylbutyramide.

Photolysis of the amide of (+)-4-methylhexanamide in presence of t-butylhypochlorite and iodine, and hydrolysis of the resulting iodine chloride complex, gave racemic 4-methyl-4-hexanolactone, in line with the stepwise radical mechanism and against a direct insertion mechanism

(4.51). The yields obtained in the lactonisations parallel the strengths of the C—H bonds broken in the hydrogen abstraction step. Abstraction of benzylic or tertiary hydrogen gives high yields. Abstraction of secondary hydrogen affords satisfactory yields but, as in the photolysis of nitrites, abstraction of primary hydrogen is difficult and gives poor yields.

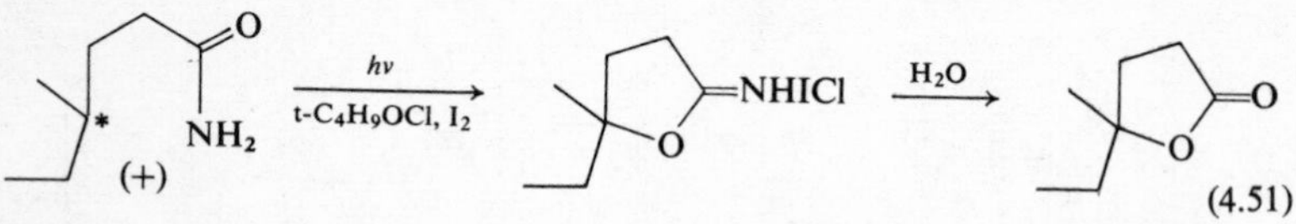

(+) $\xrightarrow[t\text{-}C_4H_9OCl,\ I_2]{h\nu}$ =NHICl $\xrightarrow{H_2O}$ racemic (4.51)

In sterically favourable cases, disubstitution at the site of reaction may occur, and the product is then an anhydride. Thus, the amide (49) when treated under the usual conditions, was converted into the anhydride (50) instead of the expected γ-lactone (Baldwin, Barton, Dainis and Pereira, 1968) (4.52).

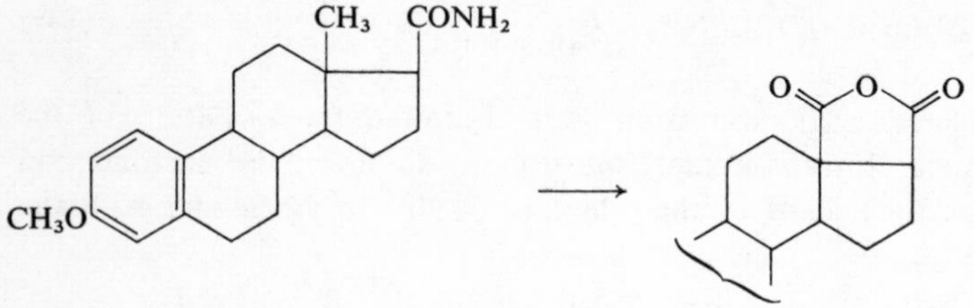

(49) → (50) (4.52)

Photochemical lactonisation of iodoamides has been used in the synthesis of a number of lactones related to diterpenes. The octahydrophenanthrene derivative (51) on irradiation in presence of lead tetraacetate and iodine gave a mixture of two lactones. In this reaction the γ-lactone was the main product (4.53). In the kaurane series, in contrast, the amide (52) gave mainly the δ-lactone (4.54). This is presumably due to steric congestion at the C-6 methylene group in the kauranes, favouring attack at the C-10 methyl group through a seven-membered cyclic transition state.

γ-Lactones can also be conveniently obtained from carboxylic acids by photolysis of the *N*-chloroamides, and this method offers certain advantages, particularly with the lower aliphatic acids where the iodoamide method gives poor yields (Beckwith and Goodrich, 1965). Chloro-

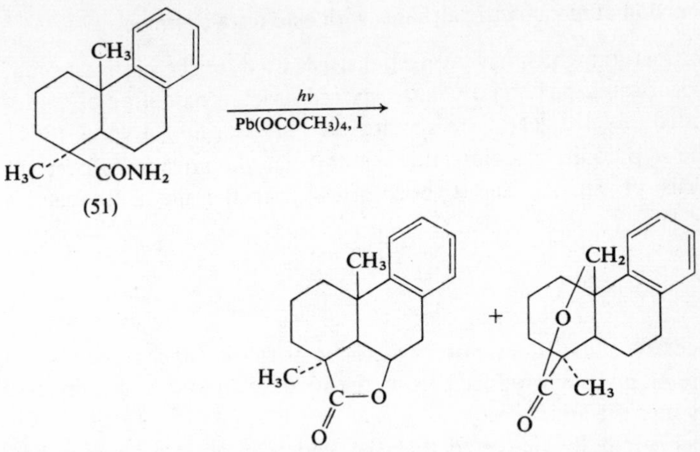

$$\text{(51)} \xrightarrow[\text{Pb(OCOCH}_3)_4,\ \text{I}]{h\nu} \text{lactone } (8\text{–}12\%) + \text{lactone } (2\text{–}3\%) \quad (4.53)$$

amides are readily obtained from the amides by action of t-butyl hypochlorite, and irradiation of a range of examples, followed by hydrolysis, afforded γ-lactones in yields of 20–60 per cent. δ-Lactones were also produced, but always in minor amounts. It is presumed that γ-chloroamides are intermediates and that the reaction path is similar to that suggested for the iodoamide reaction. Interestingly, similar transformations could be effected, in comparable yields, by reaction of the *N*-chloroamides with a catalytic amount of cuprous chloride.

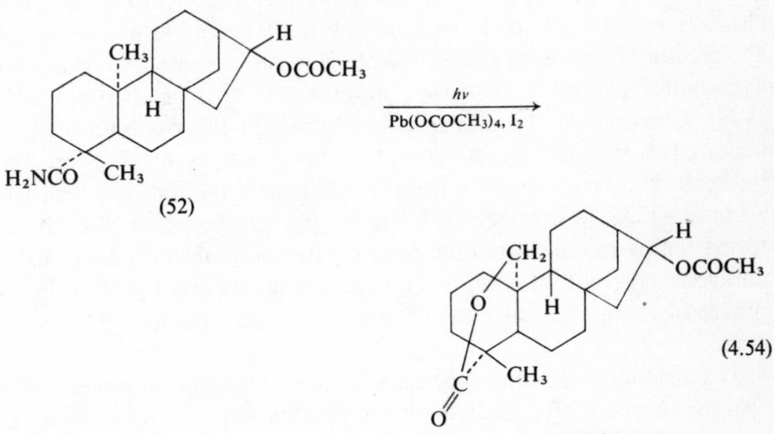

$$\text{(52)} \xrightarrow[\text{Pb(OCOCH}_3)_4,\ \text{I}_2]{h\nu} \text{lactone } (40\%) \quad (4.54)$$

4.4. Reaction of monohydric alcohols with lead tetra-acetate

Another method which has been extensively used for the preparation of tetrahydrofurans, particularly in the steroid series, is oxidation of monohydric alcohols with lead tetra-acetate (see Heusler and Kalvoda, 1964). Like the hypoiodite reaction, this method has the advantage over the photolysis of nitrites and hypochlorites that the unstable reactive

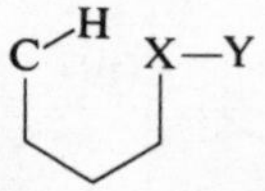

C–H X—Y

intermediate above does not have to be prepared and isolated in a separate step. It is produced from the alcohol *in situ* and converted directly into the alkoxy radical.

It was originally suggested that the lead tetra-acetate reaction proceeded via an electron-deficient oxonium ion intermediate, but it is now believed that it represents yet another example of an alkoxy radical rearrangement. The initial reaction is formation of the triacetoxy lead alkoxide

$$R\text{—}OH + Pb(OCOCH_3)_4 \rightleftharpoons RO\text{—}Pb(OCOCH_3)_3 + HO_2CCH_3 \qquad (4.55)$$

which is easily split thermally or photolytically to give the corresponding alkoxy radical. Formation of the lead alkoxide is dependent on steric effects in the alcohol, and tertiary or strongly hindered secondary alcohols react only slowly.

The steric requirements for the formation of tetrahydrofurans from alcohols with lead tetra-acetate are substantially the same as for the abstraction of hydrogen in the homolysis of nitrites, hypochlorites and hypoiodites (compare Bošnjak, Andrejević, Čeković and Mihailović, 1972). Cleavage of the alkoxide leads through the oxy radical to a favoured six-membered cyclic transition state which in certain cases may be oxidised directly to the tetrahydrofuran with the aid of a lead triacetate radical. In other cases, 1,5-hydrogen transfer occurs, but direct formation of an ether from the resulting carbon radical is energetically unfavourable, and conversion into the tetrahydrofuran is thought to proceed by oxidation to a carbonium ion, which subsequently cyclises (4.56).

This reaction has been extensively applied in the steroid series, and it also provides a useful method for the preparation of simple alkyltetrahydrofurans from primary and secondary aliphatic alcohols. Thus,

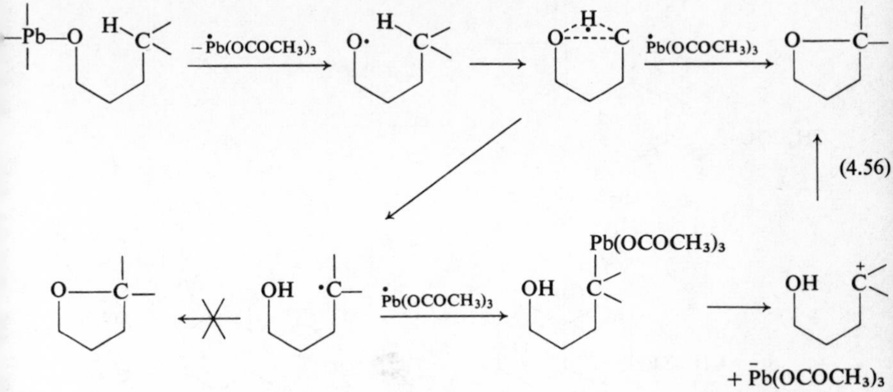

(4.56)

1-heptanol, treated with lead tetra-acetate in boiling cyclohexane gives 2-propyltetrahydrofuran in 50 per cent yield, accompanied by a small amount of 2-ethyltetrahydropyran and 2-heptanol is converted into a mixture of *cis*- and *trans*-2-ethyl-5-methyltetrahydrofuran. The *cis* and *trans* forms were obtained in nearly equal amounts, showing that the oxidation of aliphatic alcohols with lead tetra-acetate is not stereoselective.

Similar conversions of primary and secondary alcohols into tetrahydrofuran derivatives can be readily achieved at room temperature by irradiation of a mixture of the alcohol and lead tetra-acetate in benzene solution, or by irradiation in presence of silver oxide and bromine or mercuric oxide and iodine (Mihailović, Čeković and Stanković, 1969).

In the steroid series the products in general are similar to those obtained in the hypoiodite reaction (4.57) (see p. 252). However, in some cases

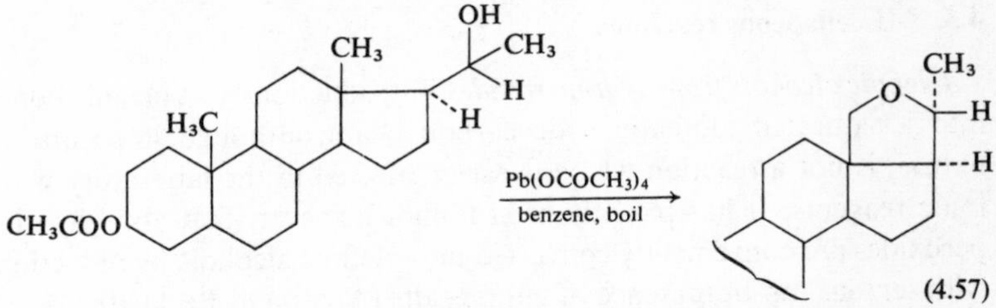

(4.57)

the yields are less good. For example, the 6β-alcohol (53) with lead tetra-acetate and iodine is converted into the ether (54) in high yield but with lead tetra-acetate alone the ketone (55) is the main product (4.58).

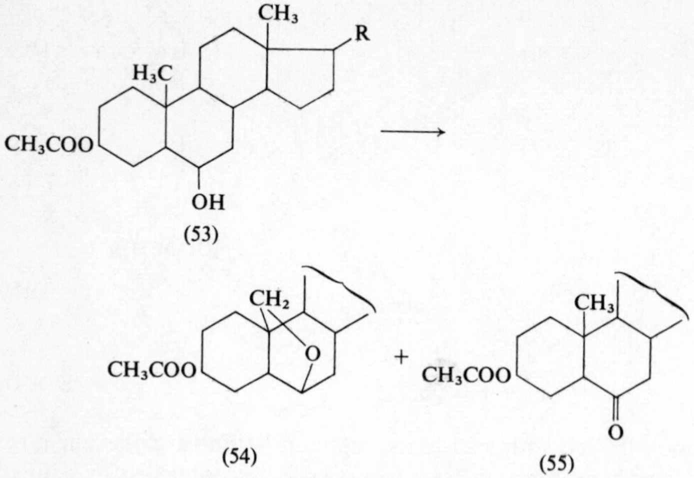

CH_3COO, H_3C, CH_3, R, OH

(53)

CH_2, O, CH_3COO + CH_3, CH_3COO, O (4.58)

(54) (55)

Products of the type of (55) are produced to a greater or lesser extent in all the reactions involving oxy radicals which have been discussed. Carbonyl-forming fragmentation reactions may also take place (4.59),

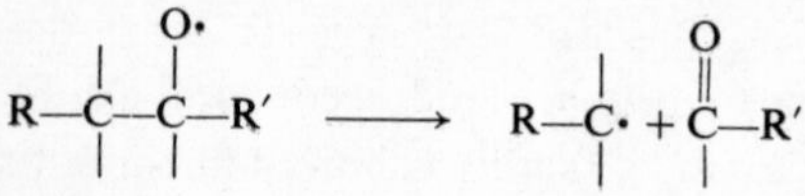

$$\mathrm{R{-}\overset{|}{\underset{|}{C}}{-}\overset{O\cdot}{\underset{|}{C}}{-}R'} \longrightarrow \mathrm{R{-}\overset{|}{\underset{|}{C}}\cdot + \overset{O}{\overset{\|}{\underset{|}{C}}}{-}R'} \qquad (4.59)$$

the extent of which is influenced by the structure of the compound. In particular, fragmentations which give rise to allyl or benzyl radicals, or to a radical adjacent to an oxygen function are strongly favoured, and in such cases hydrogen abstraction may be completely suppressed.

4.5. Miscellaneous reactions

Olefinic alcohols from hydroperoxides. Introduction of an olefinic bond into a saturated aliphatic hydrocarbon chain, although it occurs in nature, is not a reaction which is easily effected in the laboratory with ionic reagents. It has recently been found, however, that alkyl hydroperoxides are conveniently converted into olefinic alcohols by reduction with ferrous ion in presence of cupric salts (Acott and Beckwith, 1964; Čeković and Green, 1974). The key step in this reaction is once again a 1,5-intramolecular hydrogen transfer to the oxy radical produced by reduction of the hydroperoxide; subsequent oxidation of the resulting carbon radical by cupric ion with simultaneous expulsion of a proton leads to formation of the olefin. Thus, 2-methyl-2-hexyloxy radical generated by ferrous ion reduction of the hydroperoxide (56) undergoes intramolecular hydrogen transfer to give the carbon radical (57) which in presence of cupricion is oxidised to a mixture of 2-methyl-4-hexen-2-ol and (mainly) 2-methyl-5-hexen-2-ol (4.60). Similarly, the hexyl hydro-

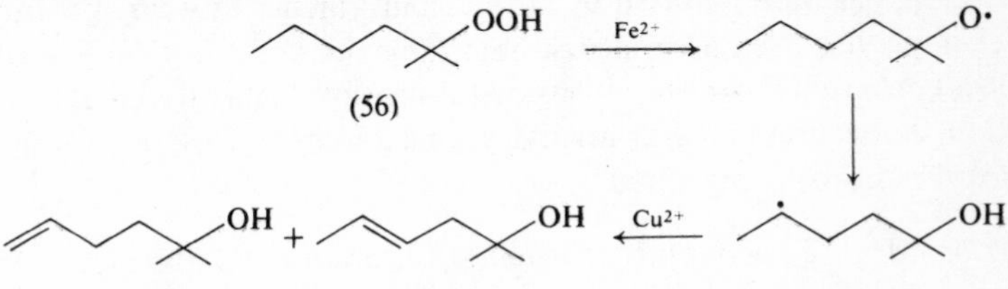

OOH $\xrightarrow{Fe^{2+}}$ O·

(56)

(4.60)

OH + OH $\xleftarrow{Cu^{2+}}$ OH

(46%) (4%) (57)

peroxide (58) affords the two olefins (59) and (60) (4.61). The same reaction can be effected by photolysis of organic nitrites in presence of cupric acetate (Čeković and Srnic, 1976).

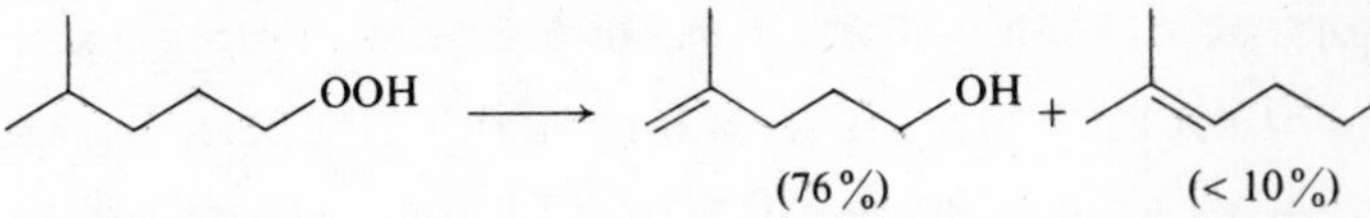

OOH ⟶ OH + OH

(76%) (< 10%) (4.61)

(58) (59) (60)

It is of interest that in all cases studied so far, the olefin with the double bond in the 4,5-position to the hydroxyl group, and not necessarily the thermodynamically more stable olefin, is the main product. This is possibly due to a directive effect of the hydroxyl group exerted through a cyclic transition state of the form (61) which models show is much less strained than the form (62) leading to the γ-olefin, the minor product of the reactions (4.62).

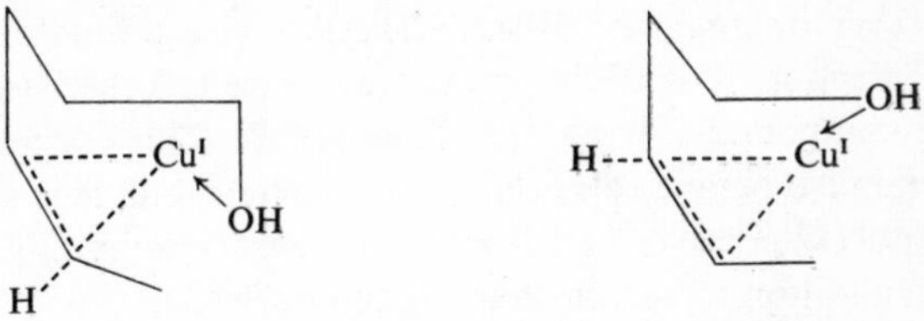

Cu^{I} OH H OH H Cu^{I} (4.62)

(61) (62)

Alcohols from aliphatic and alicyclic hydrocarbons. Another reaction which is common in nature but for which there is yet no convenient laboratory analogy entails the introduction of a hydroxyl group at the site of an unactivated aliphatic C—H bond. One method by which this transformation can be effected in the laboratory in certain cases, by photolysis of an organic nitrite in presence of oxygen, has been described on p. 248. A procedure which may be more closely related to the reaction in nature, involving oxidation of an alcohol with hydrogen peroxide and ferrous ion, has been reported by Groves and Van der Puy (1974). Only one example has been described so far: the oxidation of cyclohexanol, which is converted into *cis*-cyclohexan-1,3-diol in a highly stereoselective reaction. A reaction pathway involving a metal-bound oxidant, not free hydroxyl radicals, is suggested.

Cyclobutanols by photolysis of ketones. Intramolecular abstraction of hydrogen by an oxy radical is also involved in the formation of cyclobutanol derivatives by photolysis of aldehydes and ketones, and in the photolytic fragmentation of ketones to give olefins and ketones of smaller carbon number by cleavage between the α- and β-carbon atoms. Thus, irradiation of methyl neopentyl ketone gives acetone and 2-methylpropene (4.63). A structure in which there is a hydrogen atom

$$(CH_3)_3CCH_2COCH_3 \xrightarrow{h\nu} (CH_3)_2C{=}CH_2 + CH_3COCH_3 \qquad (4.63)$$

on the γ-carbon atom is necessary for those reactions to take place, and a mechanism involving hydrogen transfer through a diradical six-membered cyclic transition state has been proposed (4.64). In many

(4.64) [scheme: C—C(O•) … H, α H$_2$C, C H$_2$ β, C γ ⟶ C—C(OH)=CH$_2$ + CH$_2$=C<]

cases these fragmentation reactions are accompanied by another reaction leading to cyclobutanol derivatives, and although the yields obtained are small the reaction provides a practicable route to otherwise difficultly accessible compounds (Yang and Yang, 1958). Thus, 2-pentanone on irradiation in iso-octane solution affords acetone, ethylene and methylcyclobutanol (12 per cent) and 2-octanone gives 1-methyl-2-propylcyclobutanol in addition to acetone and 1-pentene (4.65).

$$CH_3CO(CH_2)_5CH_3 \xrightarrow[\text{iso-octane}]{h\nu} \text{1-methyl-2-propylcyclobutanol (CH}_3\text{, OH; C}_3\text{H}_7\text{)} \; (17\%) \qquad (4.65)$$

With 1,2-diketones, 2-hydroxycyclobutanones are obtained, and in these cases the yields are much higher owing to the stabilisation of the intermediate excited state by the α-carbonyl group.

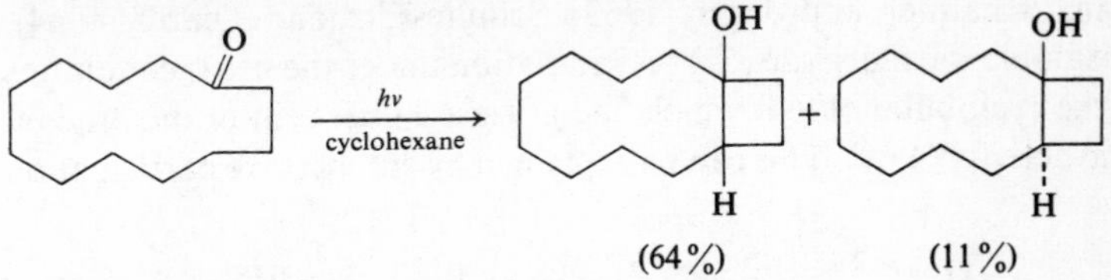

O
hν
cyclohexane
OH
H
+
OH
H
(64%)
(11%)

(4.66)

The reaction has also been effected with cyclic ketones (4.66), and has been applied in the steroid series for reaction at the unactivated C-18 methyl group through attack by a suitably situated carbonyl group at C-20 (see Schaffner, Arigoni and Jeger, 1960) (4.67).

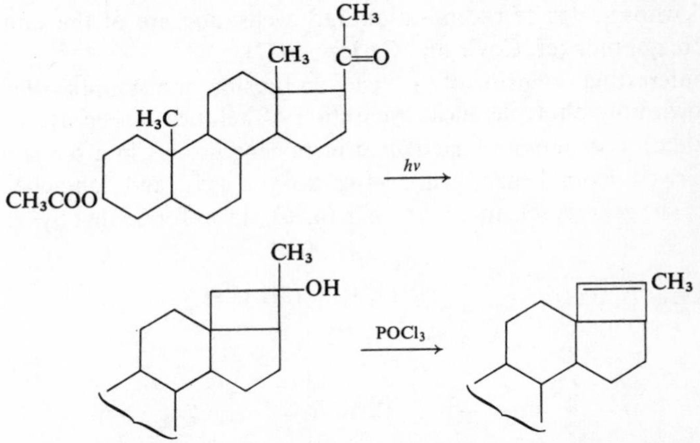

CH3
CH3
C=O
H3C
CH3COO
hν
CH3
OH
POCl3
CH3

(4.67)

These reactions are thought to take place by initial photoexcitation of the carbonyl group to a diradical, followed by 1,5-intramolecular hydrogen transfer to the oxy radical and intramolecular combination of the two carbon radicals with formation of the four-membered ring (4.68). There is evidence, however, that the reaction may not be entirely a stepwise process involving discrete radical intermediates. Experiments

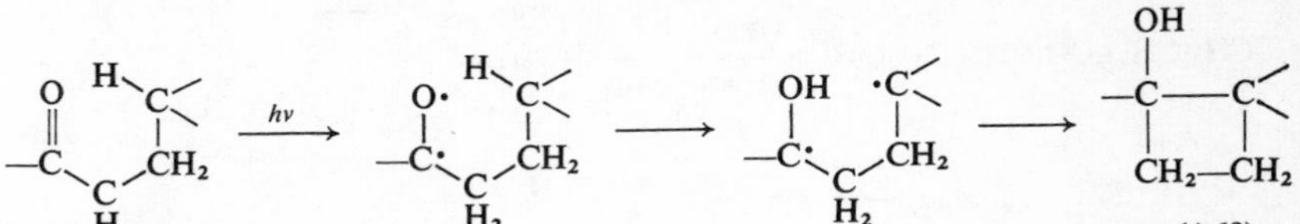

O
H
C
CH2
C
H2
hν
O·
OH
·C
CH2—CH2

(4.68)

using optically active ketones with an asymmetric carbon atom at the γ-position gave cyclobutanol derivatives which retained some optical activity, showing partial retention of configuration during cyclisation (Orban, Schaffner and Jeger, 1963; Schulte-Elte and Ohloff, 1964). For example the aldehyde (63) on irradiation under the usual conditions gave the cyclobutanol (64) which had at least 24 per cent of the original optical activity (4.69). This can be explained by competitive participation

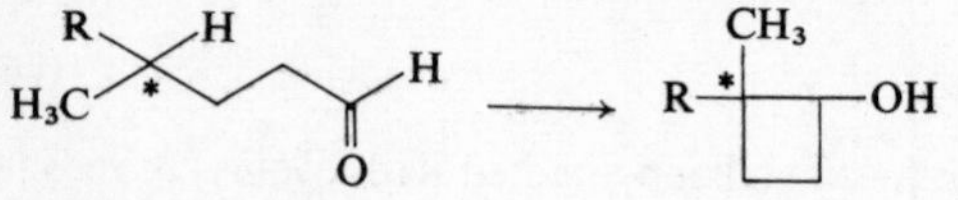

R = $(CH_3)_2CH(CH_2)_3$—

(63) → (64) (4.69)

of stepwise and concerted mechanisms, or by intervention of a short-lived diradical whose rates of racemisation and cyclisation are of the same order of magnitude (cf. Coyle and Carless, 1972).

In an interesting extension of this reaction Breslow and Winnik (1969) have shown how photochemical cyclisation of ketones can be used to effect oxidation of remote unactivated methylene groups in a series of esters derived from benzophenone-4-carboxylic acid and long-chain alcohols. The general scheme is shown in (4.70). It was found that hydro-

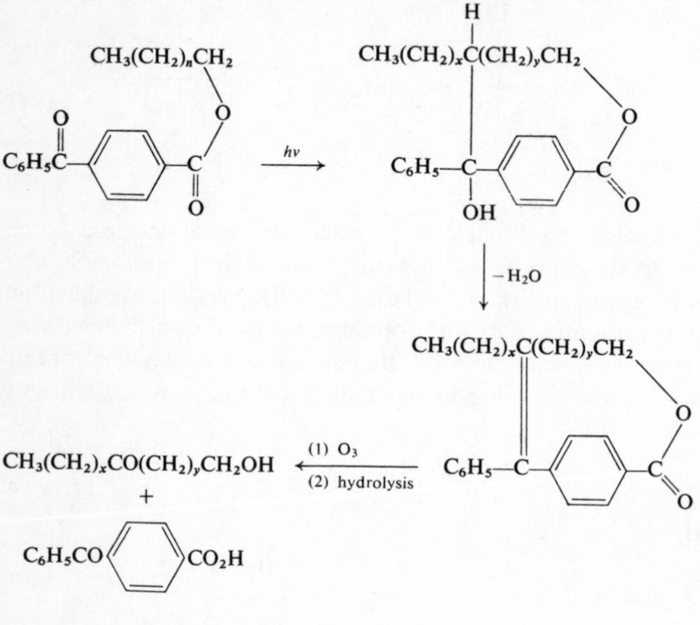

$CH_3(CH_2)_nCH_2$–O–CO–C_6H_4–CO–C_6H_5 $\xrightarrow{h\nu}$ $CH_3(CH_2)_xCH(CH_2)_yCH_2$–O–CO–$C_6H_4$–C(OH)–$C_6H_5$ (cyclic)

$\downarrow -H_2O$ (4.70)

$CH_3(CH_2)_xC(CH_2)_yCH_2$–O–CO–$C_6H_4$–C(=)–$C_6H_5$ (cyclic)

$\xrightarrow{(1)\ O_3;\ (2)\ \text{hydrolysis}}$ $CH_3(CH_2)_xCO(CH_2)_yCH_2OH$ + C_6H_5CO–C_6H_4–CO_2H

gen abstraction can occur over very large distances, much further than the six atoms common in the reactions discussed above. With the ester derived from hexadecanol there was considerable preference for reaction at C-14.

This procedure has been used with conspicuous success to introduce functional groups at unactivated positions of the steroid nucleus. Here the rigidity of the molecules allows much greater selectivity than in the experiments with long-chain alkanes (Breslow, 1972; Breslow, Baldwin, Flechtner, Kalicky, Liu and Washburn, 1973). Thus, irradiation of the benzophenone-4-propionic ester of cholestan-3α-ol (65, $n = 2$) led specifically to attack at three α-oriented axial hydrogen atoms at C-7, C-12 and C-14 to give, after further manipulation of the initial products, the 8(14)- and 14(15)-cholestenols and the ketones (66) and (67). The

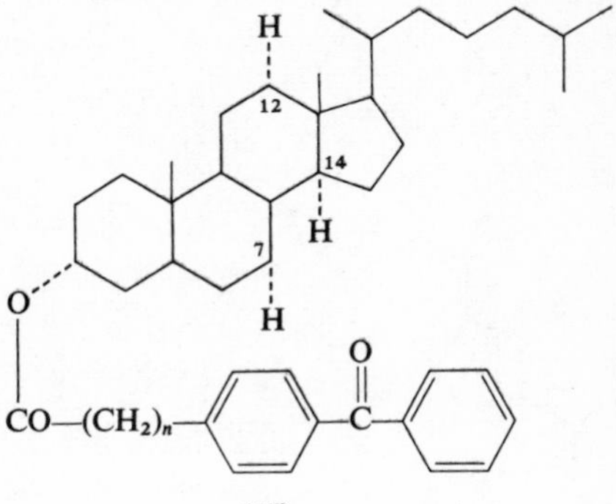

(4.71)

(65)

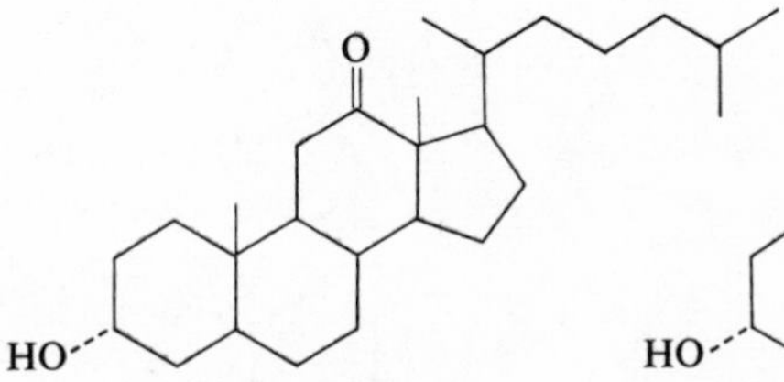

(66) 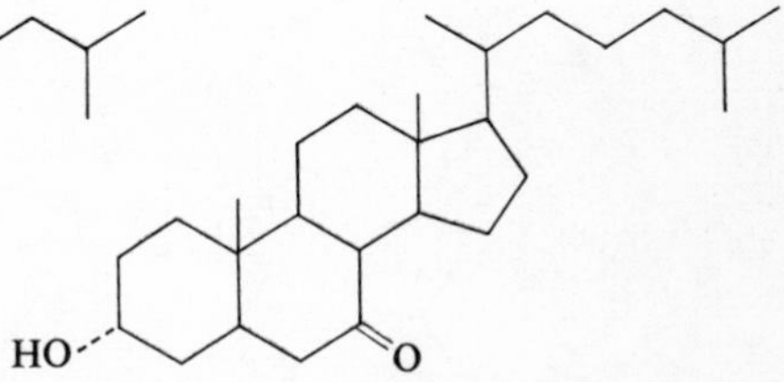 (67)

point of attack on the steroid nucleus can be controlled to some extent by varying the length and the point of attachment of the ester side chain and by manipulation of the experimental conditions. Thus, with the acetate (65, $n = 1$) the 14(15)-cholestenol (70) was obtained specifically in 55 per cent yield.

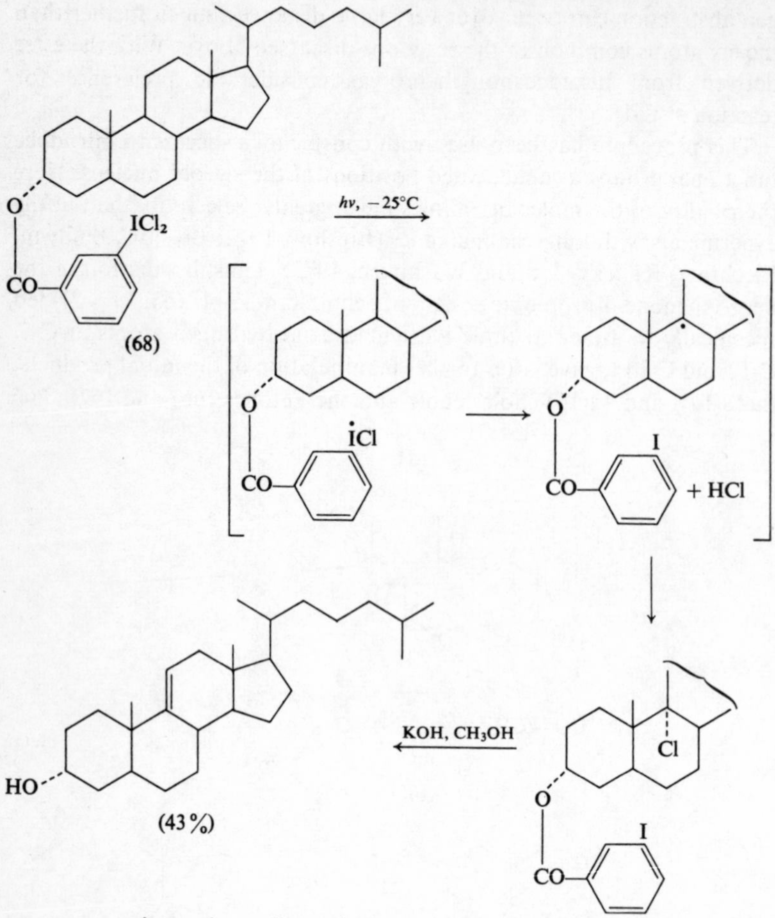

$h\nu$, −25°C

ICl_2

CO

(68)

$\dot{I}Cl$

CO

I

CO

+ HCl

KOH, CH_3OH

Cl

I

CO

HO

(43%)

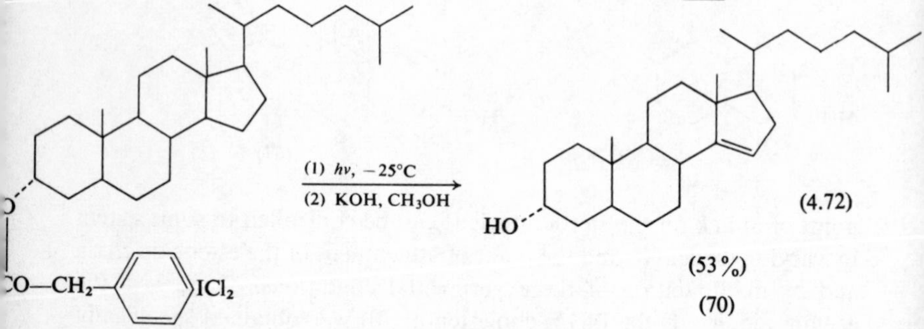

(1) $h\nu$, −25°C

(2) KOH, CH_3OH

HO

(4.72)

(53%)

(70)

CO—CH_2

ICl_2

(69)

An alternative procedure which leads to olefinic products uses a free radical chain reaction of aryliodine dichlorides (Breslow, Corcoran, Dale, Liu and Kalicky, 1974; Breslow, Corcoran and Snider, 1974). Thus the α-cholestanol ester (68), on irradiation followed by treatment with methanolic potassium hydroxide, gave the 9(11)-cholestenol selectively in 43 per cent yield. With the *p*-dichloroiodophenylacetate (69), which can 'reach' further into the steroid nucleus, reaction took place selectively at C-14 to give, after reaction with potassium hydroxide, the 14(15)-cholestenol (70).

These conversions were even more conveniently effected by irradiation of the appropriate iodoaryl ester in the presence of phenyliodine dichloride. A radical chain mechanism is postulated.

Intramolecular hydrogen abstraction by carbon radicals. There are only a few authenticated examples of intramolecular hydrogen abstraction by carbon radicals, and reactions incorporating such a step are not yet of practical synthetic importance. One example is seen in the reaction of 2-(4′-methylbenzoyl)benzenediazonium salts (71) with carbon tetrachloride and alkali, which leads to the formation of small amounts of the products (72) and (73) (4.73). Formation of the latter is attributed to

$$\text{(71)}\ \xrightarrow{CCl_4,\ NaOH}\ \text{(72)} + \text{(73)} \qquad (4.73)$$

an intramolecular transfer involving an aromatic C—H bond (De Los and Relyea, 1956) (4.74). Again, decomposition of ε-phenylcaproyl

$$\longrightarrow \xrightarrow{CCl_4} \text{(73)} \qquad (4.74)$$

peroxide in boiling cumene gives, in addition to the expected 1,10-diphenyldecane, a 15 per cent yield of 5,6-diphenyldecane formed by recombination of rearranged radicals (Grob and Kammüller, 1957)

(4.75). A six-membered cyclic transition state for the hydrogen transfer step is suggested by the observation that the next lower homologue, δ-phenylvaleryl peroxide, on decomposition in boiling benzene, gives

$$[C_6H_5(CH_2)_5CO_2]_2 \xrightarrow{\text{heat}} C_6H_5CH_2(CH_2)_3CH_2^{\bullet}$$

$$\downarrow$$

$$\left[C_6H_5\text{—}\underset{H}{C}\cdots H \cdots CH_2 \text{ (six-membered cyclic transition state)} \right] \quad (4.75)$$

$$C_6H_5\dot{C}H(CH_2)_3CH_3 \longleftarrow$$

$$\downarrow$$

$$\begin{array}{c} C_6H_5CH(CH_2)_3CH_3 \\ | \\ C_6H_5CH(CH_2)_3CH_3 \end{array}$$

tetralin, but no product derivable by an intramolecular hydrogen transfer reaction (see also Nedelec and Lefort, 1972).

5 Synthetic applications of organoboranes and organosilanes

One of the most rapidly expanding areas of organic chemistry in recent years has been in the field of organoboranes. These versatile reagents, which are readily obtained by addition of borane, BH_3 (which exists as the gaseous dimer diborane B_2H_6), to olefins or acetylenes, undergo a wide variety of reactions, many of which are of great value in synthesis (Brown, Kramer, Levy and Midland, 1974; Cragg, 1973). In addition, diborane itself is a powerful reducing agent and attacks a variety of unsaturated groups besides carbon–carbon multiple bonds (Brown and Subba Rao, 1960) (see Table 5.1).

5.1. Reductions with diborane and dialkylboranes

Diborane for reductions, or for the hydroboration reactions described later, is conveniently prepared (5.1) by reaction of boron trifluoride etherate with sodium borohydride in diglyme solution (diglyme is the dimethyl ether of diethylene glycol). In some cases the reagent is generated in the presence of the compound being reduced or hydroborated, and

$$3NaBH_4 + 4BF_3 \longrightarrow 3NaBF_4 + 2B_2H_6 \qquad (5.1)$$

reaction takes place directly. In other cases where especially pure material is required, or where it is desired to avoid reduction of the substrate by the sodium borohydride the diborane is prepared separately and distilled into tetrahydrofuran. The resulting solution of diborane, which probably contains an equilibrium mixture of the free dimer and the tetrahydrofuran–borane complex, can be standardised by a titration procedure (House, 1965).

Reaction of diborane with unsaturated groups takes place readily at room temperature, and the products shown in Table 5.1 are isolated in high yield after hydrolysis of the intermediate boron compound (see also

TABLE 5.1 *Reduction of functional groups with diborane*

Reactant	Product
—CO_2H	—CH_2OH
—CH═CH—	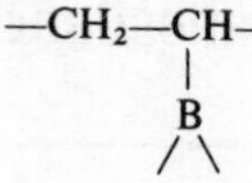 —CH_2—CH(—B<)— (before hydrolysis)
—CHO	—CH_2OH
>CO	>CHOH
—C≡N	—CH_2NH_2
—$CONR_2$	—CH_2NR_2
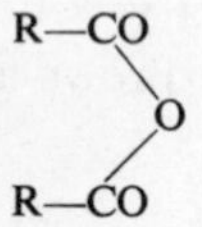 (R—CO)$_2$O	R—CH_2OH
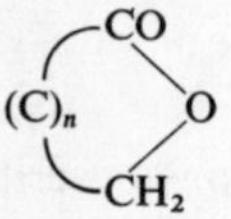 cyclic lactone: $(C)_n$ with CO—O—CH_2	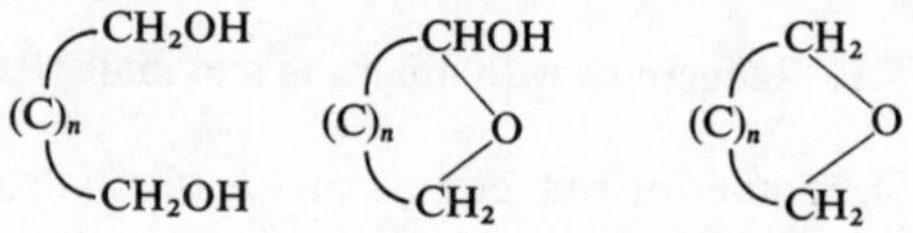 $(C)_n(CH_2OH)_2$; cyclic $(C)_n$ with CHOH—O—CH_2; cyclic ether $(C)_n$ with CH_2—O—CH_2
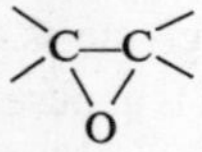 >C—C< epoxide (O bridge)	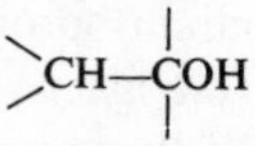 >CH—C(OH)<
—CO_2R	—CH_2OH + ROH
—COCl	no reaction
—NO_2	no reaction

Walker, 1976). Diborane reacts rapidly with water, and reactions must be effected under anhydrous conditions, best under nitrogen, since diborane itself, and the lower alkylboranes, may ignite in air.

A valuable feature of reductions with diborane is that they do not simply parallel those with sodium borohydride. This is because sodium borohydride is a nucleophile and reacts by addition of hydride ion to the more positive end of a polarised multiple bond (see p. 463), whereas diborane is a Lewis acid and attacks electron-rich centres. For example, while sodium borohydride very rapidly reduces acid chlorides to primary alcohols, the reaction being facilitated by the electron-withdrawing effect of the halogen, diborane does not react with acid chlorides under the usual mild conditions. Reduction of carbonyl groups by diborane is believed to take place by addition of the electron-deficient borane to the oxygen atom, followed by the irreversible transfer of hydride ion from boron to carbon (5.2). The inertness of acid chlorides can be

$$>C{=}O + BH_3 \rightleftharpoons >C{=}\overset{+}{O}{-}\overset{-}{B}H_3 \longrightarrow -\underset{H}{\overset{|}{C}}{-}OBH_2 \qquad (5.2)$$

ascribed to the decreased basic properties of the carbonyl oxygen resulting from the electron-withdrawing effect of the halogen. For a similar reason esters are reduced only slowly by diborane.

Remarkable is the ready reduction of carboxylic acids to primary alcohols with diborane, which can often be selectively effected in presence of other unsaturated groups. *p*-Nitrobenzoic acid, for example, is reduced to *p*-nitrobenzyl alcohol in 79 per cent yield. Reduction of carboxylic acids is believed to proceed by way of a triacyloxyborane, the carbonyl groups of which are rapidly reduced by further reaction with diborane.

The reduction of epoxides with diborane is also noteworthy since it gives rise to the *less* substituted carbinol in preponderant amount, in contrast to reduction with complex hydrides (see p. 473). The reaction is catalysed by small amounts of sodium or lithium borohydride and high yields of the alcohol are obtained (Brown and Nung Min Yoon, 1968*a*) (5.3). With 1-alkylcycloalkene epoxides the 2-alkylcycloalkanols

$$CH_3{-}\overset{CH_3}{\overset{|}{C}}{-}CH{-}CH_3 \text{ (epoxide, O bridging)} \xrightarrow[\text{THF, 0°C}]{B_2H_6,\ LiBH_4} (CH_3)_2\overset{OH}{\overset{|}{C}}CH_2CH_3 + (CH_3)_2CHCHOHCH_3$$

(25% of product) (75% of product)

1-methylcyclohexene epoxide $\xrightarrow[\text{THF, 0°C}]{B_2H_6,\ LiBH_4}$ 2-methylcyclohexanol (72% of product) + 1-methylcyclohexanol (28% of product) (5.3) produced are entirely *cis*, and this reaction thus complements the hydroboration–oxidation of cycloalkenes described on p. 283, which leads to *trans*-2-alkylcycloalkanols. Reaction with diborane in the presence of boron trifluoride has also been used for the reduction of epoxides and for the conversion of lactones and some esters into ethers (Brown and Nung Min Yoon, 1968*b*).

Reducing agents more selective than diborane itself are obtained by replacing one or two of the hydrogen atoms of the borane by bulky alkyl groups. Particularly useful reagents of this type are di(3-methyl-2-butyl)borane (disiamylborane) and 2,3-dimethyl-2-butylborane (thexylborane) which are easily obtained by action of diborane on 2-methyl-2-butene and 2,3-dimethyl-2-butene (see p. 275) (Brown, Bigley, Arora and Nung Min Yoon, 1970; Brown, Heim and Nung Min Yoon, 1972). These are much milder reducing agents than diborane, and because of the large steric requirements of the alkyl groups the rate of reaction is strongly influenced by the structure of the compound being reduced. Aldehydes and ketones are converted into the corresponding alcohols, although the reactivity of ketones varies widely with their structure. Acid chlorides, acid anhydrides, and esters do not react, and epoxides are reduced only very slowly. Carboxylic acids are not reduced; in contrast to the rapid reaction with diborane, they simply form the dialkylboron carboxylate which on hydrolysis regenerates the acid. Presumably the large size of the dialkylboron group in the carboxylate prevents further attack by the reagent on the carbonyl group.

Two useful reactions of disiamylborane, which are not paralleled in the reactions of diborane itself, are the reductions of lactones to hydroxyaldehydes and of dimethylamides to aldehydes (5.4). High yields of

$$\text{(butyrolactone)} + (C_5H_{11})_2BH \xrightarrow{\text{THF, 0°C}} \text{(2-tetrahydrofuranyl: H, } OB(C_5H_{11})_2\text{)} \xrightarrow{H_2O} \underset{CH_2OH}{CH_2}\!-\!\underset{CHO}{CH_2} \quad (73\%) \qquad (5.4)$$

$$RCON(CH_3)_2 \xrightarrow[\text{THF, 0°C}]{(C_5H_{11})_2BH} R\overset{H}{\underset{OB(C_5H_{11})_2}{C}}N(CH_3)_2 \xrightarrow{H_2O} RCHO$$

aldehyde are obtained, and reduction with disiamylborane supplements the reactions with hydridoalkoxyaluminates described on p. 475 for the conversion of carboxylic acid derivatives into aldehydes. Another reagent, 9-borabicyclo[3,3,1]nonane (p. 278), is highly selective for the reduction of $\alpha\beta$-unsaturated carbonyl compounds to allylic alcohols in the presence of other functional groups; nitro and ester groups, for example, react only slowly under the reaction conditions (Brown, Krishnamurthy and Nung Min Yoon, 1976).

A valuable feature of the alkylboranes is the high stereoselectivity which they show in the reduction of cyclic ketones to alcohols. With 2-substituted cycloalkanones and bicyclic ketones reaction takes place from the less hindered side of the carbonyl group to give the less stable of the two possible epimers. Thus, whereas reduction of 2-methylcyclohexanone with diborane gave mainly the more stable *trans*-2-methylcyclohexanone, with disiamylborane the preponderant product was the *cis* isomer, and using the bulky optically active terpenoid derivative diisopinocampheylborane as reducing agent, the yield of the *cis* isomer rose to 94 per cent. The product of the last reaction was optically active (Brown and Varma, 1974). Reduction of a series of open-chain methyl ketones, $RCOCH_3$, with this reagent gave secondary alcohols of 11–30 per cent optical purity (Brown and Bigley, 1961).

Very high stereoselectivity is shown also by certain complex trialkylborohydrides such as lithium perhydro-9*b*-boraphenalylhydride (2), which is readily obtainable from *cis,cis,trans*-perhydro-9*b*-boraphenalene (1) with lithium hydride in tetrahydrofuran (Brown and Dickason, 1970), and lithium tri-s-butylborohydride (known commercially as 'Selectride') which is best prepared and used *in situ*, by reaction of tri-s-butylborane with lithium trimethoxyaluminohydride. With the latter

$$LiAlH(OCH_3)_3 + (s\text{-}C_4H_9)_3B \xrightarrow[25°C]{THF} Li(s\text{-}C_4H_9)_3BH + Al(OCH_3)_3$$

reagent 2-methylcyclohexanone is converted into *cis*-2-methylcyclohexanol in 99 per cent yield, and even cyclic ketones with the alkyl group relatively remote from the reaction centre are reduced selectively (Brown and Krishnamurthy, 1972); attack on the carbonyl group always takes place from the equatorial direction.

5.2. Hydroboration

Undoubtedly the most important synthetic application of diborane in organic chemistry is in the preparation of alkyl- and alkenyl-boranes

(1) $\xrightarrow[\text{THF}]{\text{LiH}}$ (2) Li^+

$$\text{2-methylcyclohexanone} \xrightarrow[\text{THF, 0°C}]{Li(s\text{-}C_4H_9)_3BH} \text{2-methylcyclohexanol} \quad (99\%) \qquad (5.5)$$

$$\text{4-methylcyclohexanone} \xrightarrow[\text{THF, 0°C}]{Li(s\text{-}C_4H_9)_3BH} \text{4-methylcyclohexanol} \quad (98\%)$$

by addition to olefins and acetylenes, a process known as hydroboration (Brown, 1962) (5.6). This reaction, which is catalysed by ethers, has

$$>C{=}C< + H{-}B< \longrightarrow H{-}\overset{|}{\underset{|}{C}}{-}\overset{|}{\underset{|}{C}}{-}B<$$

$$-C{\equiv}C- + H{-}B< \longrightarrow H{-}C{=}C{-}B< \qquad (5.6)$$

been applied to a large number of olefins of widely different structures. In nearly all cases the addition proceeds rapidly and simply at room temperature and only the most hindered olefins do not react. With simple olefins (mono- and di-substituted ethylenes) all three hydrogen atoms of the borane are used, to form a trialkylborane (5.7). But if

$$CH_3CH{=}CHCH_3 \xrightarrow[25°C]{B_2H_6,\ \text{diglyme}} (CH_3CH_2\overset{CH_3}{\overset{|}{C}H})_3B \qquad (5.7)$$

the double bond carries a bulky substituent rapid reaction may proceed only to the dialkylborane stage. Similarly, trisubstituted ethylenes normally give the dialkylborane and tetrasubstituted olefins form only monoalkylboranes (5.8). This order of reactivity has been exploited in the preparation of a number of mono- and di-alkylboranes which are less reactive and more selective than diborane (see p. 278). Alkylboranes in which there is still a hydrogen atom attached to the boron usually exist in the dimeric form like diborane itself.

$$(CH_3)_3CCH{=}CHCH_3 \xrightarrow[25^\circ C]{B_2H_6,\ diglyme} [(CH_3)_3CCH_2\overset{\displaystyle CH_3}{\overset{|}{C}}H]_2BH$$

$$(CH_3)_2C{=}CHCH_3 \xrightarrow[25^\circ C]{B_2H_6,\ diglyme} [(CH_3)_2CH\overset{\displaystyle CH_3}{\overset{|}{C}}H]_2BH \qquad (5.8)$$

$$(CH_3)_2C{=}C(CH_3)_2 \xrightarrow[25^\circ C]{B_2H_6,\ diglyme} (CH_3)_2CH\underset{\displaystyle CH_3}{\underset{|}{\overset{\displaystyle CH_3}{\overset{|}{C}}}}{-}BH_2$$

Partially alkylated boranes obtained from hindered alkenes may themselves be used to hydroborate less hindered alkenes. In some cases stepwise addition to two different olefins is possible, leading to trialkylboranes with three different alkyl groups. Thexylborane (p. 272) is particularly useful in this respect. With non-terminal olefins it forms thexylmonoalkylboranes which themselves react with less hindered olefins to give trialkylboranes (cf. also p. 277).

$$\text{Thexyl}{-}BH_2 + \text{cyclohexene} \longrightarrow \text{Thexyl}{-}B(H)(C_6H_{11}) \xrightarrow{CH_2{=}CHCH_2CO_2CH_3} \text{Thexyl}{-}B(CH_2CH_2CH_2CO_2CH_3)(C_6H_{11}) \qquad (5.9)$$

Other methods for preparing organoboranes are known (Smith, 1974), but hydroboration of olefins remains the most widely employed procedure. In addition to borane itself and the alkylboranes several other hydroborating agents have been introduced, mainly with the object of preparing mono- and di-alkylboranes and trialkylboranes containing different alkyl groups. The availability of the latter type of compound would greatly extend the scope of some of the transformations of alkylboranes which are described below. Except in a small number of cases involving highly hindered olefins, mono- and di-alkylboranes are not readily obtained by direct hydroboration of olefins, but recently monoalkylboranes have been prepared by hydroboration of olefins with disubstituted boranes such as dichloroborane (cf. p. 283) (Brown and Ravindran, 1973*c*) or 'catecholborane' (3). The latter compound reacts rapidly with olefins at 100°C to give the corresponding *B*-alkyl derivatives reduction of which with lithium aluminium hydride or, better, aluminium hydride, affords the corresponding monoalkylborane. Addition of an olefin to the crude reaction product gives a mixed trialkylborane RR'_2B (Brown and Gupta, 1971*a*, 1975; Lane and Kabalka, 1976).

OH OH $\xrightarrow[\text{THF, 0°C}]{BH_3}$ (3) BH $\xrightarrow{100°C}$ B— ; $\xrightarrow{AlH_3}$ H_2B— (90%) $\xrightarrow{\text{2-butene}}$ $(s\text{-}C_4H_9)_2B$— (5.10)

Mixed trialkylboranes can also be obtained by a sequence starting from monochloroborane. This compound reacts readily with a variety of olefins to give dialkylchloroboranes in excellent yield (Brown and Ravindran, 1972). With alcohols these form dialkylborinic esters which are reduced with aluminium hydride to dialkylboranes, isolated as their complexes with aluminium methoxide. Reaction of these complexes with olefins gives the mixed trialkylboranes in good yield (Brown and Gupta, 1971*b*). A related route proceeds from a bromodialkylborane by reaction with sodium hydride in presence of an olefin (Pelter, Rowe and Smith, 1975).

A completely different approach to mixed trialkylboranes which proceeds by one-carbon homologation of a symmetrical organoborane is exemplified in (5.11) (Negishi, Yoshida, Silveira and Chiou, 1975). No 'cross-over' products were obtained when a mixture of trioctylborane and tripentylborane was used in the reaction, showing that the rearrangement is intramolecular.

Recently a route to trialkylboranes containing three different alkyl groups has been found. It depends on the discovery that thexylalkyl-

$$(C_4H_9)_3B + LiCH_2SCH_3 \xrightarrow[0°C]{THF} (C_4H_9)_3\bar{B}{-}CH_2SCH_3\ \ Li^+$$

$$\downarrow CH_3I$$

$$(C_4H_9)_2BCH_2C_4H_9\ (93\%) \longleftarrow \left[(C_4H_9)_2\bar{B}(C_4H_9){-}CH_2{-}\overset{+}{S}(CH_3)_2\right] \qquad (5.11)$$

boranes react with some olefins with displacement of the thexyl group to give dialkylboranes RR^1BH, which themselves react with less hindered olefins to form the totally 'mixed' trialkylborane RR^1R^2B (Brown, Katz, Lane and Negishi, 1975).

Directive effects. Addition of diborane to an unsymmetrical olefin could, of course, give rise to two different products, by addition of the boron atom at either end of the double bond. It is found in practice, however, that, in the absence of strongly polar neighbouring substituents, the reactions are highly selective and give predominantly the isomer in which boron is bound to the less highly substituted carbon atom (5.12).

$$CH_3CH_2C(CH_3){=}CH_2 \xrightarrow[diglyme]{B_2H_6} CH_3CH_2CH(CH_3)CH_2B\langle \quad (99\%)$$

$$CH_3CH{=}CHC(CH_3)_3 \quad \text{(58\%) at C-2, (42\%) at C-3} \qquad (CH_3)_2C{=}C(H)CH_3 \quad \text{(2\%), (98\%)} \qquad (5.12)$$

In disubstituted internal olefins there is some preference for addition of boron to the end of the double bond adjacent to the less branched alkyl substituent.

Even greater selectivity can be achieved by hydroboration with less reactive organoboranes such as di(3-methyl-2-butyl)borane (disiamylborane) and 2,3-dimethyl-2-butylborane (thexylborane), which are readily obtained by hydroboration of 2-methyl-2-butene and 2,3-dimethyl-2-butene. Thus hydroboration of 1-hexene with diborane itself gives a product containing 94 per cent of the primary and 6 per cent of the secondary borane, but with disiamylborane the primary borane is formed almost exclusively. Another very useful reagent is 9-borabicyclo[3,3,1]nonane, formed by reaction of diborane with 1,5-cyclo-octadiene in tetrahydrofuran. This is a crystalline compound, which, unlike the two reagents above, can be isolated and stored. On reaction with olefins it readily affords *B*-alkyl derivatives which undergo all the reactions of other trialkylboranes and in its reactions with unsymmetrical olefins it shows even greater selectivity than disiamylborane. *B*-alkyl and *B*-aryl derivatives of 9-borabicyclo[3,3,1]nonane can also be obtained by reaction with the corresponding organolithium compounds, thus making available derivatives, such as the *B*-phenyl compound, which cannot be prepared by way of an olefin (Brown and Rogic, 1969; Brown, Knights and Scouten, 1974) (5.13).

BH B—CH_2CH_2R

$+ B_2H_6$ $\xrightarrow{THF}$ $\xrightarrow{RCH=CH_2}$

C_6H_5Li (5.13)

BHC_6H_5 B—C_6H_5

Li^+ $\xrightarrow[0°C]{CH_3SO_3H}$ $+ H_2$

$+ CH_3SO_3Li$

Because of the large steric requirements of disiamylborane, it reacts at markedly different rates with different kinds of olefinic double bonds, and this allows some remarkably selective reactions. Thus a mixture of 18 per cent *cis*- and 82 per cent *trans*-2-pentene, by a single treatment with disiamylborane, gave *trans*-2-pentene of 98 per cent purity through preferential reaction of the *cis* isomer. Another striking example is the conversion of limonene into the monohydroborated product by preferential reaction at the disubstituted double bond (5.14).

$\xrightarrow[THF,\ 0°C]{(C_5H_{11})_2BH}$ $\xrightarrow[H_2O_2,\ NaOH]{oxidation}$ (5.14)

$CH_2B(C_5H_{11})_2$ CH_2OH

Mechanism of hydroboration. All the available evidence suggests that hydroboration is a concerted process and takes place through a cyclic four-membered transition state

$$\begin{array}{ccc} \overset{\delta-}{C} & \cdots\cdots & \overset{\delta+}{C} \\ \vdots & & \vdots \\ \underset{\delta+}{B} & \cdots\cdots & \underset{\delta-}{H} \end{array}$$

formed by addition to the double bond of a polarised B—H bond in which the boron atom is the more positive. This is supported by the stereochemistry of the reaction (p. 283) and by the directive effect of polar substituents. Thus, in allyl derivatives and in nuclear substituted styrenes the proportion of product formed by addition of boron to the α-carbon atom increases with the electronegativity of the substituent (Brown and Sharp, 1966; Brown and Gallivan, 1968) (5.15).

$$CH_2{=}CHCH_3 \quad (94\%)\ (6\%) \qquad CH_2{=}CHCH_2OC_2H_5 \quad (19\%) \qquad CH_2{=}CHCH_2Cl \quad (40\%)$$

$$p\text{-}CH_3O{-}C_6H_4{-}CH{=}CH_2 \quad (5\%) \qquad C_6H_5{-}CH{=}CH_2 \quad (18\%)\ (82\%) \qquad p\text{-}Cl{-}C_6H_4{-}CH{=}CH_2 \quad (25\%) \tag{5.15}$$

Hydroboration of substituted olefins. Hydroboration is readily effected with olefins containing many types of functional groups, thus greatly extending the usefulness of the reaction in synthesis. Where the other group is not reduced by diborane, hydroboration proceeds without difficulty, except in cases where the organoborane produced has a good leaving group in the β position, when an elimination may ensue. Thus, reaction of 5-chloro-1-pentene with diborane, or, better, disiamyl-

$$ClCH_2CH{=}CH_2 \longrightarrow Cl{-}CH_2CH_2CH_2{-}B(C_5H_{11})_2 \xrightarrow{NaOH\ (\bar{O}H)} \underset{\text{cyclopropane}}{CH_2{-}CH_2{-}CH_2}\ (80\%) \tag{5.16}$$

borane proceeds normally to give the trialkylborane, oxidation of which (see p. 288) gives the 1,5-chlorohydrin in almost quantitative yield, but the organoborane from allyl chloride and disiamylborane is converted into cyclopropane with alkali, and *gem*-dimethylcyclopropane was obtained from αα-dimethylallyl chloride in 75 per cent yield (Brown and Rhodes, 1969). In a related sequence 1,5- and 1,6-cyclodecadienes are obtained by elimination from suitably substituted decalylboranes (5.17) (Marshall, 1971).

OSO_2CH_3 CH_3 $\xrightarrow{B_2H_6,\ THF}$ OSO_2CH_3 CH_3 H $B \leftarrow \bar{O}H$ $\xrightarrow{NaOH}$ CH_3 (90%) (5.17)

In substituted olefins polar substituents may have a determining influence on the direction of addition of diborane to nearby double bonds (Brown and Keblys, 1964; Brown and Sharp, 1968). Thus, in the addition of diborane to 1-chloro-2-butene the initial product is the s-butyl derivative (4) but this immediately undergoes elimination to form the terminal olefin which, after further addition of diborane and final oxidation gives butanol (5.18). Similar results were obtained with the

$$CH_3CH{=}CHCH_2Cl \xrightarrow[THF]{B_2H_6} \underset{\underset{(4)}{B\langle}}{CH_3CH_2CHCH_2Cl} \longrightarrow \qquad (5.18)$$

$$CH_3CH_2CH{=}CH_2 \xrightarrow{B_2H_6} CH_3CH_2CH_2CH_2B\langle \longrightarrow CH_3CH_2CH_2CH_2OH$$

corresponding acetate, but with the ethyl ether there was less tendency for elimination, ethoxy being a poor leaving group, and oxidation gave the ethoxy alcohols almost quantitatively. With allyl alcohols the hydroxy group directs the boron atom to the adjacent carbon atom, and high yields of 1,2-glycols can be obtained by subsequent oxidation. In these reactions elimination is prevented by protection of the hydroxyl group as a borinate ester or by some other means (Brown and Gallivan, 1968) (5.19).

$$C_6H_5CH{=}CHCH_2OH \xrightarrow[\text{THF, 0°C}]{(C_5H_{11})_2BH} C_6H_5CH{=}CHCH_2OB(C_5H_{11})_2$$

$$\downarrow \text{(1) } B_2H_6 \text{; (2) } H_2O_2\text{, NaOH}$$

$$C_6H_5CH_2CHOHCH_2OH \quad (92\%) \qquad (5.19)$$

OH

(1) hydroboration

(2) oxidation

OH

OH

(90%)

Selective hydroboration of olefins containing functional groups which are normally reduced by diborane is more difficult, but can be achieved by use of the stoichiometric amount of diborane or, better, with disiamylborane, since addition to olefinic bonds is so much faster than to most other unsaturated groups. Thus, it has been possible to convert a series of terminally unsaturated olefinic esters into the ω-hydroxy esters by hydroboration with disiamylborane and subsequent oxidation. Disiamylborane does not reduce ester groups, and the ω-hydroxy esters were obtained in excellent yield virtually free from the isomeric secondary alcohols (5.20).

$$CH_2{=}CHCH_2CO_2C_2H_5 \xrightarrow[\text{THF, 0°C}]{(C_5H_{11})_2BH} {>}B{-}CH_2CH_2CH_2CO_2C_2H_5$$

$$\downarrow H_2O_2\text{, NaOH} \qquad (5.20)$$

$$HOCH_2CH_2CH_2CO_2C_2H_5 \quad (78\%)$$

Hydroboration of acetylenes. Acetylenes also can be hydroborated, giving either mono- or di-hydroborated products depending on the borane used and the structure of the acetylene. Hydroboration of di-substituted triple bonds proceeds readily with diborane and can be controlled to give mainly the alkenylborane. Even better results are obtained with disiamylborane or thexylborane or with catecholborane (Brown and Gupta, 1972, 1975). Oxidation of the products with alkaline hydrogen peroxide affords ketones (5.21). The same result is, of course, achieved by acid-catalysed hydration of acetylenes but in certain cases the route via hydroboration may be more regiospecific. Reaction of

$$CH_3CH_2C{\equiv}CCH_2CH_3 \xrightarrow[\text{diglyme}]{(C_5H_{11})_2BH} CH_3CH_2C(B{<}){=}CHCH_2CH_3 \xrightarrow{\text{oxidation}} \underset{(68\%)}{CH_3CH_2COCH_2CH_2CH_3} \quad (5.21)$$

terminal acetylenes with diborane yields mainly dihydroboration products, but with disiamylborane or trimethylamine-t-butylborane high yields of monohydroboration products are obtained. These compounds are synthetically useful because in them the boron atom is attached to the end of the chain and on oxidation they afford aldehydes. The series of reactions, hydroboration followed by oxidation, thus complements the usual acid-catalysed hydration of terminal acetylenes which leads to methyl ketones (Zweifel and Brown, 1963) (5.22).

$$CH{\equiv}CCH_2CH_3 + t\text{-}C_4H_9\bar{B}H_2\overset{+}{N}(CH_3)_3 \longrightarrow {>}\bar{B}(\overset{+}{N}(CH_3)_3){-}CH{=}CHCH_2CH_3 \xrightarrow{\text{oxidation}} OCHCH_2CH_2CH_3$$

$$CH{\equiv}C(CH_2)_5CH_3 \xrightarrow{(C_5H_{11})_2BH} (C_5H_{11})_2B{-}CH{=}CH(CH_2)_5CH_3 \xrightarrow{\text{oxidation}} OCH(CH_2)_6CH_3 \quad (70\%) \quad (5.22)$$

Other useful reagents for hydroboration of acetylenes are monochloroborane and dichloroborane. Monochloroborane, readily obtained by reaction of boron trichloride etherate with lithium borohydride, reacts smoothly with acetylenes in ether to give dialkenylchloroboranes. These are protonolysed to *cis*-alkenes and oxidised to aldehydes or ketones as in the examples cited above, but in addition they are usefully converted into *cis,trans*-dienes by reaction with sodium hydroxide and iodine, and this provides a very convenient route to dienes of this type from acetylenes (cf. p. 292) (Brown and Ravindran, 1973*a*).

$$C_2H_5C{\equiv}CC_2H_5 + BH_2Cl \xrightarrow[0°C]{ether} [(C_2H_5)(H)C{=}C(C_2H_5)]_2BCl \xrightarrow{I_2,\ NaOH} (C_2H_5)(H)C{=}C(C_2H_5){-}C(C_2H_5){=}C(C_2H_5)(H) \quad (85\%) \tag{5.23}$$

By reaction with dichloroborane in presence of boron trichloride acetylenes form alkenyldichloroboranes in high yield. These also can be protonolysed to *cis* olefins and oxidised to aldehydes and ketones. By reaction with water the 1-alkenyldichloroboranes obtained from terminal acetylenes are converted into 1-alkenylboronic acids which can subsequently be converted into stereochemically pure alkenyl iodides and bromides of opposite configuration (Brown and Ravindran, 1973*b*) (see p. 132). The same result is achieved using the *B*-alkenylcatechol-boranes obtained from the acetylenes with catecholborane (Lane and Kabalka, 1976).

Stereochemistry of hydroboration. Hydroboration of olefins and acetylenes is highly stereoselective and takes place by *cis* addition, in the case of olefins to the less hindered side of the multiple bond. Thus, reduction of acetylenes by hydroboration followed by hydrolysis affords *cis* olefins (see p. 287) and 1-alkylcycloalkenes, on hydroboration and oxidation, yield *trans*-2-alkylcycloalkanols almost exclusively (5.24)

$$\text{1-methylcyclohexene} \xrightarrow[diglyme]{B_2H_6} (\text{trans-2-methylcyclohexyl})_2BH \xrightarrow[H_2O_2,\ NaOH]{oxidation} \text{trans-2-methylcyclohexanol (CH}_3\text{, OH)} \tag{5.24}$$

(cf. p. 272). Both the oxidation of the boron–carbon bond to form an alcohol and hydrolysis to form a hydrocarbon have been found to occur with retention of configuration, thus establishing the *cis* stereochemistry of the original addition.

The complementary stereochemistry may sometimes be achieved by photochemically induced *cis* addition of a trialkylborane to cyclohexenes. Thus, irradiation of cyclohexene in presence of triethylborane, and oxidation of the reaction mixture gave stereoselectively *cis*-2-ethylcyclohexanol in 80 per cent yield. The reaction occurs only with cyclohexenes and is believed to proceed by way of the highly strained *trans* configuration of the olefin (Miyamoto, Isayama, Utimoto and Nozaki, 1971).

$h\nu$; $(C_2H_5)_3B$; $(C_2H_5)_2\bar{B}$—C_2H_5 ; H_2O_2, NaOH ; C_2H_5, OH ; $B(C_2H_5)_2$

(5.25)

The hydroboration–oxidation sequence has been widely used to prepare alcohols of known stereochemistry. Thus, cholesterol gives cholestan-3β,6α-diol, and pinene is converted into isopinocampheol, in each case addition taking place from the less hindered side of the double bond (5.26).

The diisopinocampheylboranes (5) obtained from (+)- and (−)-α-pinene are themselves optically active and have been used in the asymmetric synthesis of alcohols. Thus, hydroboration of *trans*-2-butene, *trans*-3-hexene and norbornene with the reagent and oxidation of the resulting organoboranes produced the corresponding alcohols with optical purities of 70–90 per cent. Similarly, a key step in an asymmetric synthesis of loganin (6), an important intermediate in the biosynthesis of indole and monoterpene alkaloids, involved asymmetric hydroboration of 5-methylcyclopentadiene with (+)- or (−)-diisopinocampheylborane. With the (+)-borane the (−)-alcohol was obtained whereas the (−)-borane gave the (+)-alcohol. Both alcohols were at least 98 per cent optically pure (Partridge, Chadha and Uskoković, 1973). The absolute configuration of diisopinocampheylborane may be deduced from the known configuration of α-pinene, and on the basis of a simple model for

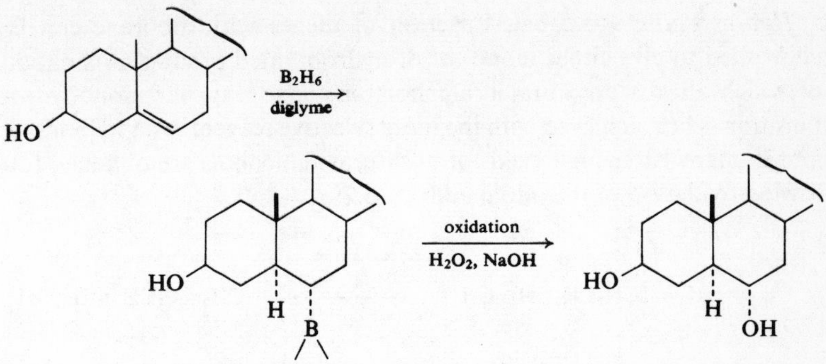

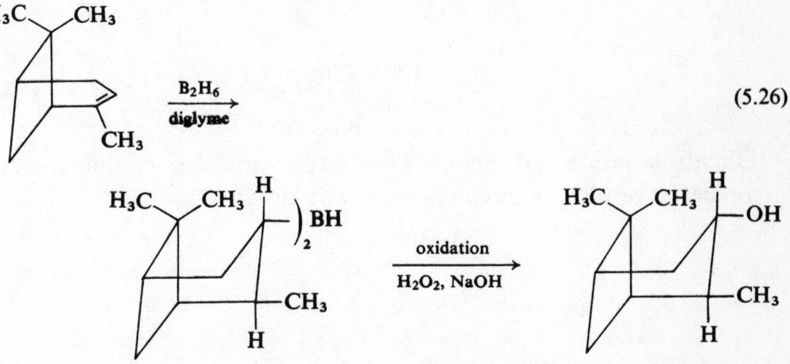

HO — B_2H_6, diglyme →

HO, H, B — oxidation, H_2O_2, NaOH → HO, H, OH

H_3C, CH_3, CH_3 — B_2H_6, diglyme → (5.26)

H_3C, CH_3, H, $)_2BH$, CH_3, H — oxidation, H_2O_2, NaOH → H_3C, CH_3, H, OH, CH_3, H

(5) (89%)

hydroboration it is possible to predict the absolute configuration of the optically active alcohols obtained. Diisopinocampheylborane has also found a use for the resolution of allenes. Thus racemic 1,3-diphenylallene treated with a limited amount of the (+)-borane gave recovered (−)-allene of moderate activity.

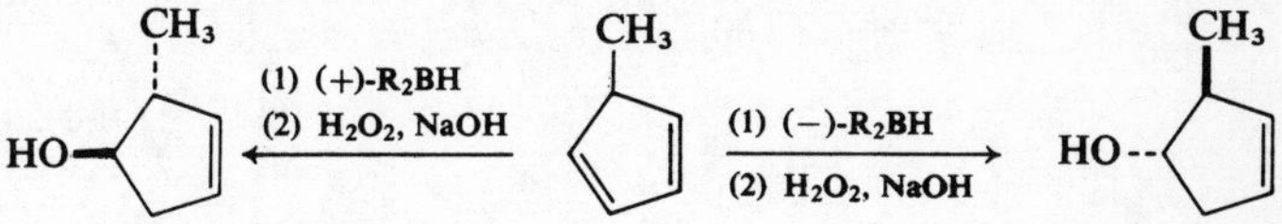

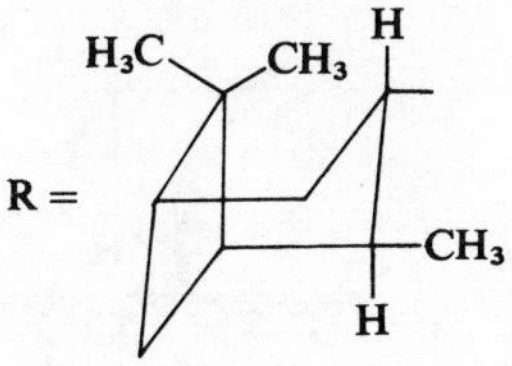

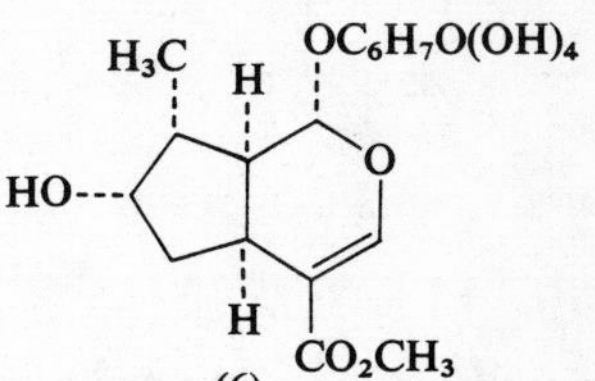

CH_3, HO ← (1) (+)-R_2BH (2) H_2O_2, NaOH — CH_3 — (1) (−)-R_2BH (2) H_2O_2, NaOH → HO, CH_3

R = H_3C, CH_3, H, CH_3, H

H_3C, H, $OC_6H_7O(OH)_4$, O, HO, H, CO_2CH_3

(6)

(5.27)

Hydroboration of dienes. Reaction of dienes with diborane can be controlled to give either mono- or di-hydroborated products, oxidation of which affords unsaturated alcohols or diols. However monohydroboration is best achieved with the more selective reagent disiamylborane, and in many cases high yields of unsaturated alcohols are obtained following oxidation of the initial adduct (5.28).

$$CH_2{=}C(CH_3)CH_2CH_2CH{=}CH_2 \xrightarrow[H_2O_2,\ NaOH]{(1)\ (C_5H_{11})_2BH\ \ (2)\ oxidation,} CH_2{=}C(CH_3)(CH_2)_3CH_2OH \quad (71\%)$$

$$\text{4-vinylcyclohexene } (CH{=}CH_2) \xrightarrow[(2)\ oxidation,\ H_2O_2,\ NaOH]{(1)\ (C_5H_{11})_2BH} \text{4-}(CH_2CH_2OH)\text{cyclohexene} \quad (72\%) \qquad (5.28)$$

Dihydroboration of dienes can be effected with excess of diborane but again better results are often obtained with disiamylborane.

$$\text{1,5-hexadiene} \xrightarrow{\text{thexyl-}BH_2} \text{borepane-B-thexyl } (7)$$

$$\text{limonene} \xrightarrow[THF,\ 0°C]{\text{thexyl-}BH_2} (8) \xrightarrow{CH_3CO_2H^*} CH_3COO\text{-B(thexyl)-menthyl} \xrightarrow{oxidation,\ H_2O_2,\ NaOH} \text{menthol (HO, } CH_3\text{, H, } H_3C\text{, } CH_3) \qquad (5.29)$$

* Preferential protonolysis of the primary organoborane.

With trienes selective hydroboration is more difficult. It has been achieved in some cases by protection of two of the double bonds by incorporation in an organometallic complex. Thus, 1,3,5-heptatriene was converted into 3,5-heptadienol by hydroboration of the tricarbonyliron complex followed by oxidation and removal of the protecting group (Mauldin, Biehl and Reeves, 1972).

Hydroboration of conjugated diynes affords a route to conjugated *cis* enynes, *αβ*-acetylenic ketones and *cis,cis* dienes (Zweifel and Polston, 1970).

In suitable cases hydroboration of dienes with thexylborane gives cyclic or bicyclic organoboranes, some of which undergo synthetically useful transformations (see p. 295) (Brown and Negishi, 1972). 1,5-Hexadiene, for example, is converted mainly into the boracycloheptane (7), and D-(+)-limonene is converted specifically into D-(−)-carvomenthol by way of the cyclic organoborane (8) (5.29).

5.3. Reactions of organoboranes

The usefulness of the hydroboration reaction in synthesis arises from the fact that the intermediate alkylboranes can be converted by further reaction into a variety of other products. On hydrolysis (protonolysis), for example, the boron atom is replaced by hydrogen, and under appropriate conditions the boranes are readily oxidised to alcohols or carbonyl compounds. A practical advantage of these reactions is that it is often unnecessary to isolate the intermediate organoborane.

Protonolysis. Protonolysis is best effected with an organic carboxylic acid (boiling propionic acid is often used), and provides a convenient non-catalytic method for reduction of carbon–carbon multiple bonds (5.30). Protonolysis is believed to take place with retention of configuration at the carbon atom concerned (Brown and Murray, 1961), and in accordance with the proposed mechanism of hydroboration, acetylenes are converted cleanly into *cis* olefins, and cyclic olefins undergo *cis* addition of hydrogen. An advantage of the hydroboration–protonolysis procedure is that it can sometimes be used for the reduction of double or triple bonds in compounds which contain other easily reducible groups. Allyl methyl sulphide, for example, is converted into methyl propyl sulphide in 78 per cent yield.

$$R_2B{-}R \;+\; C_2H_5CO_2H \xrightarrow{C_2H_5CO_2H} R_2BOCOC_2H_5 + RH \longrightarrow 3RH$$

$$C_4H_9CH{=}CH_2 \xrightarrow[\text{diglyme}]{B_2H_6} (C_4H_9CH_2CH_2)_3B \text{ (not isolated)} \xrightarrow{\text{reflux, propionic acid}} C_4H_9CH_2CH_3 \quad (91\%) \tag{5.30}$$

$$C_2H_5C{\equiv}CC_2H_5 \xrightarrow[\text{diglyme}]{(C_5H_{11})_2BH} \underset{H}{\overset{C_2H_5}{}}\!\!>C{=}C<\!\!\underset{B<}{\overset{C_2H_5}{}} \xrightarrow{\text{acetic acid, }25°C} \underset{H}{\overset{C_2H_5}{}}\!\!>C{=}C<\!\!\underset{H}{\overset{C_2H_5}{}} \quad (68\%) \text{ (almost 100\% } cis)$$

$$CH_3SCH_2CH{=}CH_2 \xrightarrow[\text{(2) propionic acid}]{\text{(1) } B_2H_6} CH_3SCH_2CH_2CH_3 \quad (78\%)$$

Oxidation. Oxidation of organoboranes to alcohols is usually effected with alkaline hydrogen peroxide although other methods can be used (Wilke and Heimback, 1962; Kabalka and Hedgecock, 1975). The reaction is of wide applicability and many functional groups are unaffected by the reaction conditions, so that a variety of substituted olefins can be converted into alcohols by this procedure (Zweifel and

$$CH_2{=}CHCH_2Cl \xrightarrow[\text{THF}]{(C_5H_{11})_2BH} (C_5H_{11})_2BCH_2CH_2CH_2Cl \xrightarrow{\text{NaOH, } H_2O_2\text{, pH 7–8}} CH_2OHCH_2CH_2Cl \quad (77\%) \tag{5.31}$$

$$C_6H_{13}C{\equiv}CH \xrightarrow[\text{(2) } H_2O_2\text{, NaOH}]{\text{(1) } (C_5H_{11})_2BH\text{, diglyme, 0°C}} C_6H_{13}CH_2CHO \quad (70\%)$$

Brown, 1963) (5.31). Several examples have already been given. A useful feature of the reaction in synthesis is that it results in over-all anti-Markownikoff addition of water to the double or triple bond, and thus complements the more usual acid-catalysed hydration (see p. 281). This follows from the fact that in the hydroboration step the boron atom adds to the less substituted carbon of the multiple bond.

Another noteworthy feature of the reaction, pointed out on p. 283, is that it leads to *cis* addition of the elements of water to the double bond. This is determined by the mechanism of the reaction and not by the stability of the product, for the thermodynamically more stable product is not invariably formed. Thus, hydroboration–oxidation of 1-methylcyclohexene affords the more stable *trans*-2-methylcyclohexanol, but 1,2-dimethylcyclohexene gives the less stable *cis*-1,2-dimethylcyclohexanol. β-Pinene similarly gives the less stable *cis*-myrtanol by addition of B—H to the less hindered side of the double bond (5.32).

$$\text{β-pinene} \xrightarrow{B_2H_6} \text{(CH}_2\text{B<, H)} \xrightarrow{H_2O_2,\ NaOH} \text{(CH}_2\text{OH, H)} \qquad (5.32)$$

Oxidation of the carbon–boron bond to form an alcohol is believed to take place with retention of configuration, and the reaction path (5.33), involving intramolecular transfer of an alkyl group from boron to oxygen has been proposed.

$$R_3B \xrightarrow{HOO^-} R_2\overset{-}{B}(R)\text{—O—OH} \longrightarrow R_2B\text{—O—R} \xrightarrow{OH^-} R_2\overset{-}{B}(OH)\text{—O—R} \longrightarrow R_2B\text{—}\bar{O} + R\text{—OH} \twoheadrightarrow 3ROH + B(OH)_3 \qquad (5.33)$$

Oxidation of organoboranes to ketones is easily accomplished with aqueous chromic acid, and the sequence provides a very convenient method for conversion of an olefin into a ketone without isolation of the corresponding alcohol (Brown and Garg, 1961).

Other useful conversions which proceed in good yield are the reaction of trialkylboranes with hydroxylamine-*O*-sulphonic acid to give primary amines (Brown, Heydkamp, Breuer and Murphy, 1964) and of dialkylchloroboranes with organic azides forming secondary amines (Brown, Midland and Levy, 1972, 1973). Primary organic bromides and iodides are also readily obtained by reaction of trialkylboranes derived from terminal olefins with bromine or iodine in presence of methanolic sodium methoxide (Brown and Lane, 1970; Brown, Rathke and Rogic, 1968).

Isomerisation and cyclisation of alkylboranes. An interesting and synthetically useful reaction of organoboranes is their ready isomerisation on gentle heating to compounds in which the boron atom is attached to the least hindered carbon of the alkyl group. Thus, tri(3-hexyl)borane on heating at 150°C for 1 h is converted into a mixture in which the major component has the boron atom attached to the terminal carbon atom (5.34). The reaction is strongly catalysed by

$$CH_3CH_2\underset{\underset{\diagup\diagdown}{B}}{\underset{|}{C}}HCH_2CH_2CH_3 \xrightarrow[(B_2H_6)]{150°C} \rangle BCH_2(CH_2)_4CH_3 \quad (90\%) \qquad (5.34)$$

diborane or other molecules containing boron–hydrogen bonds, and proceeds by a succession of eliminations and additions of borane, leading, eventually, to the most stable alkylborane with the boron atom in the least hindered position. In line with this mechanism, it is found that the boron atom can migrate readily along a straight chain of carbon atoms and past a single alkyl branch, but it cannot pass a completely substituted carbon atom (5.35). This reaction can be used to

$$CH_3\!-\!\underset{CH_3}{\overset{CH_3}{C}}\!-\!CH\!=\!\overset{CH_3}{C}\!-\!CH_3 \xrightarrow{B_2H_6} CH_3\!-\!\underset{CH_3}{\overset{CH_3}{C}}\!-\!\underset{\underset{\diagup\diagdown}{B}}{CH}\!-\!\overset{CH_3}{CH}\!-\!CH_3$$

$$\downarrow \text{100°C, 2 h} \qquad (5.35)$$

$$CH_3\!-\!\underset{CH_3}{\overset{CH_3}{C}}\!-\!CH_2\!-\!\overset{CH_3}{CH}\!-\!CH_2\!-\!B\langle$$

bring about 'contrathermodynamic isomerisation' of olefins, for after rearrangement the isomerised olefin can be displaced from the organoborane by heating it with a higher boiling olefin (Brown and Bhatt, 1966).

Oxidation of the rearranged organoboranes provides a means of converting internal olefins into primary alcohols, as illustrated in (5.36).

$$\text{1-ethylcyclohexene } (CH_2CH_3) \xrightarrow{B_2H_6} \text{2-ethylcyclohexyl-}B\langle \xrightarrow{150°C}$$

$$\text{cyclohexyl-}CH_2CH_2B\langle \ (85\%) \xrightarrow{H_2O_2,\ NaOH} \text{cyclohexyl-}CH_2CH_2OH \qquad (5.36)$$

$$(C_2H_5)_2C{=}CHCH_3 \xrightarrow{B_2H_6} (C_2H_5)_2CHCH(B\langle)CH_3$$

$$\downarrow 160°C$$

$$(C_2H_5)_2CHCH_2CH_2OH \xleftarrow[NaOH]{H_2O_2} (C_2H_5)_2CHCH_2CH_2B\langle$$

Certain organoboranes, which cannot isomerise for structural reasons, cyclise at elevated temperatures to form cyclic organoboranes. This reaction appears to be especially prone to occur when a methyl group is in a position to form a five- or six-membered boron heterocycle (5.37). Oxidation of the cyclic products gives 1,4- and 1,5-diols (Brown, Murray, Müller and Zweifel, 1966).

$$(CH_3)_3C{-}CH{=}CH{-}C(CH_3)_3 \xrightarrow{B_2H_6} (CH_3)_2C(CH_3){-}CH_2{-}CH(C(CH_3)_3){-}BH_2$$

$$\downarrow 150°C \qquad (5.37)$$

$$HOCH_2C(CH_3)_2CH_2CHOHC(CH_3)_3 \xleftarrow[NaOH]{H_2O_2} \text{cyclic } (CH_3)_2C\langle CH_2{-}CH(C(CH_3)_3){-}BH{-}CH_2 \rangle + H_2$$

Formation of cyclic compounds by pyrolysis of organoboranes has been extensively studied by Köster (1964).

5.4. Formation of carbon–carbon bonds

One of the most useful applications of organoboranes in synthesis is in the formation of carbon–carbon bonds. As will be seen, most of the reactions used involve the migration of a group from boron to carbon. Many of them proceed in high yield and under very mild conditions.

Coupling reactions. In an interesting extension of the hydroboration reaction it has been found that treatment of the preformed trialkylboranes with alkaline silver nitrate leads to coupling of the alkyl groups. Reaction takes place under mild conditions and provides a useful method for making carbon–carbon bonds (Brown, Verbrugge and Snyder, 1961). 1-Hexene, for example, is converted into dodecane in 70 per cent yield by way of trihexylborane. 'Mixed' coupling can also be achieved in reasonable yield by using an excess of one of the olefins. For example, hexylcyclopentane was obtained in 45 per cent yield from cyclopentene and excess of 1-hexene by hydroboration of the mixture of olefins followed by addition of silver nitrate and sodium hydroxide. The reaction is believed to take place by way of alkyl free radicals formed by breakdown of intermediate silver alkyls.

Another series of reactions leading to conjugated dienes from acetylenes via alkenylboranes may conveniently be discussed here, although they are not, strictly speaking, coupling reactions. Both *cis,trans* and *trans,trans* dienes can be obtained by appropriate choice of reaction. In each case the key step involves migration of an alkenyl group, with retention of configuration, from boron to an adjacent carbon atom. An analogous migration of an alkyl group from boron to oxygen was seen in the oxidation of trialkylboranes with hydrogen peroxide (p. 289). For the synthesis of *cis,trans* dienes the alkenylborane, which is easily obtained from the acetylene by hydroboration, is treated with iodine and sodium hydroxide, when the diene is obtained directly. For example, *cis,trans*-4,5-diethyl-3,5-octadiene is obtained in 68 per cent yield by way of the trialkenylborane prepared by addition of borane to 3-hexyne (Zweifel, Polston and Whitney, 1968). The reaction is analogous to Zweifel's *cis* olefin synthesis (p. 136) and is thought to take the pathway shown (5.38), in which iodide-assisted migration of the *cis* alkenyl group is followed by *trans* de-iodoboronation.

Trans,trans dienes are obtained from 1-chloroacetylenes. Hydroboration with thexylborane gives thexyl-1-chloroalkenylboranes, which react further with acetylenes to thexyldialkenylboranes. Reaction of these

C_2H_5 C_2H_5 $C=C$ H

H B R
H $C=C$ $)_3B$ C_2H_5 C_2H_5 $\xrightarrow[\text{THF, 0°C}]{I_2,\ NaOH}$ C_2H_5 $C—C$ I^+ C_2H_5 I^-

H R
C_2H_5 $C—C---B$ R I
I C_2H_5

H B
C_2H_5 $C—C$ C_2H_5 $\longleftarrow$
I R

$-OH$

H C_2H_5
C_2H_5 $C=C$ H
$C=C$
C_2H_5 C_2H_5

$R =$ C_2H_5 $C=C$ C_2H_5 H

(5.38)

$\vdash\!\!\vdash BH_2 + ClC{\equiv}CC_6H_{11} \longrightarrow \vdash\!\!\vdash B$ H
Cl $C=C$ H C_6H_{11}

1-hexyne

H C_4H_9
$C=C$ H
$\vdash\!\!\vdash B$ H
$C=C$
Cl C_6H_{11}

$\vdash\!\!\vdash B$ OCH_3
$C=C$ C_6H_{11} H $\xleftarrow[\text{THF, 0°C}]{NaOCH_3}$
H $C=C$ H
C_4H_9

H^+

(5.39)

H C_6H_{11}
H $C=C$ H
$C=C$
C_4H_9 H

with sodium methoxide forms intermediates which on protonolysis are converted into the corresponding *trans,trans* dienes; oxidation with alkaline hydrogen peroxide affords $\alpha\beta$-unsaturated ketones. Thus, *trans,trans*-1-cyclohexyl-1,3-octadiene was obtained in 53 per cent yield as shown in (5.39) (Negishi and Yoshida, 1973). The use of alkenylboranes for the stereoselective synthesis of olefins is described on pp. 132–137.

Carbonylation of organoboranes. One of the most useful reactions of organoboranes in synthesis is their reaction with carbon monoxide at 100–125°C which, under appropriate conditions, can be directed to give primary, secondary and tertiary alcohols, aldehydes, and open-chain, cyclic and polycyclic ketones (Brown, 1969, 1972).

It had earlier been observed by Hillman and others that trialkylboranes reacted with carbon monoxide in presence of water at high pressures and a temperature of 50–75°C to form compounds which could be oxidised to dialkylcarbinols. Brown and Rathke (1967) found, however, that at a slightly higher temperature a wide variety of organoboranes reacted with carbon monoxide at atmospheric pressure in diglyme solution to form intermediates which were oxidised by alkaline hydrogen peroxide to trialkylcarbinols in excellent yield (5.40). The

$$R_3B + CO \xrightarrow{\text{diglyme}} (R_3CBO)_x \xrightarrow{H_2O_2,\ NaOH} R_3COH \qquad (5.40)$$

reaction appears to be of wide applicability, and for trialkylcarbinols containing bulky groups gives much higher yields than any other method. Tricyclohexylcarbinol, for example, is obtained from cyclohexene in 85 per cent yield, whereas the Grignard method gives only 7 per cent.

The reaction obviously involves migration of alkyl groups from boron to the carbon atom of carbon monoxide, and this was shown to occur intramolecularly by the fact that carbonylation of an equimolecular mixture of triethylborane and tributylborane gave, after oxidation, only triethylcarbinol and tributylcarbinol; no 'mixed' carbinols were formed. Similarly, dicyclohexyloctylborane gave only dicyclohexyloctylcarbinol. The stepwise mechanism (5.41), involving three successive intramolecular transfers has been proposed. If the carbonylation is conducted in presence of a small amount of water, migration of the third alkyl group is inhibited. Oxidation of the intermediate thus produced then gives the dialkyl ketone instead of the trialkyl carbinol. Alkaline hydrolysis is said (Brown, 1969) to lead to the secondary

$$R_3B + CO \rightleftharpoons R_3\overset{-}{B}\overset{+}{C}O \rightleftharpoons R_2B{-}\underset{\displaystyle O}{\overset{\|}{C}}{-}R$$

$$R_2B\underset{\displaystyle O}{\overset{\|}{C}}R \longrightarrow RB\underset{\backslash\,O\,/}{-}CR_2 \qquad (5.41)$$

$$RB\underset{\backslash\,O\,/}{-}CR_2 \longrightarrow OBCR_3$$

alcohol, but no experimental details or examples of this transformation appear to have been published. The water apparently converts the bora-epoxide into the corresponding hydrate which less easily undergoes transfer of the third alkyl group (5.42). Yields obtained are generally

$$RB\underset{\backslash\,O\,/}{-}CR_2 \xrightarrow[\text{fast}]{H_2O} \underset{OH}{\underset{|}{RB}}{-}\underset{OH}{\underset{|}{CR_2}} \xrightarrow{\text{slow}} (HO)_2BCR_3 \qquad (5.42)$$

high, and the sequence provides a very convenient synthetic route to ketones. 1-Octene, for example, was smoothly converted into dioctyl ketone in 80 per cent yield, and cyclopentene gave dicyclopentyl ketone (90 per cent).

The method can be extended to the synthesis of unsymmetrical ketones by using 'mixed' organoboranes (see pp. 275–277). The 2,3-dimethyl-2-

$$\underset{CH_3\ \ CH_3}{\overset{CH_3\ \ CH_3}{CH{-}C{-}BH_2}} \xrightarrow{\text{olefin A}} \underset{CH_3\ \ CH_3}{\overset{CH_3\ \ CH_3}{CH{-}C{-}B}}\!\!\begin{matrix}R_A\\H\end{matrix} \xrightarrow{\text{olefin B}} \underset{CH_3\ \ CH_3}{\overset{CH_3\ \ CH_3}{CH{-}C{-}B}}\!\!\begin{matrix}R_A\\R_B\end{matrix} \qquad (5.43)$$

butyl (thexyl) group shows an exceptionally low aptitude for migration, and carbonylation of trialkylboranes containing a thexyl group (5.43) in the presence of water, followed by oxidation, leads to high yields of the ketone R_ACOR_B. Because of the bulky nature of the thexyl group, carbonylation of these compounds requires rather more vigorous conditions than usual and generally has to be effected under pressure (5.44). The presence of functional groups in the olefin does not interfere with the reaction, and the procedure can be used to synthesise a ketone from almost any two olefins (Brown and Negishi, 1967*a*).

Dienes similarly yield cyclic ketones, and in a notable extension of the reaction bicyclic ketones have been prepared, as illustrated below for the stereospecific conversion of cyclohexanone into the thermodynamically disfavoured *trans*-perhydroindanone (Brown and Negishi,

$$(CH_3)_2C{=}CH_2 + \text{thexyl}{-}BH_2 \longrightarrow \text{thexyl}{-}B(H)CH_2CH(CH_3)_2 \xrightarrow{CH_2{=}CHCH_2CO_2C_2H_5} \text{thexyl}{-}B(CH_2CH(CH_3)_2)(CH_2)_3CO_2C_2H_5 \xrightarrow[(2)\ H_2O_2,\ NaOH]{(1)\ CO,\ H_2O,\ 50°C,\ 70\ atm} (CH_3)_2CHCH_2CO(CH_2)_3CO_2C_2H_5\ (84\%) \quad (5.44)$$

1967*b*; 1968) (5.45). *Trans*-1-decalone is obtained similarly from 1-allylcyclohexene, and the reaction appears to have considerable generality for the synthesis of *trans* fused polycyclic ketones with a carbonyl group adjacent to the *trans* ring junction. The high stereospecificity

$$CH_2CH{=}CH_2\,|\,CH_2CH{=}CH_2 \longrightarrow \text{cycloheptanone}\ (65\%)$$

$$\text{cyclohexanone} \xrightarrow{\text{2 steps}} \text{1-vinylcyclohexene} \xrightarrow{\text{thexyl}{-}BH_2} \text{trans-fused bicyclic thexylborane} \xrightarrow[(2)\ H_2O_2,\ NaOH]{(1)\ CO,\ H_2O,\ 50°C,\ 70\ atm} \text{trans-hydrindan-1-one} \quad (5.45)$$

of these reactions, leading exclusively to the *trans* fused compounds, is a result of the mechanism of hydroboration which requires *cis* addition of the boron–hydrogen bond to the olefinic double bond.

The carbonylation reaction can also be modified to produce aldehydes and primary alcohols. In the presence of lithium hydridotrimethoxyaluminate, the rate of reaction of carbon monoxide with organoboranes at atmospheric pressure is greatly increased, and the products, on oxidation with buffered hydrogen peroxide, afford aldehydes. Alkaline hydrolysis, on the other hand, gives the corresponding primary alcohol. The reaction is believed to proceed by reduction of one of the intermediates R_3BCO or R_2BCOR (see p. 295) by the complex hydride thus precluding migration of further alkyl groups (5.46). Cyclohexene, for

$$R_3B + CO + LiAlH(OCH_3)_3 \longrightarrow R_2B{-}\underset{OAl(OCH_3)_3Li}{\overset{H}{\underset{|}{\overset{|}{C}}}}{-}R \qquad (5.46)$$

$$\xrightarrow{H_2O_2} RCHO + 2ROH \qquad \xrightarrow{H_2O,\ NaOH} RCH_2OH$$

example, was readily converted into cyclohexanecarboxaldehyde and cyclohexylcarbinol.

A disadvantage of this direct procedure is that only one of the three alkyl groups of the trialkylborane is converted into the required derivative, so that even under ideal conditions a maximum yield of only 33 per cent is possible. This difficulty has been overcome by hydroboration with 9-borabicyclo[3,3,1]nonane (see p. 277). Reaction of the resulting *B*-alkylborabicyclo-nonane with carbon monoxide and lithium hydridotrimethoxyaluminate results in preferential migration of the B-alkyl group, and high yields of aldehydes containing a variety of functional groups have been obtained from variously substituted olefins by this method (5.47). In some cases, where the olefin contains a reducible

$$\text{cyclohexene} \longrightarrow \text{cyclohexane-CHO} \quad (81\%) \qquad (5.47)$$

$$CH_2{=}CHCH_2CO_2C_2H_5 \longrightarrow OCH(CH_2)_3CO_2C_2H_5 \quad (83\%)$$

substituent, better yields of the required aldehyde are obtained with the weaker reducing agent lithium hydridotri-t-butoxyaluminate (Brown, Knights and Coleman, 1969; Brown and Coleman, 1969). Dienes can be converted into the corresponding diols or dialdehydes, and by taking advantage of selective hydroboration with 9-borabicyclo[3,3,1]-nonane it is sometimes possible to effect the conversion of diene into an unsaturated aldehyde.

The reactions are highly stereoselective, the original boron–carbon bond being replaced by a carbon–carbon bond with retention of the original stereochemistry. 1-Methylcyclopentene, for example, was converted entirely into *trans*-2-methylcyclopentylmethanol.

A very useful alternative to the carbonylation route to ketones and trialkylmethanols from alkylboranes has been described by Pelter and his co-workers (Pelter, Smith, Hutchings and Rowe, 1975; Pelter, Hutchings, Smith and Williams, 1975). In this sequence the key intermediate is a cyanoborate and the carbon atom which eventually becomes the carbonyl group of the ketone or the 'methanol' carbon of the trialkylmethanol originates not in carbon monoxide but in cyanide ion. If a solution of a trialkylborane is added to a suspension of sodium cyanide in tetrahydrofuran or diglyme, the cyanide dissolves and a trialkylcyanoborate is formed. Addition of one molar equivalent of benzoyl chloride or, better, trifluoroacetic anhydride to the cooled solution induces two successive migrations of alkyl groups from boron to the adjacent carbon atom of the cyanide, forming the cyclic organoborane intermediate (9), oxidation of which without isolation affords the ketone in high yield (5.48). As in

$$\text{ThexBH}_2 \xrightarrow{\text{olefin A}} \text{ThexB(R}^1\text{)H} \xrightarrow{\text{olefin B}} \text{ThexBR}^1\text{R}^2 \xrightarrow[\text{diglyme}]{\text{NaCN}} \text{Thex}\overset{-}{\text{B}}\text{R}^1\text{R}^2\text{—C}{\equiv}\text{N} \xrightarrow[-78^\circ\text{C}\to 0^\circ\text{C}]{(\text{CF}_3\text{CO})_2\text{O},}$$

$$\left[\text{ThexB(R}^2\text{)—C(R}^1\text{)=N—C(=O)CF}_3\right] \longrightarrow \underset{(9)}{\text{ThexB}\langle\text{C(R}^1\text{R}^2\text{)—N=C(CF}_3\text{)—O}\rangle} \xrightarrow[\text{NaOH}]{\text{H}_2\text{O}_2} \text{R}^1\text{COR}^2 \qquad (5.48)$$

the carbonylation reaction, maximum utilisation of the alkyl groups on the borane is achieved by using thexylborane as the hydroborating agent; the thexyl group does not migrate and mixed ketones are easily obtained from two different olefins by this technique.

If an excess of trifluoroacetic anhydride is employed at the rearrangement step, the reaction takes a different course; a third migration from boron to carbon ensues and oxidation of the product affords a trialkylmethanol.

$$\text{RB(oxazoline, } R_2, CF_3) \xrightarrow[0^\circ C]{\text{trifluoroacetic anhydride}} \text{R}\overset{-}{B}(O_2CCF_3)\text{–}\overset{+}{N}(COCF_3) \text{ ring } \longrightarrow CF_3CO_2B(CR_3)\text{–}N=C(CF_3)\text{–}O \xrightarrow[NaOH]{H_2O_2} R_3COH$$

(5.49)

Both sequences give high yields of products with complete retention of configuration of the migrating groups. Some examples are given in (5.50).

$$\text{cyclopentene} + CH_2{=}CH(CH_2)_6I \longrightarrow \text{cyclopentyl–}CO(CH_2)_8I \quad (76\%)$$

$$\longrightarrow \quad (80\%) \qquad (5.50)$$

$$\text{1-methylcyclopentene} \longrightarrow (\text{2-methylcyclopentyl})_3COH \quad (85\%)$$

Another procedure which is said to be particularly suitable for the preparation of trialkylmethanols containing tertiary or other hindered alkyl groups makes use of the reaction between trialkylboranes and dichloromethyl methyl ether in the presence of lithium triethylcarboxide, which results in the transfer of all three groups from boron to a carbon atom of the ether; oxidation of the product affords the corresponding tertiary alcohol in high yield (Brown, Katz and Carlson, 1973). The precise mechanism of this reaction is not yet known.

In the syntheses of trialkylmethanols described above, all three alkyl groups of the methanol originate in the trialkylborane. An alternative approach in which two alkyl groups of the methanol originate in the trialkylborane and the third in an acyl anion equivalent (p. 44) is effected by treatment of the trialkylborane successively, in the same vessel, with a 1-lithio-1,1-bis(phenylthio)alkane, mercuric chloride and alkaline hydrogen peroxide (5.51). Excellent yields of the tertiary alcohols are

$$\text{(}t\text{-Bu)B(C}_6\text{H}_{11})_2 + \text{iso-C}_3\text{H}_7\text{—}\overset{-}{\text{C}}(\text{SC}_6\text{H}_5)_2\ \text{Li}^+ \xrightarrow[20^\circ\text{C}]{\text{THF}} \left[(t\text{-Bu})\overset{-}{\text{B}}(\text{C}_6\text{H}_{11})_2\text{—C(SC}_6\text{H}_5)_2\text{—C}_3\text{H}_7\text{-iso}\right]\text{Li}^+$$

$$\longrightarrow (t\text{-Bu})\text{B(C}_6\text{H}_{11})\text{—C(C}_6\text{H}_{11})(\text{SC}_6\text{H}_5)\text{—C}_3\text{H}_7\text{-iso} \xrightarrow[-78^\circ\text{C}]{\text{HgCl}_2} (t\text{-Bu})\text{B(Cl)—C(C}_6\text{H}_{11})_2\text{—C}_3\text{H}_7\text{-iso}$$

$$\xrightarrow{\text{H}_2\text{O}_2,\ \text{NaOH}} \text{HO—C(C}_6\text{H}_{11})_2\text{—C}_3\text{H}_7\text{-iso} \quad (98\%) \tag{5.51}$$

obtained. The overall result is synthetically equivalent to the reaction of a Grignard reagent with a ketone or an ester but the present method has the advantage that it is applicable in the presence of functional groups which would be attacked by Grignard reagents (Hughes, Pelter and Smith, 1974). By using bis(phenylthio)methyl-lithium and trialkylboranes, dialkylmethanols are obtained (Hughes, Pelter, Smith, Negishi and Yoshida, 1976).

Trialkylmethanols may also be obtained from alkynyltrialkylborates. Warming these with aqueous hydrochloric acid leads to migration of two alkyl groups from boron to carbon, producing a t-alkylborane which affords the t-alcohol on oxidation (Midland and Brown, 1975).

Reactions of trialkylalkynylborates. Trialkylalkynylborates, which are easily obtained by reaction of lithium acetylides with trialkylboranes, serve as intermediates for the synthesis of acetylenes, ketones, dienes and diynes. The reactions again involve migration of a group from boron to carbon, induced by attack of an electrophile on the triple bond of the alkynyl substituent. Substituted acetylenes are obtained in excellent yield by reaction of the borates with iodine (Suzuki, Miyaura, Abiko, Itoh, Brown, Sinclair and Midland, 1973; Midland, Sinclair and Brown, 1974). The same transformation can, of course, often be effected by

$$(C_4H_9)_3B + LiC{\equiv}CC_6H_5 \xrightarrow{THF} (C_4H_9)_2\overset{-}{B}(C_4H_9){-}C{\equiv}CC_6H_5\ Li^+ \quad (I{-}I)$$

$$\downarrow I_2,\ -78°C \qquad (5.52)$$

$$C_4H_9C{\equiv}CC_6H_5\ (98\%) \longleftarrow (C_4H_9)_2B(C_4H_9)C{=}C(C_6H_5)I$$

direct alkylation of the alkali metal acetylide, but the present procedure can be used to introduce substituents bearing functional groups which might react with the acetylide in the direct reaction. The mechanism of the reaction is thought to involve electrophilic attack of iodine on the triple bond with migration of an alkyl group from boron to the adjacent carbon and subsequent deiodoboration with regeneration of the triple bond. That is to say migration is induced by generation of a partial positive charge on the adjacent carbon rather than by nucleophilic attack on the boron atom as in the rearrangements discussed above. This reaction was extended by Pelter, Smith and Tabata (1975) to provide an excellent new synthesis of symmetrical diynes.

$$(C_6H_{11})_2BBr + 2LiC{\equiv}CC_4H_9 \xrightarrow{THF} [(C_6H_{11})_2\overset{-}{B}(C{\equiv}CC_4H_9)_2]Li^+$$

$$\downarrow I_2,\ -78°C \qquad (5.53)$$

$$C_4H_9C{\equiv}CC{\equiv}CC_4H_9\quad (92\%)$$

Trans enynes are also readily obtained, reduction of which gives *cis,trans* dienes (Negishi, Lew and Yoshida, 1973).

Terminal alkynes R—C≡CH may also be converted into ketones or αβ-unsaturated ketones via derived borates. Thus, alkylation or protonation of lithium alkynyltrialkylborates followed by oxidation gives ketones RR^1CHCOR^2 in good yield. The nature of the substituent groups R, R^1 and R^2 can be varied widely and the sequence provides a versatile route to α-substituted ketones (Pelter, Harrison and Kirkpatrick, 1973).

$$(C_6H_{13})_3B + LiC{\equiv}CC_4H_9 \longrightarrow [(C_6H_{13})_3\bar{B}{-}C{\equiv}CC_4H_9]\ Li^+ \xrightarrow{(CH_3)_2SO_4,\ \text{diglyme},\ -78^\circ C} (C_6H_{13})_2B(C_6H_{13})C{=}C(C_4H_9)CH_3 \xrightarrow[NaOH]{H_2O_2} C_6H_{13}COCH(C_4H_9)CH_3 \ (84\%) \quad (5.54)$$

Hydrolysis of the intermediates with acetic acid, instead of oxidation, gives olefins in excellent yields, although as mixtures of geometrical isomers. However, alkylation of thexyldialkylalkynylborates followed by protonolysis of the resulting alkenylboranes provides a good stereoselective route to trisubstituted olefins from acetylenes; the groups derived from the alkylating agent and from the borane end up *cis* to each other in the olefin produced (Pelter, Subramanyan, Laub, Gould and Harrison, 1975).

Reaction with α-bromoketones and α-bromoesters. Organoboranes react readily with α-bromoketones and α-bromoesters in presence of potassium t-butoxide, or, better, the hindered potassium 2,6-di-t-butylphenoxide to give the corresponding α-alkyl or α-aryl derivatives (Brown, Nambu and Rogic, 1969). 2-Ethylcyclohexanone, for example, is readily obtained from triethylborane and 2-bromocyclohexanone. The reaction path (5.55) has been suggested; the key step again involves migration of a group

$$RCOCH_2Br + \text{base} \longrightarrow RCO\bar{C}HBr \xrightarrow{BR^1_3} RCOCH(Br)\bar{B}R^1_3 \longrightarrow RCOCH(R^1)BR^1_2 + Br^- \xrightarrow{C_2H_5OH} RCOCH_2R^1 + R^1_2BOC_2H_5 \quad (5.55)$$

from boron to the adjacent carbon atom with expulsion of bromide ion (see Prager and Reece, 1975).

The usefulness of the direct reaction using trialkylboranes is limited by the fact that organoboranes containing highly branched groups do not react and, in addition, only one of the three alkyl groups in the trialkylborane takes part in the reaction. Both of these difficulties can be circumvented by using a B-alkyl-9-borabicyclo[3,3,1]nonane reagent, instead of the trialkylborane. By this procedure good yields of α-substituted ketones are obtained from α-bromoketones (Brown, Nambu and Rogic, 1969).

Mono- and di-halogenoacetates undergo similar reactions, providing a useful alternative to the malonic ester route to substituted acetic acids. Again best yields are obtained using 9-aryl- or 9-alkyl-borabicyclo[3,3,1]-nonanes as starting materials (Brown and Rogic, 1969). Thus, isobutylene was converted into ethyl 4-methylpentanoate in 53 per cent yield by reaction of the borane with ethyl bromoacetate in the presence of potassium t-butoxide and 1-methylcyclopentene similarly gave ethyl *trans*-2-methylcyclopentylacetate (5.56).

$CH_2CH(CH_3)_2$ B (9-borabicyclo[3,3,1]nonane) + $BrCH_2CO_2C_2H_5$ →[t-C_4H_9OK, t-C_4H_9OH] $(CH_3)_2CHCH_2CH_2CO_2C_2H_5$ (5.56)

1-methylcyclopentene (CH_3) →[B_2H_6] 2-methylcyclopentyl-B (CH_3) →[$BrCH_2CO_2C_2H_5$, t-C_4H_9OK, t-C_4H_9OH] *trans*-2-methylcyclopentyl $CH_2CO_2C_2H_5$ (CH_3)

The reaction can be extended to dibromoacetates, and can be controlled to yield either α-bromoacetates or disubstituted acetates. With one equivalent of base, only one bromine is replaced and an α-bromo ester results; with two equivalents of base the dialkylated ester is formed. Moreover, the dialkylation can be effected in two successive steps, allowing the introduction of two different organic groups into the acetic acid moiety (Brown and Rogic, 1969).

Reaction with diazo compounds. Another route from organoboranes to ketones and esters depends on the nucleophilic addition of a diazo compound to the organoborane. Diazoketones, for example, react with organoboranes to form products which yield ketones on alkaline hydrolysis. The mechanism (5.57) has been suggested, in which migration of R

$$R_3B + N_2CHCOCH_3 \longrightarrow R_3\bar{B}CH(N_2^+)COCH_3 \xrightarrow{-N_2} R_2BCH(R)COCH_3 \xrightarrow{\text{hydrolysis}} RCH_2COCH_3 \quad (5.57)$$

from boron to carbon is facilitated by expulsion of nitrogen. The reaction appears to be fairly general and yields of 36–89 per cent are claimed. Diazo-esters and -nitriles react similarly to give the corresponding homologated esters and nitriles (Hooz, Gunn and Kono, 1971).

Free radical α-bromination. Free radical bromination of trialkylboranes with bromine in presence of light or with *N*-bromosuccinimide leads readily to the formation of α-bromo derivatives. It has been estimated that reaction at the α-carbon–hydrogen bond of triethylborane is favoured over that at the secondary carbon–hydrogen bond of propane by some 60 kJ mol^{-1}. This enhanced reactivity presumably arises from stabilisation of the free radical at the α-carbon by overlap of the unpaired electron with the vacant *p*-orbital of the boron. In any

(5.58)

$(CH_2)_3CO_2C_2H_5$ — B — cyclopentyl $\xrightarrow{Br_2,\ h\nu}$ H_2O: → B, $(CH_2)_3CO_2C_2H_5$, Br (on cyclopentane carbon) $\xrightarrow{H_2O}$ OH–B, $(CH_2)_3CO_2C_2H_5$ (on cyclopentane carbon) $\xrightarrow{H_2O_2,\ NaOH}$ HO, $(CH_2)_3CO_2C_2H_5$ (cyclopentane)

event the reaction provides another method for making carbon–carbon bonds from trialkylboranes, for if the reaction is conducted in the presence of water, migration of an alkyl group from boron to carbon takes place, with expulsion of bromide ion giving a tertiary alcohol. Again the thexyl group is used as an inert blocking group to achieve the union of two other residues (5.58).

Carbocyclic structures can also be produced, by ring contraction of boracyclanes (Yamamoto and Brown, 1973).

BH_2

B

Br_2, $h\nu$ (5.59)

H

(1) H_2O

(2) H_2O_2, NaOH

OH

(60%)

H Br B $\ddot{O}H_2$

It is to be noted that in this approach to tertiary alcohols all the carbon atoms are derived from the trialkylborane; there is no 'extra' carbon as in the sequences described above.

Carbanionic migration from boron to carbon – summary. Nearly all the reactions discussed so far take place by migration of a group from boron to an adjacent carbon or hetero-atom, induced by nucleophilic attack on the electron deficient boron of an alkylborane or by electrophilic attack at an unsaturated substituent of a borate. They all proceed in high yield with retention of configuration of the migrating group. The main reactions are summarised in Table 5.2.

Free radical reactions of organoboranes. There is another smaller group of reactions of organoboranes which proceed by a free radical pathway, and some of these are useful in synthesis (Brown and Midland, 1972). One is the autoxidation of trialkylboranes, which can be controlled to give either hydroperoxides or alcohols. The initial product R_2BO_2R

TABLE 5.2 *Reactions involving migration from boron to an adjacent atom*

$$R_3B + {}^{-}OOH \longrightarrow R_2\overset{-}{B}(R)\text{—}O\text{—}OH \longrightarrow R_2B\text{—}OR \longrightarrow ROH$$

$$R_3B + \bar{O}\text{—}\overset{+}{N}(CH_3)_3 \longrightarrow R_2\overset{-}{B}(R)\text{—}O\text{—}\overset{+}{N}(CH_3)_3 \longrightarrow R_2B\text{—}OR \longrightarrow ROH$$

$$R_3B + CO \longrightarrow R_2\overset{-}{B}(R)\text{—}\overset{+}{C}O \longrightarrow R_2B\text{—}C(R){=}O \longrightarrow RCOR,\ R_3COH$$

$$R_3B + CN^- \longrightarrow R_2\overset{-}{B}(R)\text{—}C{\equiv}N\ (X^+) \xrightarrow{X^+} R_2B\text{—}C(R){=}NX \twoheadrightarrow RCOR$$

$$R_3B + \bar{C}H(Br)CO_2C_2H_5 \longrightarrow R_2\overset{-}{B}(R)\text{—}CH(Br)CO_2C_2H_5 \longrightarrow R_2B\text{—}CH(R)CO_2C_2H_5 \longrightarrow RCH_2CO_2C_2H_5$$

$$\text{⊢+—}B(R)(CHR^1R^2) \xrightarrow{Br_2,\ h\nu} \text{⊢+—}B(R)(CR^1R^2Br)\ (H_2O{:}) \longrightarrow \text{⊢+—}B(OH)\text{—}CRR^1R^2 \longrightarrow RR^1R^2COH$$

$$R_2\overset{-}{B}(R)\text{—}C{\equiv}C\text{—}R^1 + R^2\text{—}X \longrightarrow R_2B\text{—}C(R){=}C(R^1)(R^2) \longrightarrow R\text{—}C{\equiv}C\text{—}R^1\ (\text{for } R^2\text{—}X = I_2),\quad RCOCHR^1R^2$$

TABLE 5.2—(*Continued*)

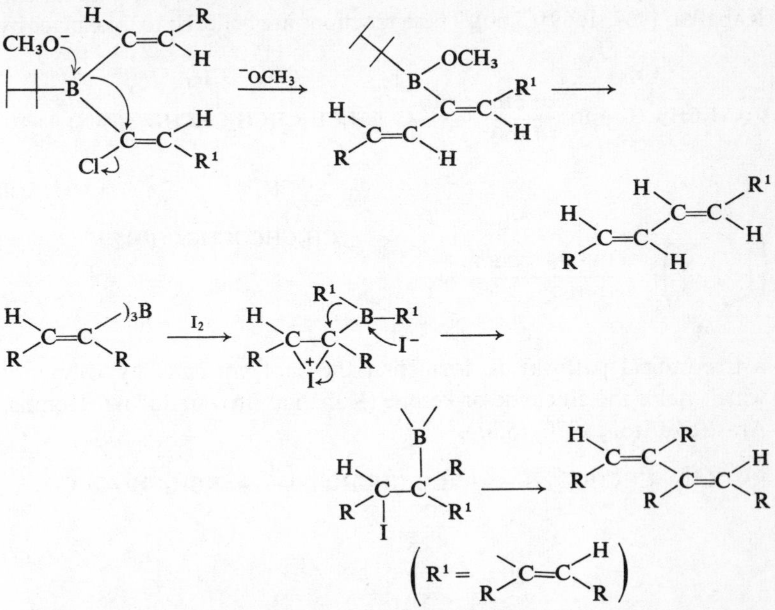

may react with a second molecule of oxygen to given an intermediate, $RB(O_2R)_2$, which is oxidised by hydrogen peroxide to a mixture of an alcohol and the corresponding hydroperoxide in the ratio 1:2. The

$$R_3B + O_2 \longrightarrow R\cdot + R_2BO_2\cdot$$

$$R\cdot + O_2 \longrightarrow RO_2\cdot$$

$$RO_2\cdot + BR_3 \longrightarrow R_2BO_2R + R\cdot$$

hydroperoxide can be isolated by extraction of the reaction solution with aqueous potassium hydroxide (Brown and Midland, 1971). By restricting the amount of oxygen a different intermediate is formed which affords the alcohol on alkaline hydrolysis.

Another useful free radical reaction of organoboranes is the 1,4-addition to $\alpha\beta$-unsaturated carbonyl compounds.

1,4-Addition to $\alpha\beta$-unsaturated carbonyl compounds. Trialkylboranes undergo a remarkably fast addition to many $\alpha\beta$-unsaturated carbonyl compounds, such as acrolein, methyl vinyl ketone and 2-methylenecycloalkanones, to form aldehydes and ketones in which the chain has been extended by three or more carbon atoms (Brown, Rogic, Rathke and Kabalka, 1967, 1969) (5.60). These reactions are believed to take place by

$$(CH_3CH_2\overset{\overset{\displaystyle CH_3}{|}}{C}H\text{—})_3B \xrightarrow[(2)\ H_2O]{(1)\ CH_2=\overset{\overset{\displaystyle CH_3}{|}}{C}CHO} CH_3CH_2\overset{\overset{\displaystyle CH_3}{|}}{C}HCH_2\overset{\overset{\displaystyle CH_3}{|}}{C}HCHO \quad (96\%)$$

$$(C_6H_{11})_3B \xrightarrow[(2)\ H_2O]{(1)\ CH_2=\overset{\overset{\displaystyle CH_3}{|}}{C}COCH_3} C_6H_{11}CH_2\overset{\overset{\displaystyle CH_3}{|}}{C}HCOCH_3 \quad (100\%) \qquad (5.60)$$

a free-radical pathway to form first the enol borinate, hydrolysis of which yields the aldehyde or ketone (Kabalka, Brown, Suzuki, Honma, Arase and Itoh, 1970) (5.61).

$$R\cdot + CH_2=CHCHO \longrightarrow RCH_2\dot{C}HCHO \longleftrightarrow RCH_2CH=CHO\cdot$$

$$\downarrow R_3B \qquad (5.61)$$

$$RCH_2CH_2CHO \xleftarrow{H_2O} RCH_2CH=CHOBR_2 + R\cdot$$

It is found in practice that only aldehydes and ketones containing the group $CH_2=\overset{|}{C}-\overset{|}{C}=O$ undergo the spontaneous reaction easily. If there is one or more alkyl substituent on the terminal carbon, as in *trans*-crotonaldehyde or cyclohexenone, spontaneous reaction does not take place, supposedly as a result of a shorter radical chain length and inefficient initiation. In presence of catalytic amounts of diacetyl peroxide or of oxygen, however, or by irradiation of the reaction mixture with ultraviolet light, all of which favour the formation of radicals from the trialkylborane, these 'inert' compounds too undergo rapid addition of trialkylboranes to form enol borinates which are readily hydrolysed to the carbonyl compounds (5.62). The sequence provides a most useful and versatile method for the synthesis of aldehydes and ketones from practically any combination of olefin and conjugated aldehyde or ketone (Brown and Kabalka, 1970). $\alpha\beta$-Unsaturated methyl ketones are obtained similarly from trialkylboranes and acetylacetylene (Suzuki *et al.*, 1970).

$$(C_2H_5)_3B + CH_3CH{=}CHCOCH_3 \xrightarrow[25^\circ C,\ 24\ h]{\text{diacetyl peroxide, diglyme, } H_2O} C_2H_5CH(CH_3)CH_2COCH_3 \quad (88\%)$$

$$(C_6H_{11})_3B + CH_3CH{=}CHCHO \xrightarrow[\text{iso-}C_3H_7OH,\ H_2O]{h\nu,\ 25^\circ C} C_6H_{11}CH(CH_3)CH_2CHO \quad (100\%)$$

(5.62)

$$(C_2H_5)_3B + \text{cyclopent-2-enone} \xrightarrow[25^\circ C]{\text{cat. } O_2,\ \text{THF}} \text{3-}C_2H_5\text{-cyclopentanone} \quad (68\%)$$

5.5. New applications of organo-silicon compounds in synthesis

Boron is not the only inorganic element whose derivatives are of value in organic synthesis. Reagents based on silicon, phosphorus and sulphur are now being used increasingly and a number of examples of their application have already been given. The three hetero-elements are less electronegative than their first row counterparts and in consequence they form comparatively weak bonds to carbon and comparatively strong bonds to electronegative elements such as oxygen and fluorine (see Table).

Electronegativities and bond strengths (Cotton and Wilkinson, 1972)

	C	Si	P	S
Electronegativity (Allred–Rochow scale)	2·5	1·7	2·1	2·4
Bond strength/kJ mol^{-1}	H—C (416) C—C (356) O—C (336)	H—Si (323) C—Si (301) O—Si (368) F—Si (582)	H—P (322) C—P (264) O—P (380)*	H—S (347) C—S (272) O—S (250)*

* Estimated.

In general, therefore, it is energetically profitable to replace a bond from silicon, phosphorus or sulphur to carbon with a bond to an electronegative element such as oxygen or fluorine, and this can often be exploited to remove the hetero-atom from the product of a reaction after it has performed its synthetic function.

Two other properties of the hetero-elements which are important synthetically are that they can stabilise a negative charge on an adjacent carbon atom, for example tetramethylsilane can be metallated, and they readily undergo nucleophilic substitution (5.63). The latter reaction

$$(CH_3)_4Si \xrightarrow[\text{TMEDA}]{C_4H_9Li} (CH_3)_3SiCH_2Li$$

$$SiCl_4 \xrightarrow{CH_3MgBr} CH_3SiCl_3 \longrightarrow (CH_3)_2SiCl_2 \longrightarrow (CH_3)_3SiCl \longrightarrow (CH_3)_4Si \qquad (5.63)$$

TMEDA = tetramethylethylenediamine

is used to introduce the hetero-elements into organic compounds. Both of these effects can be ascribed, at least partly, to the presence of empty $3d$ orbitals in the hetero-atoms. They can interact with $2p$ orbitals on adjacent carbon atoms and they may also aid nucleophilic substitution by allowing the nucleophile to attach itself to the hetero-atom before the leaving group departs.

Silicon is generally employed in organic synthesis in the form of a trialkylsilyl functional group, R_3Si, often the trimethylsilyl group, $(CH_3)_3Si$, and in this form it plays three main roles: (*a*) as a protecting group and for trapping of reactive intermediates, (*b*) for oxygen-capture with elimination to form olefins, and (*c*) in the stabilisation of α-carbanions and β-carbonium ions.

Protection of functional groups. The trimethylsilyl group is well known as a protecting group for amino, hydroxyl and terminal alkynyl groups (Klebe, 1972). The derivatives are easily made by reaction of a suitable metallic derivative of the compound with chlorotrimethylsilane. Gener-

$$\text{iso-}C_3H_7(CH_3)_2Si\text{—}Cl + R\text{—}\ddot{O}H \longrightarrow R\text{—}O\text{—}Si(CH_3)_2C_3H_7\text{-iso} \quad (H^+,\ :OH_2)$$

$$\xrightarrow{CH_3CO_2H,\ H_2O} R\text{—}OH + \text{iso-}C_3H_7(CH_3)_2SiOH \qquad (5.64)$$

ally, however, trimethylsilyl ethers are too susceptible to solvolysis in protic media to be broadly useful in synthesis. The dimethylisopropylsilyl ethers are more stable and require dilute acid for hydrolysis. They have the added advantage that the bulky dimethylisopropylsilyl group may be used to mask nearby functional groups, allowing selective reaction elsewhere in the molecule. Thus, prostaglandin E_2 was converted into prostaglandin E_1 by selective reduction of one olefinic bond in the bis(dimethylisopropylsilyl) derivative (10).

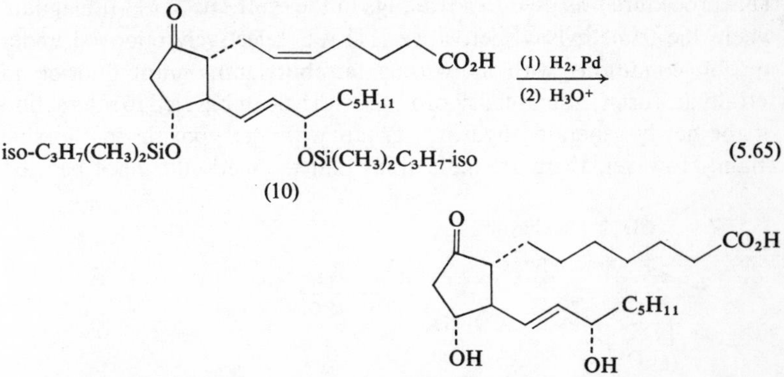

O

CO_2H (1) H_2, Pd

(2) H_3O^+

C_5H_{11}

iso-$C_3H_7(CH_3)_2SiO$ $OSi(CH_3)_2C_3H_7$-iso (5.65)

(10)

O

CO_2H

C_5H_{11}

OH OH

Even better results have been obtained with the t-butyldimethylsilyl protecting group (Corey and Venkateswarlu, 1972). These derivatives are prepared by reaction of the alcohol with t-butylchlorodimethylsilane in the presence of imidazole, presumably by way of *N*-t-butyldimethylsilylimidazole, the conjugate acid of which would be a very reactive silylating agent (nucleophilic attack on silicon aided by a good leaving group). The t-butyldimethylsilyl derivatives are stable to aqueous or alcoholic base under the conditions normally used for the hydrolysis of acetates and they are resistant to hydrogenolysis with hydrogen and palladium, which cleaves benzyl ethers, and to mild chemical re-

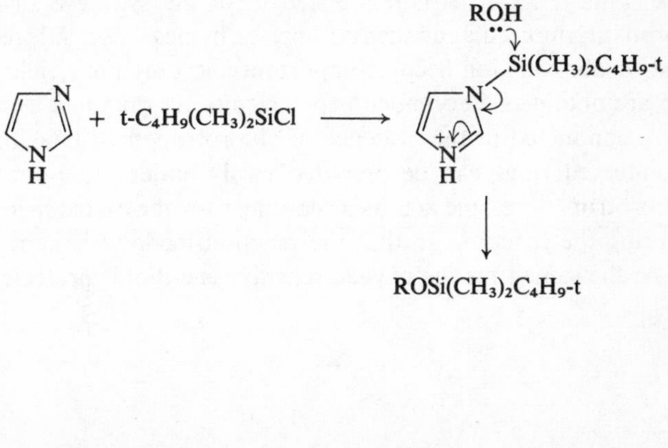

ROH

$Si(CH_3)_2C_4H_9$-t

N N

N + t-$C_4H_9(CH_3)_2SiCl$ ⟶ N

H H (5.66)

$ROSi(CH_3)_2C_4H_9$-t

duction, for example with zinc and methanol, conditions generally employed for regeneration of alcohols from trichloroethyl ethers. They can thus be used for the selective protection of hydroxyl groups while other groups in the molecule are progressively unmasked. On the other hand the t-butyldimethylsilyl group can itself be selectively removed by treatment of the derivative with tetra-butylammonium fluoride (cleavage of a silicon–oxygen bond by nucleophilic attack of fluoride ion and formation of an even stronger silicon–fluorine bond). This procedure was used to advantage in the synthesis of (±)-fumagillin, where the trimethylsilyl derivative (11) was selectively removed under aprotic conditions with anhydrous tetrabutylammonium fluoride in tetrahydrofuran; attempted hydrolysis with acid or base led to interaction of the newly generated hydroxyl group with the epoxide in the side chain. However, there are indications that the method cannot be used

CH_2O–$Si(CH_3)_3$ / Br / F^- / CO_2CH_3 / H O

(11)

$\xrightarrow[\text{THF, 25°C}]{(C_4H_9)_4\overset{+}{N}\ F^-}$

CH_2OH / Br / CO_2CH_3 / H O

(5.67)

where there is an opportunity for acyl switching (Dodd, Golding and Ionannu, 1975).

Chlorotrimethylsilane is a very useful adjunct in the acyloin condensation. As is well known, this reaction provides a valuable and widely used method for the synthesis of medium-sized ring compounds from diesters (Finley, 1964). It is less effective for the synthesis of smaller rings or in intermolecular condensations, for in these cases side reactions induced by alkoxide ion become important and only poor yields of the acyloin are obtained. Very much better results are obtained if the reactions are conducted in the presence of chlorotrimethylsilane and even four-membered rings can be prepared easily under these conditions. The chlorotrimethylsilane acts as a scavenger for the alkoxide ion liberated during the reaction, so that the reaction medium is kept neutral (5.68). At the same time the oxygen-sensitive ene-diol is protected as its bis-trimethylsilyl derivative which may be isolated and purified before hydrolysis to the acyloin (Rühlmann, Seefluth and Becker, 1967; Bloomfield, 1968). Thus, diethyl suberate was converted into the seven-membered cyclic acyloin in 75 per cent yield and the four-membered ring compound (12) was obtained in 90 per cent yield; in the absence of chlorotrimethylsilane, only poor yields were obtained in both reactions.

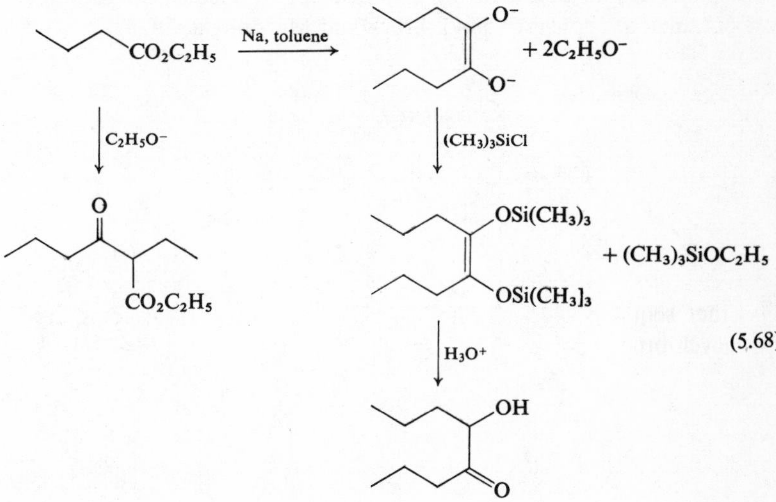

Na, toluene; $CO_2C_2H_5$; $+ 2C_2H_5O^-$; O^-; $C_2H_5O^-$; $(CH_3)_3SiCl$; $OSi(CH_3)_3$; $+ (CH_3)_3SiOC_2H_5$; H_3O^+; OH; O

(5.68)

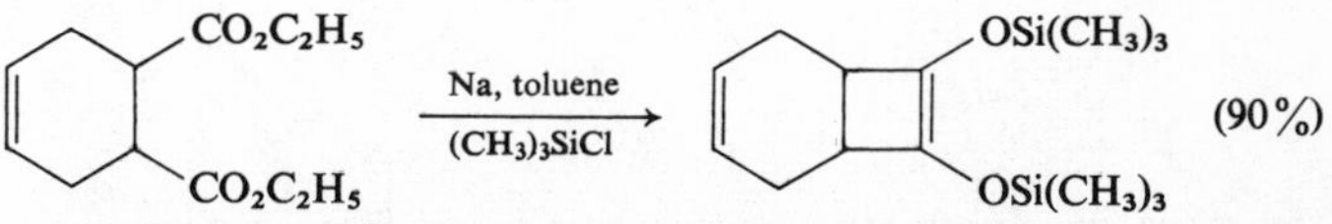

Na, toluene, $(CH_3)_3SiCl$; $CO_2C_2H_5$; $OSi(CH_3)_3$; (90%)

(12)

Trimethylsilyl enol ethers can also be used with advantage for the generation of specific enolate anions from unsymmetrical ketones (cf. p. 20) (Stork and Hudrlik, 1968; House, Czuba, Gall and Olmstead, 1969). Reaction of the mixture of enolates formed from an unsymmetrical ketone by action of a base with chlorotrimethylsilane gives a mixture of the two trimethylsilyl enol ethers which can generally be separated into its components by fractional distillation or preparative scale gas–liquid chromatography. Cleavage of the purified silyl ethers with methyllithium or with lithium amide in liquid ammonia (Benkley and Heathcock, 1975) regenerates the specific lithium enolates, which retain their structural integrity as long as the reaction medium is kept free of protons which could lead to equilibration. A useful alternative route to specific trimethylsilyl enol ethers by thermal rearrangement of trimethylsilyl esters of β-keto acids has been described by Coates, Sandefur and Smillie (1975).

Trialkylsilyl enol ethers of cyclic ketones react with dichlorocarbene to form 2-trialkylsiloxy-1,1-dichlorocyclopropane derivatives which undergo ring expansion on treatment with acid. (±)-Muscone, for example, was obtained as shown in (5.69) (Stork and Macdonald, 1975).

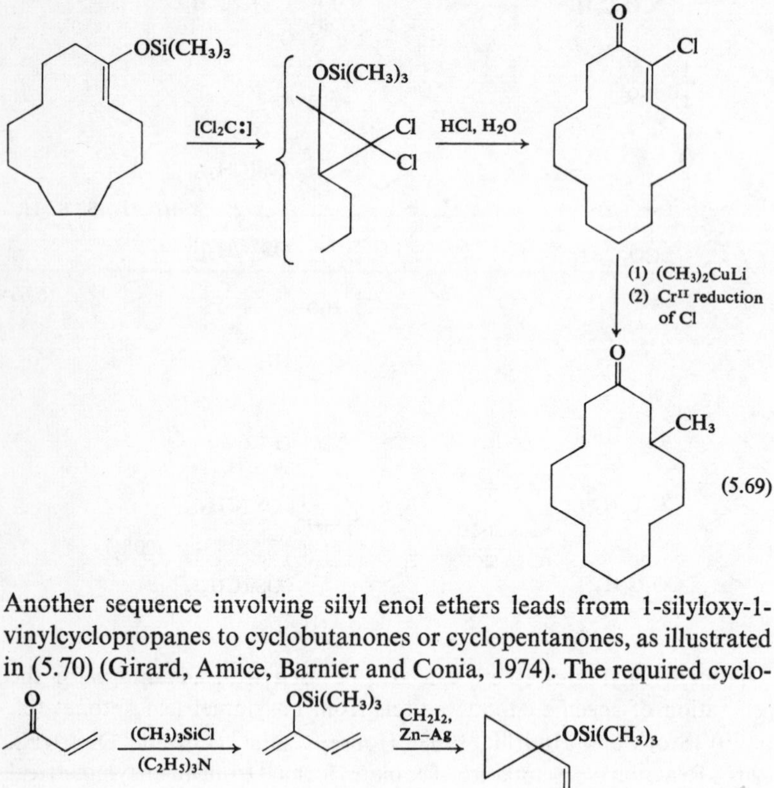

$OSi(CH_3)_3$ $[Cl_2C:]$ $OSi(CH_3)_3$ Cl Cl HCl, H_2O O Cl

(1) $(CH_3)_2CuLi$
(2) Cr^{II} reduction of Cl

O CH_3

(5.69)

Another sequence involving silyl enol ethers leads from 1-silyloxy-1-vinylcyclopropanes to cyclobutanones or cyclopentanones, as illustrated in (5.70) (Girard, Amice, Barnier and Conia, 1974). The required cyclo-

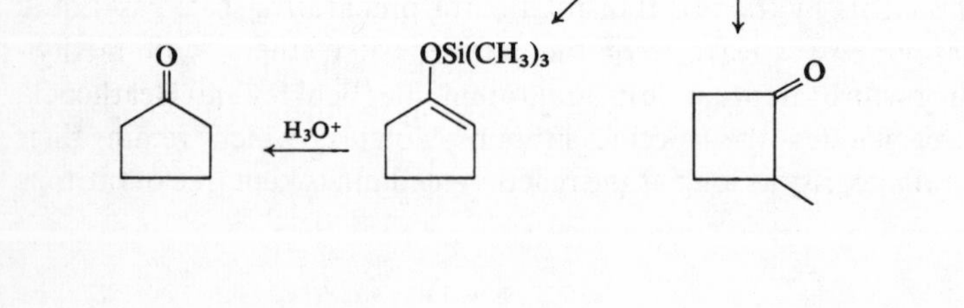

O $(CH_3)_3SiCl$ $(C_2H_5)_3N$ $OSi(CH_3)_3$ CH_2I_2, Zn–Ag $OSi(CH_3)_3$

Δ, 350°C H⁺

(5.70)

O H_3O^+ $OSi(CH_3)_3$ O

propanes are easily obtained from silyl enol ethers of $\alpha\beta$-unsaturated ketones by the Simmons–Smith reaction (p. 77).

Trimethylsilyl cyanide. Another useful silicon reagent is trimethylsilyl cyanide, which is easily prepared from chlorotrimethylsilane and metallic cyanides. It reacts readily with aldehydes and ketones in the presence of catalytic amounts of Lewis acids or of cyanide ion, to give the trimethylsilyl ethers of the corresponding cyanhydrins in high yield (Evans, Carroll and Truesdale, 1974). Even normally unreactive ketones react readily

$CN^- +$ C=O $\xrightarrow{(CH_3)_3SiCN}$ C(CN)O$^-$ + $(CH_3)_3Si$—CN $\longrightarrow$ C(CN)OSi$(CH_3)_3$ + CN^- (5.71)

with trimethylsilyl cyanide due to the formation of the strong Si—O bond which displaces the equilibrium in favour of the derivative, and the reaction provides a valuable alternative to the base-catalysed addition of hydrogen cyanide to carbonyl compounds which often gives only poor yields. Tetralone, for example, is reported not to form a cyanhydrin, but it gives a trimethylsilyl derivative in excellent yield. The silylated cyanohydrins can be hydrolysed to α-hydroxy acids (Corey, Crouse and Anderson, 1975) and on reduction with lithium aluminium hydride they afford the corresponding β-amino alcohols in excellent yield. This

Tetralone $\xrightarrow[ZnI_2,\ 25°C]{(CH_3)_3SiCN}$ NC OSi$(CH_3)_3$ (95%) $\xrightarrow[(2)\ H_3O^+]{(1)\ LiAlH_4}$ H_2NCH_2 OH

NC OSi$(CH_3)_3$ $\xrightarrow{\text{hydrolysis}}$ HOOC OH (5.72)

sequence provides a better route to these valuable intermediates (they are used in the ring expansion of cycloalkanones) than the classical methods involving addition of hydrogen cyanide or nitromethane to the carbonyl compound. $\alpha\beta$-Unsaturated carbonyl compounds and *p*-quinones react exlusively at the carbonyl group; no 1,4-addition compounds are formed. This contrasts markedly with the base-catalysed addition of

$$\text{p-benzoquinone} \xrightarrow[CN^-]{(CH_3)_3SiCN} \text{NC, OSi(CH}_3)_3\text{ adduct} \xrightarrow[(2)\ H_2O]{(1)\ CH_3Li} \text{NC, OSi(CH}_3)_3\text{; HO, CH}_3 \xrightarrow{AgF,\ THF,\ H_2O} \text{4-hydroxy-4-methylcyclohexadienone} \xrightarrow{BF_3\ \text{etherate}} \text{methylhydroquinone} \quad (5.73)$$

$$\text{(CH}_3)_3\text{SiO, CN adduct} \xrightarrow[(2)\ NH_4Cl,\ H_2O]{(1)\ \text{prenyl bromide, Mg, ether}} \text{(CH}_3)_3\text{SiO, CN; HO adduct} \xrightarrow{NaF,\ H_2O} [\text{4-hydroxy-4-(1,1-dimethylallyl)cyclohexadienone}] \xrightarrow{[3,3]\text{-sigmatropic}} \text{prenylhydroquinone}$$

hydrogen cyanide to these compounds which generally gives the 1,4-addition products. In the reaction with quinones selective protection of one of the carbonyl groups can be achieved; subsequent transformations lead to *p*-quinols, in yields much superior to those obtained by any other method, and to substituted quinones.

The sequence has been exploited in a new approach to biologically important isoprenylquinones. Reaction of the protected quinol with an allylic Grignard reagent and removal of the protecting group with sodium fluoride or silver fluoride, is followed by immediate Cope rearrangement (not dienone–phenol rearrangement this time) accompanied by allylic inversion, to give the isoprenylhydroquinone or the quinone directly, depending on the reaction conditions and the structure of the original quinone (5.73) (Evans and Hoffman, 1976).

Promising reagents containing both silicon and sulphur or silicon and phosphorus are being developed. One of these, methyl trimethylsilyl sulphide, $CH_3SSi(CH_3)_3$, reacts with aldehydes and ketones under mild conditions to form dimethylthioketals in high yield without the requirement of an acid catalyst (Evans, Grimm and Truesdale, 1975). With $\alpha\beta$-unsaturated aldehydes and ketones, however, conjugate addition takes place. Conjugate addition to $\alpha\beta$-unsaturated carbonyl compounds is also a feature of the reactions of the new reagent, $(CH_3)_3Si—\overset{+}{P}(C_6H_5)_3Cl^-$, obtained by reaction of chlorotrimethylsilane with triphenylphosphine. The products are readily converted, in a Wittig reaction with another carbonyl compound, into $\alpha\beta$-unsaturated carbonyl compounds, formed, in effect, by joining together two electrophilic species, represented by benzaldehyde and methyl vinyl ketone in the example in (5.74) (Evans, Truesdale, Grimm and Nesbitt, 1977).

$(C_6H_5)_3P$, $(CH_3)_3SiCl$, 25°C → $(C_6H_5)_3\overset{+}{P}$ ... CH_3 Cl^- / $OSi(CH_3)_3$

(1) C_4H_9Li
(2) C_6H_5CHO (5.74)

C_6H_5 ... CH_3 (O) ← H_3O^+ — C_6H_5CH ... CH_3 / $OSi(CH_3)_3$

(75%)

Elimination from β-hydroxysilanes. The Peterson olefin synthesis and related processes. Silicon, with its empty 3*d* orbitals, can stabilise a negative charge on an adjacent carbon atom, although it cannot do so as well as sulphur. α-Silyl carbanions can be obtained from α-chlorosilanes by reaction with magnesium or lithium or by halogen–metal exchange with an alkyl-lithium, by addition of alkyl-lithium reagents to vinylsilanes or by metallation of weakly acidic silanes with butyl-lithium (see Chan and Chang, 1974). These α-silyl carbanions react readily with carbonyl

$$(CH_3)_3SiCH_2Cl \xrightarrow{Mg} (CH_3)_3SiCH_2MgCl$$

$$(CH_3)_3SiCH{=}CH_2 \xrightarrow{RLi} (CH_3)_3SiCH(Li)CH_2R \qquad (5.75)$$

$$(CH_3)_3SiCH_2SCH_3 \xrightarrow{C_4H_9Li} (CH_3)_3SiCH(Li)SCH_3$$

compounds, forming lithio derivatives of β-hydroxysilanes which eliminate the trimethylsilyloxy group spontaneously in tetrahydrofuran solution or by treatment with acetyl chloride to give olefins in good yield. The

$$(CH_3)_3SiCH_2C_6H_5 \xrightarrow[TMEDA]{C_4H_9Li} (CH_3)_3SiCH(Li)C_6H_5 \xrightarrow[\text{ether–pentane, } 0^\circ\rightarrow35^\circ C]{C_6H_5COC_6H_5} (C_6H_5)_2C(OLi)-CH(Si(CH_3)_3)C_6H_5$$

$$\downarrow$$

$$\left[(C_6H_5)_2C(-O-)-CHC_6H_5(-Si(CH_3)_3-)\right] \longrightarrow (CH_3)_3SiOLi + (C_6H_5)_2C{=}CHC_6H_5 \ (77\%)$$

$$(CH_3)_3SiOLi \xrightarrow{H_3O^+} (CH_3)_3SiOSi(CH_3)_3 \ (13)$$

$$CH_3SCH(Li)Si(CH_3)_3 \xrightarrow[\text{ether–pentane, } 0^\circ C\rightarrow35^\circ C]{C_6H_5COC_6H_5} CH_3SCH(Si(CH_3)_3)-C(OLi)(C_6H_5)_2 \longrightarrow CH_3SCH{=}C(C_6H_5)_2 \qquad (5.76)$$

TMEDA = tetramethylethylenediamine

sequence provides a valuable alternative to the Wittig olefin synthesis (p. 106) and has the practical advantage that the by-product hexamethyldisiloxane (13) is volatile and easier to remove from the reaction product than triphenylphosphine oxide formed in the Wittig reaction. The major driving force for the reaction (now usually known as the Peterson reaction) is provided by the formation of the strong silicon–oxygen bond. It is assumed that the elimination takes place by way of a four-membered cyclic transition state, although there is no direct evidence for this (Peterson, 1968; Chan and Chang, 1974).

The reaction can also be used for the preparation of $\alpha\beta$-unsaturated esters and is a favourable alternative to the phosphonate modification of the Wittig reaction for this, since the silicon reagents are, in general, more reactive than phosphorus ylids. Even easily enolisable ketones which often give poor yields in the Wittig reaction, react readily with the silicon reagents (Hartzell, Sullivan and Rathke, 1974; Shimoji, Taguchi, Oshima, Yamamoto and Nozaki, 1974).

$$(CH_3)_3SiCH_2CO_2C_2H_5 \xrightarrow[-76°C]{(C_6H_{11})_2NLi,\ THF} (CH_3)_3Si\underset{\displaystyle Li}{\underset{|}{C}}HCO_2C_2H_5 \qquad (5.77)$$

$$\xrightarrow[THF,\ -78°C \rightarrow 25°C]{\text{cyclohexanone}} \text{cyclohexylidene}{=}CHCO_2C_2H_5 \quad (95\%)$$

When the double bond to be generated in the Peterson reaction is part of a strained system, elimination from the β-hydroxysilane does not take place easily under the usual conditions, but it can be effected by treating the corresponding β-halogenosilane with fluoride ion (5.78). Elimination from β-halogenosilanes, promoted by fluoride ion, is an efficient route to strained alkenes; allenes, cyclopropenes and cyclopropanones can all be prepared by this route (Chan and Massuda, 1975).

As ordinarily effected the Peterson reaction gives mixtures of *cis* and *trans* olefins where this is possible. However, it appears that this is due to the fact that the β-hydroxysilane formed in the condensation step is a mixture of the *threo* and *erythro* isomers. The actual elimination is highly stereoselective, and with a pure diastereomer of the hydroxysilane elimination can be controlled to give either the *cis* or the *trans* olefin. Under basic conditions *syn* elimination takes place so that, for example, the *threo* hydroxysilane (14) gives mainly the *trans* octene, while under

$(C_6H_5)_3Si$ $=CH_2$ Li $\xrightarrow{C_6H_5CHO}$ $(C_6H_5)_3Si$ $=CH_2$ HO— C_6H_5 $\xrightarrow{SOCl_2}$ $(C_6H_5)_3Si$ F^- $=CH_2$ Cl— C_6H_5 $\xrightarrow{(C_2H_5)_4\overset{+}{N}\ F^-}$ H, C_6H_5 >C=C=CH_2 (59%)

CH_2=CH—$Si(CH_3)_3$ $\xrightarrow{[Cl_2C:]}$ Cl Cl cyclopropane $Si(CH_3)_3$ F^- $\xrightarrow[\text{diglyme}]{CsF}$ Cl cyclopropene (5.78)

acid conditions the elimination is *anti* and *cis* octene is formed (5.79). Similarly the *erythro* hydroxysilane gives the *cis* octene with base and the *trans* octene with acid (Hudrlik and Peterson, 1975; Hudrlik, Peterson and Rona, 1975).

$(CH_3)_3Si$, C_3H_7, H, OH, C_3H_7, H (14)

$\xrightarrow[\text{(syn elimination)}]{\text{KH, THF}}$ H, C_3H_7 / C_3H_7, H (96%; 95% *trans*)

$\xrightarrow[\text{(anti elimination)}]{BF_3\text{ etherate, }CH_2Cl_2}$ H, H / C_3H_7, C_3H_7 (99%; 94% *cis*)

(5.79)

The diastereomeric β-hydroxysilanes are obtained from the appropriate $\alpha\beta$-epoxysilanes by reaction with lithium organocuprates (see p. 64). Reaction occurs selectively at the carbon bearing the silicon atom, to give the *erythro* hydroxysilane from the *cis* epoxide and the *threo* hydroxysilane from the *trans* epoxide. The series of reactions therefore furnishes a method for the stereoselective synthesis of olefins from 1-alkenylsilanes.

So far, apparently, only the synthesis of *cis*- and *trans*-disubstituted olefins has been reported although it is claimed that the method is suitable for the stereoselective synthesis of trisubstituted olefins as well.

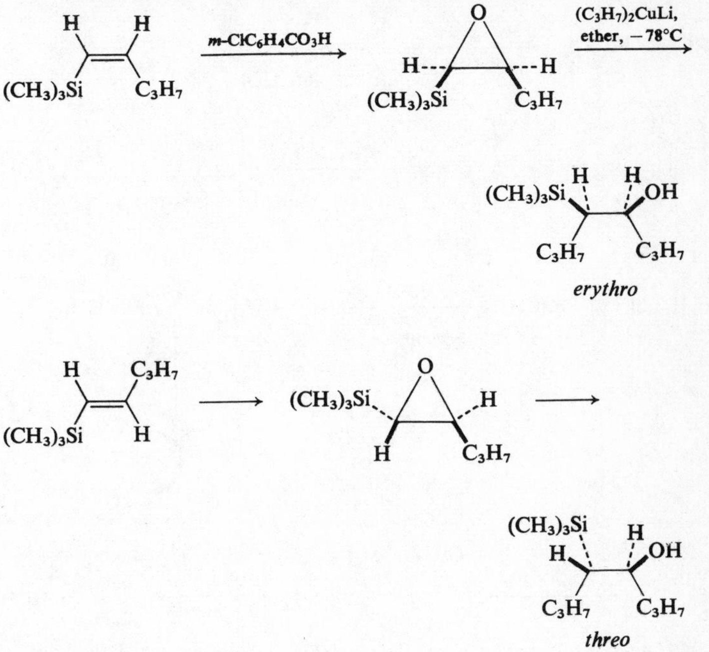

m-$ClC_6H_4CO_3H$; $(C_3H_7)_2CuLi$, ether, −78°C

erythro (70%)

(5.80)

threo (82%)

1-Alkenylsilanes can be made by several routes, from acetylenes, from 1-alkenyl halides or from carbonyl compounds using the Peterson reaction (5.81). β-Hydroxylsilanes can also be obtained by reaction of carbonyl compounds with α-silyl carbanions, themselves formed by action of butyl-lithium on monosubstituted methyl α-silylmethyl selenides (Dumont and Krief, 1976).

The Peterson reaction has been used to prepare ketene thioacetals by reaction of aldehydes or ketones with 2-lithio-2-trimethylsilyl-1,3-dithiane, which is readily available from 1,3-dithiane. The ketene thioacetals are useful synthetic intermediates (see p. 49). On hydrolysis they give carboxylic acids, allowing the conversion $RR^1CO \rightarrow RR^1CHCO_2H$, and reduction followed by hydrolysis gives aldehydes (5.82); alkylation before hydrolysis leads to ketones, $RR^1CO \rightarrow RR^1CHCOR^2$ (Carey and Court, 1972).

$$RC{\equiv}CH \xrightarrow[H_2PtCl_6]{HSi(CH_3)_3} \text{(E)-RCH=CHSi(CH_3)_3}$$

$$RC{\equiv}CH \xrightarrow{(CH_3)_3SiCl,\ (C_2H_5)_3N} RC{\equiv}CSi(CH_3)_3 \xrightarrow[\text{catalyst}]{H_2,\ \text{Lindlar}} \text{(Z)-RCH=CHSi(CH_3)_3}$$

$$\text{C=C-Hal.} \xrightarrow{Li} \text{C=C-Li} \xrightarrow{(CH_3)_3SiCl} \text{C=C-Si(CH_3)_3} \qquad (5.81)$$

$$(CH_3)_3SiCH_2Si(CH_3)_3 \xrightarrow[\text{hexamethylphosphoramide}]{C_4H_9Li} (CH_3)_3SiCH(Li)Si(CH_3)_3 \xrightarrow{C_6H_5COC_6H_5} (C_6H_5)_2C{=}CHSi(CH_3)_3$$

$$\text{2-Li-2-Si(CH}_3)_3\text{-1,3-dithiane} \xrightarrow[\text{THF, 0°C}]{C_6H_5COC_6H_5} (C_6H_5)_2C{=}\text{(1,3-dithian-2-ylidene)} \xrightarrow[(2)\ \text{hydrolysis}]{(1)\ \text{reduction}} (C_6H_5)_2CHCHO \ (70\%) \qquad (5.82)$$

Annelation with silylated reagents. The ability of silicon to stabilise an adjacent carbanion has been exploited in an improved method for the annelation of ketones, that is for building a ring system on to an already existing carbon framework.

Annelation of 2-alkylcyclohexanones with methyl vinyl ketone and its homologues is an important route to fused polycyclic systems, for example in the synthesis of steroids. Reaction at the less substituted α-

$$\text{2-R-cyclohexanone} + CH_2{=}CHCOCH_3 \longrightarrow \text{2-R-2-(3-oxobutyl)cyclohexanone (15)} \longrightarrow \text{R-substituted octalone} \qquad (5.83)$$

carbon of the ketone can be effected using the corresponding enamine (p. 29), but the basic conditions required for reaction at the more substituted α-carbon generally result in polymerisation of the vinyl ketone, and low yields of the desired products are often obtained. Other difficulties are caused by the fact that the Michael adduct (15) and the original ketone have similar acid strengths and reactivities so that competitive reaction of the product with the vinyl ketone can ensue, and in reactions involving the thermodynamically less stable enolate equilibration to the more stable enolate can take place by interaction with (15), with the result that, in addition to the poor yields, mixtures of structural isomers may be obtained. Many of these difficulties can be avoided by using α-silylated vinyl ketones. These derivatives react readily with cyclohexanones under aprotic conditions, which lead to maximum polymerisation with methyl vinyl ketone itself, to give high yields of condensation products. Thus, reaction of the lithium enolate of cyclohexanone with the silylated vinyl ketone (16) and cyclisation of the product with sodium methoxide in methanol, with simultaneous cleavage of the triethylsilyl group, gave the octalone (17) in 80 per cent yield (5.84). The corresponding reaction with methyl vinyl ketone itself gave less than 5 per cent of the Michael addition product.

OLi + $Si(C_2H_5)_3$ (16) $\xrightarrow[-78°C \to 0°C]{THF}$

$Si(C_2H_5)_3$ $\xrightarrow{NaOCH_3, CH_3OH}$ (17) (80%) (5.84)

The trialkylsilyl group stabilises the anion of the Michael adduct relative to that of the starting ketone, thus facilitating the forward reaction and at the same time discourages equilibration of positionally unstable enolates of the starting material. The steric effect of the large trialkylsilyl group may also play a part by preventing further unwanted reactions at the α-carbon of the Michael adduct (Stork and Ganem, 1973; Boeckmann, 1974; Stork and Singh, 1974).

Another much used method for introducing a 3-ketoalkyl group at the α-position of a cyclohexanone to give a product suitable for cyclisation entails alkylation with an alkyl halide bearing a functional group at

C-3 which is eventually transformed into the 3-keto group. This procedure has hitherto suffered from the disadvantage that several steps were often required to bring about the desired transformation. This difficulty can be avoided by alkylation with an allyl halide carrying a trialkylsilyl substituent on one of the olefinic carbon atoms. The resulting alkenylsilanes, which are in effect masked carbonyl groups, are smoothly converted into 3-ketoalkyl groups by epoxidation and rearrangement with acid as described below (5.85). The exact nature of the side chain introduced can be controlled by proper choice of the silylated allyl halide (Stork and Jung, 1974).

CH_3 O^- → $ICH_2CH{=}C(CH_3)Si(CH_3)_3$, THF, 25°C → CH_3 O $Si(CH_3)_3$

(1) $m\text{-}ClC_6H_4CO_3H$
(2) HCO_2H (5.85)

CH_3 H O (82%) ← KOH, C_2H_5OH ← CH_3 O O (83%)

1-*Alkenylsilanes and αβ-epoxysilanes.* 1-Alkenylsilanes are valuable in synthesis because they can be converted into carbonyl compounds by way of the corresponding epoxides. The alkenylsilanes themselves can be

$Si(CH_3)_3$ → O $Si(CH_3)_3$ → O (5.86)

prepared in a number of ways, from acetylenes, from alkenyl halides or from carbonyl compounds using the Peterson reaction, as described on p. 321. Oxidation with peroxy acids gives the corresponding epoxides, and these are converted into carbonyl compounds on treatment with acid under mild conditions. Because silicon can stabilise a positive charge on a β-carbon atom (see p. 326) opening of the epoxide ring takes place in one direction only to give the product with the carbonyl group at the carbon atom which was originally attached to silicon. Yields are generally high. Thus, methyl neopentyl ketone was obtained from the appropriate alkenylsilane as shown in (5.87) (Stork and Colvin, 1971; Gröbel and Seebach, 1974).

$(CH_3)_3Si$ C=CH_2 —m-$ClC_6H_4CO_3H$→ $(CH_3)_3Si$ O
↓ CH_3OH, H_2SO_4
CH_3 OH $(CH_3)_3Si$ HO— $\overset{+}{C}H_2$ ← HO ← O= (74%) (5.87)

Si$(CH_3)_3$ H_2C= Li —CuI, THF, −20°C→ Si$(CH_3)_3$)$_2$CuLi —O, −10°C→ O Si$(CH_3)_3$
↓ ketal + m-$ClC_6H_4CO_3H$ (5.88)
O ← −H_3O^+ — O O Si$(CH_3)_3$ O

Si$(CH_3)_3$ H H)$_2$CuLi —O etc.→ O CHO

The reaction of epoxysilanes with organocuprates to give β-hydroxysilanes has already been referred to (p. 320). β-Hydroxysilanes are also formed by reduction of the epoxides with lithium aluminium hydride. As in the cuprate reactions, attack of the reagent takes place at the carbon bearing the silicon atom and not at the less substituted carbon as in the reduction of ordinary epoxides. It is suggested that the empty 3*d* orbitals on silicon may play some part in directing attack on the adjacent carbon atom (Eisch and Trainor, 1963).

Another useful synthetic sequence leading to 1,4- and 1,5-diketones by way of epoxysilanes is illustrated in (5.88). The sequence results, in effect, in the conjugate addition of acyl anions, $RCO^{(-)}$, or enolate anions, $RCOCH_2^{(-)}$, to $\alpha\beta$-unsaturated carbonyl compounds. It has the further practical advantage that the second carbonyl group is introduced in a masked form, so that the two carbonyl groups can be differentiated in subsequent reactions (Boeckman and Bruza, 1974).

Si(CH3)3 Si(CH3)3 O
H H
CuLi ≡ ≡
2
R R R

(5.89)

Si(CH3)3 Si(CH3)3 O
R CuLi ≡ R CH⁻ ≡ R CH2⁻
2

Stabilisation of carbonium ions by a β-silyl group. A noteworthy feature of the chemistry of organosilanes is the ability of the silicon atom to stabilise a positive charge on a β-carbon atom (compare Eaborn, 1973). This is reflected, for example, in the fact that trimethylsilylbenzene undergoes electrophilic substitution at the site of the substituent 10 000 times faster than benzene itself. This effect has been used to control site selectivity in some aromatic electrophilic substitution reactions. Thus, bromination of *o*- and *p*-trimethylsilylbenzoic acid takes place at the site of the trimethylsilyl group, *ortho* and *para* to the *meta*-directing carboxylic acid group, aided partly by the readiness with which the trimethylsilyl group is cleaved from a carbon atom next to a carbonium ion centre on attack by a nucleophile, and partly by stabilisation of the carbonium ion by the silicon through hyperconjugation or by bridging (Koenig and Weber, 1973).

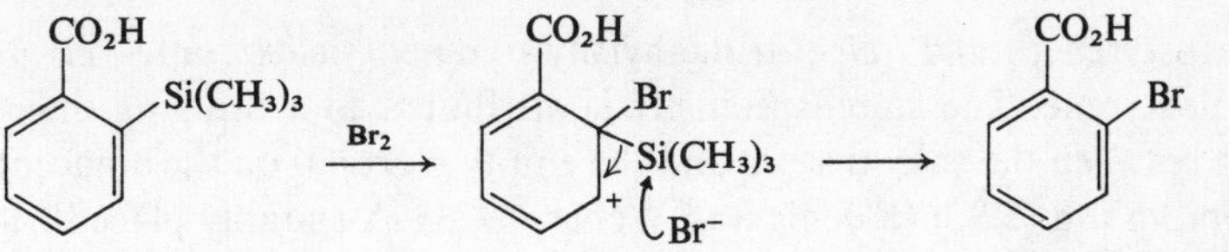

(5.90)

This effect has been used recently to direct electrophilic substitution in alkenes. In alkenylsilanes electrophilic substitution is directed to the carbon atom carrying the trimethylsilyl group (5.91). The reactions are

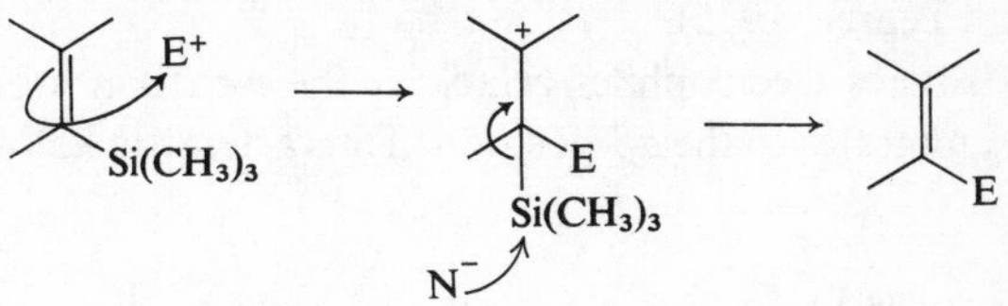

(5.91)

E^+ = an electrophile

stereospecific as well as site specific and take place with complete retention of configuration of the double bond. Thus, *trans*-β-trimethylsilylstyrene reacts with deuterium chloride to give virtually only *trans*-β-

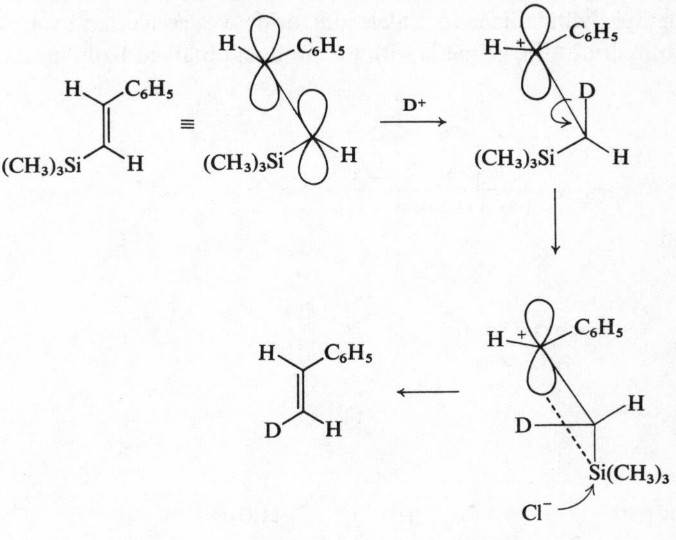

(5.92)

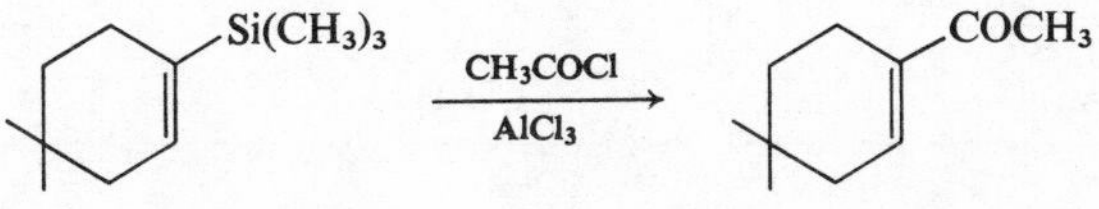

(18) (19) (77%)

deuteriostyrene, and *cis*-β-trimethylsilylstyrene yields only *cis*-β-deuteriostyrene. The stereospecificity is attributed to a bridging interaction between the silicon atom and the empty *p* orbital on the adjacent carbonium ion (5.92) (Koenig and Weber, 1973). A suitably placed trimethylsilyl group can be used similarly to direct the course of aliphatic Friedel–Crafts reactions. Thus, acetylation of the trimethylsilylcyclohexene (18) gave only the acetyl derivative (19). None of the product formed by attack at the other end of the double bond was detected (Fleming and Pearce, 1975).

With allylsilanes electrophiles attack at the γ-carbon atom and the double bond migrates to the $\alpha\beta$-position. Thus artemisia ketone (21) was

$(CH_3)_3Si$ α β γ E^+ N^- ⟶ E (5.93)

obtained in 90 per cent yield by action of senecioyl chloride and aluminium chloride on trimethylisopentenylsilane (20) (Pillot, Dunogues and Calas, 1976), and the adduct (22), obtained by Diels–Alder addition of 1-trimethylsilylbutadiene to maleic anhydride was converted by acid into the anhydride (23), isomeric with the product obtained from maleic

$Si(CH_3)_3$ (20) —(1) O, Cl; $AlCl_3$, CH_2Cl_2, $-60°C$; (2) H_2O, NH_4Cl⟶ O (21) (90%)

CO, O, CO, $Si(CH_3)_3$ (22) —p-$CH_3C_6H_4SO_3H$; C_6H_6, reflux⟶ CO, O, CO (23)

CO_2H, CO_2H, $Si(CH_3)_3$ (24) —CH_3CO_3H, ether; 25°C⟶ HO, CO_2H, CO_2H (25) (5.94)

anhydride and butadiene itself. Similarly, reaction of the dicarboxylic acid (24) with unbuffered peroxyacetic acid gave the allyl alcohol (25), by way of an intermediate epoxide (Carter and Fleming, 1976).

6 Oxidation

For practical purposes most organic chemists mean by 'oxidation' either addition of oxygen to the substrate, as in the conversion of ethylene into ethylene oxide, removal of hydrogen as in the oxidation of ethanol to acetaldehyde or removal of one electron as in the conversion of phenoxide anion to the phenoxy radical.

Of the wide variety of agents available for the oxidation of organic compounds, probably the most widely used are potassium permanganate and derivatives of hexavalent chromium. Permanganate, a derivative of heptavalent manganese, is a very powerful oxidant. Its reactivity depends to a great extent on whether it is used under acid, neutral or basic conditions. In acid solution it is reduced to the divalent manganous ion Mn^{2+} with net transfer of five electrons ($Mn^{VII} \rightarrow Mn^{II}$), while in neutral or basic media manganese dioxide is usually formed, corresponding to a three electron change ($Mn^{VII} \rightarrow Mn^{IV}$). Permanganate is generally used in aqueous solution and this restricts its usefulness since not many organic compounds are sufficiently soluble in water and only a few organic solvents are resistant to the oxidising action of the reagent. Solutions in acetic acid, t-butanol or dry acetone or pyridine can sometimes be employed.

The situation has recently been transformed by the discovery that in the presence of certain tetra-alkylammonium or phosphonium salts, known as phase-transfer catalysts, or of the macrocyclic ether dicyclohexano-18-crown-6 (1), potassium permanganate dissolves in benzene, and the resulting solutions of 'purple benzene' are excellent reagents for the oxidation of a variety of organic substrates. Thus, in presence of the crown ether (1) olefins, alcohols, aldehydes and alkylbenzenes are rapidly oxidised to carboxylic acids in high yield at room temperature. Toluene gives benzoic acid in 78 per cent yield and α-pinene is converted into pinonic acid in 90 per cent yield (Sam and Simmons, 1972; Landini, Montanari and Pirisi, 1974). Similarly, heptanoic acid is obtained from 1-octene in 90 per cent yield with aqueous permanganate in benzene in presence of cetyltrimethylammonium chloride (Starks, 1971). Under basic conditions, olefins are smoothly converted into 1,2-diols (6.1); the yields obtained in the latter reactions are much higher than those usually encountered in the oxidation of olefins to glycols with permanganate.

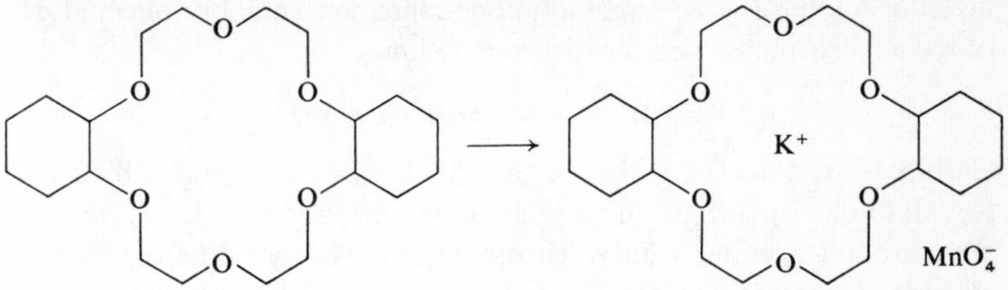

(1) → (2) K^+ MnO_4^-

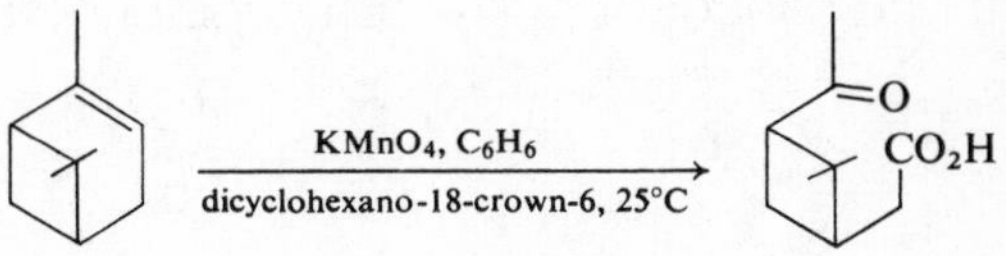

$$\xrightarrow[\text{dicyclohexano-18-crown-6, 25°C}]{KMnO_4,\ C_6H_6} \quad CO_2H \quad (90\%) \qquad (6.1)$$

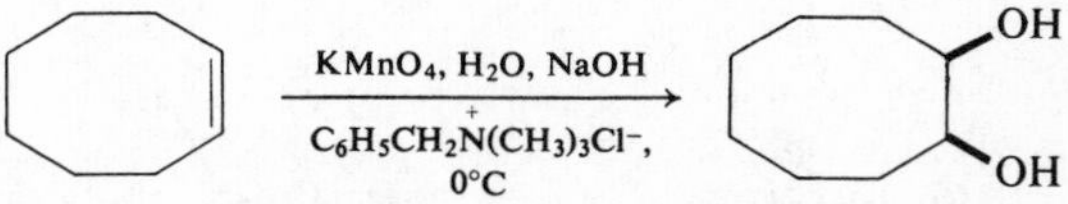

$$\xrightarrow[C_6H_5CH_2\overset{+}{N}(CH_3)_3Cl^-,\ 0°C]{KMnO_4,\ H_2O,\ NaOH} \quad OH,\ OH \quad (50\%)$$

The catalytic action of the quaternary salts is believed to be due to the ability of their organic-soluble cations to transfer anions (e.g. MnO_4^-) from the aqueous into the organic phase. With the crown ether (1) the organic-soluble complex (2) is formed (Dockx, 1973; Dehmlow, 1974).

Chromic acid, a derivative of hexavalent chromium, is one of the most versatile of the available oxidising agents, and reacts with almost all types of oxidisable groups. The reactions can often be controlled to yield largely one product, and for this reason chromic acid oxidation is a useful process in synthesis. In oxidations chromium is reduced from the hexavalent to the trivalent state ($Cr^{VI} \rightarrow Cr^{III}$) with production of a chromic salt. The commonest reagents are chromium trioxide and sodium or potassium dichromate. Chromium trioxide is a polymer which dissolves in water with depolymerisation to form chromic acid.

$$(CrO_3)_n + H_2O \longrightarrow HO-\underset{\underset{O}{\|}}{\overset{\overset{O}{\|}}{Cr}}-OH$$

It is commonly employed in solution in dilute sulphuric acid, sometimes containing acetic acid to aid dissolution of the substrate; a comparable solution is obtained by adding sodium or potassium dichromate to aqueous sulphuric acid. These solutions contain an equilibrating mixture of the acid chromate and the dichromate ions

$$2HCrO_4^- \rightleftharpoons H_2O + Cr_2O_7^{2-}$$

Chromium trioxide may also be used in solution in acetic anhydride, t-butanol or in pyridine. In these solutions the reactive species present are chromyl acetate, t-butyl chromate and the pyridine–chromium trioxide complex, as shown (6.2).

$$(CrO_3)_n + CH_3COOCOCH_3 \longrightarrow CH_3COO\overset{\overset{\displaystyle O}{\|}}{\underset{\underset{\displaystyle O}{\|}}{Cr}}OCOCH_3$$

$$(CrO_3)_n + (CH_3)_3COH \longrightarrow (CH_3)_3CO\overset{\overset{\displaystyle O}{\|}}{\underset{\underset{\displaystyle O}{\|}}{Cr}}OC(CH_3)_3 \qquad (6.2)$$

$$(CrO_3)_n + C_5H_5N \longrightarrow \bar{O}-\overset{\overset{\displaystyle O}{\|}}{\underset{\underset{\displaystyle O}{\|}}{Cr}}-\overset{+}{N}C_5H_5$$

The commonest lower oxidation state of chromium is Cr^{III} and most oxidations with chromic acid lead to this state by a net transfer of three electrons. However, each stage in the oxidation of most organic compounds involves transfer of only two electrons, and it is evident that most reactions must give rise to Cr^{V} or Cr^{IV} species as intermediates, as exemplified in the following scheme for the oxidation of a secondary alcohol.

$$Cr^{VI} + R_2CHOH \longrightarrow R_2C{=}O + 2H^+ + Cr^{IV}$$

$$Cr^{IV} + Cr^{VI} \longrightarrow 2Cr^{V}$$

$$Cr^{V} + R_2CHOH \longrightarrow R_2C{=}O + 2H^+ + Cr^{III}$$

The Cr^{IV} and Cr^{V} ions are themselves powerful oxidants and may give rise to unwanted side-reactions and to products different from those formed in the original oxidation by Cr^{VI}.

6.1. Oxidation of hydrocarbons

Paraffin hydrocarbons. Under vigorous conditions both chromic acid and permanganate attack alkanes, but the reaction is of little synthetic use for usually mixtures of products are obtained in low yield. The reaction is of importance in the Kuhn–Roth estimation of methyl groups. This depends on the fact that a methyl group is rarely attacked (the relative rates of oxidation of primary, secondary and tertiary C—H bonds are 1:110:7000) and is eventually converted into acetic acid. The usual method is to boil the substance with chromic acid in aqueous sulphuric acid and determine the amount of steam volatile acid formed (almost entirely acetic acid) by titration with alkali. The sesquiterpene hydrocarbon cadinene, for example, under these conditions gives three molecules of acetic acid, one from each of the ring methyl substituents and one other from the isopropyl group (6.3). Under less vigorous

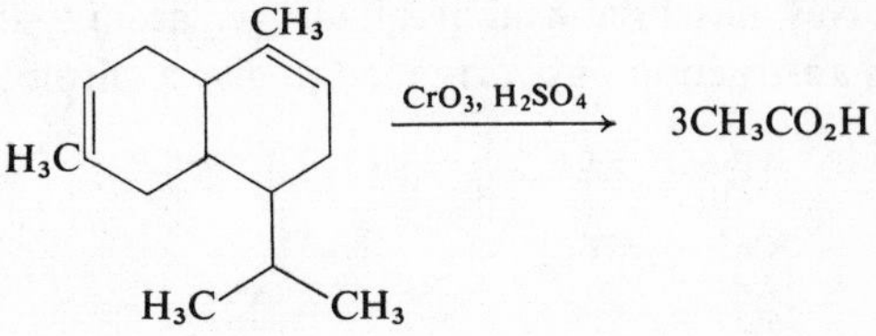

$$\xrightarrow{CrO_3,\ H_2SO_4} 3CH_3CO_2H \qquad (6.3)$$

conditions intermediate oxidation products can sometimes be isolated, and this may be useful in degradative work.

Aromatic hydrocarbons. In the absence of activating hydroxyl or amino substituents benzene rings are only slowly attacked by chromic acid or permanganate, but alkyl side chains are degraded with formation of benzene carboxylic acids (6.4). This is a useful method for the preparation of benzene carboxylic acids, and also for determining the orientation pattern of unknown polyalkylbenzenes which are degraded to benzene polycarboxylic acids of known orientation. Nuclear hydroxyl or amino substituents, if present, must be converted into their methyl ethers or acetyl derivatives, for otherwise they activate the ring to attack, and quinones or, with excess of reagent, carbon dioxide and water are formed. With side chains longer than methyl, initial attack is thought to take place at the benzylic carbon atom. This is suggested by the fact that t-butylbenzene is very resistant to oxidation and ethylbenzene gives some acetophenone as well as benzoic acid. The rate-determining step in these chromic acid oxidations is known to be cleavage of the benzylic

$$p\text{-}NO_2C_6H_4CH_3 \xrightarrow[\text{boil}]{Na_2Cr_2O_7,\ \text{aq. } H_2SO_4} p\text{-}NO_2C_6H_4CO_2H \quad (86\%)$$

$$4\text{-}CH_3C_5H_4N \xrightarrow[\text{boil}]{\text{aq. } KMnO_4} 4\text{-}HO_2CC_5H_4N \quad (50\%) \qquad (6.4)$$

$$C_6H_5C_8H_{17} \xrightarrow[\text{aq. } H_2SO_4]{CrO_3,\ CH_3COOH} C_6H_5CO_2H$$

C—H bond, but it is uncertain at present whether attack on the benzylic hydrogen involves initial removal of a hydrogen atom to form a radical, or a concerted abstraction of hydride ion to give a chromate (6.5).

$$R_3C{-}H + Cr^{VI} \longrightarrow R_3C\cdot + Cr^{V}$$

$$R_3C{-}H + (HO)_2CrO_2 \longrightarrow R_3C{-}O{-}Cr^{IV} \qquad (6.5)$$

The conversion of a methyl group attached to a benzene ring into the formyl group can be achieved by oxidation with chromium trioxide in acetic anhydride in the presence of a strong acid, or with a solution of chromyl chloride in carbon disulphide or carbon tetrachloride (the Étard reaction) (6.6). The success of the first reaction is due to the initial formation of the di-acetate which protects the aldehyde group against further oxidation. In the Étard reaction a complex of composition

$$p\text{-}CH_3C_6H_4CH_3 \xrightarrow[H_2SO_4]{CrO_3,\ \text{acetic anhydride}} p\text{-}(CH(OCOCH_3)_2)_2C_6H_4 \xrightarrow[\text{aq. } H_2SO_4]{\text{hydrolysis}} p\text{-}OHCC_6H_4CHO \quad (52\%)$$

$$p\text{-}BrC_6H_4CH_3 \xrightarrow[CCl_4]{CrO_2Cl_2} \text{complex} \xrightarrow[H_2O]{\text{hydrolysis}} p\text{-}BrC_6H_4CHO \quad (80\%) \qquad (6.6)$$

1 hydrocarbon: $2CrO_2Cl_2$ is first formed and is converted into the aldehyde by treatment with water. Another useful reagent for this purpose is ceric ion which readily oxidises aromatic methyl groups to aldehyde in acidic media. The aldehyde group is not oxidised further and, in a polymethyl compound only one methyl group is oxidised under normal conditions (Syper, 1966; Dust and Gill, 1970). Mesitylene, for example, gave 3,5-dimethylbenzaldehyde quantitatively, and acet-*p*-toluidide was converted into *p*-acetamidobenzaldehyde (94 per cent).

Unsaturated hydrocarbons. Oxidation of unsaturated hydrocarbons may take place both at the double bond and at the adjacent allylic positions and important synthetic reactions of both types, involving oxidations with permanganate, peroxy-acids, ozone and other reagents are discussed below. With chromic acid, mixtures of products are often produced, and for this reason oxidation of olefinic compounds with chromic acid is of limited value in synthesis. For example, with chromic acid in acetic acid cyclohexene gives a mixture of cyclohexenone and adipic acid, although with straight chain alkenes allylic oxidation appears to be a relatively minor reaction. With acetic acid as solvent epoxides are also frequently produced, and with aqueous acid more complex mixtures including rearranged products may result (Wiberg, 1965). Chromyl chloride oxidises unsaturated hydrocarbons to (saturated) aldehydes and ketones in good to excellent yields, although sometimes oxidation is accompanied by molecular rearrangement. 2-Phenylpropene for example, gives 2-phenylpropanal in 60 per cent yield, but 2,3-dimethyl-2-butene rearranges to give 3,3-dimethyl-2-butanone (Freeman and Yamachika, 1972).

Allylic oxidation of olefins to give $\alpha\beta$-unsaturated carbonyl compounds can sometimes be effected with the chromium trioxide–pyridine complex (Collins' reagent; see p. 342), but if there is more than one allylic methyl or methylene group present mixtures of products are generally obtained (Dauben, Lorber and Fullerton, 1969). Selenium dioxide has also been widely used for the allylic oxidation of olefins, giving allylic alcohols, $\alpha\beta$-unsaturated carbonyl compounds or a mixture of the two depending on the experimental conditions (Rabjohn, 1949; Jerussi, 1970). For example, with selenium dioxide in ethanol 1-methylcyclohexene is converted into a mixture of 2-methyl-2-cyclohexenol and 2-methyl-2-cyclohexenone, but in water the ketone is formed exclusively in 90 per cent yield. The allylic alcohols can be easily oxidised further to the $\alpha\beta$-unsaturated carbonyl compounds (see p. 343). A very useful application of this reaction is in the oxidation of aliphatic 1,1-dimethylalkenes to the corresponding *trans* allylic alcohols or $\alpha\beta$-unsaturated aldehydes by selective attack on the *trans* methyl group (Bhalerao and Rapaport, 1971). Thus, 2-methyl-2-heptene is converted almost exclusively into the *trans* allylic alcohol (3) and 2,7-dimethyl-2,6-octadiene gives the bis-aldehyde (4). The high selectivity shown in these reactions is a consequence of the mechanism, which involves an initial 'ene' addition of selenious acid (from SeO_2 and water) to the less hindered side of the olefin followed by dehydration and [2,3]-sigmatropic rearrangement of the resulting allylseleninic acid (Arigoni, Vasella, Sharpless and Jensen, 1973) (6.7).

0·5 mol SeO_2, 95% C_2H_5OH, reflux → CH_2OH (3)

2 mol SeO_2, 95% C_2H_5OH, reflux → OHC … CHO (4)

HO, OH, Se, H, O, H, CH_2 — 'ene' reaction → HO, OH, H, Se, OH — $-H_2O$ → OH, H, Se^+, O^-

[2,3]-sigmatropic rearrangement

H, CH_2OH ← H_2O — H, CH_2OSeOH (6.7)

Allylic amination of olefins can be effected by selenium or, better, sulphur reagents of the type TosylN=S=NTosyl (Sharpless and Hori, 1976). Reactions take place easily at room temperature without allylic rearrangement. For example methylenecyclohexane is converted into the 2-sulphonamido derivative (5) in 84 per cent yield. The reactions are believed to take place by the same sequence of ene reaction and [2,3]-sigmatropic rearrangement which has been established for oxidations by selenium dioxide. The sulphenamide intermediates are easily cleaved to the corresponding allylic sulphonamides with trimethyl phosphite in methanol at room temperature.

NTos, S, NTos, CH_2, H; CH_2Cl_2, 25°C; NHTos, S, NTos; CH_2, SNHTos, NTos; $(CH_3O)_3P$, CH_3OH, 25°C; CH_2, NHTos (84%) (5) (6.8)

Tos ≡ CH_3–C_6H_4–SO_2

An indirect method for the conversion of an olefin into an allylic alcohol with a shift in the position of the double bond proceeds from the olefin epoxide by reaction, first, with phenylselenide anion to give the corresponding β-hydroxyselenide. This is not isolated but is oxidised directly with excess of hydrogen peroxide to the unstable selenoxide which decomposes spontaneously to the *trans* allylic alcohol with elimination of the selenium (6.9) (cf. p. 100) (Sharpless and Lauer, 1973). Remarkably, elimination from the selenoxides always takes place away from the hydroxy group; no more than traces of the alternative ketonic products have ever been isolated. The reactions proceed in high yield under mild non-basic conditions, and in some situations this sequence would clearly be preferable to base-catalysed isomerisation, which has also been employed for conversion of epoxides into allylic alcohols (see p. 373).

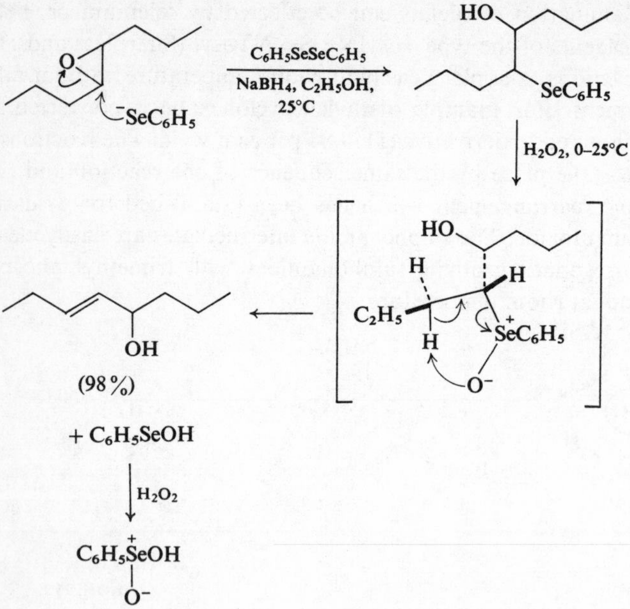

$$C_6H_5SeSeC_6H_5 \quad NaBH_4, C_2H_5OH, 25°C$$

$^{-}SeC_6H_5$ HO SeC$_6$H$_5$

H_2O_2, 0–25°C

HO H H C_2H_5 H $\overset{+}{Se}C_6H_5$ O^- (6.9)

OH (98%)

+ C_6H_5SeOH

H_2O_2

$C_6H_5\overset{+}{Se}OH$ with O^-

6.2. Oxidation of alcohols

Chromic acid. One of the most important uses of chromic acid in synthesis is in the oxidation of alcohols, and particularly in the oxidation of secondary alcohols to ketones. This reaction is commonly effected with a solution of the alcohol and aqueous acidic chromic acid in acetic acid, or with aqueous acidic chromic acid in a heterogeneous mixture. If no complicating structural features are present high yields of ketone are usually obtained.

Oxidation of primary alcohols to aldehydes with acidic solutions of chromic acid is usually less satisfactory because the aldehyde is easily oxidised further to the carboxylic acid and, more importantly, because under the acidic conditions the aldehyde reacts with unchanged alcohol to form a hemiacetal which is rapidly oxidised to an ester (6.10). Satis-

$$C_3H_7CH_2OH \xrightarrow[H_2O,\ H_2SO_4,\ <20°C]{Na_2Cr_2O_7} C_3H_7CHO \quad (6.10)$$

$$C_3H_7CHO \xrightleftharpoons{C_3H_7CH_2OH} C_3H_7CHOHOC_4H_9$$

$$C_3H_7COOC_4H_9 \longleftarrow C_3H_7CHOHOC_4H_9$$

factory yields of aldehyde can be obtained in favourable cases by removing the aldehyde from the reaction medium by distillation as it is formed or, better, by use of the pyridine–chromium trioxide complex (p. 341) or of t-butyl chromate as oxidising agent. The latter reagent is prepared by careful addition of chromium trioxide to t-butanol and is reported to give excellent yields of aldehydes by oxidation of primary alcohols in petroleum ether solution (Wiberg, 1965).

In general tertiary alcohols are unaffected by chromic acid, but tertiary 1,2-diols are rapidly cleaved, provided they are sterically capable of forming cyclic chromate esters. *cis*-1,2-Dimethylcyclopentandiol, for example, is oxidised 17×10^3 times faster than the *trans* isomer (6.11).

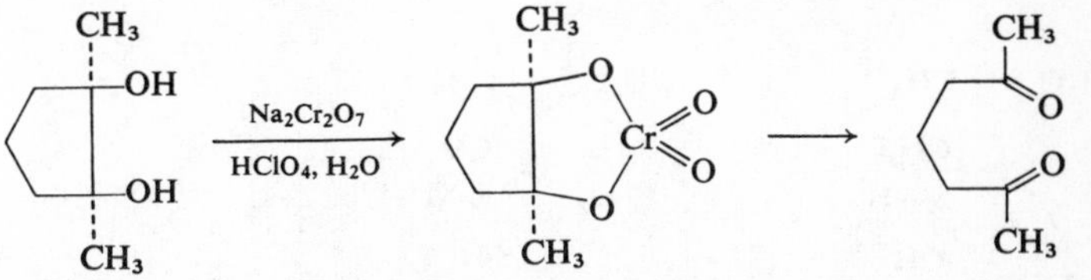

(6.11)

Oxidation of alcohols by chromic acid is believed to take place by initial formation of a chromate ester, followed by breakdown of the ester, as shown for oxidation of isopropanol. Whether proton abstraction from the ester takes place by an intermolecular or intramolecular process is uncertain (6.12).

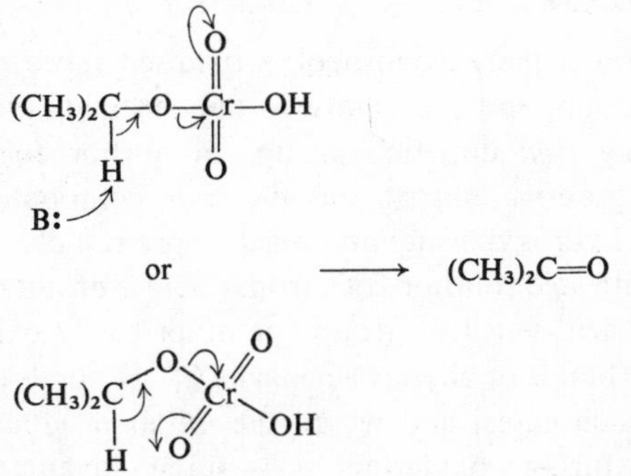

(6.12)

With unhindered alcohols the initial reaction to form the chromate ester is fast, and the subsequent decomposition of the ester is the rate-controlling step. Where formation of the ester results in steric overcrowding, ester decomposition is accelerated because steric strain is relieved in going from reactant to product. In extreme cases the initial esterification may become rate-determining. In the cyclohexane series it is found that axial hydroxyl groups are generally oxidised more rapidly than equatorial by a factor of about 3, presumably because of 1,3-diaxial interactions in the axial ester. This has been used in the determination of configurations of steroidal alcohols. Relative oxidation rates for a number of epimeric alcohols are given in (6.13). In the

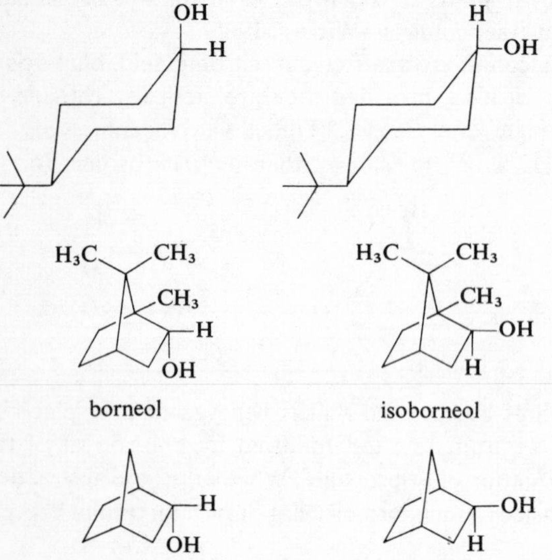

		Relative rates of oxidation with chromic acid
(axial OH)	(equatorial OH)	3:1
borneol	isoborneol	1:2
endo-norborneol	*exo*-norborneol	2·5:1

(6.13)

borneol–isoborneol pair, isoborneol is oxidised more rapidly because the hydroxyl group, and even more so the chromate ester, is sterically crowded by the *gem* dimethyl group. In norborneol, however, the *gem* dimethyl group is absent and the *endo* compound now has the more hindered hydroxyl group and reacts more rapidly.

Oxidation with acid solutions of chromic acid is unsuitable for alcohols which contain acid-sensitive groups or other easily oxidisable groups such as olefinic bonds or allylic or benzylic C—H bonds elsewhere in the molecule. In such cases, and where the initial product of reaction is susceptible to further oxidation, it is often advantageous to effect reaction in the presence of an immiscible solvent such as benzene or ether, which serves to reduce contact between the organic compounds and the acidic oxidising solution (compare Brown, Garg and Kwang-Ting Liu, 1971). Another method which often allows selective oxidation of a hydroxyl group in such molecules is by dropwise addition of the stoichiometric amount of a solution of chromium trioxide in aqueous sulphuric acid (the Jones reagent) to a cooled (0–20°C) solution of the alcohol in acetone (Bowers, Halsall, Jones and Lemin, 1953). Over-oxidation is thus lessened or prevented, and selective oxidation of unsaturated secondary alcohols to unsaturated ketones without appreciable oxidation or rearrangement of double bonds can often be achieved in good yield. Primary alcohols may give either aldehydes or carboxylic acids. In many cases the 'end point' is easily observed by the persistence of the red colour of the chromic acid after addition of the theoretical amount of oxidant (6.14).

$$CH_3CHOHC{\equiv}C(CH_2)_3CH_3 \xrightarrow[CH_3COCH_3]{CrO_3,\ H_2SO_4} CH_3COC{\equiv}C(CH_2)_3CH_3 \quad (80\%)$$

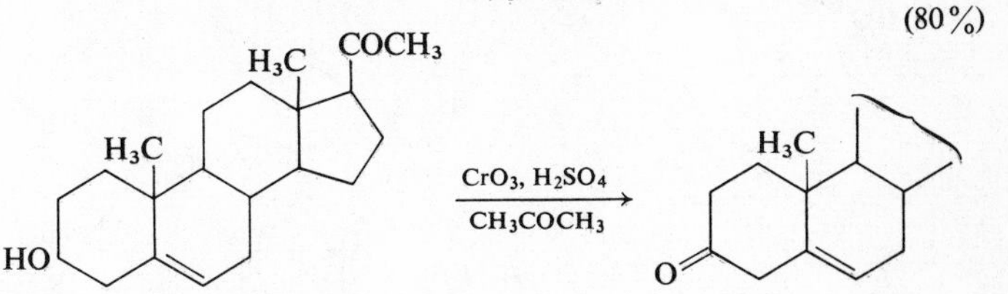

H_3C COCH$_3$ H_3C HO $\xrightarrow[CH_3COCH_3]{CrO_3,\ H_2SO_4}$ H_3C O (89%) (6.14)

$$CH{\equiv}C{-}CH{=}CHCH_2OH \longrightarrow CH{\equiv}C{-}CH{=}CHCO_2H \quad (60\%)$$

Chromium trioxide–pyridine complex. A useful reagent for oxidation of alcohols that contain acid-sensitive functional groups is the chromium trioxide–pyridine complex $CrO_3.2C_5H_5N$ (Poos, Arth, Beyler and Sarett, 1953). As originally described the reagent was prepared by addition of chromium trioxide to pyridine (caution: never the other way round) to form a pyridine solution of the complex, or, more simply, by adding a concentrated aqueous solution of chromium trioxide to pyridine. Reaction was effected at room temperature in pyridine solution by addition of the alcohol to the previously prepared reagent. Under

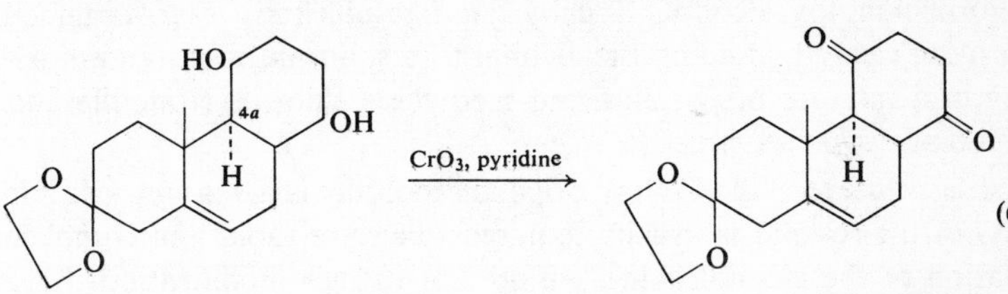

HO 4a OH H O O (6) $\xrightarrow{CrO_3,\ pyridine}$ O O H O O (90%) (6.15)

these conditions secondary alcohols are converted into the corresponding carbonyl compounds in good yield. Acid-sensitive protecting groups are unaffected. The tricyclic diol (6), for example, is converted into the dione without cleavage of the ketal, migration of the double bond or epimerisation at C-4*a*, (6.15). Polyhydroxy compounds can sometimes be selectively oxidised at one position by protection of the other hydroxyl groups by acetal formation, as in the example (6.16).

HO, CH_3, HO, HO — CH_3CHO, H^+ → H_3C, O, CH_3, H, O, HO

↓ CrO_3, pyridine

H_3C, O, CH_3, H, O, O

(6.16)

The reaction is generally less successful with primary alcohols. Allylic and benzylic alcohols are readily converted into the corresponding aldehydes, but yields of aldehydes from non-allylic primary alcohols by the original procedure are capricious. However, a modification by Collins, Hess and Frank (1968) gives high yields in a rapid reaction at room temperature. They made and isolated separately the complex $CrO_3.2C_6H_5N$ and effected oxidation with a solution of this material in methylene chloride under anhydrous conditions. By this procedure 1-heptanol gave heptanal in 80 per cent yield (Collins and Hess, 1972) and 5,9-dimethyl-5,9-decadienal was obtained in 92 per cent yield from the corresponding alcohol. Equally good results have been obtained even more easily by adding the alcohol to a solution of chromium trioxide in a mixture of pyridine and methylene chloride (Ratcliffe and Rodehorst, 1970; Ratcliffe, 1976).

A disadvantage of the Collins' original procedure is that a considerable excess of the reagent is usually required to ensure rapid and complete oxidation of the alcohol, and a number of further modifications have been introduced to overcome this. Thus, good results were obtained by oxidation of a wide range of alcohols in methylene chloride with the complex of chromium trioxide and 3,5-dimethylpyrazole (Corey and Fleet, 1973) or, better, with pyridinium chlorochromate, $C_5H_5\overset{+}{N}HCrO_3Cl^-$ (Corey and Suggs, 1975). The latter reagent is easily prepared by addition of pyridine to a solution of chromium trioxide in hydrochloric acid, and, when used in small excess in methylene chloride solution it gives yields of aldehydes and ketones at least as good as those obtained using Collins' reagent. For the oxidation of compounds containing acid-sensitive protecting groups the slightly acidic nature of the reagent is masked by addition of powdered sodium acetate (6.17). A solution of

$$\xrightarrow[NaOCOCH_3,\ CH_2Cl_2]{C_5H_5\overset{+}{N}HCrO_3Cl^-} \qquad (6.17)$$

CH_2OH → CHO (85%)

chromyl chloride in pyridine is also said to be efficacious for the oxidation of primary alcohols to aldehydes, but it is less selective than Collins' reagent for oxidation of allylic alcohols (Sharpless and Akashi, 1975).

Manganese dioxide and argentic oxide. Another useful mild reagent for the oxidation of primary and secondary alcohols to carbonyl compounds is manganese dioxide (Evans, 1959). The advantage of this reagent is that it is specific for allylic and benzylic hydroxyl groups, and reaction takes place under mild conditions (room temperature) in a neutral solvent (water, benzene, petroleum ether, chloroform). The general technique is simply to stir a solution of the alcohol in the solvent with the manganese dioxide for some hours. The manganese dioxide has to be specially prepared to obtain maximum activity. The best method appears to be by reaction of manganous sulphate with potassium permanganate in alkaline solution; the hydrated manganese dioxide obtained is highly active, but whether the actual oxidising agent is manganese dioxide itself or some other manganese compound adsorbed on the surface of the dioxide is not clear at present.

Olefinic and acetylenic bonds are unaffected by the reagent as illustrated in the examples (6.18), and the reaction has been widely used for oxidation of polyunsaturated alcohols in the carotenoid and vitamin A series. Hydroxyl groups adjacent to acetylenic bonds or cyclopropane

$$CH{\equiv}CCH{=}C(CH_3){-}CH_2OH \xrightarrow[CH_3COCH_3]{MnO_2} CH{\equiv}CCH{=}C(CH_3){-}CHO \quad (35\%)$$

$$\xrightarrow[CHCl_3]{MnO_2} \quad (80\%) \qquad (6.18)$$

rings are also easily oxidised, but under ordinary conditions saturated alcohols are not attacked (although they may be under more vigorous conditions), allowing selective oxidation of activated hydroxyl groups in appropriate cases (6.19).

$$(CH_3O)_2C_6H_3{-}CHOHCH_2CH_2OH \xrightarrow[CH_3COCH_3,\,25°C]{MnO_2} (CH_3O)_2C_6H_3{-}COCH_2CH_2OH \quad (94\%) \qquad (6.19)$$

$$\xrightarrow[\text{room temp.}]{MnO_2,\ CHCl_3} \quad (62\%)$$

Although in general oxidation of allylic primary alcohols with manganese dioxide takes place without significant further oxidation to carboxylic acids, in the presence of cyanide ions and an alcohol, very high yields of carboxylic esters can be obtained from the corresponding αβ-unsaturated aldehydes (Corey, Gilman and Ganem, 1968). Thus, in methanol solution, cinnamaldehyde is converted into methyl cinnamate in >95 per cent yield, and geranial gives methyl geranate in 85–95 per cent yield. Reaction is thought to proceed through the cyanhydrin (6.20). An important feature of the reaction is that oxidation takes place without any *cis–trans* isomerisation of the αβ-olefinic bond.

$$\begin{array}{ccc} >C{=}C<_{CHO} & \xrightarrow[ROH]{CN^-} & >C{=}C<_{CH(CN)OH} \\ & & \downarrow MnO_2 \\ >C{=}C<_{COOR} + HCN & \xleftarrow{ROH} & >C{=}C<_{C(=O)-CN} \end{array} \quad (6.20)$$

The traditional method of oxidising an aldehyde to a carboxylic acid using alkaline silver(I) oxide (Ag_2O) is relatively unsatisfactory for $\alpha\beta$-unsaturated aldehydes, since appreciable *cis-trans* $\alpha\beta$-isomerisation and other base-catalysed side reactions can occur.

Non-conjugated aldehydes are not converted into esters by the action of cyanide ion and manganese dioxide in methanol, even though cyanhydrin formation occurs readily. But with the more powerful reagent argentic oxide, AgO, prepared by action of alkaline permanganate on silver nitrate, oxidation to the corresponding *carboxylic acid* takes place readily with both conjugated and non-conjugated aldehydes. Thus, cinnamaldehyde with argentic oxide and sodium cyanide in methanol led to cinnamic acid in >90 per cent yield, and the non-conjugated aldehyde cyclohexene-3-carboxaldehyde was smoothly transformed into cyclohexene-3-carboxylic acid. Cinnamaldehyde cyanhydrin was likewise converted into cinnamic acid, and the catalytic effect of cyanide ions was indicated by the fact that little oxidation of cinnamaldehyde to the acid occurred in the absence of cyanide. The difference between the two reagents, manganese dioxide and argentic oxide, may be due to the presence of nucleophilic hydroxide or oxide ions on the surface of the argentic oxide, which strongly catalyse heterogeneous hydrolysis of the acyl cyanide in this system.

An even simpler method for the conversion of aldehydes into carboxylic acids, which is especially applicable to non-conjugated aldehydes, is oxidation with argentic oxide in tetrahydrofuran–water at 25°C under neutral conditions. Dodecenal and cyclohexenyl-3-carboxaldehyde, for example, were oxidised to the corresponding carboxylic acids in >90 per cent yield by this method. The oxidation of conjugated aldehydes is slower under these conditions, and cyanide-catalysed oxidation is preferable in such cases (Corey, Gilman and Ganem, 1968).

Oxidation of $\alpha\beta$-unsaturated aldehydes to $\alpha\beta$-unsaturated carboxylic acids can also be effected in high yield with hydrogen peroxide and selenium dioxide in t-butanol solution (Smith and Holm, 1957).

Silver carbonate. An excellent reagent for oxidising primary and secondary alcohols to aldehydes and ketones in high yield under mild and essentially neutral conditions is silver carbonate precipitated on celite (Fétizon and Golfier, 1968). The reaction is easily effected in boiling benzene and the product is recovered, usually in a high state of purity, by simply filtering off the spent reagent and evaporating off the solvent. Other functional groups are unaffected. Under these conditions nerol, for example, is converted into neral in 95 per cent yield. Highly hindered hydroxyl groups are not attacked, allowing selective oxidation in appropriate cases as in the first example below, where attack on the C-6 hydroxyl group is prevented by the steric effect of the *gem* dimethyl group at C-4. Primary alcohols are oxidised more slowly than secondary which are themselves much less reactive than benzylic and allylic alcohols, and in acetone or methanol solution selective oxidation of benzylic or allylic hydroxyl groups can be effected (6.21). The reaction is apparently not ionic for the vinylcyclopropyl-

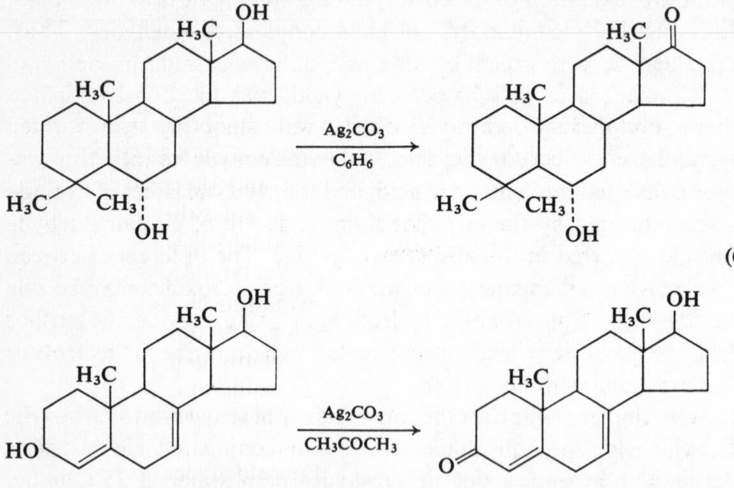

H_3C OH H_3C H_3C CH_3 OH $\xrightarrow[C_6H_6]{Ag_2CO_3}$ H_3C O H_3C H_3C CH_3 OH

(6.21)

H_3C OH H_3C HO $\xrightarrow[CH_3COCH_3]{Ag_2CO_3}$ H_3C OH H_3C O

methanol (7) is oxidised to the aldehyde without opening of the cyclopropane ring. A concerted mechanism has been suggested (6.22) (Fétizon, Golfier and Mourgues, 1972).

The behaviour of diols is different and depends on their structure. Butan-1,4-diols, pentan-1,5-diols and hexan-1,6-diols are converted, in an interesting reaction, into the corresponding γ-, δ- and ε-lactones (6.23). Other diols give hydroxy-ketones or -aldehydes. Vicinal diols are oxidised to α-hydroxy-ketones and β-glycols give β-hydroxy-ketones,

$$\text{(7) cyclopropane-}CH_2OH \xrightarrow[C_6H_6]{Ag_2CO_3} \text{cyclopropane-}CHO$$

$$(6.22)$$

$$\text{H–}\ddot{O}\text{–CH}\cdots Ag^+\ Ag^+\ (^-O)_2C{=}O \longrightarrow \text{H–}\overset{+}{O}(\text{Ag})\text{–CH} \quad Ag^+ \quad CO_3^{2-} \longrightarrow \text{>C=}\overset{+}{O}\text{–H} + 2Ag + CO_3^{2-} + H^+$$

$$HOCH_2(CH_2)_3CH_2OH \xrightarrow[C_6H_6]{Ag_2CO_3} \text{H}_2\text{C(CH}_2\text{)}_3\text{CO–O (lactone)} \quad (95\%)$$

$$(6.23)$$

$$\text{(CH}_3\text{)}_2\text{C(O–CH(H)CH}_2\text{OH)}_2 \xrightarrow[C_6H_6]{Ag_2CO_3} \text{bicyclic lactone}$$

usually in high yield (6.24). Butan-1,3-diol forms 1-hydroxy-3-butanone exclusively; no significant amount of hydroxy-aldehyde is found, in line with the observation that secondary alcohols are oxidised more

$$\text{cyclohexane-1,2-diol} \xrightarrow[C_6H_6]{Ag_2CO_3} \text{2-hydroxycyclohexanone} \quad (45\%) \qquad (6.24)$$

$$HOCH_2CH_2CHOHCH_3 \xrightarrow[C_6H_6]{Ag_2CO_3} HOCH_2CH_2COCH_3 \quad (80\%)$$

rapidly than primary with this reagent. Hydroxy-aldehydes are obtained from hexan-1,7-diol and octan-1,8-diol (6.25) (Fétizon, Golfier and Louis, 1975).

$$CH_2OH(CH_2)_5CH_2OH \xrightarrow[\text{boiling } C_6H_6]{Ag_2CO_3} CHO(CH_2)_5CH_2OH \qquad (6.25)$$

The reagent also oxidises some nitrogen-containing functional groups. Hydrazones, for example, are rapidly converted into diazoalkanes and hydroxylamines give nitrones (Fétizon, Golfier, Milcent and Papadakis,

1975). It is very effective for the important oxidative coupling of phenols. In neutral media it effects selective coupling of hindered phenols, giving the corresponding dipheno- and stilbene-quinones in high yield. Hydroquinones and catechols are oxidised to the corresponding *p*- and *o*-quinones in excellent yield and *o*-aminophenols form phenoxazones (Balogh, Fétizon and Golfier, 1971).

Oxidation via alkoxysulphonium salts. A number of methods for oxidising primary and secondary alcohols to aldehydes and ketones by action of a base on the derived alkoxysulphonium salts differ from each other mainly in the way in which the alkoxysulphonium salt is obtained from the alcohol. The conditions of reaction are mild and high yields of carbonyl compounds are generally obtained. One of the earliest procedures involved reaction of the alcohol with dimethyl sulphoxide and dicyclohexylcarbodiimide in presence of phosphoric acid or pyridinium trifluoroacetate as a proton source (Epstein and Sweat, 1967). This method has been used to oxidise a number of sensitive compounds in

CH_3 CH_3 O O CH_3SO_2O— —OH O O CH_3 CH_3 $\xrightarrow[\text{dimethyl sulphoxide, phosphoric acid}]{\text{dicyclohexylcarbodiimide}}$ CH_3 CH_3 O O CH_3SO_2O— =O O O CH_3 CH_3 (74%) (6.26)

O HN CH_3 O N CH_2OH O $OCOCH_3$ $\xrightarrow[\text{dimethyl sulphoxide, phosphoric acid, room temp.}]{\text{dicyclohexylcarbodiimide}}$ O HN CH_3 O N CHO O $OCOCH_3$ (90%)

the steroid, alkaloid and carbohydrate series. Thus, the 'isolated' hydroxyl group of the mannitol derivative in (6.26) was smoothly oxidised to the ketone in 74 per cent yield, and 3′-*O*-acetyl-thymidine was converted into the 5′-aldehyde in 90 per cent yield, with no trace of the carboxylic acid.

The mechanism of the oxidation has been elucidated by tracer experiments and is thought to involve initial formation of a sulphoxide–carbodiimide adduct which reacts with the alcohol to give an alkoxysulphonium ion. This then undergoes proton abstraction to form an ylid which collapses to the ketone and dimethyl sulphide by an intramolecular concerted process (6.27) (Moffatt, 1971).

$$C_6H_{11}{-}N{=}C{=}N{-}C_6H_{11} + CH_3SOCH_3 \xrightarrow{H^+} C_6H_{11}{-}NH{-}C(=N{-}C_6H_{11}){-}O{-}\overset{+}{S}(CH_3)_2$$

$$\xrightarrow{RCHOHR} (CH_3)_2\overset{+}{S}{-}O{-}CHR_2 + (C_6H_{11}NH)_2CO \qquad (6.27)$$

$$(CH_3)_2\overset{+}{S}{-}O{-}CHR_2 \xrightarrow{base} \overset{-}{C}H_2(CH_3)\overset{+}{S}{-}O{-}C(H)R_2 \longrightarrow (CH_3)_2S + O{=}CR_2$$

A related method uses dimethyl sulphoxide and acetic anhydride or, better, methanesulphonic anhydride in hexamethylphosphoramide (Albright, 1974). Here the anhydride acts as the 'activating' group. The mechanism established by tracer studies again involves an ylid (6.28). Methyl thiomethyl ethers, formed by alternative breakdown of the alkoxysulphonium ion are often obtained as by-products in oxidations with dimethyl sulphoxide (cf. 6.28), but appear to be formed in only small amounts in the methanesulphonic anhydride procedure. Thus, testosterone gave 70 per cent of the androsten-3,17-dione and 30 per cent of methylthio compound with the acetic anhydride–dimethyl sulphoxide reagent, but with methanesulphonic anhydride the diketone was obtained in 99 per cent yield at –20°C (6.29). The method is widely used for the oxidation of hydroxyl groups in the carbohydrate series (Brimacombe, 1969).

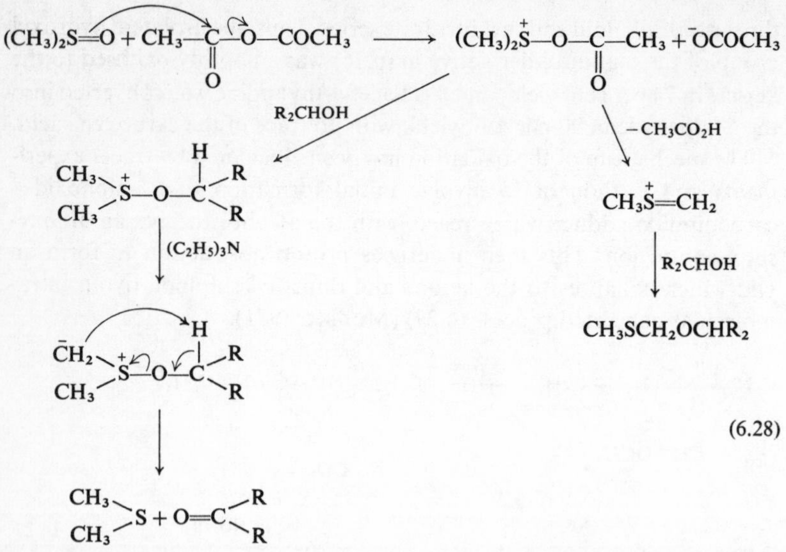

$$(CH_3)_2S{=}O + CH_3{-}\underset{\underset{O}{\|}}{C}{-}O{-}COCH_3 \longrightarrow (CH_3)_2\overset{+}{S}{-}O{-}\underset{\underset{O}{\|}}{C}{-}CH_3 + \bar{O}COCH_3$$

R_2CHOH; $-CH_3CO_2H$

$(CH_3)_2\overset{+}{S}{-}O{-}CHR_2 \xrightarrow{(C_2H_5)_3N} \bar{C}H_2(CH_3)\overset{+}{S}{-}O{-}CHR_2 \longrightarrow (CH_3)_2S + O{=}CR_2$

$CH_3\overset{+}{S}{=}CH_2 \xrightarrow{R_2CHOH} CH_3SCH_2OCHR_2$

(6.28)

OH, H_3C, H_3C, O; O, H_3C; OCH_2SCH_3, H_3C

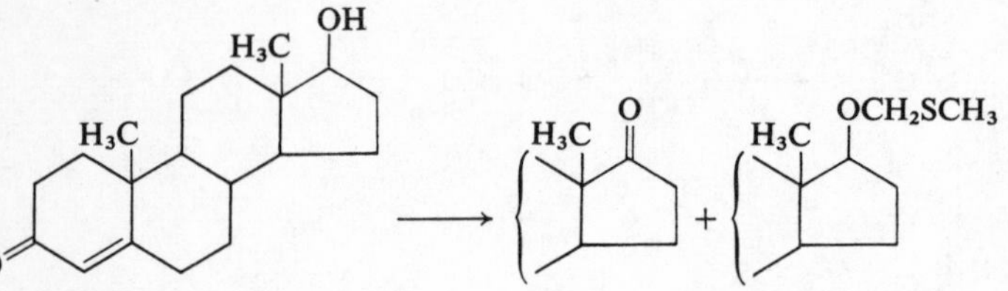

(6.29)

A useful alternative approach makes use of the complexes formed from a methyl sulphide with chlorine or *N*-chlorosuccinimide (Corey and Kim, 1972). These complexes react readily with alcohols to form the corresponding alkoxysulphonium salts which decompose in the usual way in presence of triethylamine to give the carbonyl compound (6.30). Tracer experiments have shown that the transformation again involves an intermediate ylid (McCormick, 1974). Very good results are claimed under mild conditions, but the method is unsuitable for allylic or benzylic alcohols, which form chlorides in high yield rather than carbonyl compounds. A valuable application is in the oxidation of secondary-tertiary-1,2-diols to α-ketols without rupture of the carbon–carbon bond (Corey and Kim, 1974). This unusual transformation depends on the fact that straightforward oxidation of the secondary alcohol function proceeds through a five-membered transition state and is thus favoured over glycol cleavage which requires a seven-membered transition state (6.30).

$$\text{cyclohexyl-}CH_2OH + \text{succinimido-}N\text{—}\overset{+}{S}(CH_3)_2\ \overset{-}{Cl} \xrightarrow[-25°C]{\text{toluene}} \text{cyclohexyl-}CH_2O\overset{+}{S}(CH_3)_2\ Cl^- \xrightarrow{(C_2H_5)_3N} \text{cyclohexyl-}CHO\ (93\%)$$

$$\xrightarrow[(2)\ (C_2H_5)_3N]{(1)\ Cl_2,\ C_6H_5SCH_3,\ CH_2Cl_2,\ -25°C} \quad (80\%) \qquad (6.30)$$

Alkoxysulphonium salts convertible into carbonyl compounds on treatment with base have also been obtained by reaction of primary and secondary alcohols with the complex $(CH_3)_2\overset{+}{S}OCl\ Cl^-$, obtained from dimethyl sulphoxide and chlorine (Corey and Kim, 1973) and by reaction of the alcohol chloroformate with dimethyl sulphoxide (Barton and Forbes, 1975).

Related to these reactions is the oxidation of alkyl halides and toluene-*p*-sulphonates to carbonyl compounds with dimethyl sulphoxide. This useful reaction is simply effected by warming the halide or sulphonate in dimethyl sulphoxide, generally in presence of a proton acceptor such as sodium bicarbonate or collidine. Oxidation never proceeds beyond the carbonyl stage and other functional groups are generally unaffected (Epstein and Sweat, 1967). The reaction has been applied to phenacyl halides, benzyl halides, primary sulphonates and iodides and a limited number of secondary sulphonates. Primary bromides and chlorides give only poor yields, but may be converted *in situ* into the corresponding sulphonates and oxidised without purification (Kornblum, Jones and Anderson, 1959). Much improved yields of aldehydes from primary bromides can be obtained at room temperature by conducting the reaction in presence of silver ion and triethylamine (Epstein and Ollinger, 1970; Ganem and Boeckman, 1974). With secondary sulphonates and halides elimination becomes an important side reaction and the reaction is less useful with such compounds. Where elimination is not possible good yields of ketones are obtained. It is thought that this reaction also probably proceeds through an ylid in most cases.

A method for converting secondary–tertiary 1,2-diols into α-ketols was described above. Another procedure which allows the conversion of primary–secondary diols into hydroxy aldehydes has recently been reported by Boeckman and Ganem (1974). It depends on the fact that on reaction of a primary–secondary diol with triphenylphosphine dibromide in dimethylformamide at 0°C, the primary hydroxyl group is converted into the bromide, whereas the secondary group gives the formate ester. The primary bromide may then be converted into the aldehyde by reaction with dimethyl sulphoxide and triethylamine in the presence of silver ion as described above. Using this procedure, 1,4-dihydroxypentane was converted into the formate of 4-hydroxypentanal in 55 per cent yield.

Other methods. Among a number of other useful methods for selective oxidation of primary and secondary alcohols to aldehydes and ketones under mild conditions are the Oppenauer oxidation, oxidation with lead tetra-acetate and catalytic oxidation with oxygen and platinum.

$$CH_3CHOHCH{=}CHCH{=}C(CH_3)CH{=}CH_2 \xrightarrow[\text{boiling benzene}]{\text{Al t-butoxide, acetone,}} CH_3COCH{=}CHCH{=}C(CH_3)CH{=}CH_2 \quad (80\%) \qquad (6.31)$$

H$_3$C ... HO $\xrightarrow[CH_3COCH_3]{Al(OC_3H_7\text{-iso})_3}$ H$_3$C ... O

The Oppenauer oxidation (Djerassi, 1951) with aluminium alkoxides in acetone is the reverse of the Meerwein–Pondorff–Verley reduction discussed on p. 462. It has been widely used in the steroid series, particularly for the oxidation of allylic secondary hydroxyl groups to $\alpha\beta$-unsaturated ketones (6.31). $\beta\gamma$-Double bonds generally migrate into conjugation with the carbonyl group under the conditions of the reaction. The aluminium alkoxide serves only to form the aluminium alkoxide of the alcohol which is then oxidised through a cyclic transition state at the expense of the acetone. By use of excess of acetone the equilibrium is forced to the right (6.32).

$$(CH_3)_2C{=}O + \text{Al}{-}O{-}CHR_2 \rightleftharpoons (CH_3)_2CH{-}O{-}\text{Al} + R_2C{=}O \qquad (6.32)$$

Lead tetra-acetate in refluxing benzene, hexane or chloroform is a good reagent for oxidation of primary and secondary alcohols to the corresponding aldehyde or ketone, provided there is no δ C—H group in the molecule. In this circumstance high yields of tetrahydrofuran derivatives are obtained (see p. 258). In pyridine solution lead tetra-acetate oxidises a variety of primary and secondary alcohols to the carbonyl compounds in good yield at room temperature whether they contain δ C—H groups or not. Both allylic and saturated alcohols are oxidised and cyclisation products appear not to be formed (Partch, 1964) (6.33).

$$CH_3(CH_2)_3CH_2OH \xrightarrow[\text{pyridine}]{Pb(OCOCH_3)_4} CH_3(CH_2)_3CHO \quad (70\%)$$

$$CH_3CHOHCH_2CH_2CHOHCH_3 \longrightarrow CH_3COCH_2CH_2COCH_3 \quad (89\%)$$

$$C_6H_5CH{=}CHCH_2OH \longrightarrow C_6H_5CH{=}CHCHO \quad (91\%) \qquad (6.33)$$

Catalytic oxygenation with a platinum catalyst and molecular oxygen is another valuable method for oxidation of primary and secondary hydroxyl groups under mild conditions. With primary alcohols the reaction can be regulated to give aldehydes or acids. Double bonds, in general, are not affected and unsaturated alcohols such as tiglic alcohol

(2-methyl-2-buten-1-ol) can be oxidised catalytically to the unsaturated aldehyde (Heyns and Paulsen, 1963) (6.34).

$$CH_3(CH_2)_{10}CH_2OH \xrightarrow[C_7H_{16},\ \frac{1}{2}\,h]{O_2,\ Pt} CH_3(CH_2)_{10}CHO \quad (77\%)$$

$$\xrightarrow{2\,h} CH_3(CH_2)_{10}COOH \quad (96\%) \qquad (6.34)$$

$$CH_3CH{=}C(CH_3){-}CH_2OH \xrightarrow[C_7H_{16}]{O_2,\ Pt} CH_3CH{=}C(CH_3){-}CHO$$

In general, primary hydroxyl groups are attacked before secondary, and in cyclic secondary alcohols axial groups appear to react before equatorial (but see Heyns, Soldat and Köll, 1975). The method has been widely used in the carbohydrate series to effect selective oxidation of specific hydroxyl groups (Brimacombe, 1969). Thus, L-sorbose is oxidised at 30°C to 2-keto-L-gulonic acid, an intermediate in a synthesis of ascorbic acid (6.35).

$$\begin{array}{c} CH_2OH \\ | \\ C(OH) \\ | \\ HO{-}C{-}H \\ | \\ H{-}C{-}OH \\ | \\ HO{-}C{-}H \\ | \\ CH_2 \end{array} \text{(ring O)} \xrightarrow{O_2,\ Pt} \begin{array}{c} COOH \\ | \\ C(OH) \\ | \\ HO{-}CH \\ | \\ H{-}C{-}OH \\ | \\ HO{-}C{-}H \\ | \\ CH_2 \end{array} \text{(ring O)} \qquad (6.35)$$

Other useful reagents are the *N*-halo-imides, *N*-bromosuccinimide and *N*-bromo- and *N*-chloro-acetamide. In aqueous acetone or aqueous dioxan they readily oxidise many primary and secondary alcohols to aldehydes and ketones (Filler, 1963). In general, secondary alcohols are oxidised more readily than primary, and among cyclic secondary alcohols, axial hydroxyl groups are oxidised more easily than equatorial ones. This has led to the extensive use of these reagents for selective oxidation of nuclear hydroxyl groups in the steroid series. Thus, aetiocholan-$3\alpha,11\alpha,17\beta$-triol with *N*-bromo-acetamide in aqueous acetone is converted into aetiocholan-11α-ol-3,17-dione by oxidation of the two axial hydroxyl groups (6.36). By contrast, chromic acid gave the 3,11,17-trione. Benzylic or allylic alcohols are particularly easily oxidised.

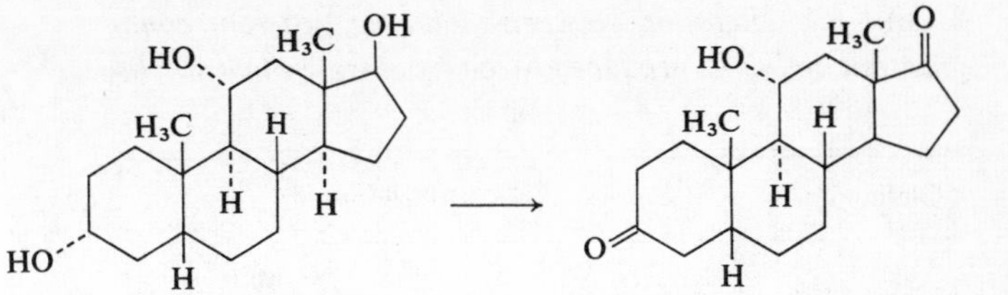

(6.36)

Recently Rees and Storr (1969) have shown that 1-chlorobenzotriazole, which is readily obtained from benzotriazole with sodium hypochlorite, is an excellent reagent for the oxidation of alcohols to carbonyl compounds under mild conditions. Oxidation of alcohols to aldehydes and ketones in neutral solution at room temperature has also been effected in good yield with 4-phenyl-1,2,4-triazoline-3,5-dione (Cookson, Stevens and Watts, 1966).

Steroidal allylic alcohols have also been selectively oxidised to the corresponding $\alpha\beta$-unsaturated ketones in excellent yield with the high-potential quinone 2,3-dichloro-5,6-dicyanobenzoquinone (6.37). Saturated alcohols are unaffected.

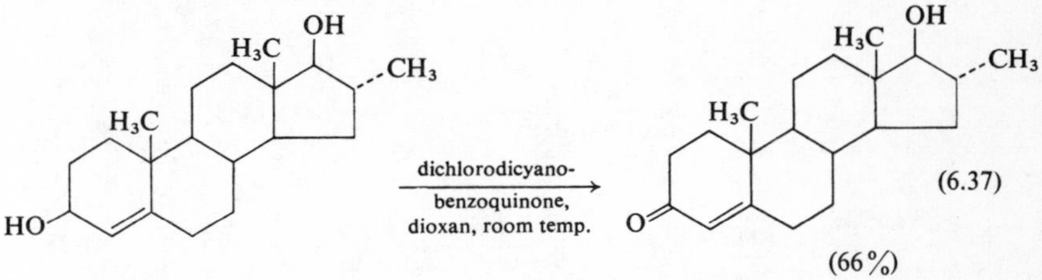

dichlorodicyano-benzoquinone, dioxan, room temp. (6.37)

(66%)

6.3. Oxidation of olefinic bonds

Perhydroxylation. Perhydroxylation of olefinic double bonds is useful in degradation and synthesis, and can be accomplished stereospecifically with a number of different reagents. For *cis*-hydroxylation the best

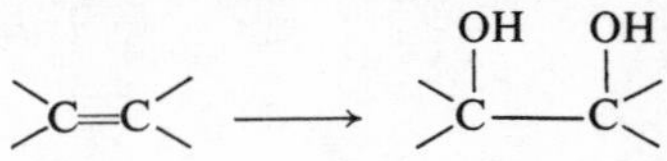

$$>C{=}C< \longrightarrow >C(OH){-}C(OH)<$$

methods are reaction with potassium permanganate, with osmium tetroxide or with iodine and moist silver acetate. The most important method of *trans*-hydroxylation is reaction with peroxy-acids, but the Prévost reaction, that is the action of iodine and silver acetate under anhydrous conditions, is also useful.

TABLE 6.1 *Relation between olefins of different configuration and diols produced by* cis- *and* trans- *hydroxylation*

Olefin	Perhydroxylation	
	cis	*trans*
cis	*cis, meso, erythro*	*trans, racemic, threo*
trans	*trans, racemic, threo*	*cis, meso, erythro*

The olefin may, of course, have either the *cis* or *trans* configuration and by *cis*- or *trans*-hydroxylation can give rise to isomeric diols which may be described by the terms *cis* or *trans*, *erythro* or *threo*, or *meso* or *racemic*, depending on the nature of the other groups attached to the glycol system. The relation between the olefins and the diols produced is shown in Table 6.1.

Potassium permanganate. Oxidation with permanganate is a widely used method for *cis*-perhydroxylation of olefins, but needs careful control

Cyclohexene $\xrightarrow[H_2O,\ NaOH,\ 5°C]{KMnO_4}$ cyclohexane-1,2-diol (OH, OH) (38%)

Norbornene $\xrightarrow[H_2O,\ NaOH,\ 0°C]{KMnO_4,\ t\text{-}C_4H_9OH}$ norbornane-2,3-diol (OH, OH) (40%)

Norbornene $\xrightarrow[H_2O,\ neutral]{KMnO_4}$ cyclopentane-1,3-dicarbaldehyde (CHO, CHO) (60%)

(6.38)

oleic acid, $CH_3(CH_2)_7$(H)C=C(H)$(CH_2)_7CO_2H$ $\xrightarrow[H_2O,\ NaOH,\ 10°C]{KMnO_4}$ $CH_3(CH_2)_7$–C(H)(OH)–C(H)(OH)–$(CH_2)_7CO_2H$

erythro-9,10-dihydroxystearic acid (80%)

oleic acid $\xrightarrow{KMnO_4,\ H_2O,\ H_2SO_4,\ pH\ 9}$ $CH_3(CH_2)_7C{-}C(CH_2)_7CO_2H$ (one carbon bearing H and OH, the other =O)

(75%, mixture of two products)

to avoid over-oxidation. Best results are obtained in alkaline solution, using water or aqueous organic solvents (acetone, ethanol or t-butanol); in acid or neutral solution α-ketols or even cleavage products are formed. The method is particularly suitable for perhydroxylation of unsaturated acids, which dissolve in the alkaline solution (6.38), but in many other cases only poor yields of the diols are obtained because of the insolubility of the substrate in the aqueous oxidising medium. Greatly improved yields can be obtained by effecting the oxidations in presence of a phase-transfer catalyst (cf. p. 330). Thus, oxidation of *cis*-cyclo-octene with aqueous alkaline permanganate in presence of benzyltrimethylammonium chloride gave *cis*-cyclo-octan-1,2-diol in 50 per cent yield; in absence of the catalyst the yield was only 7 per cent.

These reactions are believed to proceed through the formation of cyclic manganese esters. The mode of addition of the hydroxyl groups is shown to be *cis* by the conversion of maleic acid into *meso*-tartaric acid and of fumaric acid into (±)-tartaric acid, and the cyclic ester mechanism is supported by studies with ^{18}O which show that transfer of

$$\text{RHC=CHR} \xrightarrow[(\text{Mn}^{\text{VII}})]{\text{MnO}_4^-} \text{RHC(O)–CHR(O)}\!>\!\bar{\text{M}}\text{nO}_2 \; (\text{Mn}^{\text{V}})$$

$$\text{cyclic ester } (\text{Mn}^{\text{V}}) \xrightarrow{\text{MnO}_4^-} \text{RHC(O)–CHR(O)}\!>\!\text{MnO}_2 \; (\text{Mn}^{\text{VI}}) \xrightarrow{\text{H}_2\text{O:}} \text{RHC(OMnO}_2^-\text{)–C(=O)R} \; (\text{Mn}^{\text{IV}}) \xrightarrow{\text{H}_2\text{O}} \text{RHC(OH)–C(=O)R} + \text{MnO}_2$$

$$\text{cyclic ester } (\text{Mn}^{\text{V}}) \xrightarrow{\text{OH}^-} \text{RHC(OH)–CHR(OMnO}_3^{2-}) \; (\text{Mn}^{\text{V}}) \xrightarrow{\text{H}_2\text{O}} \text{RHC(OH)–CHR(OH)} \qquad (6.39)$$

oxygen from permanganate to the substrate occurs. Competition between ring-opening of the cyclic ester by hydroxyl ion and further oxidation by permanganate accounts for the effect of pH on the distribution of products (Stewart, 1965) (6.39).

Osmium tetroxide. Reaction with osmium tetroxide is another good method for *cis*-hydroxylation of double bonds, but is suitable only for small-scale work with valuable compounds because of the expense and toxicity of the reagent; for example it has been much used in the steroid hormone series (Gunstone, 1960; Fieser and Fieser, 1967) (6.40).

CH_3, CH_3 — (1) OsO_4, pyridine, ether; (2) hydrolysis, KOH, mannitol → CH_3, OH, OH, CH_3 (70%) (6.40)

H_3C, O, H_3C, HO — (1) OsO_4, pyridine; (2) aq. $NaHSO_3$ → H_3C, HO, HO, OH (86%)

Reaction is accelerated by tertiary bases, especially pyridine, and pyridine is often added to the reaction medium. Brightly coloured complexes in which osmium is co-ordinated with two molecules of base separate in almost quantitative yield (6.41). Hydrolysis of the osmate esters to the

$$\mathrm{>C{=}C<} + OsO_4 \xrightarrow{\text{pyridine}} \text{(—C—O, —C—O)Os(=O)}_2\text{(Pyr.)}_2 \xrightarrow{H_2O} \text{—C—OH, —C—OH} \qquad (6.41)$$

diols is effected by treatment with sodium hydroxide and mannitol, sodium sulphite or hydrogen sulphide.

Osmium tetroxide has also been used as a catalyst in conjunction with other oxidising agents, notably with chlorates and with hydrogen peroxide. Thus, the furan–maleic anhydride adduct on treatment with hydrogen peroxide in t-butanol in the presence of a catalytic amount of osmium tetroxide (the Milas reagent) gave the *exo* diol in 50 per cent yield. Similarly 2-pentenoic acid, with aqueous barium chlorate and a catalytic amount of osmium tetroxide was converted into the corresponding diol. In these reactions the initial osmate ester is oxidatively hydrolysed by the oxidising agent with regeneration of osmium tetroxide which continues the reaction, so that a small amount suffices. A disadvantage of these procedures is that in certain cases considerable amounts of over-oxidation products (ketols and compounds resulting from cleavage of the carbon–carbon bond) are formed. It has recently been found, however, that this difficulty can be largely avoided by effecting the reaction in alkaline solution in the presence of t-butyl hydroperoxide as oxidising agent. An added advantage of this procedure is that it is effective for oxidation of tri- and tetra-substituted olefins, which are often completely resistant to the perchlorate or hydrogen peroxide reagents (Sharpless and Akashi, 1976). Even better yields of *cis* 1,2-glycols are obtained by oxidation of olefins with tertiary amine oxides and catalytic osmium tetroxide (van Rheenen, Kelly and Cha, 1976).

A promising method for converting olefins into vicinal hydroxytoluene-*p*-sulphonamides by reaction with chloramine T in presence of a catalytic amount of osmium tetroxide has been reported by Sharpless, Chong and Oshima (1976). The sulphonylimido osmium compound (8) is presumed to be the effective reagent, and is continuously regenerated during the reaction. The sulphonamides are readily converted into the *cis* α-hydroxyamines by cleavage with sodium in liquid ammonia.

TosNClNa + RCH=CHR $\xrightarrow[\text{t-}C_4H_9OH,\ 60°C]{1\%\ OsO_4}$ HO(R)CH–CH(R)NHTos

Na, liq. NH_3 (6.42)

HO(R)CH–CH(R)NH_2

(8) $O_2Os(O)=NTos$

Tos ≡ $CH_3C_6H_4SO_2$

Oxidation with iodine and silver carboxylates. Many of the difficulties attending the oxidation of olefins to 1,2-glycols with other reagents can be avoided by using Prévost's reagent – a solution of iodine in carbon tetrachloride together with an equivalent of silver acetate or silver benzoate. Under anhydrous conditions this oxidant directly yields the diacyl derivative of the *trans* glycol (Prévost conditions), while in presence of water the monoester of the *cis* glycol is obtained (Woodward conditions). Thus, cyclohexene on treatment with iodine and silver benzoate in boiling carbon tetrachloride under anhydrous conditions gives the dibenzoate of *trans*-1,2-dihydroxycyclohexane. With iodine and silver acetate in moist acetic acid, however, the monoacetate of *cis*-1,2-dihydroxycyclohexane is formed. Similarly, oleic acid on oxidation under Prévost conditions gives *threo*-9,10-dihydroxystearic acid, while by the Woodward procedure the *erythro* isomer results (Gunstone, 1960).

The value of these reagents is due to their specificity and to the mildness of the reaction conditions; free iodine, under the conditions used, hardly affects other sensitive groups in the molecule. Reaction

$>C{=}C<$ + I–I Ag^+ $^-OCOCH_3$ ⟶ cyclic iodonium ion ($^-OCOCH_3$ attacks) + AgI

↓

$O{=}C(CH_3)–O–C–C–I$ ⟶(Ag^+) cyclic acetoxonium ion (9) ⟷ resonance form with $^+O{=}C$; attack by $^-OCOCH_3$

(9) → $AcO–C–C–OAc$ (trans diacetate)

(9) →(H_2O) $CH_3C(OH)(O–C)(O–C)$ ⟶ $CH_3C(=O)–O–C–C–OH$

(6.43)

proceeds through formation of an iodonium cation which, in presence of acyloxy and silver ions, forms the resonance stabilised cation (9) (6.43). Attack on the cation by acetate ion in a bimolecular process gives the *trans* diacyl compound. In the presence of water, however, a hydroxy acetal is formed which affords the *cis* hydroxy-acyloxy compound. With conformationally rigid molecules the *cis* diol obtained by the Woodward method may not have the same configuration as that obtained with osmium tetroxide. In an alternative procedure which avoids the use of expensive silver salts the *trans* iodoacetates are obtained in high yield from the alkene by reaction with iodine and thallium(I) acetate. Solvolysis in wet acetic acid under reflux affords the corresponding *cis* hydroxy-acetates, while in dry acetic acid the *trans* diacetate is obtained (Cambie, Hayward, Roberts and Rutledge, 1974).

Oxidation with peroxy-acids. Oxidation of olefins with peroxy-acids gives rise to epoxides (oxiranes) or to *trans* 1,2-glycols, depending on the experimental conditions (Swern, 1953).

The peroxy-acids most commonly used in the laboratory are perbenzoic, perphthalic, performic, peracetic and trifluoroperacetic acid. Recently, *m*-chloroperbenzoic acid has come into use; it is commercially available and is an excellent reagent for epoxidation of olefinic double bonds. It is more stable than the other peroxy-acids and has even been used at an elevated temperature (ethylene dichloride at 90°C), in the presence of a radical inhibitor, to effect the epoxidation of unreactive compounds (Kishi, Aratani, Tanino, Fukuyama and Goto, 1972). The other reagents are rather unstable and are generally prepared freshly before use. Perbenzoic acid is obtained from benzoyl peroxide and sodium methoxide or by reaction of benzoic acid with hydrogen peroxide in methanesulphonic acid; the sulphonic acid serves both as the solvent and an acid catalyst. Performic and peracetic acids are often prepared *in situ*, and not isolated, by action of hydrogen peroxide on the acids.

$$\mathrm{RCOOH + H_2O_2 \rightleftharpoons H_2O + RCO_3H}$$

With acetic acid the equilibrium is attained only slowly, and sulphuric acid is usually added as a catalyst to hasten formation of the peroxy-acid. Solutions prepared in this way from acetic acid and formic acid are widely used for the preparation of *trans* glycols from olefins. Trifluoroperacetic acid, the most powerful of the common reagents, is obtained mixed with trifluoroacetic acid by action of hydrogen peroxide on trifluoroacetic anhydride (6.44). The use of insoluble resins containing aromatic peroxy-acid residues for effecting epoxidation of olefins has been explored by Harrison and Hodge (1976). Yields of epoxides are variable but the method has the advantage that the products are easily isolated by filtering off the spent reagent.

$$(CF_3CO)_2O + H_2O_2 \xrightarrow{CH_2Cl_2} CF_3CO_2H + CF_3CO_3H \qquad (6.44)$$

It is probable that reaction of all peroxy-acids with olefins gives rise to the epoxide in the first place, but unless proper conditions are chosen the epoxide may be converted directly into an acyl derivative of the α-glycol. The best reagents for the preparation of the epoxide appear to be the three aromatic peroxy-acids, although solutions of peracetic acid buffered with sodium acetate to neutralise the sulphuric acid used in their preparation, are also frequently employed (6.45).

CH_3 $\xrightarrow[CHCl_3]{C_6H_5CO_3H}$ CH_3 O (75%)

(6.45)

C_6H_5 H C=C H C_6H_5 $\xrightarrow[CH_3CO_2Na,\ 25°C]{CH_3CO_3H,\ CH_3CO_2H}$ C_6H_5 O H C—C H C_6H_5

CH_3 CH_3 $\xrightarrow[\text{dioxan, }0°C]{m\text{-}ClC_6H_4CO_3H}$ CH_3 H H CH_3 O

Reaction is believed to take place by electrophilic attack of the peroxy-acid on the double bond, as illustrated in the equation (6.46) (Lee

C=C O H O O—C—R ⟶ C—C O + H O C—R O (6.46)

and Uff, 1967). In accordance with this mechanism the rate of epoxidation is increased by electron-withdrawing groups in the peroxy-acid (for example, trifluoroperacetic acid is more reactive than peracetic acid) or electron-donating substituents on the double bond. Terminal mono-olefins react only slowly with most peroxy-acids, but the rate of reaction increases with the degree of alkyl substitution. 1,2-Dimethyl-1,4-cyclohexadiene for example, reacts entirely at the tetrasubstituted double bond (6.47). On the other hand, conjugation of the olefin with other

$$\text{1,2-dimethyl-1,4-cyclohexadiene} \xrightarrow[\text{CHCl}_3]{m\text{-ClC}_6\text{H}_4\text{CO}_3\text{H}} \text{epoxide (80\%)} \quad (6.47)$$

unsaturated groups reduces the rate of epoxidation because of delocalisation of the π electrons. $\alpha\beta$-Unsaturated acids or esters, for example, require the strong reagent trifluoroperacetic acid for successful reaction. These reactions are carried out in presence of a buffer to prevent hydrolysis of the oxide by trifluoroacetic acid. Equally good or better results are obtained more conveniently with *m*-chloroperbenzoic acid at an elevated temperature (cf. p. 361). With $\alpha\beta$-unsaturated ketones (6.48),

$$(CH_3)_2C{=}CHCOCH_3 \xrightarrow{CH_3CO_3H} (CH_3)_2\overset{O}{C{-}}CHCOCH_3 + (CH_3)_2\overset{O}{C{-}}CHOCOCH_3 \quad (6.48)$$

(20% of product) (80% of product)

reaction is complicated by simultaneous, or even preferential, Baeyer–Villiger type oxidation at the carbonyl group (see p. 395). Epoxides of $\alpha\beta$-unsaturated aldehydes and ketones are best made by action of nucleophilic reagents such as hydrogen peroxide or t-butyl hydroperoxide in alkaline solution. The reaction is believed to take the course shown in (6.49). Thus, acrolein is readily converted into glycidaldehyde by this method, and cyclohexenone gives 2,3-epoxycyclohexanone.

$$R_2C{=}CH{-}\overset{O}{\overset{\|}{C}}{-}R + {}^{-}O{-}OH \rightleftharpoons R{-}\underset{HO-O}{\overset{R}{C}}{-}CH{=}\overset{O^-}{C}{-}R \longrightarrow R{-}\overset{R}{C}{-}CH{-}CO{-}R \ (\text{epoxide, O bridging C–CH}) \quad (6.49)$$

Epoxidations with peroxy-acids are highly stereoselective and take place by *cis* addition to the olefinic bond. This follows from the results of numerous experiments and has been shown unequivocally by X-ray diffraction and infrared studies of the products obtained by epoxidation of oleic and elaidic acid with perbenzoic and peracetic acid. Thus, oleic acid gave *cis*-9,10-epoxystearic acid as shown (6.50), whereas elaidic acid gave the isomeric *trans* compound.

$$CH_3(CH_2)_7CH{=}CH(CH_2)_7CO_2H \ (cis) \xrightarrow{C_6H_5CO_3H} cis\text{-}CH_3(CH_2)_7\overset{O}{CH{-}CH}(CH_2)_7CO_2H$$

$$CH_3(CH_2)_7CH{=}CH(CH_2)_7CO_2H \ (trans) \xrightarrow{C_6H_5CO_3H} trans\text{-}CH_3(CH_2)_7\overset{O}{CH{-}CH}(CH_2)_7CO_2H \qquad (6.50)$$

With conformationally rigid cyclic olefins the reagent usually approaches from the less hindered side of the double bond, as illustrated in the epoxidation of norbornene shown in (6.51), but with flexible

$$\text{norbornene} \xrightarrow[CHCl_3]{C_6H_5CO_3H} \text{exo-epoxide } (94\%) + \text{endo-epoxide } (6\%) \qquad (6.51)$$

molecules it may be more difficult to predict the stereochemical outcome (cf. Berti, 1973). Where there is a polar substituent in the allylic position this may influence the direction of attack by the peroxy-acid (Henbest, 1965). Thus, whereas 2-cyclohexenyl acetate gives mainly the *trans* epoxide as expected, the free hydroxy compound gives the *cis* epoxide in 80 per cent yield under the same conditions (6.52). Similarly 1-cholestene forms the α-epoxide while with 1-cholesten-3β-ol the β-epoxide is obtained. The rate of reaction is faster with the hydroxy compounds and it has been suggested that hydrogen bonding causes association of the reactants in an orientation favourable for *cis* epoxidation.

Oxidation with a peroxy-acid is not, of course, the only way to convert an olefin into an epoxide. The formation of epoxides by reaction of olefins with hydroperoxides and a molybdenum or vanadium catalyst has been known for some time but has not been much used. It has been found recently, however, that these reagents show remarkable reactivity towards allylic alcohols, and provide a very useful and highly selective

OCOCH$_3$ → $C_6H_5CO_3H$, C_6H_6, 0°C → OCOCH$_3$, O

OH → $C_6H_5CO_3H$, C_6H_6, 0°C → OH, O

(6.52)

monoperphthalic acid, ether → O

HO → monoperphthalic acid, ether → O, HO

method for epoxidation of these compounds (Sharpless and Michaelson, 1973; Tanaka, Yamamoto, Nozaki, Sharpless, Michaelson and Cutting, 1974). Thus, reaction of geraniol with t-butyl hydroperoxide in boiling benzene in presence of a catalytic amount of vanadium acetylacetonate gave the 2,3-epoxide almost exclusively in 93 per cent yield (6.53). Peroxy-acids, in contrast, gave variable amounts of the 6,7-epoxide. The reaction has been exploited in a useful method for transposition of the hydroxy group and the double bond in allylic alcohols (6.53). (–)-*cis*-Carveol, for example, was converted into (+)-*cis*-carveol in 57 per cent yield (Yasuda, Yamamoto and Nozaki, 1976).

These reagents are more stereoselective in their action than peroxy-acids. Thus, the acyclic allylic alcohol (10) with t-butyl hydroperoxide and vanadium acetylacetonate gave the *erythro* epoxide almost exclusively, whereas with *m*-chloroperbenzoic acid a mixture of the *erythro* and *threo* isomers was produced. Similarly, in reactions of the new reagents with cyclic olefinic alcohols the *syn* effect of the hydroxyl groups is even more marked than it is in reactions with peroxy-acids (cf. p. 364). The precise course of these reactions is not clear, but the rate accelerations and high stereoselectivities observed suggest a mechanism in which the hydroxyl group is co-ordinated to the metal in the rate-determining step.

$$\text{geraniol} \xrightarrow[\text{VO(acac)}_2 \text{ catalyst, } C_6H_6\text{, reflux}]{t\text{-}C_4H_9OOH} \text{2,3-epoxygeraniol} \quad (93\%)$$

$$-\overset{|}{C}{=}\overset{|}{C}-\underset{OH}{\overset{|}{C}}- \xrightarrow[(2)\ CH_3SO_2Cl,\ (C_2H_5)_3N]{(1)\ t\text{-}C_4H_9OOH,\ VO(acac)_2} -\overset{|}{C}\overset{O}{-}\overset{|}{C}-\underset{OSO_2CH_3}{\overset{|}{C}}- \xrightarrow{\text{Na, liq. } NH_3} -\underset{OH}{\overset{|}{C}}-\overset{|}{C}{=}\overset{|}{C}-$$

$$\text{(−)-}cis\text{-carveol} \longrightarrow \text{(+)-}cis\text{-carveol} \qquad (6.53)$$

(−)-*cis*-carveol (+)-*cis*-carveol

$$\mathbf{(10)} \xrightarrow{t\text{-}C_4H_9OOH,\ VO(acac)_2} \text{epoxide} \ (98\%\ erythro)$$

$$\text{cyclohex-3-enol} \xrightarrow[C_6H_6\text{, reflux}]{t\text{-}C_4H_9OOH,\ Mo(CO)_6 \text{ catalyst}} \text{epoxy alcohol} \ (90\%;\ 98\%\ cis) \qquad (6.54)$$

Epoxides may also be obtained by action of base on bromohydrins, and these are themselves conveniently prepared from olefins by reaction with *N*-bromosuccinimide in aqueous dimethoxyethane. This method has found its most useful application in the selective epoxidation of terminal double bonds in acyclic polyolefins (van Tamelen and Sharpless, 1967). Squalene, for example, was converted selectively into the 2,3-monoepoxide, and *Cecropia* juvenile hormone was obtained in 52 per cent yield from the corresponding triene (Corey, Katzenellenbogen, Gilman, Roman and Erickson, 1968). The selectivity shown in these reactions is highly dependent on the oxidising agent and on the solvent. It is believed that in aqueous dimethoxyethane the polyolefin adopts a coiled conformation which protects the internal double bonds from attack. Free hypobromous acid is apparently not involved in the formation of the bromohydrins for it forms different products in reaction with squalene.

Another useful reagent is peroxybenzimidic acid, $C_6H_5C(=NH)—O—OH$, formed *in situ* by reaction of benzonitrile with hydrogen peroxide; it has the advantage that it can be used under mildly alkaline conditions (Payne, 1962). Thus, 2-allylcyclohexanone was readily converted into the corresponding epoxide with the alkaline reagent, whereas with peracetic acid Baeyer–Villiger ring expansion intervened. It may also show useful stereochemical preferences (cf. Woodward *et al.*, 1973).

Epoxides are synthetically useful because they react with a variety of reagents with opening of the epoxide ring (Parker and Isaacs, 1959; Dittius, 1965). Hydrolysis with dilute mineral acids affords *trans* 1,2-diols and with carboxylic acids monoesters of *trans* 1,2-diols are obtained. Consequently, epoxidations with peroxy-acids carried out in presence of an excess of the corresponding carboxylic acid or a mineral acid, frequently yield hydroxy-esters or diols derived from the initially formed epoxide (6.55). This is especially the case for reactions run in formic acid and for reaction mixtures which contain a strong mineral acid.

H_2O_2, HCO_2H → [O] → OCOH, OH (6.55)

H_2O_2, CH_3CO_2H, H_2SO_4 → [O] → OH, OH

Another useful reaction of epoxides is reduction with lithium aluminium hydride to form alcohols (6.56). In general reaction takes place at the less substituted carbon atom to give the more substituted carbinol (see. p. 473).

$$C_2H_5CH—CH_2 \text{ (O bridging)} \xrightarrow[\text{ether}]{LiAlH_4} C_2H_5CHOHCH_3 \quad (100\%) \qquad (6.56)$$

De-oxygenation of epoxides to olefins can be effected by a number of methods. Two of these involve selenium reagents. Reaction of the epoxide with triphenylphosphine selenide and trifluoroacetic acid (Clive and Denyer, 1973) or with potassium selenocyanate (Behan, Johnstone and Wright, 1975) gives the corresponding olefin directly with retention of configuration of the substituent groups on the epoxide. Both conversions are believed to proceed by extrusion of selenium from the derived episelenide (6.57).

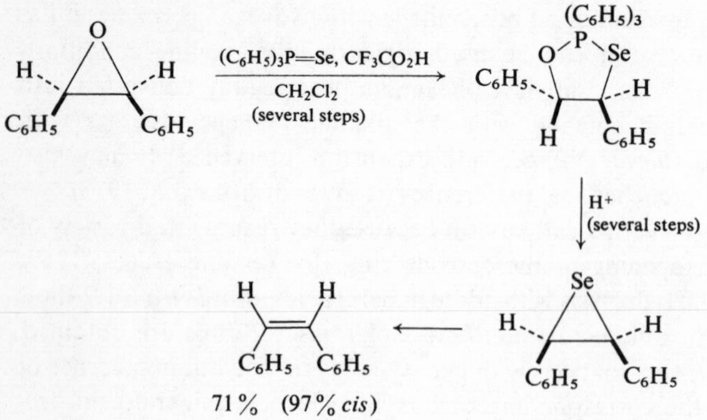

$(C_6H_5)_3P{=}Se$, CF_3CO_2H; CH_2Cl_2 (several steps)

H^+ (several steps)

71% (97% *cis*)

(6.57)

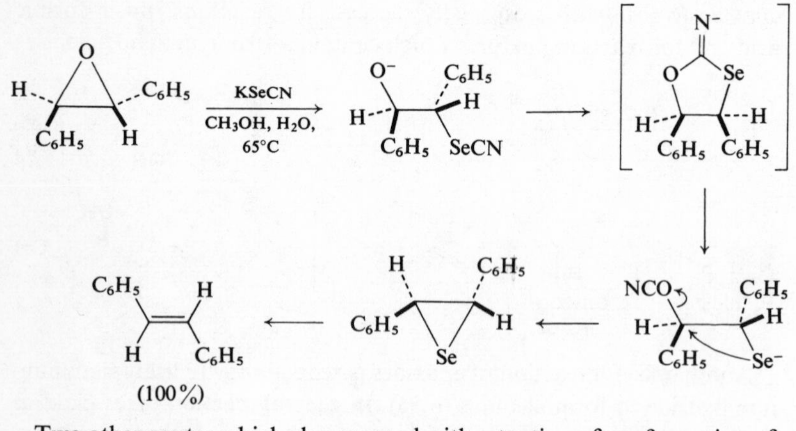

KSeCN; CH_3OH, H_2O, 65°C

(100%)

Two other routes which also proceed with retention of configuration of substituent groups on the epoxide make use of organometallic reagents. In the first, the epoxide is deoxygenated in a rapid reaction at room temperature with lower-valent tungsten chlorides produced *in situ* by reduction of the hexachloride with an alkyl lithium. *trans*-Cyclodecene oxide, for example, is converted into *trans*-cyclodecene in 94 per cent yield with 95 per cent retention of stereochemistry (Sharpless, Umbreit, Nieh and Flood, 1972). The second method uses sodium (cyclopentadienyl)dicarbonylferrate, which is readily prepared by reduction of $[\eta^5\text{-}C_5H_5Fe(CO)_2]_2$ with sodium amalgam. This reagent reacts rapidly with epoxides at or below room temperature in tetrahydrofuran solution to give alkoxides (as 12) which are converted without isolation into the complexes (13) by reaction with two equivalents of fluoroboric or hexafluorophosphoric acid; the olefin is liberated by reaction with sodium iodide in acetone. The total sequence is highly stereoselective.

$$\bar{Fp} + \text{epoxide} \xrightarrow{\text{THF}} Fp\text{–}CH_2CH_2\text{–}O^- \;(12) \xrightarrow{H^+} Fp\text{–}CH_2CH_2\text{–}OH \xrightarrow{H^+} [Fp\text{–}\|]^+ \;(13) \xrightarrow{\text{NaI}} \| \qquad (6.58)$$

$$\bar{Fp} \equiv [\eta^5\text{-}C_5H_5Fe(CO)_2]^-$$

cis- and *trans*-2-Butene epoxides, for example, were converted into *cis*- and *trans*-2-butenes respectively in 90 per cent yield and with more than 98 per cent retention of configuration. Terminal epoxides are more rapidly reduced than internal ones, and there is promise that the method may be useful for selective deoxygenation of one epoxide group in a molecule containing several (Giering, Rosenblum and Tancrede, 1972). Another method using titanium(II) salts has been described but is not stereoselective (McMurry and Fleming, 1975).

A number of methods are also available for the conversion of epoxides into olefins with *inversion* of configuration. In one, reaction of the epoxide with lithium diphenylphosphide in tetrahydrofuran followed by hydrogen peroxide in acetic acid leads, by a single inversion, to a β-hydroxy-diphenylphosphine oxide. The latter, on treatment with sodium hydride in dimethylformamide undergoes a stereospecific *syn* elimination of diphenylphosphinate to give the olefin (Bridges and Whitham, 1974). The procedure provides the basis of an excellent method for the inversion of olefins. *cis*-1-Methylcyclo-octene, for example, is converted into the *trans* isomer in 63 per cent yield. Some mono-, di-, and tri-substituted epoxides have also been deoxygenated in good yield with *inversion* of stereochemistry by reaction with trimethylsilylpotassium, generated *in situ* from hexamethyldisilane and potassium methoxide (6.59). Since olefins can be epoxidised with retention of configuration the sequence forms another method for inversion of configuration of olefins. *trans*-3-Hexene, for example, was converted into the *cis* isomer in 99 per cent yield and with more than 99 per cent stereospecificity (Dervan and Shippey, 1976).

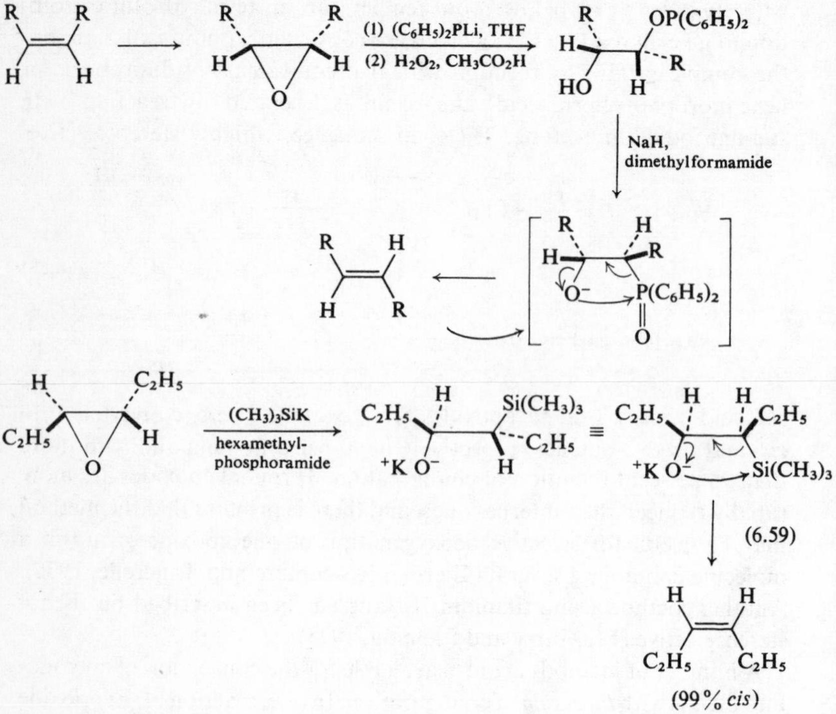

R R H H → R R H H O —(1) $(C_6H_5)_2PLi$, THF (2) H_2O_2, CH_3CO_2H→ R H HO H $OP(C_6H_5)_2$ R

NaH, dimethylformamide

[R H H R O^- $P(C_6H_5)_2$ O] → R H H R

H C_2H_5 C_2H_5 H O —$(CH_3)_3SiK$ hexamethylphosphoramide→ C_2H_5 H $Si(CH_3)_3$ C_2H_5 ^+K O^- H ≡ C_2H_5 H H C_2H_5 ^+K O^- $Si(CH_3)_3$

(6.59)

H H C_2H_5 C_2H_5

(99% *cis*)

Ring-opening reactions of epoxides take place in most cases with inversion of configuration at the carbon atom attacked, resulting in overall *trans* addition to the double bond of the olefin. Thus hydrolysis with dilute mineral acid affords the *trans* glycol (6.60), and the sequence of epoxidation followed by hydrolysis of the epoxide represents a widely used method for the preparation of *trans* diols from olefins, and is a

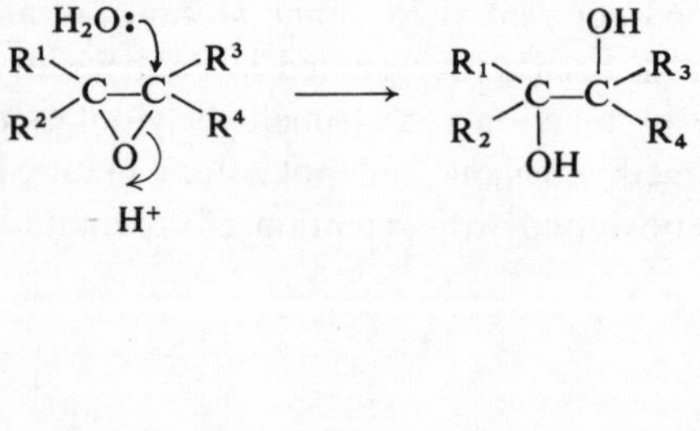

H_2O: R^1 R^2 C—C R^3 R^4 O H^+ → R_1 R_2 C—C OH OH R_3 R_4 (6.60)

convenient alternative to the Prévost route discussed on p. 360. *trans*-1-Methylcyclopentan-1,2-diol, for example, is readily obtained by hydrolysis of 1-methylcyclopentene epoxide (6.61).

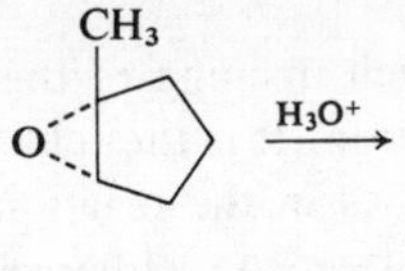

CH_3 H_3O^+ → CH_3 HO H OH (6.61)

Similarly reduction of 1,2-dimethylcyclopentene epoxide affords *trans*-1,2-dimethylcyclopentanol.

In conformationally rigid cyclohexane derivatives it is found that reactions leading to opening of the epoxide ring take place with formation of the *trans* diaxial product, rather than the *trans* diequatorial one, in preponderant amount. Thus, acid-catalysed hydrolysis of 2,3-epoxy-*trans*-decalin gives a 90 per cent yield of the *trans* diaxial diol as the only isolated product (6.62). Numerous other examples are known in the

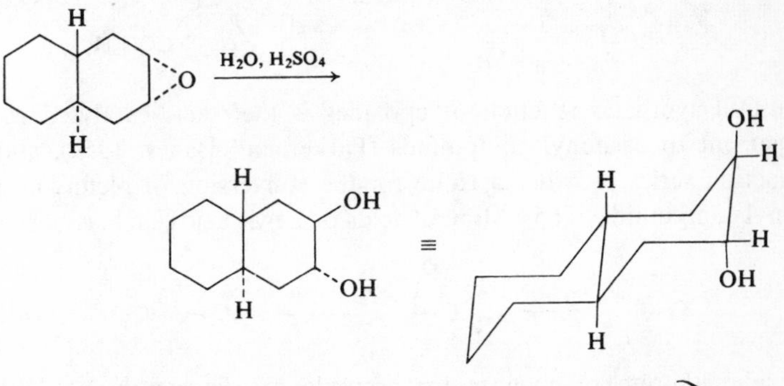

H O H H_2O, H_2SO_4 →

H OH OH H ≡ H OH H H OH H (6.62)

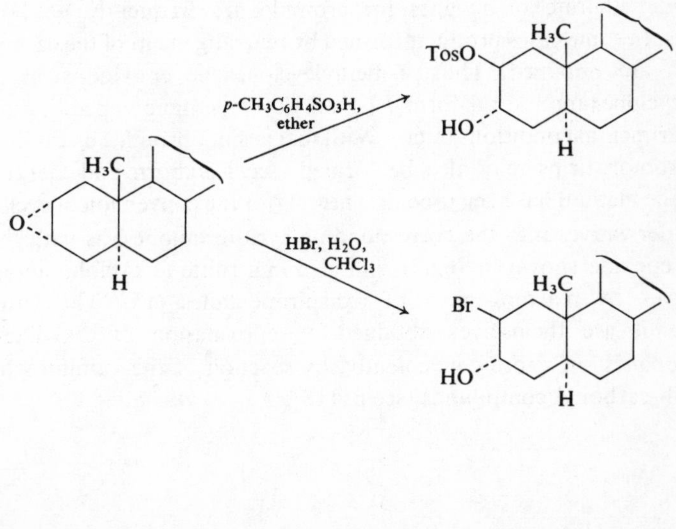

H_3C O H; p-$CH_3C_6H_4SO_3H$, ether → H_3C TosO HO H; $HBr, H_2O, CHCl_3$ → H_3C Br HO H (6.63)

steroid series. Reaction of 2α,3α-epoxycholestane with toluene-*p*-sulphonic acid, for example, gives the 3α-hydroxy-2β-tosyloxy derivative (diaxial) and reaction with hydrobromic acid leads to the diaxial bromohydrin (6.63).

A few cases are known in which opening of the epoxide ring takes place with retention of configuration. It is thought that these reactions may involve a double inversion, as in the reaction of dypnone oxide with hydrochloric acid (6.64). In other cases either retention or inversion of configuration may be observed depending on the conditions.

$$C_6H_5,\ H_3C\text{-}C(\overset{+}{O}H)C(H)(COC_6H_5) \xrightarrow{\text{inversion}} [C_6H_5, H_3C, Cl^-, H, OH, \overset{+}{O}, C_6H_5] \xrightarrow{\text{inversion}} C_6H_5\text{-}C(Cl)(CH_3)\text{-}C(H)(OH)COC_6H_5 \qquad (6.64)$$

A useful synthetic reaction of epoxides is their acid-catalysed rearrangement to carbonyl compounds (Parker and Isaacs, 1959), and the reaction series provides a route for the conversion of olefins into carbonyl compounds (6.65). Mineral acids or Lewis acids such as boron

$$>C{=}C< \longrightarrow >C\overset{O}{-}C< \longrightarrow -\overset{|}{\underset{|}{C}}-\overset{|}{C}{=}O \qquad (6.65)$$

trifluoride etherate or magnesium bromide are frequently used as catalysts. In some cases products formed by rearrangement of the carbon skeleton are obtained. Thus, 1-methylcyclohexene epoxide gives 2-methylcyclohexanone or 1-formyl-1-methylcyclopentane depending on the experimental conditions (6.66). With magnesium bromide as catalyst *trans* bromohydrins may also be formed (see Rickborn and Gerkin, 1968). The method has been used in synthesis for the conversion of cyclohexene derivatives into the corresponding cyclohexanones, as with the steroid epoxide shown in (6.67) and also in a route to cyclobutanone derivatives by rearrangement of oxaspiropentanes (15). The latter compounds are themselves obtained by epoxidation of alkylidene cyclopropanes or, more conveniently, by reaction of the sulphur ylid (14) with carbonyl compounds (see p. 118).

CH_3 ... $\xrightarrow[60°C]{MgBr_2,\ ether}$ ... CH_3 ... + ... $COCH_3$

$MgBr_2$, ether, 0°C ... CH_3 ... CHO (6.66)

H_3C ... O ... H $\xrightarrow[C_6H_6,\ 25°C]{BF_3,\ etherate}$ H_3C ... H ... O (6.67)

quantitative

$(C_6H_5)_2\overset{+}{S}$ (14) $\xrightarrow{(C_6H_5)_2C{=}O}$ C_6H_5 C_6H_5 O (15) $\xrightarrow[C_6H_6]{LiClO_4}$ C_6H_5 C_6H_5 O (90%)

Epoxides also undergo a number of base-catalysed rearrangements (cf. Crandall and Luan-Ho Chang, 1967). The most useful synthetically is the rearrangement to allylic alcohols induced by strong non-nucleophilic bases such as lithium diethylamide (Crandall and Crawley, 1973). Reaction proceeds by abstraction of a proton from a carbon adjacent to the epoxide ring, and with unsymmetrical epoxides is highly regioselective, leading to the less substituted double bond. Thus, 2-pentene oxide gives 1-penten-3-ol and 2-methyl-3-heptene oxide leads exclusively to *trans*-2-methyl-4-hepten-3-ol (Rickborn and Thummel, 1969; but see Yasuda, Tanaka, Oshima, Yamamoto and Nozaki, 1974). With acyclic compounds there is high selectivity for the formation of the *trans* olefin from both *cis* and *trans* epoxides. In the cyclic series the reaction proceeds by a *syn* elimination (6.68), and it is believed that this course probably holds for reactions in the acyclic series as well (Thummel and Rickborn, 1970).

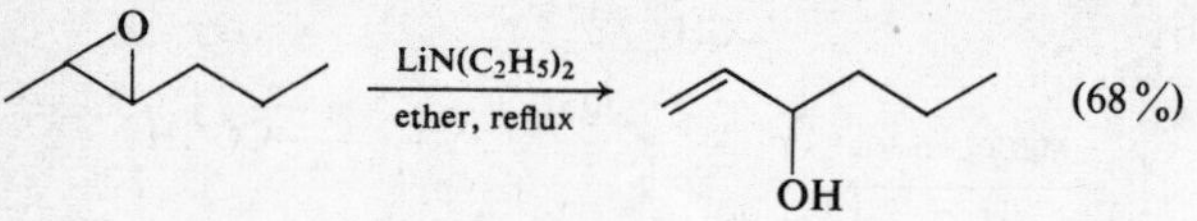

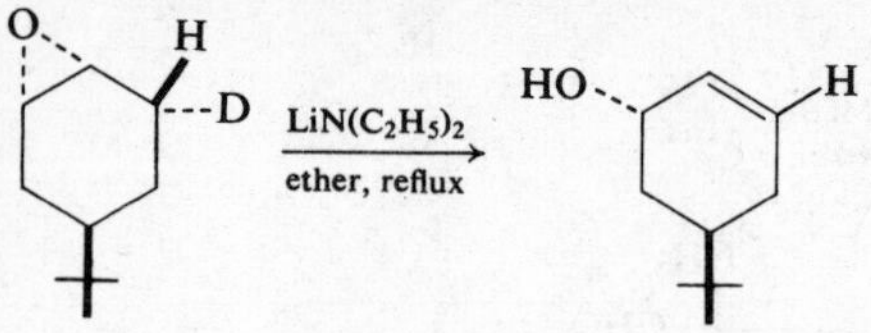

$\xrightarrow[\text{ether, reflux}]{LiN(C_2H_5)_2}$ (68%)

OH

$\xrightarrow[\text{ether, reflux}]{LiN(C_2H_5)_2}$ (6.68)

$\longrightarrow$ via

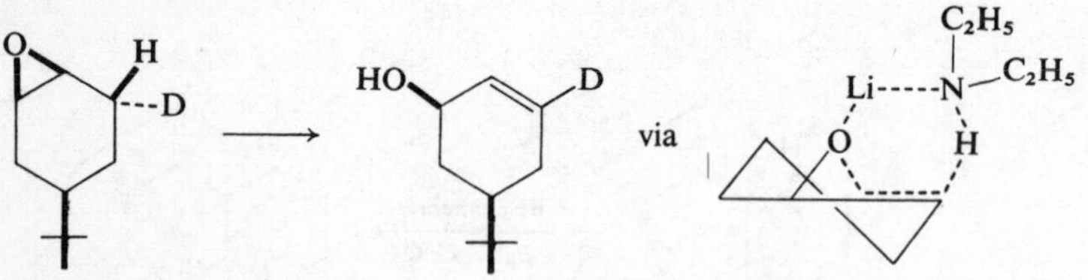

Ozonolysis. Ozonolysis, that is reaction of an olefin with ozone followed by splitting of the resulting ozonide, is a very convenient method for oxidative cleavage of olefinic double bonds.

Recent physical measurements have shown that the ozone molecule is a resonance hybrid of the structures shown in (6.69). It is an electro-

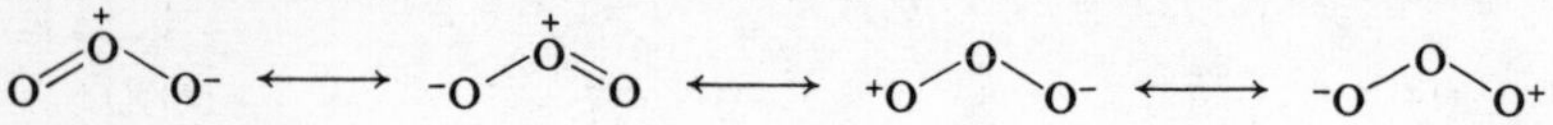

$$O{=}\overset{+}{O}{-}O^- \longleftrightarrow {}^-O{-}\overset{+}{O}{=}O \longleftrightarrow {}^+O{-}O{-}O^- \longleftrightarrow {}^-O{-}O{-}O^+ \qquad (6.69)$$

philic reagent and reacts with olefinic double bonds to form ozonides which can be cleaved oxidatively or reductively to carboxylic acids, ketones or aldehydes; the nature of the products formed depends on the method used and on the structure of the olefin (Bailey, 1958). The reaction is usually carried out by passing a stream of oxygen containing 2–10 per cent of ozone into a solution or suspension of the compound in a suitable solvent, such as methylene chloride or methanol, at or below room temperature. Oxidation of the crude ozonisation product, without isolation, by hydrogen peroxide or other reagent, leads normally to

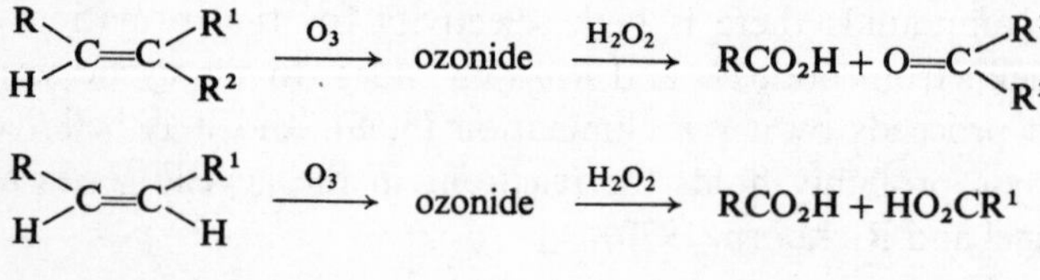

$$\begin{array}{l} RHC{=}CR^1R^2 \xrightarrow{O_3} \text{ozonide} \xrightarrow{H_2O_2} RCO_2H + O{=}CR^1R^2 \\ RHC{=}CHR^1 \xrightarrow{O_3} \text{ozonide} \xrightarrow{H_2O_2} RCO_2H + HO_2CR^1 \end{array} \qquad (6.70)$$

carboxylic acids or ketones or both, depending on the degree of substitution of the olefin (6.70). Thus oleic acid gives azelaic and pelargonic acids (6.71).

$$CH_3(CH_2)_7CH{=}CH(CH_2)_7CO_2H \xrightarrow[\text{(2) oxidation}]{\text{(1) } O_3} CH_3(CH_2)_7CO_2H + HO_2C(CH_2)_7CO_2H \quad (6.71)$$

Reductive decomposition of the crude ozonide leads to aldehydes and ketones (6.72). Various methods of reduction have been used including

$$\text{R(H)C}{=}\text{CR}^1\text{R}^2 \xrightarrow{O_3} \text{ozonide} \xrightarrow{\text{reduction}} \text{RCHO} + \text{O}{=}\text{CR}^1\text{R}^2 \quad (6.72)$$

catalytic hydrogenation and reduction with zinc and acids or with triethyl phosphite, but, in general, yields of aldehydes have not been high. Reaction with dimethyl sulphide in methanol has been found to give excellent results (Pappas and Keaveney, 1966) and this reagent appears to be superior to all others previously used. Reaction takes place under neutral conditions and the reagent is highly selective; nitro and carbonyl groups, for example, elsewhere in the molecule are

$$CH_2{=}CH(CH_2)_5CH_3 \xrightarrow{O_3} \text{ozonide} \xrightarrow{CH_3SCH_3,\ CH_3OH} CHO(CH_2)_5CH_3 + CH_2O \ (75\%)$$

$$\text{4-vinylpyridine} \xrightarrow[\text{(2) } CH_3SCH_3]{\text{(1) } O_3} \text{pyridine-4-carboxaldehyde} + CH_2O \quad (6.73)$$

not affected (6.73). The procedure hinges on the fact that hydroperoxides are rapidly and cleanly reduced to alcohols by sulphides (6.74). Ozonisation of an olefin in methanol solution gives rise to a hydroperoxide (see p. 376) which is reduced in the same way by dimethyl sulphide to the hemi-acetal as shown (6.75).

$$(CH_3)_3C{-}O{-}OH \xrightarrow{(CH_3)_2S} (CH_3)_3COH + (CH_3)_2SO \quad (6.74)$$

$$\rangle C(OCH_3)(O{-}OH) \xrightarrow{(CH_3)_2S} \rangle C(OCH_3)(OH) \longrightarrow \rangle C{=}O + \begin{matrix} CH_3OH \\ (CH_3)_2SO \end{matrix}$$

$$\text{1,2-dimethylcyclopentene} \xrightarrow[\text{tetracyanoethylene}]{O_3,\ CH_3CO_2H,\ -70°C} CH_3CO(CH_2)_3COCH_3\ (71\%) + O\langle C(CN)_2{-}C(CN)_2 \qquad (6.75)$$

Criegee and Gunther (1963) have found that carbonyl compounds can be obtained directly from olefins by reaction of ozone with a 1:1 mixture of the olefin and tetracyanoethylene in acetic acid (6.75). Similar results were obtained by reaction in methylene chloride solution in presence of a small amount of pyridine (Conia and Leriverend, 1960). Under these conditions cyclobutanone, for example, was obtained directly from methylenecyclobutane in high yield.

Largely through the work of Criegee (1975) it now seems clear that most normal ozonolyses proceed by formation of a primary ozonide which decomposes to give a zwitterion and a carbonyl compound. The fate of the zwitterion depends on its structure, on the structure of the carbonyl compound and on the solvent. In an inert (non participating) solvent, if the carbonyl compound is reactive, it may react with the zwitterion to form an ozonide (16); otherwise the zwitterion may dimerise to the peroxide (17) or give ill-defined polymers. In nucleophilic solvents such as methanol or acetic acid, however, hydroperoxides of the type of (18) are formed (6.76). Strong evidence for the intermediacy

$$R_2C{=}CR_2 \xrightarrow{O_3} \text{primary ozonide (1,2,3-trioxolane, } R_2C{-}CR_2\text{)}$$

$$\Big\Downarrow$$

$$R_2C{=}\overset{+}{O}{-}\overset{-}{O} \longleftrightarrow R_2\overset{+}{C}{-}O{-}\overset{-}{O} + O{=}CR_2 \qquad (6.76)$$

$$\xrightarrow{CH_3OH} R_2C(OCH_3){-}O{-}OH\ (18) \qquad \xrightarrow{C_5H_{12}} \text{ozonide } R_2C\langle O{-}O,\,O\rangle CR_2\ (16) \ \text{or}\ \text{dimeric peroxide } R_2C\langle O{-}O,\,O{-}O\rangle CR_2\ (17)$$

of a zwitterion was found by Criegee in the ozonolysis of tetramethylethylene. In an inert solvent the cyclic peroxide and acetone were obtained. But when formaldehyde was added to the reaction mixture the known ozonide of isobutene was isolated. In the first case the intermediate zwitterion has dimerised; but in the second it has reacted preferentially with the highly reactive carbonyl compound (6.77). Criegee's general

$$(CH_3)_2C(O{-}O)_2C(CH_3)_2 \xleftarrow[C_5H_{12}]{O_2} (CH_3)_2C{=}C(CH_3)_2 \xrightarrow{O_3,\ C_5H_{12};\ CH_2O} (CH_3)_2C\langle O{-}CH_2{-}O{-}O\rangle \qquad (6.77)$$

scheme is supported by much recent experimental work, although it appears that in some cases the exact mechanism is sensitive to a number of experimental factors (Murray, 1968; Fliszar and Carles, 1969; Klopman and Joiner, 1975).

Ozonolysis is widely used both in degradative work and in synthesis for the preparation of aldehydes, ketones and carboxylic acids from olefins. Thus, ω-aldehydic acids can be obtained from cycloalkanone enol ethers (6.78). Cleavage of cyclic ketones without loss of carbon

$$\text{cyclo-}[(CH_2)_n{-}C(=O){-}CH_2] \xrightarrow{HC(OC_2H_5)_3} \text{cyclo-}[(CH_2)_n{-}C(OC_2H_5)_2{-}CH_2] \xrightarrow[\Delta]{p\text{-}CH_3C_6H_4SO_3H} \text{cyclo-}[(CH_2)_n{-}C(OC_2H_5){=}CH] \xrightarrow[(2)\ H_2,\ Pd]{(1)\ O_3} OHC{-}(CH_2)_n{-}CO_2C_2H_5 \qquad (6.78)$$

$n = 3, 4, 5, 6$

can also be achieved by ozonolysis of the benzylidene or furfurylidene derivatives or, better, of derived trimethylsilyl ethers (6.79) (Clark and Heathcock, 1976).

$$\text{2-methylcyclohexanone} \xrightarrow[\text{(2) } (CH_3)_3SiCl]{\text{(1) } LiN(iso\text{-}C_3H_7)_2} \text{1-}OSi(CH_3)_3\text{-6-}CH_3\text{-cyclohexene} \xrightarrow[\text{(2) } NaBH_4]{\text{(1) } O_3,\ CH_3OH} HOCH_2(CH_2)_3CH(CH_3)CO_2H \quad (94\%)$$

$$\text{2-methylcyclohexanone} \xrightarrow[\text{(2) } (CH_3)_3SiCl]{\text{(1) } NaH} \text{1-}OSi(CH_3)_3\text{-2-}CH_3\text{-cyclohexene} \xrightarrow[\text{(2) } (CH_3)_2S]{\text{(1) } O_3,\ CH_3OH} HO_2C(CH_2)_4COCH_3 \quad (90\%) \tag{6.79}$$

With some substrates ozonolysis follows an 'abnormal' course. Olefins which are sterically hindered on one side of the double bond often give rise to epoxides, or rearrangement products thereof, containing the same number of carbon atoms as the starting material. In the example (6.80) the epoxide is formed, supposedly, by way of a π-complex which, because of steric hindrance, can stabilise itself more easily by losing a molecule of oxygen than by collapsing to the 'normal' zwitterion.

$$(CH_3)_3C\text{—}C(CH_3)_2\text{—}C(C(CH_3)_3)\text{=}CH_2 \longrightarrow (CH_3)_3C\text{—}C(CH_3)_2\text{—}\overset{O}{C(C(CH_3)_3)\text{—}CH_2} \tag{6.80}$$

$\alpha\beta$-Unsaturated ketones or acids generally give products containing fewer than the expected number of carbon atoms. Thus, the tricyclic $\alpha\beta$-unsaturated ketone shown in (6.81) is converted into the keto acid

$$\text{tricyclic }\alpha\beta\text{-unsaturated ketone } (H_3C,\ H_3C,\ CH_3) \xrightarrow[\text{(2) } H_2O_2]{\text{(1) } O_3,\ CH_3CO_2H} \text{bicyclic ketone with } CH_2CH_2CO_2H\ (H_3C,\ H_3C,\ CH_3) \tag{6.81}$$

with loss of a carbon atom. The following general mechanism involving a release of electrons by O—H heterolysis has been suggested to account for these results (6.82). Abnormal results have also been observed in the ozonolysis of allylic alcohols, and amines (Bailey, 1958).

$$R{-}CO{-}C(R^1){=}C< \xrightarrow[H_2O]{O_3} HO{-}C(R)(OH){-}C(R^1)\text{(ozonide)} \longrightarrow HO{-}C(R){=}O + {}^{-}O{-}C(R^1){=}O + O{=}C< \qquad (6.82)$$

$$>C{=}C(R)CO_2H \xrightarrow{O_3} \text{ozonide} \longrightarrow >C{=}O + HO{-}C(R){=}O + CO_2$$

A useful alternative to ozonolysis is oxidation with periodate in presence of a catalytic amount of potassium permanganate (6.83). With this reagent double bonds are cleaved to give ketones and aldehydes or, more usually, carboxylic acids formed by further oxidation of the aldehydes (Lemieux and von Rudloff, 1955). Thus, citronellal is oxidised to acetone and 3-methyladipic acid in high yield, and the unsaturated alcohol shown (6.83) is cleaved without attack on the acetoxy and primary alcohol groups. Reaction is conducted at pH 7–8, and

$$\text{citronellal (CHO)} \xrightarrow[H_2O,\ t\text{-}C_4H_9OH]{NaIO_4,\ KMnO_4} \text{3-methyladipic acid } (CO_2H,\ CO_2H) + CH_3COCH_3$$

$$CH_3(CH_2)_5CH(OCOCH_3)CH_2CH{=}CH(CH_2)_5CH_2OH \qquad (6.83)$$

$$\downarrow NaIO_4,\ KMnO_4,\ K_2CO_3;\ H_2O,\ t\text{-}C_4H_9OH$$

$$CH_3(CH_2)_5CH(OCOCH_3)CH_2CO_2H + HO_2C(CH_2)_5CH_2OH$$

under these conditions the α-hydroxyketone or the 1,2-diol formed from the olefin with permanganate is cleaved by the periodate to carbonyl compounds. The permanganate is reduced only to the manganate stage at this pH and is re-oxidised by the periodate, which itself does not attack the double bond. Only catalytic amounts of permanganate are thus needed.

Similar results are obtained with a combination of periodate and osmium tetroxide, and this technique has the advantage that it does not proceed beyond the aldehyde stage. It produces the same result as ozonolysis followed by reductive cleavage of the ozonide. Cyclohexene, for example, is converted into adipaldehyde in 77 per cent yield. The osmium tetroxide oxidises the olefin to the diol, which is cleaved by the periodate. Catalytic amounts of osmium tetroxide are sufficient because the periodate oxidises the reduced osmium back to the tetroxide.

Another excellent method for cleaving olefinic double bonds is by action of ruthenium tetroxide in combination with sodium periodate (see below). Carboxylic acids are usually produced from disubstituted olefins with this reagent.

6.4. Photosensitised oxidation of olefins

Irradiation of dilute solutions of olefins and conjugated dienes in presence of oxygen and a sensitiser gives rise to hydroperoxides and cyclic peroxides (6.84) (Adams, 1971; Kearns, 1971; Denny and Nickon, 1973). The

$$\underset{\alpha\ \ \beta\ \ \gamma}{\overset{\text{H}}{-\text{C}-\text{C}=\text{C}-}} \xrightarrow[\text{sensitiser}]{O_2,\ h\nu} \underset{\alpha\ \ \beta\ \ \gamma}{\overset{\text{OOH}}{-\text{C}=\text{C}-\text{C}-}} \xrightarrow{\text{reduction}} \underset{\alpha\ \ \beta\ \ \gamma}{\overset{\text{OH}}{-\text{C}=\text{C}-\text{C}-}} \qquad (6.84)$$

$$\text{(α-terpinene)} \xrightarrow[\text{sensitiser}]{O_2,\ h\nu} \text{(ascaridole, cyclic O–O peroxide)}$$

addition to dienes is analogous to the Diels–Alder reaction and is discussed further on p. 172.

Photosensitised oxygenation of mono-olefins has been widely studied and provides a convenient method for introducing oxygen in a highly specific fashion into such compounds (Schenck, 1957; Gollnick, 1968). The first products of the reaction are allylic hydroperoxides which may be isolated if desired or reduced directly to the corresponding allylic alcohols. For example, α-pinene is converted into *trans*-pinocarveyl hydroperoxide which on reduction gives *trans*-pinocarveol, and 2-methyl-2-pentene affords, after reduction of the hydroperoxides, a mixture of two methylpentenols corresponding to attack at each end of the double bond (6.85).

OOH; O_2, $h\nu$, sensitiser; Na_2SO_3, aq. CH_3OH; OH; O_2; OOH

(6.85)

(1) O_2, $h\nu$, sensitiser; (2) reduction; OH; + HO

(49% of product) (51% of product)

The oxidations are conducted in solution in benzene, pyridine or a lower aliphatic alcohol. Common sensitisers are organic dyes such as fluorescein derivatives, methylene blue and certain porphyrin derivatives. Essentially no reaction takes place if the olefin lacks an allylic hydrogen atom or if sensitiser, light or oxygen is excluded. In every case the oxygen molecule adds to one carbon of the double bond and a hydrogen atom from the allylic position migrates to the oxygen with concomitant shift of the double bond. The reaction is generally held to proceed by a concerted cyclic mechanism, analogous to that postulated for the 'ene' reaction (p. 224) (6.86) (Foote, 1971), but there is some evidence that in some cases it may proceed in two steps, possibly involving

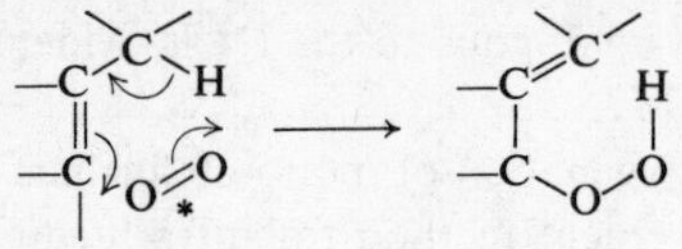

(6.86)

an intermediate peroxirane (as 20). For example, the stereospecific reaction of the chiral deuterated olefin (19) with oxygen and light in presence of a sensitiser showed no isotope effect. The two hydroperoxides (21) and (22) were formed in almost equal amounts, suggesting that the reaction involves an irreversible addition of singlet oxygen to the double bond in the rate-determining step, followed by abstraction of hydrogen or deuterium (6.87) (Stephenson, McClure and Sysak, 1973). A concerted 'ene' mechanism might have been expected to give more of the hydroperoxide (22) by operation of the normal deuterium isotope effect (compare 6.86) (but see Gorman, 1973).

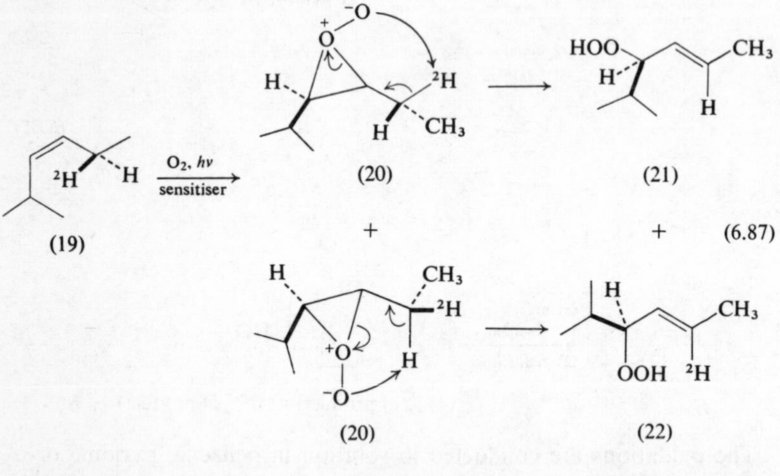

(6.87)

The reaction should be distinguished from the familiar *radical* oxidation of olefins which can also be initiated by sensitisers (such as benzophenone) which abstract hydrogen and which may also give rise to allylic hydroperoxides. It is only in exceptional cases, however, that the products of these autoxidations are the same as those of photo-oxygenation. Autoxidation of α-pinene, for example, leads to verbenyl hydroperoxide and not to pinocarveyl hydroperoxide. The photo-sensitised reactions do not involve free radical intermediates, as autoxidations do. This is shown, for example, by the fact that the alcohol (23) obtained as one product from photosensitised oxygenation of (+)-

limonene is optically active (6.88). A (symmetrical) allylic free-radical intermediate (24) would have given a racemic product.

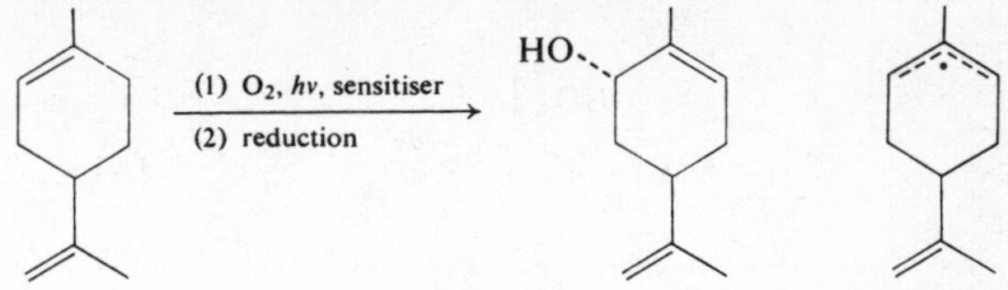

(1) O_2, $h\nu$, sensitiser
(2) reduction

HO

(23) (24) (6.88)

optically pure

Photosensitised oxygenation is useful in synthesis for the specific introduction of an oxygenated functional group at the site of a double bond, and has been employed in the synthesis of a number of natural products. Thus, the compound (25), an analogue of the diterpene alkaloid garryfoline, was readily obtained by photo-oxygenation of an olefinic precursor (6.89). Again, a key step in the synthesis of the stereo-

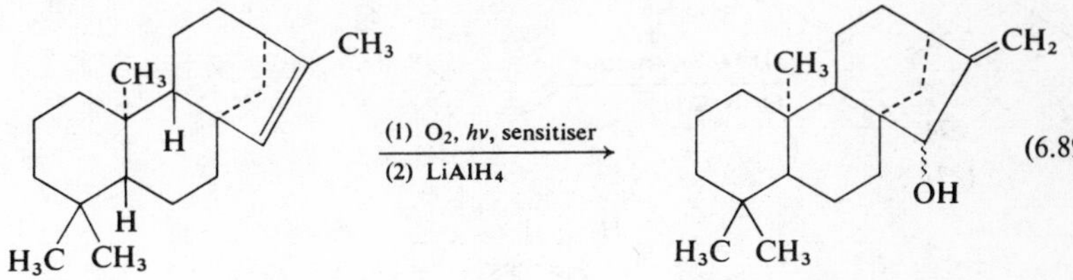

CH_3 CH_3 H H H_3C CH_3

(1) O_2, $h\nu$, sensitiser
(2) $LiAlH_4$

CH_3 CH_2 OH H_3C CH_3 (6.89)

(25) (46%)
(two epimers formed)

isomeric 'rose-oxides' (26), the active principles of rose scent, was photo-oxygenation of citronellol (6.90). An interesting reaction is the conversion of the triene (27) into a hydroxy-allene as well as the more conventional Diels–Alder type addition product (6.91). A similarly disposed allene unit occurs in certain carotenoids and it is possible that it may be formed in a similar manner in nature.

Photo-oxygenation of allylic secondary alcohols gives rise to $\alpha\beta$-epoxy-ketones. Thus, 4-cholesten-3β-ol led stereospecifically to the α-epoxy-ketone, while the 3α-ol gave the epimeric β-epoxide (6.92). From the known *cis* stereochemistry of the reaction (see below) the hydroperoxides (28 and 29) are expected intermediates; collapse of these to the epoxy-ketones accounts for the observed stereospecificity. The $\alpha\beta$-unsaturated ketone 4-cholesten-3-one was also formed in both reactions in an

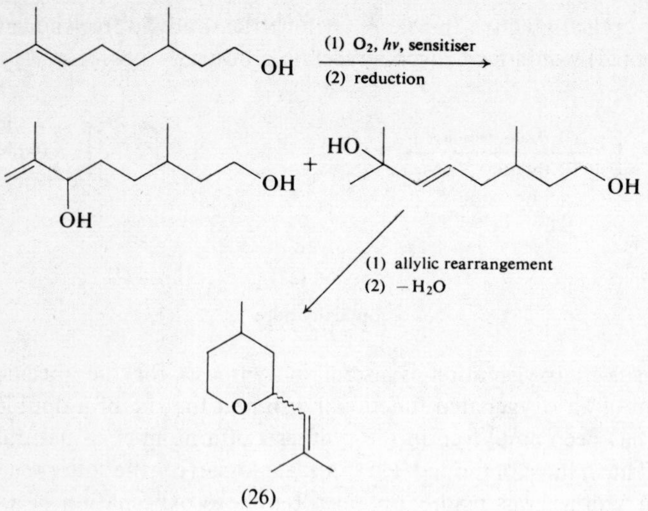

(1) O_2, $h\nu$, sensitiser
(2) reduction

OH + HO OH (6.90)

(1) allylic rearrangement
(2) $-H_2O$

O

(26)

CH_3

(1) O_2, $h\nu$, sensitiser
(2) $NaBH_4$

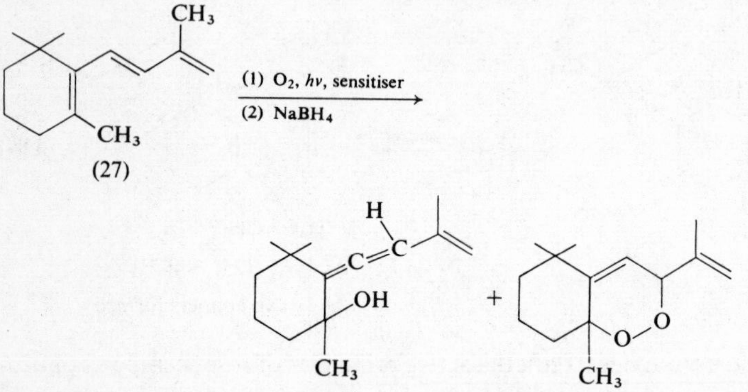

CH_3
(27)

H C=C OH + O O (6.91)

CH_3 CH_3

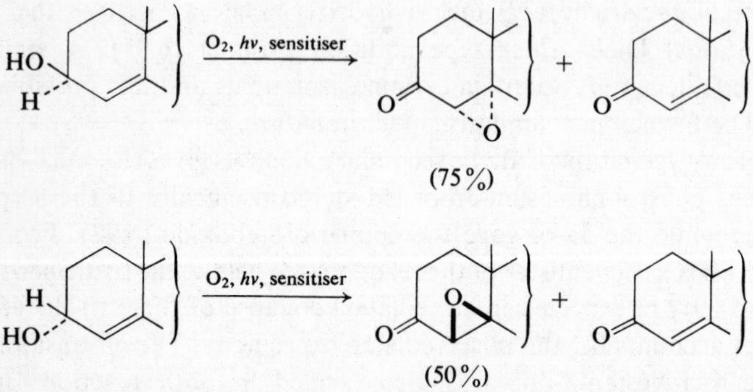

HO H O_2, $h\nu$, sensitiser O O + O

(75%) (6.92)

H HO O_2, $h\nu$, sensitiser O O + O

(50%)

amount dependent on the sensitiser used. This has been shown to be due to the differing ratios of the two forms of singlet oxygen produced (Kearns, Hollins, Khan, Chambers and Radlick, 1967).

As already indicated photo-oxygenations in many cases show a high degree of stereoselectivity. Experiments using steroidal olefins have revealed that, in agreement with the proposed cyclic reaction path (p. 381), it is a *cis* reaction, that is the new C—O bond is formed on the same side of the molecule as the C—H bond which is broken. This is shown, for example, by photo-oxygenation of the deuterated cholesterols (6.93).

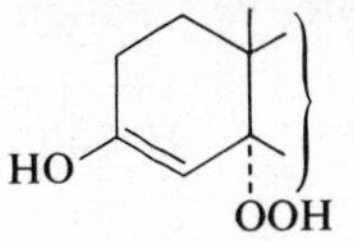

(28) 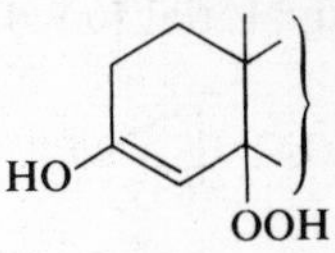 (29)

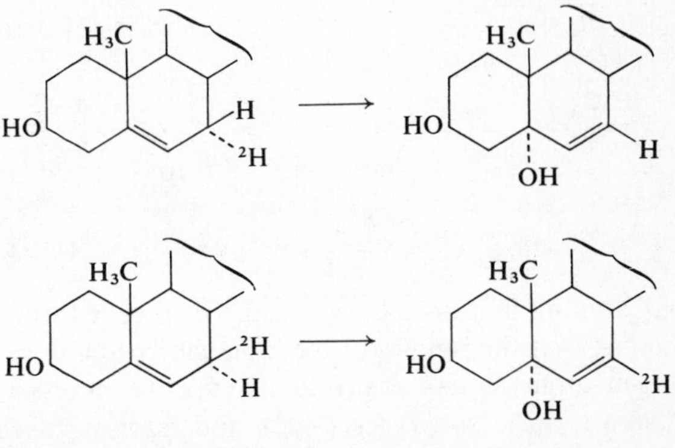

(6.93)

The 7α-deutero compound gave, after reduction of the hydroperoxide, a diol which retained only 8 per cent of the original deuterium, whereas the diol from the 7β-deutero compound retained 95 per cent of the deuterium. In both these reactions approach of the activated oxygen takes place from the less hindered α-side of the steroid molecule, with migration in the two cases of the *cis* ^{2}H or H from C-7 to oxygen.

In all these sensitised photo-oxygenations attack of the activated oxygen on the π-orbital of the double bond takes place in a direction perpendicular to the olefinic plane, and the allylic hydrogen atom must be suitably oriented to allow transfer to oxygen. This requirement is best met when the C—H bond is perpendicular to the plane of the double bond, for this favours overlap of the developing empty *p*-orbital on the γ carbon atom with the π-orbital of the double bond (6.94). A consequence

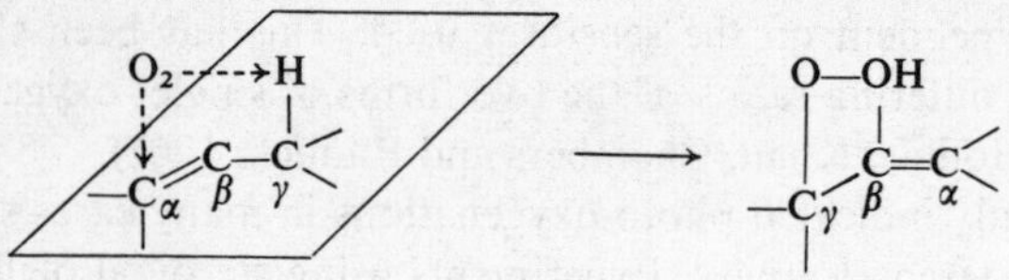

(6.94)

of this is that in cyclohexenoid systems, a *quasi* axial hydrogen atom should be better disposed for reaction than a *quasi* equatorial one. In agreement, 5α-cholest-3-ene, with a *quasi* axial C-5—H bond, is much more readily oxygenated than is 5β-cholest-3-ene, in which the C-5—H is *quasi* equatorial (6.95). Again, in the photo-oxygenation of α-pinene

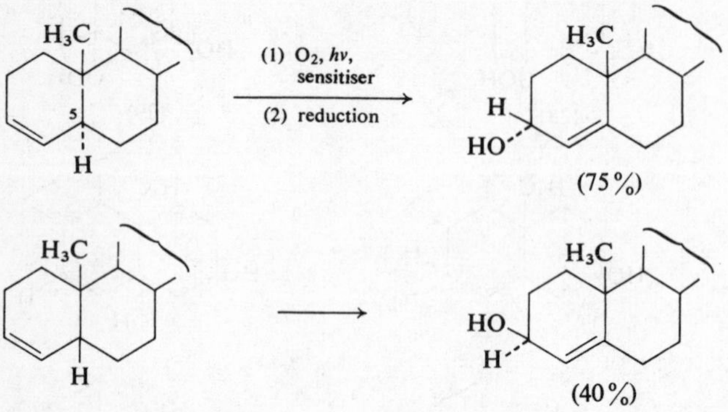

(1) O_2, *hν*, sensitiser (2) reduction

(75%)

(6.95)

(40%)

the small amount of *trans*-2-hydroperoxy-Δ^3-pinene formed is explained by assuming that the rigidity of the α-pinene system does not allow the α-hydrogen atom at C-4 easily to assume the necessary orientation for efficient transfer to oxygen (6.96), and reaction takes place preferentially at the other end of the double bond with abstraction of an

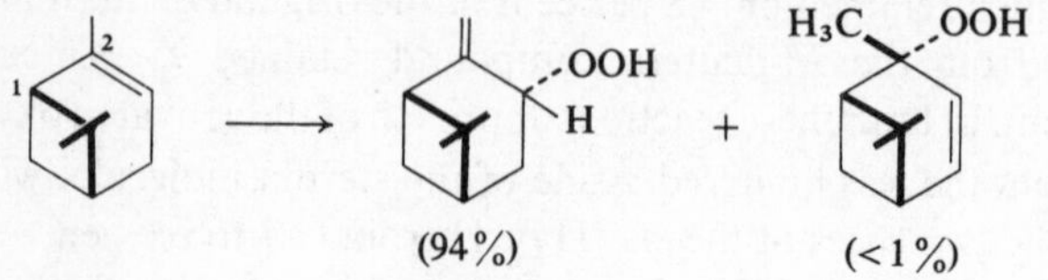

(94%) (<1%)

(6.96)

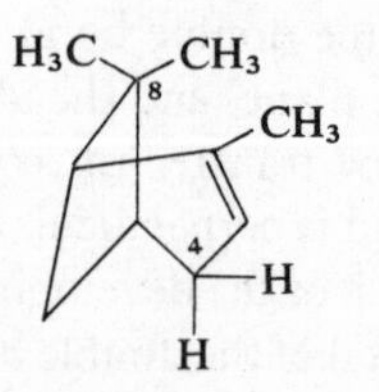

allylic hydrogen atom from the methyl group. The exclusive formation of products by oxygenation from the α-face of the molecule (i.e. *trans* to the *gem*-dimethylmethylene bridge) is due to the steric effect of the C-8 methyl group which prevents attack on the β-face of the double bond.

Consideration of the stereoelectronic course of the reaction also serves to explain some apparently anomalous results obtained with open chain olefins. 1,1-Dimethyl-2-isopropylethylene, for example, affords the secondary hydroperoxide in more than 95 per cent yield; almost none of the tertiary hydroperoxide is formed. The most stable conformation of the olefin is (30) in which the hydrogen atom at C-3 almost eclipses the double bond. The alternative conformation (31) necessary for photo-oxygenation at C-1, in which the C-3—H bond is perpendicular to the double bond, is disfavoured by steric interaction between a C-3-methyl group and the dimethyl vinyl group. Consequently very little tertiary hydroperoxide is formed (6.97). In contrast, 1,1-dimethylbutene

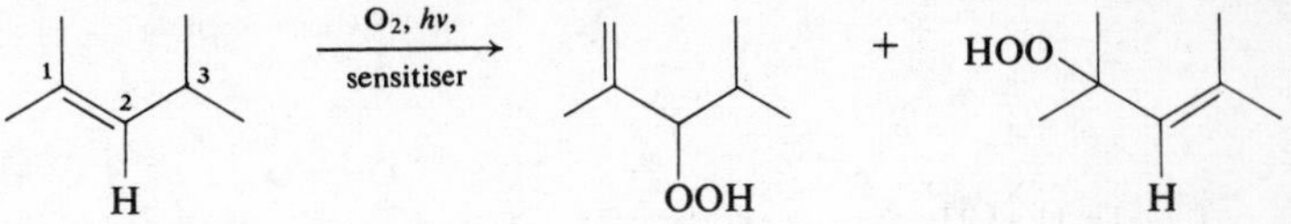

O_2, $h\nu$, sensitiser

OOH (>95%) + HOO ... H (<5%) (6.97)

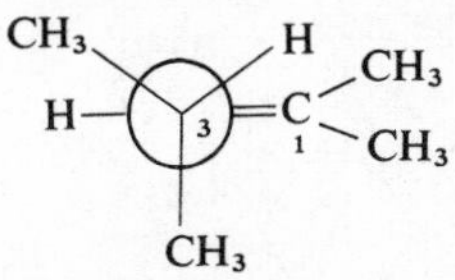

(30) 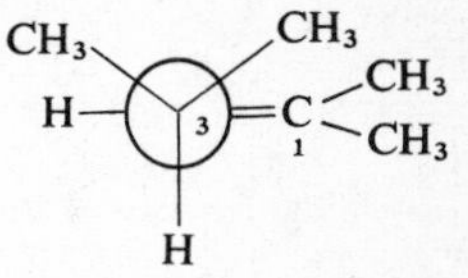 (31)

gives roughly equal amounts of secondary and tertiary hydroperoxides because the conformation (32) appropriate to tertiary peroxide formation is no longer destabilised by steric factors (6.98).

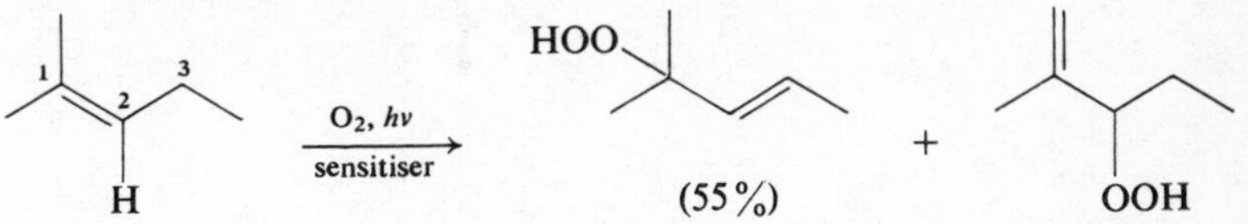

O_2, $h\nu$, sensitiser

HOO (55%) + OOH (45%)

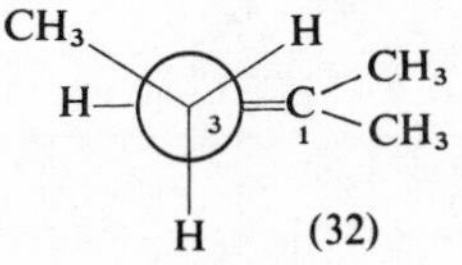

(32) (6.98)

The reactive species in these reactions is believed to be singlet oxygen, formed by energy transfer from the light-energised triplet sensitiser to oxygen or an oxygen-sensitiser complex (cf. Wagner and Hammond, 1968). In line with the suggestion that singlet oxygen is involved it has been shown (Foote, 1968) that chemically generated singlet oxygen, obtained by reaction of hypochlorites with hydrogen peroxide, gives product distributions on oxidation of olefins which are identical with those obtained in sensitised photo-oxygenations. The sensitisers used in the photo-oxidations are very bulky molecules and would be expected to exert a strong influence on the stereochemistry of the reaction intermediate if they were present in the transition state. No such effect is

$$(C_6H_5)_2CHCHO \longrightarrow (C_6H_5)_2C{=}CH{-}N(\text{piperidino}) \xrightarrow{O_2,\ h\nu,\ \text{sensitiser}} \text{(33)} \longrightarrow (C_6H_5)_2C{=}O + O{=}CH{-}N(\text{piperidino}) \qquad (6.99)$$

(33)

$$\text{1-morpholinocyclohexene} \xrightarrow{\text{singlet } O_2,\ CH_3OH} \text{dioxetane intermediate (H :B)} \longrightarrow \text{2-hydroxy-2-morpholinocyclohexanone} \longrightarrow \text{cyclohexane-1,2-dione (75\%)}$$

observed, and this would seem to tell against the suggestion that the active species is an oxygen–triplet sensitiser complex.

In many sensitised photo-oxygenations carbonyl compounds are produced in addition to hydroperoxides. These are thought to arise by decomposition of oxetanes (as 33) formed by addition of oxygen to the double bond. The reaction is favoured by the presence of a vinylic hetero-atom as in enamines, which undergo the reaction readily. Acyclic enamines give a carbonyl compound and an amide in good yield, and the reaction provides a method for cleaving carbonyl compounds between the carbonyl group and the α-carbon atom. With enamines of cyclic ketones the reaction takes a different course giving either a 1,2-diketone or its enamine derivate; in methanol solution at room temperature the diketone is obtained directly (6.99) (Wasserman and Terao, 1975).

6.5. Oxidation of ketones

Oxidation of ketones with chromic acid or permanganate under conditions vigorous enough to bring about reaction usually leads to rupture of the carbon chain adjacent to the carbonyl group with formation of carboxylic acids, and is of little value in synthesis. More important are controlled methods of oxidation leading to αβ-unsaturated ketones, α-ketols (acyloins) or lactones without disruption of the molecule.

αβ-Unsaturated ketones. Conversion of ketones into αβ-unsaturated ketones has been effected by bromination–dehydrobromination (cf. Stotter and Hill, 1973) and, in certain cases, by oxidation with selenium dioxide (Rabjohn, 1949) or high potential quinones such as chloranil or 5,6-dichloro-2,3-dicyano-*p*-benzoquinone (Stechl, 1975), but none of these methods is entirely satisfactory. Yields are sometimes not good and the reactions lack selectivity.

A new method which gives excellent yields of αβ-unsaturated compounds under mild conditions, and which promises to be of wide application, has been described recently by Sharpless, Lauer and Teranishi (1973) and by Reich, Reich and Renga (1973). It provides another example of the application of organoselenium compounds in organic synthesis. The sequence proceeds from the α-phenylseleno carbonyl compound (as 34), which is itself readily obtained from the carbonyl compound by reaction with phenylselenyl chloride at room temperature, or from the corresponding enolate anion and the selenyl halide or diphenyl diselenide −78°C. By oxidation with hydrogen peroxide or sodium periodate the selenide is converted into the corresponding selenoxide which immediately undergoes *syn* β-elimination to form the *trans* $\alpha\beta$-unsaturated ketone in high yield (cf. p. 100). Functional groups such as alcoholic hydroxyls, ester groups and olefinic double bonds appear not to interfere. Thus, propiophenone is converted in 89 per cent yield into phenyl vinyl ketone, an olefin which is difficult to obtain by other means because of its ready polymerisation and susceptibility to nucleophilic attack, and 4-acetoxycyclohexanone gives 4-acetoxycyclohexenone. The procedure can also be used to make $\alpha\beta$-unsaturated esters and lactones from the saturated precursors, as shown in (6.100).

C_6H_5 … O → (1) $LiN(iso\text{-}C_3H_7)_2$, THF, −78°C; (2) C_6H_5SeBr, −78°C → SeC_6H_5, C_6H_5, O (34)

(34) → $NaIO_4$, CH_3OH, H_2O → [$C_6H_5\overset{+}{Se}$—$\overset{-}{O}$, C_6H_5, H, C H_2, O]

→ 15–25°C → C_6H_5, O (89%) + $[C_6H_5SeOH]$

(6.100)

O, O → (1) $LiNR_2$, THF, −78°C; (2) C_6H_5SeCl; (3) H_2O_2 → O, O (56%)

CO_2CH_3 → (1) $LiN(iso\text{-}C_3H_7)_2$, THF, −78°C; (2) $C_6H_5SeSeC_6H_5$; (3) $NaIO_4$, CH_3OH, H_2O → CO_2CH_3 (80%)

The sequence provides a method (*a*) for converting $\alpha\beta$-unsaturated ketones into β-alkyl derivatives by alkylation with an organocuprate (cf. p. 66) and reaction of the intermediate copper enolate with phenylselenyl bromide, and (*b*) for the synthesis of enediones from 1,3-dicarbonyl compounds, a transformation which is difficult to bring about by other means (6.101) (Reich, Renga and Reich, 1974).

$$C_6H_5COCH{=}CHCH_3 \xrightarrow[(2)\ C_6H_5SeBr]{(1)\ (CH_3)_2CuLi} C_6H_5COCH(SeC_6H_5)CH(CH_3)_2 \xrightarrow{H_2O_2} C_6H_5COCH{=}C(CH_3)_2 \quad (85\%)$$

$$\text{2-}CO_2C_2H_5\text{-cyclohexanone} \xrightarrow[(3)\ H_2O_2,\ CH_2Cl_2]{(1)\ NaH,\ THF \quad (2)\ C_6H_5SeCl} \text{2-}CO_2C_2H_5\text{-cyclohex-2-enone} \quad (90\%) \qquad (6.101)$$

The reaction of enolates with phenylselenyl halides is very fast even at −78°C, and the kinetically generated enolates react without rearrangement to the more stable isomer. 2-Methylcyclohexanone, for example, gave only 6-methyl-2-cyclohexenone.

α-Ketols and 1,2-diketones. Oxidation of ketones at the α-carbon atom to give α-hydroxyketones (acyloins) or 1,2-diketones, is a synthetically useful transformation. Direct conversion of ketones into α-acetoxy derivatives can be effected in some cases by reaction with mercuric acetate or lead tetra-acetate, but these methods are not very selective and might well be inappropriate in cases where there were other functional groups in the molecule. Fortunately some other more selective methods are available.

In a strongly basic medium, such as a solution of potassium t-butoxide in t-butanol, ketones react rapidly with molecular oxygen to form α-hydroperoxides which, in some cases, can be reduced to the corresponding acyloin. Sometimes the hydroperoxide can be isolated and separately reduced with zinc and acetic acid, but better yields are often obtained by carrying out the oxidation in presence of triethyl phosphite which reduces the hydroperoxide directly to the acyloin. The method is useful for the introduction of a 17-α-hydroxyl substituent into 20-ketosteroids and has been used to introduce hydroxyl substituents at selected positions in the synthesis of a number of other natural products. The reactions appear to be highly stereoselective, in the examples which have been studied the hydroperoxy and derived hydroxyl substituents have the same stereochemistry as the hydrogen atoms which are replaced (Gardner, Carlon and Gnoj, 1968; see also Wasserman and Lipshutz, 1975).

$COCH_3$ CH_3 (1) O_2, t-C_4H_9OH, t-C_4H_9OK (2) Zn, CH_3CO_2H $COCH_3$ OH CH_3

(78%)

(6.102)

RN H $COCH_3$ $CONH_2$ O_2, t-C_4H_9OH, t-C_4H_9OK $P(OC_2H_5)_3$, −20°C RN HO $COCH_3$ $CONH_2$

(96%)

This method, although useful, has the disadvantage that cleavage products are often formed, and where the α-hydroperoxide bears an α-hydrogen atom an α-diketone is likely to be produced by base-catalysed elimination, resulting in poor yields of the acyloin. An alternative procedure which avoids these difficulties uses the readily available molybdenum peroxide MoO_5.pyridine.hexamethyl phosphoramide complex (Vedejs, 1974). This reagent reacts readily with enolates at temperatures between −70 and −40°C to form a Mo^{VI} ester which, after treatment with water, affords the α-hydroxy carbonyl compound in good yield without contamination by oxidative cleavage products. Ketones, esters and lactones with an enolisable methylene or methine group are all readily converted into α-hydroxy compounds by this route. With unsymmetrical ketones a single acyloin is formed with high selectivity by way of the kinetically generated enolate. 2-Phenylcyclohexanone, for example, is converted exclusively into *trans*-2-hydroxy-6-phenylcyclohexanone (6.103).

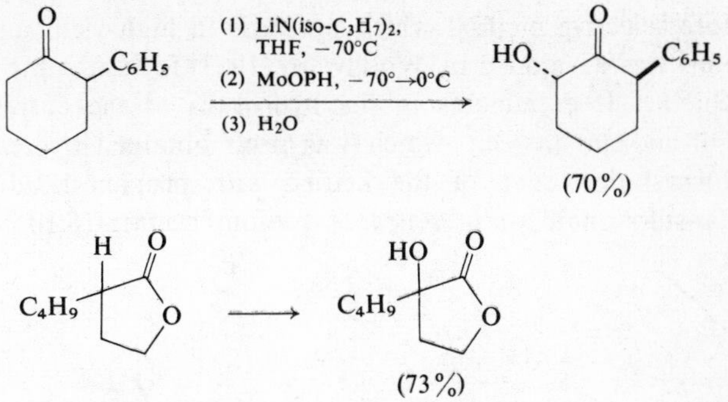

(1) $LiN(iso\text{-}C_3H_7)_2$, THF, −70°C
(2) MoOPH, −70°→0°C
(3) H_2O

(70%)

(6.103)

(73%)

MoOPH ≡ MoO_5. Pyridine.hexamethylphosphoramide complex

Direct oxidation of aldehydes and ketones to 1,2-dicarbonyl compounds can be effected with selenium dioxide (Jerussi, 1970). Acetophenone, for example, is converted into phenylglyoxal in 70 per cent yield by warming with selenium dioxide in aqueous dioxan. The mechanism of this reaction has been in doubt, but Sharpless and Gordon (1976) have produced evidence that it proceeds by way of a β-ketoseleninic acid (35), formed by electrophilic attack of selenious acid on the enol, followed by a Pummerer-like rearrangement to a short-lived selenine. (6.104). The

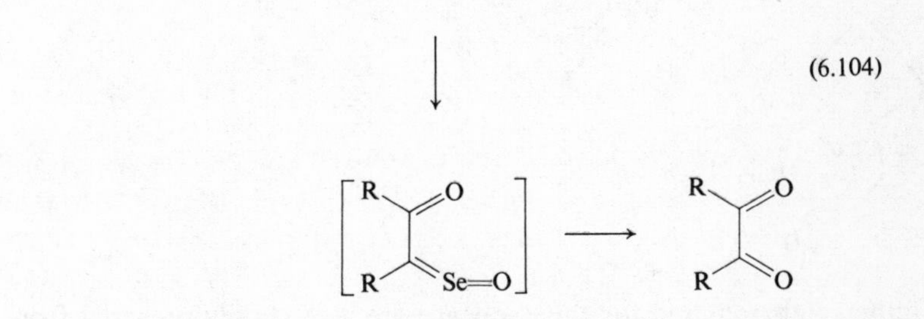

H_2SeO_3
($SeO_2 + H_2O$)

syn elimination

(35)

(6.104)

formation of $\alpha\beta$-unsaturated carbonyl compounds, which are often encountered in selenium dioxide oxidations of aldehydes and ketones, is readily explained by β-elimination from the ketoseleninic acid (cf. p. 100).

A more selective method which proceeds in high yield under mild conditions was developed by Woodward (1963) for use in his synthesis of colchicine. It culminates in the hydrolysis of the corresponding spiro-α-dithiaketal (as 36), which was itself obtained by reaction of the α-formyl derivative of the ketone with propane-1,3-dithiol di-toluene-*p*-sulphonate in presence of sodium acetate (6.105). The α-

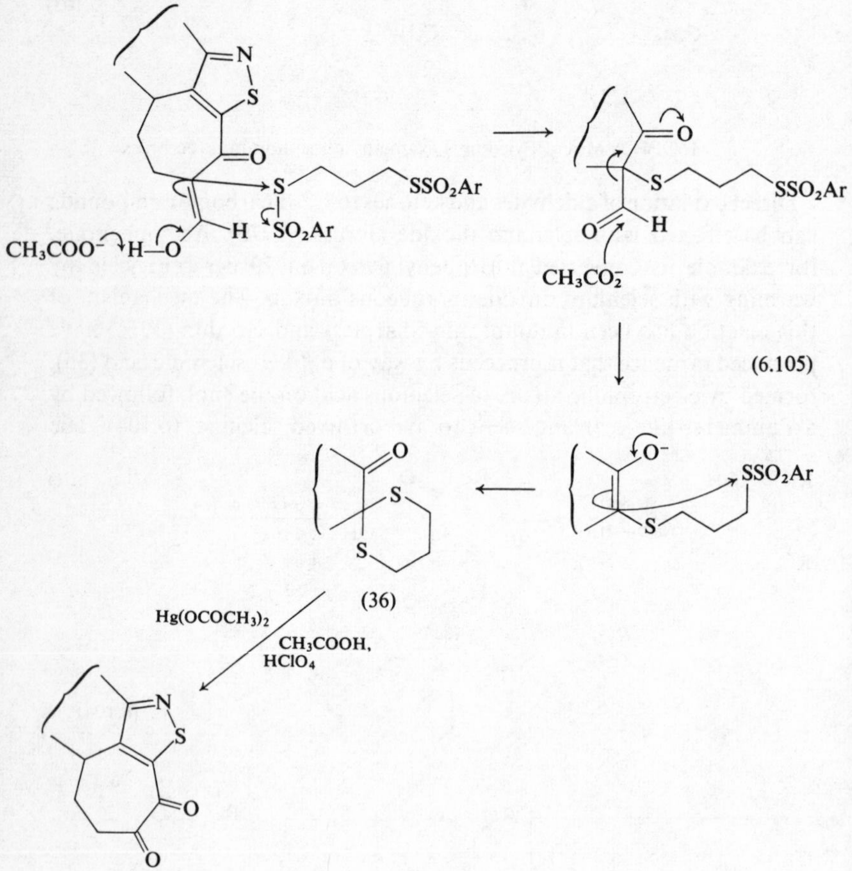

(6.105)

(36)

dithiaketals required for this sequence are also readily prepared from enamines. 2,2-(Trimethylenedithia)cyclohexanone, for example, is obtained from the pyrrolidine enamine of cyclohexanone and propane-1,3-dithiol di-toluene-*p*-sulphonate in 45 per cent yield (Woodward, Pachter and Scheinbaum, 1974).

α-Dithiaketals are useful intermediates for effecting controlled cleavage of cyclic ketones, giving open-chain products with residues at different oxidation levels at the two ends of the chain, by reaction with potassium hydroxide in t-butanol (Marshall and Seitz, 1974).

CHOH

O

S

S

O

KOH, C_4H_9OH

H

S

C

S

CO_2H

(93%)

(6.106)

6.6. Baeyer–Villiger oxidation of ketones

On oxidation with peroxy-acids, ketones are converted into esters or lactones. This reaction was discovered in 1899 by Baeyer and Villiger who found that reaction of a number of alicyclic ketones, including menthone, with Caro's acid (permonosulphuric acid) led to the formation of lactones (6.107). Better yields are obtained with organic peroxy-acids such as

O

HO_2SO_3H

O

O

(6.107)

peracetic acid in acetic acid containing sulphuric acid, or one of the stronger peroxy-acids such as permaleic or trifluoroperacetic acid. With the latter acid a buffer such as disodium hydrogen phosphate is often employed to prevent transesterification by reaction of the newly formed ester with the trifluoroacetic acid always present in the reaction medium. Very good results in difficult cases have also been obtained with solutions of potassium persulphate in 50 per cent sulphuric acid (Deno, Billups, Kramer and Lastomirsky, 1970). The reaction occurs under mild conditions and has been widely used both in degradative work and in synthesis. It is applicable to open chain and cyclic ketones as well as to aromatic ketones, and has been used to prepare a variety of steroidal and terpenoid lactones, as well as medium- and large-ring lactones which are otherwise virtually unobtainable. It also provides a route to alcohols from ketones, by hydrolysis of the esters formed (Hassall, 1957; Lee and Uff, 1967), (6.108). With unsaturated ketones Baeyer–Villiger reaction

$$CH_3COCH_2CH_3 \xrightarrow[Na_2HPO_4,\ CH_2Cl_2]{CF_3CO_3H,\ CF_3CO_2H} CH_3COOC_2H_5 \quad (72\%)$$

$$C_6H_5COCH_3 \xrightarrow[CHCl_3]{C_6H_5CO_3H} C_6H_5OCOCH_3 \xrightarrow{\text{hydrolysis}} C_6H_5OH$$

$$\text{cyclopentanone} \xrightarrow[CF_3CO_2H]{CF_3CO_3H} \delta\text{-valerolactone} \quad (88\%)$$

(6.108)

often takes place in preference to oxidation of the double bond, with formation of an unsaturated lactone (6.109).

$$\text{bicyclic enone} \xrightarrow[CHCl_3]{CH_3CO_3H} \text{unsaturated lactone} \quad (56\%)$$

$$\text{2-}CHC_6H_5\text{-cyclopentanone} \xrightarrow[CH_3CO_2H,\ CH_3CO_2K]{CH_3CO_3H} \text{6-}CHC_6H_5\text{-lactone} \quad (50\%)$$

(6.109)

An unsymmetrical ketone could obviously give rise to two different products in this reaction. Cyclohexyl phenyl ketone, for example, on reaction with peracetic acid gives both cyclohexyl benzoate and phenyl cyclohexanecarboxylate by migration of both the cyclohexyl and the phenyl group. It is found that the relative ease of migration of different groups in the reaction is in the order

t-alkyl > cyclohexyl ~ s-alkyl ~ benzyl ~ ~ phenyl > primary alkyl > methyl

That is, in the alkyl series migratory aptitudes are in the series tertiary > secondary > primary; among benzene derivatives migration is facilitated by electron-releasing *para* substituents, and hindered by electron-withdrawing ones. The methyl group shows the least tendency to migrate, so that methyl ketones always give acetates in the Baeyer–Villiger reaction. Phenyl *p*-nitrophenyl ketone gives only phenyl *p*-nitrobenzoate, and s-butyl methyl ketone is converted into s-butyl acetate. It has been suggested that the ease of migration is related to the ability of the migrating group to accommodate a partial positive charge in the transition state (see below), but it seems that in some cases steric effects may be involved, and the experimental conditions may influence the result as well. Thus, while 1-methylnorcamphor affords the expected lactone on oxidation with peracetic acid, the product obtained from camphor itself depends on the conditions. With buffered peracetic acid a mixture containing a considerable proportion of the 'abnormal' product, formed by migration of the primary carbon, is obtained; but in presence of sulphuric acid only the 'abnormal' product is formed (6.110). Epicamphor gives only the 'abnormal' product in 94 per cent yield (Sauers and Ahearn, 1961; see also Meinwald and Frauenglass, 1960).

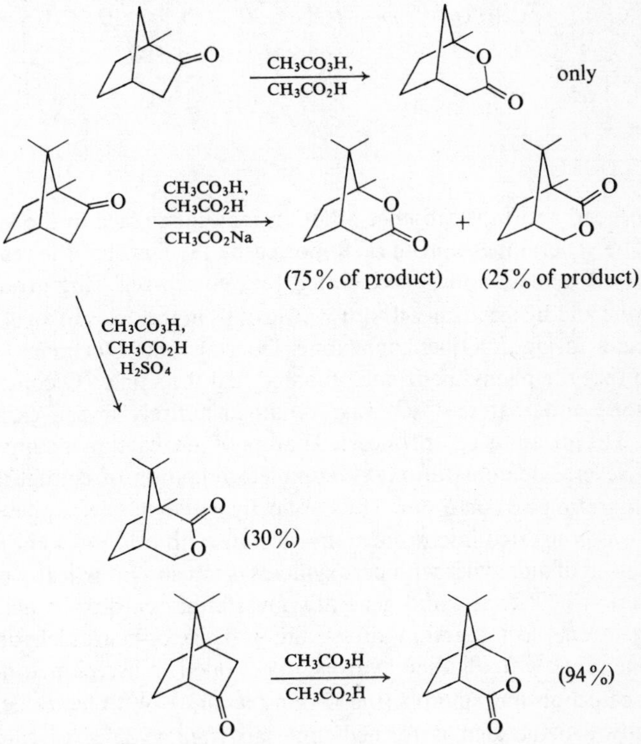

CH_3CO_3H, CH_3CO_2H → only

CH_3CO_3H, CH_3CO_2H, CH_3CO_2Na → (75% of product) + (25% of product)

CH_3CO_3H, CH_3CO_2H, H_2SO_4 → (30%)

(6.110)

CH_3CO_3H, CH_3CO_2H → (94%)

The reaction is thought to take place by a concerted intramolecular process involving migration of a group from carbon to electron-deficient oxygen, possibly by way of a cyclic transition state (Smith, 1963; Lee and Uff, 1967) (6.111). In the presence of a strong acid there may be addition of peroxy-acid to the protonated ketone, but additional acid

$$R^1R^2C{=}O + CH_3CO_3H \overset{H^+}{\rightleftharpoons} R^1R^2C(\overset{+}{O}H{-}H{\cdots}O{=}C(CH_3){-}O{-}O) \rightleftharpoons \left[R^1R^2C(O{\cdots}H{\cdots}O{=}C(CH_3)){\cdots}O{\cdots}O\right] \longrightarrow R^1C(\overset{+}{O}H)OR^2 + CH_3CO_2H \longrightarrow R^1COOR^2 \qquad (6.111)$$

is not needed and in its absence addition may take place to the ketone itself. The general mechanism is supported by the fact that the reaction is catalysed by acid and is accelerated by electron-releasing groups in the ketone and by electron-withdrawing groups in the acid. In an elegant experiment using [^{18}O]benzophenone Doering and Dorfman (1953) showed that the phenyl benzoate obtained had the same ^{18}O content as the ketone and that the ^{18}O was contained entirely in the carbonyl oxygen. The intramolecular concerted nature of the reaction is supported also by several demonstrations of complete retention of configuration in the migrating carbon atom. Thus optically active methyl α-phenethyl ketone was converted into α-phenethyl acetate with no loss of chirality.

Oxidation of aldehydes with peroxy-acids is not so synthetically useful as oxidation of ketones, and generally gives either carboxylic acids or formate esters. But reaction of *o*- and *p*-hydroxy-benzaldehydes or -acetophenones with alkaline hydrogen peroxide is a useful method for making catechols and quinols (the Dakin reaction). With benzaldehyde itself only benzoic acid is formed, but salicylaldehyde gives catechol

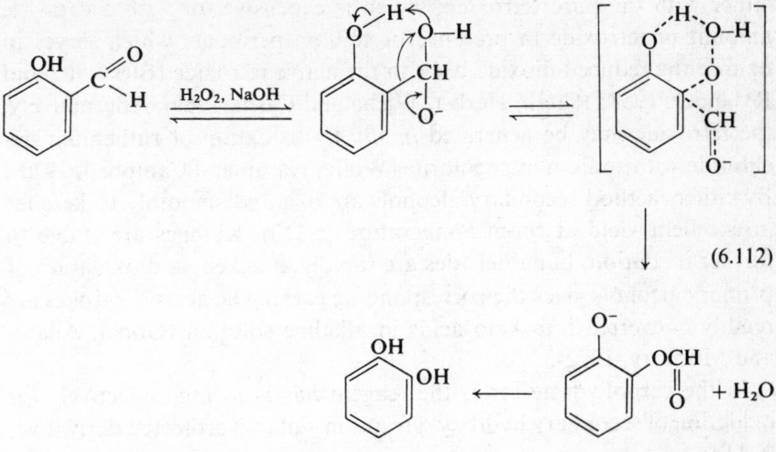

(6.112)

almost quantitatively, and 3,4-dimethylcatechol was obtained by oxidation of 2-hydroxy-3,4-dimethylacetophenone, possibly by way of a cyclic transition state similar to that postulated for the Baeyer–Villiger reaction (Lee and Uff, 1967) (6.112).

6.7. Oxidations with ruthenium tetroxide and nickel peroxide

Ruthenium tetroxide and nickel peroxide, two very powerful oxidising agents, are used for oxidation of a variety of functional groups at room temperature. Reactions with ruthenium tetroxide are generally effected in solution in carbon tetrachloride (it attacks most other organic solvents)

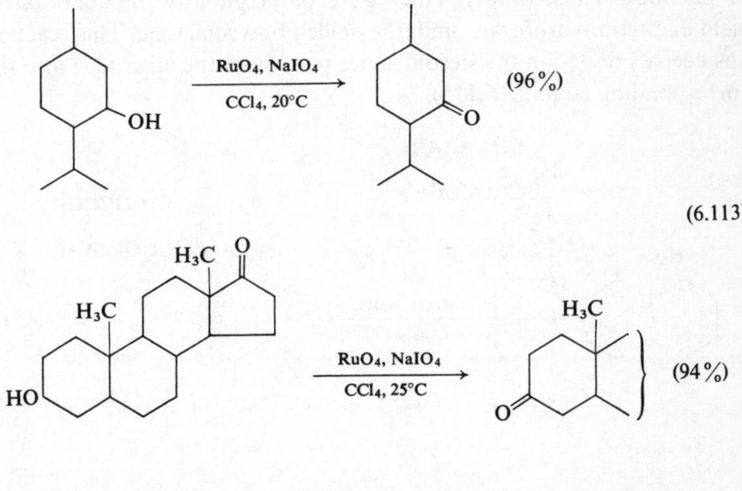

RuO_4, $NaIO_4$; CCl_4, 20°C (96%)

(6.113)

RuO_4, $NaIO_4$; CCl_4, 25°C (94%)

either with the pure tetroxide, which is expensive, or with a catalytic amount of tetroxide in presence of sodium periodate which serves to oxidise the reduced dioxide back to the active tetroxide (Birkowitz and Rylander, 1958; Piatak, Herbst, Wicha and Caspi, 1969). Alternatively, the tetroxide may be generated *in situ* by oxidation of ruthenium trichloride with sodium hypochlorite (Wolfe, Hasan and Campbell, 1970). By either method secondary alcohols are oxidised smoothly to ketones in excellent yield at room temperature (6.113). Ketones are stable to further oxidation, but aldehydes are rapidly attacked, and oxidation of primary alcohols gives the corresponding carboxylic acids. Lactones are readily converted into keto acids in alkaline solution (Gopal, Adams and Moriarty, 1972).

In the carbohydrate series the reagent has been used effectively for oxidation of secondary hydroxyl groups in suitably protected derivatives (6.114).

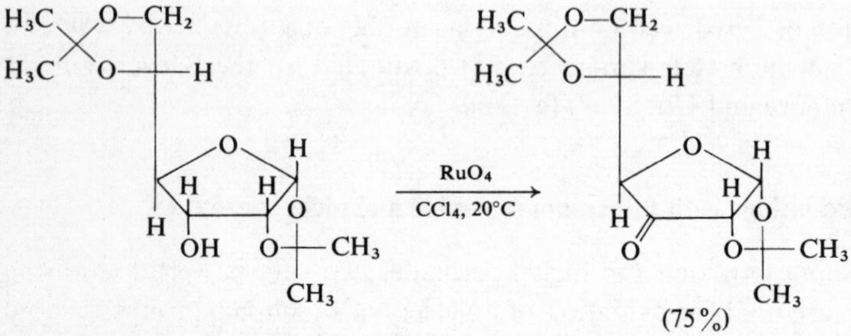

$$\xrightarrow[\text{CCl}_4,\ 20^\circ\text{C}]{\text{RuO}_4} \quad (75\%) \qquad (6.114)$$

A remarkable reaction is the smooth conversion of ethers into esters or lactones. Thus, dibutyl ether gave butyl butyrate in quantitative yield and tetrahydrofuran similarly yielded butyrolactone. This reaction has been exploited in the steroid series to convert the ether (37) into the corresponding lactone (6.115).

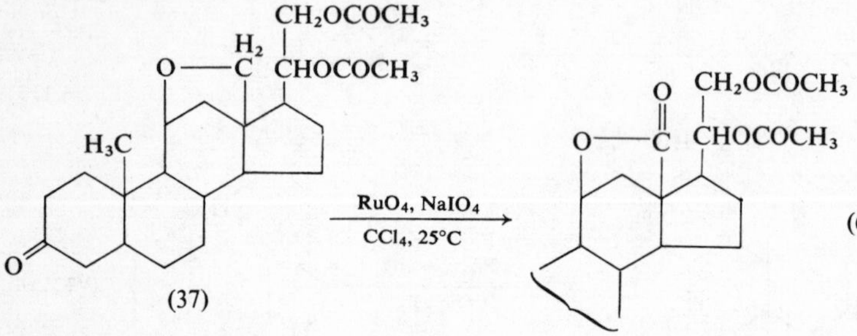

(37)

$$\xrightarrow[\text{CCl}_4,\ 25^\circ\text{C}]{\text{RuO}_4,\ \text{NaIO}_4} \qquad (6.115)$$

Unlike osmium tetroxide, ruthenium tetroxide does not react with olefinic double bonds to form diols. Instead, cleavage takes place to give carboxylic acids or ketones, and oxidation with ruthenium tetroxide and sodium periodate provides a convenient alternative to ozonolysis. Ruthenium tetroxide–sodium periodate is found, for example, to be an excellent reagent for cleavage of the diphenylethylenes obtained in the Barbier–Wieland degradation sequence (Stork, Meisels and Davies, 1963) (6.116).

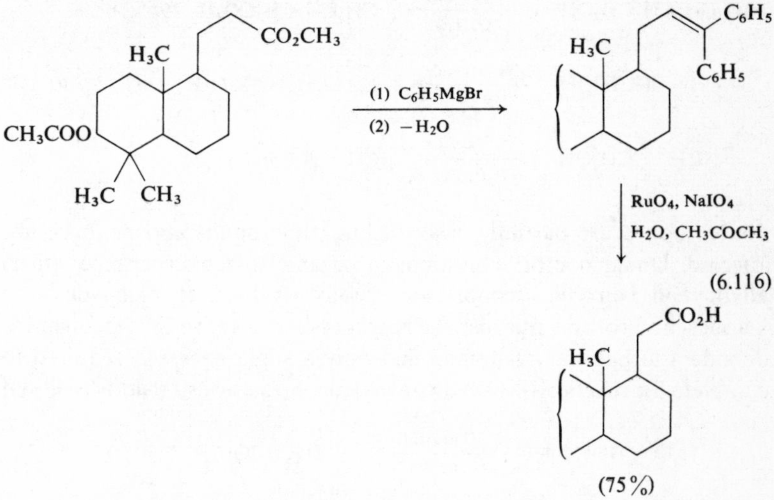

CO_2CH_3 H_3C CH_3COO H_3C CH_3 (1) C_6H_5MgBr (2) $-H_2O$ C_6H_5 H_3C C_6H_5

RuO_4, $NaIO_4$ H_2O, CH_3COCH_3 (6.116)

CO_2H H_3C (75%)

$\alpha\beta$-Unsaturated ketones are degraded in the same way as with ozone. Thus, testosterone acetate gives the ring-A cleaved product (6.117). Even benzene rings are attacked. Benzene itself is reported to react with ruthenium tetroxide with explosive violence.

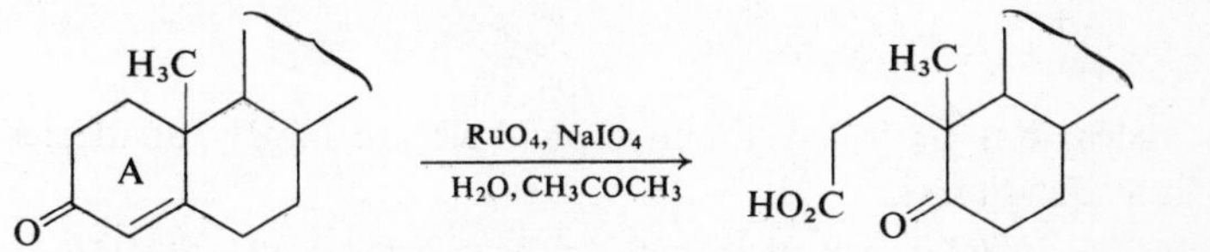

H_3C A O RuO_4, $NaIO_4$ H_2O, CH_3COCH_3 H_3C HO_2C O (6.117)

Nickel peroxide has also proved useful for the oxidation of a variety of functional groups (George and Balachandran, 1975). It is obtained as a black powder, insoluble in water and organic solvents, by action of alkaline sodium hypochlorite on nickel sulphate (Nakagawa, Konaka and Nakata, 1962). The composition of the reagent has been given as $NiO_{2.77}H_{2.85}$. It is generally used in a little excess of the stoichiometric amount and the 'available oxygen' is determined for each batch of reagent by iodometry. In many of its reactions it resembles manganese dioxide, but is a more powerful oxidant.

The oxidising power for alcohols of nickel peroxide is affected by the alkalinity of the solvent and the temperature. In aqueous alkaline solution saturated aliphatic primary alcohols are oxidised to the corresponding carboxylic acids in good yield (6.118). In some cases isolated olefinic

$$CH_3CH_2CH_2CH_2OH \xrightarrow[30°C,\ H_2O]{\text{Ni peroxide}} CH_3CH_2CH_2CO_2H \quad (94\%)$$

$$C_6H_5CH{=}CHCH_2OH \xrightarrow[H_2O,\ 50°C]{\text{Ni peroxide}} C_6H_5CH{=}CHCO_2H \quad (70\%) \qquad (6.118)$$

$$CH{\equiv}CCH_2OH \xrightarrow[H_2O,\ 5°C]{\text{Ni peroxide}} CH{\equiv}CCO_2H \quad (50\%)$$

double bonds are partially cleaved, but triple bonds appear to be unaffected. Under neutral conditions in organic solvents (benzene, ether) allylic and benzylic alcohols are readily oxidised to aldehydes and ketones, and for this purpose the reagent is more active than manganese dioxide and has the advantage that only a slight excess is required to complete the reaction (6.119). Propynyl alcohols are also readily oxidised

$$C_6H_5COCHOHC_6H_5 \xrightarrow[C_6H_6,\ 50°C]{\text{Ni peroxide}} C_6H_5COCOC_6H_5 \quad (98\%)$$

$$CH_2OH \xrightarrow[C_6H_6,\ 50°C]{\text{Ni peroxide}} CHO \quad (80\%) \qquad (6.119)$$

to aldehydes. But saturated aliphatic alcohols are largely unaffected under these conditions.

In aprotic solvents α-glycols and α-hydroxy-acids are cleaved to aldehydes and ketones, and α-keto-alcohols and -acids similarly give carboxylic acids in aqueous alkaline solution.

By oxidation at –20°C in an ethereal solution of ammonia, aromatic and allylic aldehydes can be converted into amides in good yield. At higher temperatures yields of amide decrease and nitriles are the main products. The reaction is thought to take the pathway shown in (6.120).

$$\mathrm{R{-}CHO} \xrightarrow{NH_3} \mathrm{R{-}C(NH_2)(H){-}OH} \xrightarrow{-H_2O} \mathrm{R{-}CH{=}NH}$$

$$\mathrm{R{-}C(NH_2)(H){-}OH} \xrightarrow{\text{oxidation}} \mathrm{RCONH_2} \qquad \mathrm{R{-}CH{=}NH} \xrightarrow{\text{oxidation}} \mathrm{R{-}C{\equiv}N} \qquad (6.120)$$

Other useful reactions which can be effected in good yield with nickel peroxide are the oxidative coupling of phenols and the oxidation of hydrazones to diazo compounds. Thus, 2,6-di-t-butylphenol is converted into tetra-t-butyldiphenoquinone, and benzophenone hydrazone gives diphenyldiazomethane. Studies using electron spin resonance spectroscopy have shown that oxidation of 2,6-di-t-butylphenol takes place by way of the phenoxy radical, and it is believed that all oxidations with nickel peroxide proceed by a free radical mechanism probably induced by free hydroxyl radicals on the catalyst surface (Konaka, Terabe and Kuruma, 1969). Isotopic studies suggest that the conversion of benzhydrol into benzophenone takes the course shown in (6.121). The

$$C_6H_5CHOHC_6H_5 \longrightarrow C_6H_5\dot{C}(OH)C_6H_5 \longrightarrow C_6H_5\dot{C}(O\cdot)C_6H_5 \longrightarrow C_6H_5COC_6H_5 \qquad (6.121)$$

formation of free radicals serves to explain the ready dimerisation of diphenylacetonitrile with nickel peroxide, and the conversion of chloroform into hexachloroethane.

6.8. Oxidations with thallium(III) nitrate

Some useful oxidations of alkenes and alkynes can be effected with thallium(III) nitrate (McKillop and Taylor, 1973). The effectiveness of this reagent in oxidations is due to the energetically favourable conversion $Tl^{3+} + 2e \rightarrow Tl^{+}$ ($-1{\cdot}25$ V), and to the weakness of the carbon–thallium bond (105–125 kJ mol^{-1}) which is easily cleaved heterolytically to form carbonium ion intermediates. Furthermore, nitrate ion is not very nucleophilic so that solvent may participate selectively as the nucleophile in the reactions.

Thallium(III) nitrate reacts with alkenes in methanol solution to give carbonyl compounds or 1,2-glycol dimethyl ethers. Particularly useful is the reaction with cyclic alkenes which leads by an oxidative rearrangement to ring-contracted products (McKillop, Hunt, Kienzle, Bigham and Taylor, 1973). Thus, cyclohexene reacts almost instantaneously at room temperature in methanol solution to give the dimethylacetal of cyclopentanecarboxaldehyde, and cycloheptene is similarly converted into the dimethylacetal of cyclohexanecarboxaldehyde. The reaction is believed to take the course shown in (6.122). This reaction takes place

$$\text{cyclohexene} \xrightarrow[CH_3OH]{Tl(ONO_2)_3} \text{(cation)}\ Tl(ONO_2)_2 \xrightarrow{CH_3OH} \text{2-}OCH_3\text{-cyclohexyl-}Tl(ONO_2)_2$$

$$\equiv \text{(chair: 2-}OCH_3\text{, 3, }Tl(ONO_2)_2) \longrightarrow \left[{}^{+}CHOCH_3\text{-cyclopentane}\right] \xrightarrow{CH_3OH} CH(OCH_3)_2\text{-cyclopentane}$$

$$CH_3(CH_2)_7CH{=}CH_2 \xrightarrow[CH_3OH]{Tl(ONO_2)_3} CH_3(CH_2)_7COCH_3 \quad (28\%)$$

$$+ \quad CH_3(CH_2)_7CH(OCH_3){-}CH_2(OCH_3) \quad (52\%) \qquad (6.122)$$

most easily with six- and seven-membered rings in which the migrating ring bond, at C-2–C-3, is, or can easily become, *trans* anti-parallel to the carbon–thallium bond. With aliphatic olefins and cyclic olefins in

$$CH_3O\text{-}C_6H_4\text{-}CH{=}CH_2 \xrightarrow[HNO_3,\ H_2O]{Tl(ONO_2)_3} CH_3O\text{-}C_6H_4\text{-}CH(OH)\text{-}CH_2\text{-}Tl(ONO_2)_2 \longrightarrow CH_3O\text{-}C_6H_4\text{-}CH_2CHO \quad (75\%) \qquad (6.123)$$

which a similar conformationally favourable pathway is not possible rearrangement does not take place to any appreciable extent and glycol dimethyl ethers are the main products (6.122).

Styrenes also form 1,2-dimethoxy derivatives when treated with thallium(III) nitrate in methanol, but in dilute nitric acid reaction takes a different course and high yields of arylacetaldehydes are formed. The transformation again involves a rearrangement (6.123). With alkynes the product obtained depends on the structure of the alkyne (McKillop, Oldenziel, Swann, Taylor and Robey, 1973). Diarylalkynes react with two equivalents of thallium(III) nitrate in acid solution or in methanol to give benzils in high yield (6.124), but with dialkylalkynes reaction stops at the acyloin, which does not enolise under the weakly acidic

$$\mathrm{Ar{-}C{\equiv}C{-}Ar'} \xrightarrow[\mathrm{H_3O^+}]{\mathrm{Tl(ONO_2)_3}} \mathrm{(Ar)(HO)C{=}C(Ar')Tl(ONO_2)_2} \longrightarrow \mathrm{Ar{-}C({=}O){-}CH(Tl(ONO_2)_2){-}Ar'} \ (\mathrm{H_2O:})$$

$$\longrightarrow \mathrm{Ar{-}C({=}O){-}CH(OH){-}Ar'} \underset{\mathrm{H^+}}{\rightleftharpoons} \mathrm{Ar{-}C(OH){=}C(OH){-}Ar'} \longrightarrow \mathrm{Ar{-}C{=}C{-}Ar'\ (O{-}Tl(ONO_2){-}O)}$$

$$\longrightarrow \mathrm{Ar{-}CO{-}CO{-}Ar'} \quad (90\% \text{ for } \mathrm{Ar} = p\text{-tolyl}, \mathrm{Ar'} = p\text{-}\mathrm{CH_3OC_6H_4}) \tag{6.124}$$

conditions employed. Alkylarylalkynes in acid solution give mixtures of products, but in methanol a smooth rearrangement takes place to give methyl esters of α-alkylarylacetic acids, and the reaction provides a very effective route to these valuable intermediates. Arylacetic esters are also obtained by oxidation of acetophenones with thallium(III) nitrate in acidified methanol, and this procedure is a convenient alternative to the Willgerodt reaction (McKillop, Swann and Taylor, 1973).

$$C_6H_5-C\equiv C-CH_3 \xrightarrow[CH_3OH]{Tl(ONO_2)_3} (C_6H_5)(CH_3O)C=C(CH_3)Tl(ONO_2)_2 \;(+\,CH_3OH)$$

$$\downarrow \qquad (6.125)$$

$$\underset{(80\%)}{C_6H_5-CH(CH_3)-CO_2CH_3} \xleftarrow{H_2O} (CH_3O)_2C=C(C_6H_5)(CH_3)$$

With monoalkylacetylenes yet another reaction takes place. They are oxidised with two equivalents of thallium(III) nitrate in acid solution to carboxylic acids with loss of the terminal carbon atom. 1-Octyne, for example, gave heptanoic acid in 80 per cent yield. This novel reaction is believed to go through the α-ketol which undergoes *trans* esterification with the second molecule of the nitrate.

$$R-C\equiv CH \xrightarrow[H_3O^+]{Tl(ONO_2)_3} \left[R-C(OH)=CH-Tl(ONO_2)_2 \longrightarrow RCOCH_2-Tl(ONO_2)_2 \;\; H_2O: \right]$$

$$\downarrow$$

$$H_2O: \;\; RC(=O)CH_2-O-Tl(ONO_2)_2 \xleftarrow{Tl(ONO_2)_3} RCOCH_2OH$$

$$\downarrow \qquad (6.126)$$

$$R-C(OH)(O^-)-CH_2-O-Tl(ONO_2)_2 \longrightarrow RCO_2H + CH_2O$$

7 Reduction

There must be few organic syntheses of any complexity which do not involve a reduction step at some stage. Reduction is here used in the sense of addition of hydrogen to an unsaturated group such as an olefinic bond, a carbonyl group or an aromatic nucleus, or addition of hydrogen with concomitant fission of a bond between two atoms, as in the reduction of a disulphide to a thiol or of an alkyl halide to a hydrocarbon.

Reductions are generally effected either chemically or by catalytic hydrogenation, that is by the addition of molecular hydrogen to the compound under the influence of a catalyst. Each method has its advantages. In many reductions either method may be used equally well. Complete reduction of an unsaturated compound can generally be achieved without undue difficulty, but the aim is often selective reduction of one group in a molecule in the presence of other unsaturated groups. Both catalytic and chemical methods of reduction offer considerable scope in this direction, and the method of choice in a particular case will often depend on the selectivity required and on the stereochemistry of the desired product.

7.1. Catalytic hydrogenation

Of the many methods available for reduction of organic compounds catalytic hydrogenation is one of the most convenient. Reaction is easily effected simply by stirring or shaking the substrate with the catalyst in a suitable solvent, or without a solvent if the substance being reduced is a liquid, in an atmosphere of hydrogen in an apparatus which is arranged so that the uptake of hydrogen can be measured. At the end of the reaction the catalyst is filtered off and the product recovered from the filtrate, often in a high state of purity. The method is easily adapted for work on a micro scale, or on a large, even industrial, scale. In many cases reaction proceeds smoothly at or near room temperature and at atmospheric or slightly elevated pressure. In other cases high temperatures (100–

200°C) and pressures (100–300 atmospheres) are necessary, requiring special high pressure equipment. Detailed descriptions of equipment suitable for hydrogenations under different conditions are given by Schiller (1955) and by Augustine (1965).

Catalytic hydrogenation may result simply in the addition of hydrogen to one or more unsaturated groups in the molecule, or it may be accompanied by fission of a bond between atoms. The latter process is known as hydrogenolysis.

Most of the common unsaturated groups in organic chemistry, such as $>C{=}C<$, $-C{\equiv}C-$, $>C{=}O$, $-COOR$, $-C{\equiv}N$, $-NO_2$ and aromatic and heterocyclic nuclei can be reduced catalytically under appropriate conditions, although they are not all reduced with equal ease. Certain groups, notably allylic and benzylic hydroxyl and amino groups and carbon–halogen and carbon–sulphur single bonds, readily undergo hydrogenolysis, resulting in cleavage of the bond between carbon and the hetero-atom. This may be advantageous in certain circumstances. Thus, much of the usefulness of the benzyloxycarbonyl protecting group in peptide chemistry is due to the ease with which it can be removed by hydrogenolysis over a palladium catalyst, as in the conversion of (1) into (2) (7.1).

$$p\text{-}HOC_6H_4CH_2\underset{(1)}{\overset{NHCO_2CH_2C_6H_5}{CH}}CO_2CH_3 \xrightarrow[HCl]{H_2,\ Pd,\ C_2H_5OH} p\text{-}HOC_6H_4CH_2\underset{(2)}{\overset{NH_2HCl}{CH}}CO_2CH_3 + C_6H_5CH_3 + CO_2 \qquad (7.1)$$

An alternative procedure which is sometimes advantageous is 'catalytic transfer hydrogenation' in which hydrogen is transferred to the substrate from another organic compound, usually a hydroaromatic compound or an alcohol (Brieger and Nestrick, 1974). Reaction is easily effected simply by warming the substrate and hydrogen donor together in presence of a catalyst, usually palladium.

The catalyst. Many different catalysts have been used for catalytic hydrogenations; they are mainly finely divided metals, metallic oxides or sulphides. The most commonly used in the laboratory are the platinum metals (platinum, palladium and, to a lesser extent rhodium and ruthenium), nickel and copper chromite. The catalysts are not specific and with the exception of copper chromite may be used for a variety of different reductions. The most widely used are the platinum metal catalysts and much of what can be accomplished by hydrogenation is best done with these catalysts (Rylander, 1967). They are exceptionally active catalysts and promote the reduction of most functional groups under mild conditions, with the notable exception of the carboxyl, carboxylic ester and amide groups. They are used either as the finely divided metal or, more commonly, supported on a suitable carrier such as asbestos, activated carbon, alumina or barium sulphate. In general, supported metal catalysts, since they have a larger surface area, are more active than the unsupported metal, but the activity is influenced strongly by the support and by the method of preparation, and this provides a means of preparing catalysts of varying activity. Platinum is very often used in the form of its oxide PtO_2, 'Adams' catalyst', which is reduced to metallic platinum by hydrogen in the reaction medium.

The platinum metal catalysts are usually prepared by reduction of a metallic salt, in presence of the support if a supported catalyst is wanted. Detailed instructions are given by Kimmer (1955). It has been found that very effective catalysts can be obtained by reduction of various metal salts with either sodium borohydride (Brown and Brown, 1962) or a trialkylsilane (Eaborn, Pant, Peeling and Taylor, 1969). Salts of platinum, palladium, rhodium and ruthenium are reduced to the finely divided metals which may be used directly as catalysts or adsorbed on carbon. Reaction of excess of the sodium borohydride with acid provides a source of hydrogen, allowing direct hydrogenation of easily reducible functional groups such as nitro and unhindered olefin (7.2). The selectivity of hydrogenation is increased and hydrogenolysis is

CH_3 CH_3 CH_2 —(1) $PdCl_2$, $NaBH_4$, C_2H_5OH; (2) CH_3CO_2H, $-25°C$→ CH_3 CH_3 CH_3 H (94%)

(7.2)

OH —Pd, $NaBH_4$, C_2H_5OH; CH_3CO_2H, $-25°C$→ OH CH_2CH_3

suppressed by conducting the reactions at low temperature (Brown, C. A. 1969).

Most platinum metal catalysts, with the exception of Adams' catalyst, are stable and can be kept for many years without appreciable loss of activity, but they are deactivated by many substances, particularly compounds of divalent sulphur. For best results in a hydrogenation, therefore, particularly with platinum catalysts, it is necessary to use pure materials and pure solvents. On the other hand, catalytic activity is often increased by addition of small amounts of promoters, usually platinum or palladium salts or mineral acid. The activity of platinum catalysts derived from platinum oxide is often markedly increased in presence of mineral acid; this may simply be due to neutralisation of alkaline impurities in the catalyst.

For hydrogenations at high pressure the most common catalysts are Raney nickel and copper chromite. Raney nickel is a porous, finely divided nickel obtained by treating a powdered nickel–aluminium alloy with sodium hydroxide. It is generally used at high temperatures and pressures, but with the more active catalysts many reductions can be effected at atmospheric pressure and normal temperature. Nearly all unsaturated groups can be reduced with Raney nickel but it is most frequently used for reduction of aromatic rings and hydrogenolysis of sulphur compounds (see p. 489). When freshly prepared it contains 25–100 ml adsorbed hydrogen per gram of nickel. The more adsorbed hydrogen the more active is the catalyst, and Raney nickel catalysts of graded activities can be obtained by variation of the preparative procedure (Augustine, 1965; Kimmer, 1955). Raney nickel catalysts are alkaline and may only be used for hydrogenations which are not adversely affected by basic conditions. They are deactivated by acids. Another useful catalyst which is said to resemble Raney nickel in activity is nickel boride, which is easily prepared by reduction of nickel salts in aqueous solution with sodium borohydride (Brown and Brown, 1963).

Copper chromite, $CuCr_2O_4$, is prepared by thermal decomposition of copper ammonium chromate; a more active catalyst is obtained by adding barium nitrate to the reaction mixture. It is a relatively inactive catalyst and is only effective at high temperatures (100–200°C) and pressures (200–300 atm). It does not reduce aromatic rings under these conditions as Raney nickel does, and is principally used for reduction of esters to alcohols and of amides to amines (7.3).

Hydrogenation using copper chromite is not so frequently used now because of the introduction of hydride reducing agents (p. 467)

$$C_6H_5CH_2CO_2Et \xrightarrow[25^\circ C,\ 200\ atm]{H_2,\ CuCr_2O_4} C_6H_5CH_2CH_2OH$$

$$\text{ethyl 5-oxopyrrolidine-2-carboxylate} \xrightarrow[150^\circ C,\ 300\ atm]{H_2,\ CuCr_2O_4} \text{5-(hydroxymethyl)pyrrolidin-2-one } (CH_2OH) \xrightarrow[250^\circ C,\ 300\ atm]{H_2,\ CuCr_2O_4} \text{(bicyclic diamine, N, N)} \qquad (7.3)$$

which can accomplish the same reactions under much milder conditions, but it is still useful for reactions on a large scale.

Selectivity of reduction. Many hydrogenations proceed satisfactorily under a wide range of conditions, but where a selective reduction is wanted, conditions may be more critical.

The choice of catalyst for a hydrogenation is governed by the activity and selectivity required. Selectivity is a property of the metal, but it also depends to some extent on the activity of the catalyst and on the reaction conditions. In general, the more active the catalyst the less discriminating it is in its action, and for greatest selectivity reactions should be run with the least active catalyst and under the mildest possible conditions consistent with a reasonable rate of reaction. The rate of a given hydrogenation may be increased by raising the temperature, by increasing the pressure or by an increase in the amount of catalyst used, but all these factors may result in a decrease in selectivity. For example, hydrogenation of ethyl benzoate with copper chromite catalyst under the appropriate conditions leads to benzyl alcohol by reduction of the ester group, while Raney nickel gives ethyl hexahydrobenzoate by selective attack on the benzene ring (7.4). At higher temperatures, however,

$$C_6H_5CH_2OH \xleftarrow[160^\circ C,\ 250\ atm]{H_2,\ CuCr_2O_4} C_6H_5CO_2C_2H_5 \xrightarrow[50^\circ C,\ 100\ atm]{H_2,\ Raney\ Ni} C_6H_{11}CO_2C_2H_5 \qquad (7.4)$$

the selective activity of the catalysts is lost and mixtures of the two products and toluene are obtained from both reactions.

Both the rate and, sometimes, the course of a hydrogenation may be influenced by the solvent used. The commonest solvents are methanol,

TABLE 7.1 *Approximate order of reactivity of functional groups in catalytic hydrogenation*

Functional group	Reduction product
$R—COCl$	$R—CHO$, $R—CH_2OH$
$R—NO_2$	$R—NH_2$
$R—C{\equiv}C—R$	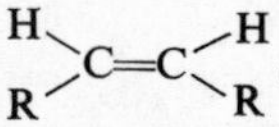 $\mathrm{(H)(R)C{=}C(H)(R)}$, RCH_2CH_2R
$R—CHO$	$R—CH_2OH$
$R—CH{=}CH—R$	$R—CH_2CH_2—R$
$R—CO—R$	$R—CHOH—R$, $R—CH_2—R$
$C_6H_5CH_2OR$	$C_6H_5CH_3 + ROH$
$R—C{\equiv}N$	$R—CH_2NH_2$
Polycyclic aromatic hydrocarbons	Partially reduced products
$R—COOR'$	$R—CH_2OH + R'OH$
$R—CONHR'$	$R—CH_2NHR'$
	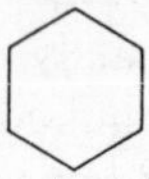
$R—CO_2^- Na^+$	inert

H.O. House, *Modern synthetic reactions*, copyright 1965, 1972, W. A. Benjamin, Inc., Menlo Park, California.

ethanol and acetic acid. Many hydrogenations over platinum metal catalysts are favoured by strong acids. For example, reduction of β-nitrostyrene in acetic acid–sulphuric acid is rapid and affords 2-phenylethylamine in 90 per cent yield; but in absence of sulphuric acid reduction is slow and the yield of amine poor.

Not all functional groups are reduced with equal ease. Table 7.1, due to House (1965), shows the approximate order of decreasing ease of catalytic hydrogenation of a number of common groups. This order is not invariable and is influenced to some extent by the structure of the compound being reduced and by the catalyst employed. In general, groups near the top of the list can be selectively reduced in the presence of groups near the bottom, but reduction of groups at the bottom in presence of the more reactive groups at the top is more difficult. For example, reduction of an unsaturated ester or ketone to a saturated ester or ketone is, in most cases, readily accomplished by hydrogenation over palladium or platinum, but selective reduction of the carbonyl group to form an unsaturated alcohol is difficult to achieve by catalytic hydrogenation and is generally effected by chemical reduction. Similarly, nitrobenzene is easily converted into aniline, but selective reduction to nitrocyclohexane is not possible (but see p. 428).

Reduction of functional groups: olefins. Hydrogenation of olefinic double bonds takes place easily and in most cases can be effected under mild conditions. Only a few highly hindered olefins are resistant to hydrogenation, and even these can generally be reduced under more vigorous conditions. Platinum and palladium are the most frequently used catalysts. Both are very active and the preference is determined by the nature of other functional groups in the molecule and by the degree of selectivity required; platinum usually brings about a more exhaustive reduction. Raney nickel may also be used in certain cases. Thus cinnamyl alcohol is reduced to the dihydro compound with Raney nickel in ethanol at 20°C, and 1,2-dimethylcyclohexene with hydrogen and platinum oxide in acetic acid is converted mainly into *cis*-1,2-dimethylcyclohexane (cf. p. 423) (7.5).

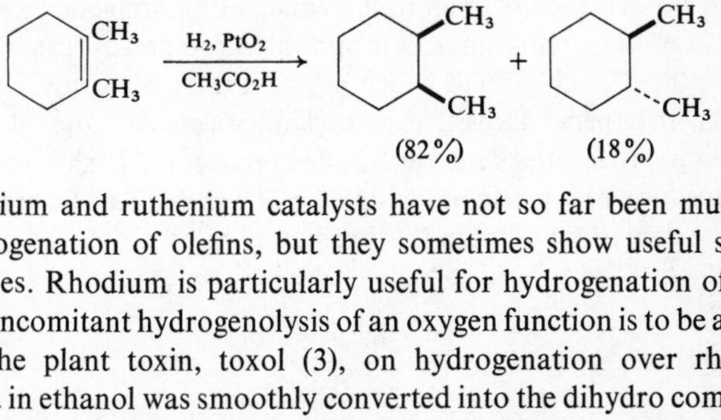

CH_3 CH_3 $\xrightarrow[CH_3CO_2H]{H_2,\ PtO_2}$ CH_3 CH_3 (82%) + CH_3 CH_3 (18%) (7.5)

Rhodium and ruthenium catalysts have not so far been much used in hydrogenation of olefins, but they sometimes show useful selective properties. Rhodium is particularly useful for hydrogenation of olefins when concomitant hydrogenolysis of an oxygen function is to be avoided. Thus, the plant toxin, toxol (3), on hydrogenation over rhodium–alumina in ethanol was smoothly converted into the dihydro compound; with platinum and palladium catalysts, on the other hand, extensive hydrogenolysis took place and a mixture of products was formed (7.6).

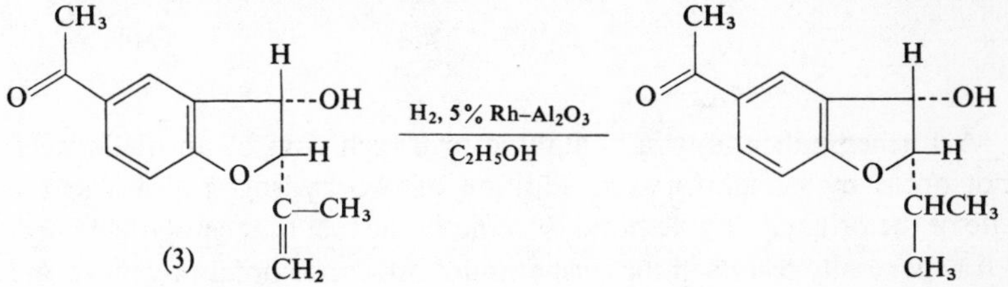

CH_3 O H OH H O C–CH_3 CH_2 (3) $\xrightarrow[C_2H_5OH]{H_2,\ 5\%\ Rh–Al_2O_3}$ CH_3 O H OH H O $CHCH_3$ CH_3 (7.6)

The ease of reduction of an olefin decreases with the degree of substitution of the double bond, and this sometimes allows selective reduction of one double bond in a polyolefin. Thus, limonene can be converted into *p*-menthene in almost quantitative yield by hydrogenation over platinum oxide if the reaction is stopped after absorption of one molecule of hydrogen (7.7). In contrast, the isomeric $\Delta^{1(7),8}$-*p*-

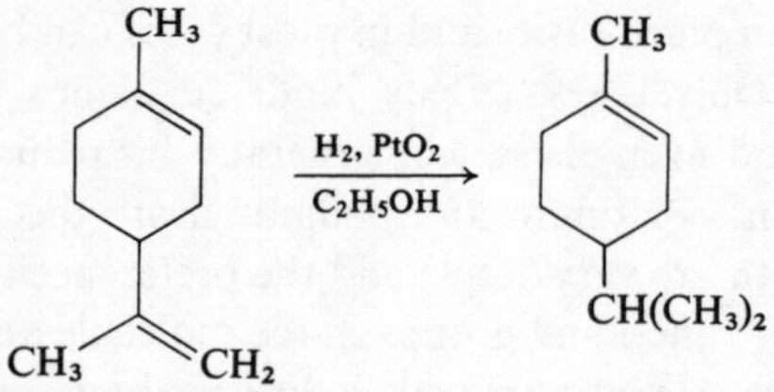

$$\text{limonene} \xrightarrow[C_2H_5OH]{H_2,\ PtO_2} \text{p-menthene}\ (\text{CH}_3,\ \text{CH(CH}_3)_2) \qquad (7.7)$$

menthadiene, in which both double bonds are disubstituted, gives only the completely reduced product.

Catalytic hydrogenation of olefins over platinum metal catalysts is often accompanied by migration of the double bond, but unless tracers are used or special products result, or the new bond is resistant to hydrogenation, no evidence of migration remains on completion of the reduction. A not uncommon result in suitable structures is formation of a tetrasubstituted double bond which resists further reduction. Thus, the tetracyclic triterpene derivative methyl dihydromasticadienol acetate, with platinum oxide and deuterium in deuteroacetic acid gave the isomer (4); the location of the original double bond was deduced from the position of the deuterium atom in this product (7.8).

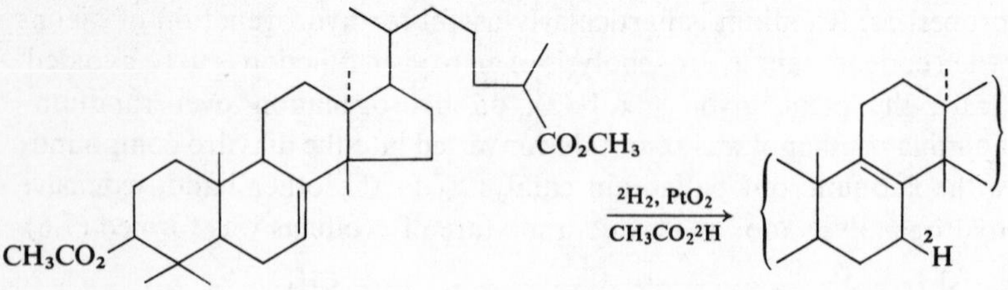

$$\text{(CH}_3\text{CO}_2\text{, CO}_2\text{CH}_3) \xrightarrow[CH_3CO_2{}^2H]{{}^2H_2,\ PtO_2} \text{(4)}\ ({}^2\text{H}) \qquad (7.8)$$

A further indication that catalytic hydrogenation of an olefin need not occur by straightforward addition of two hydrogen atoms at the site of the original double bond is found in the fact that catalytic deuteration generally results in the formation of products containing more and fewer than two atoms of deuterium per molecule. Deuteration of 1-hexene, for example, over platinum oxide was found by Smith and Burwell (1962) to afford a mixture of products containing from one to six deuterium atoms per molecule. Heterogeneous catalytic deuteration of olefins cannot therefore be safely used to prepare deuterated compounds for mechanistic or biosynthetic studies.

Selective reduction of olefinic double bonds in compounds containing other unsaturated groups can usually be accomplished, except in the presence of triple bonds, aromatic nitro groups and acyl halides. Palladium is usually the best catalyst. Thus, 2-benzylidenecyclopentanone is readily converted into 2-benzylcyclopentanone with hydrogen and palladium in methanol; with a platinum catalyst, benzylcyclopentanol is formed. Unsaturated nitriles and aliphatic nitro compounds are also reduced selectively at the double bond with a palladium catalyst.

With palladium and platinum catalysts hydrogenation of allylic or vinylic alcohols, ethers and esters, is often accompanied by hydrogenolysis of the oxygen function, and this reaction is facilitated by acids. Hydrogenolysis can often be avoided with rhodium or ruthenium catalysts.

Hydrogenation of acetylenes. Catalytic hydrogenation of acetylenes takes place in a stepwise manner, and both the olefin and the paraffin can be isolated. Complete reduction of acetylenes to the saturated compound is easily accomplished over platinum, palladium or Raney nickel. A complication which sometimes arises, particularly with platinum catalysts, is the hydrogenolysis of propargylic hydroxyl groups (7.9).

$$C_6H_{10}(OH)-CH_2-C{\equiv}C-C_6H_{10}(OH) \xrightarrow{H_2,\ Pt} C_6H_{10}(OH)-CH_2CH_2CH_2-C_6H_{11} \qquad (7.9)$$

More useful from a synthetic point of view is the partial hydrogenation of acetylenes to *cis* olefins. This reaction can be effected in high yield with a palladium–calcium carbonate catalyst which has been partially deactivated by addition of lead acetate (Lindlar's catalyst) or quinoline. It is aided by the fact that the more electrophilic acetylenic compounds are adsorbed on the electron-rich catalyst surface more strongly than the corresponding olefins. An important feature of these reductions is their high stereoselectivity. In most cases the product consists very largely of the thermodynamically less stable *cis* olefin and partial catalytic hydrogenation of acetylenes provides one of the most convenient routes to *cis* disubstituted olefins. Thus stearolic acid on reduction over Lindlar's catalyst in ethyl acetate solution affords a product containing 95 per cent of the *cis* olefin oleic acid (7.10). Partial reduction of acetylenes

$$CH_3(CH_2)_7C{\equiv}C(CH_2)_7CO_2H \xrightarrow{H_2,\ \text{Lindlar catalyst}} \underset{H}{\overset{CH_3(CH_2)_7}{}}\!\!>C{=}C<\!\!\underset{H}{\overset{(CH_2)_7CO_2H}{}} \qquad (7.10)$$

with Lindlar's catalyst has been invaluable in the synthesis of carotenoids and many other natural products with *cis* disubstituted double bonds.

Hydrogenation of aromatic compounds. Reduction of aromatic rings by catalytic hydrogenation is more difficult than that of most other functional groups, and selective reduction is not easy. Generally, at least one ring is completely reduced, for the olefinic bonds in a partially reduced product would be more easily reduced than the aromatic compound itself. The commonest catalysts are platinum and rhodium, which can be used at ordinary temperatures, and Raney nickel or ruthenium which require high temperatures and pressures.

Benzene itself can be reduced to cyclohexane with platinum oxide in acetic acid solution. Derivatives of benzene such as benzoic acid, phenol or aniline, are reduced more easily. For large-scale work the most convenient method is hydrogenation over Raney nickel at 150–200°C and 100–200 atm. A typical case is the reduction of β-naphthol in alcohol to the tetrahydro compounds (5) and (6) (7.11). Under alkaline

[β-naphthol, OH] $\xrightarrow[\text{90°C, 275 atm}]{H_2,\ \text{Raney Ni},\ C_2H_5OH}$ [OH] (86%) (5) + [OH] (7%) (6) (7.11)

conditions the alcohol (6) becomes the major product of the reaction. Hydrogenation of phenols, followed by oxidation of the resulting cyclohexanols is a convenient method for the preparation of substituted cyclohexanones. Good yields of cyclohexanones may also be obtained directly by selective hydrogenation of phenols. Thus 2,4-dimethylcyclohexanone is obtained by hydrogenation of 2,4-dimethylphenol over a palladium–carbon catalyst (7.12).

$$\text{2,4-dimethylphenol (OH, } CH_3\text{, } CH_3\text{)} \xrightarrow{H_2,\ Pd\text{–}C} \text{2,4-dimethylcyclohexanone (O, } CH_3\text{, } CH_3\text{)} \qquad (7.12)$$

Reduction of benzene derivatives carrying oxygen or nitrogen functions in benzylic positions is complicated by the easy hydrogenolysis of such groups, particularly over palladium catalysts. Preferential reduction of the benzene ring in these compounds is best achieved with ruthenium, or preferably with rhodium, catalysts which can be used under mild conditions. Thus mandelic acid is readily converted into hexahydromandelic acid over rhodium–alumina in methanol solution, whereas, with palladium, hydrogenolysis to phenylacetic acid is the main reaction (7.13).

$$C_6H_5CHOHCO_2H \xrightarrow{H_2,\ Rh\text{–}Al_2O_3} C_6H_{11}CHOHCO_2H$$

$$C_6H_5CHOHCO_2H \xrightarrow{H_2,\ Pd\text{–}C} C_6H_5CH_2CO_2H \qquad (7.13)$$

With polycyclic aromatic compounds it is often possible, by varying the conditions, to obtain either partially or completely reduced products. Thus, naphthalene can be converted into the tetrahydro or decahydro-compound over Raney nickel depending on the temperature. With anthracene and phenanthrene the 9,10-dihydro compounds are obtained by hydrogenation over copper chromite, although, in general, aromatic rings are not reduced with this catalyst. To obtain more fully hydrogenated compounds more powerful catalysts must be used.

Hydrogenation of aldehydes and ketones. Hydrogenation of the carbonyl group of aldehydes and ketones is easier than that of aromatic rings but not so easy as that of most olefinic double bonds.

For aliphatic aldehydes and ketones reduction to the alcohol is usually effected under mild conditions over platinum or the more active forms of Raney nickel. Ruthenium is also an excellent catalyst for reduction of aliphatic aldehydes and can be used to advantage with aqueous solutions (7.14). Palladium is not very active for hydrogenation of aliphatic carbonyl compounds.

$$\underset{\text{OH}}{\overset{\text{CH}_3}{\text{CH}_3\text{C}}}\!-\!\underset{\text{O}}{\overset{\|}{\text{C}}}\!-\!\text{CH}_3 \xrightarrow[\text{C}_2\text{H}_5\text{OH, 20°C}]{\text{H}_2\text{, PtO}_2} \underset{\text{OH}}{\overset{\text{CH}_3}{\text{CH}_3\text{C}}}\!-\!\text{CHOHCH}_3 \quad (88\%)$$

$$\begin{array}{c} \text{CHO} \\ \text{H}-\!\!-\text{OH} \\ \text{HO}-\!\!-\text{H} \\ \text{H}-\!\!-\text{OH} \\ \text{H}-\!\!-\text{OH} \\ \text{CH}_2\text{OH} \end{array} \xrightarrow[\text{125°C, 70 atm}]{\text{H}_2\text{, Ru–C, water}} \begin{array}{c} \text{CH}_2\text{OH} \\ \text{H}-\!\!-\text{OH} \\ \text{HO}-\!\!-\text{H} \\ \text{H}-\!\!-\text{OH} \\ \text{H}-\!\!-\text{OH} \\ \text{CH}_2\text{OH} \end{array} \quad (7.14)$$

Selective hydrogenation of a carbonyl group in presence of olefinic double bonds is difficult and in most cases is best effected with hydride reducing agents or by the Meerwein–Pondorff method (see p. 462). If the double bonds are highly substituted, selective reduction of the carbonyl group can sometimes be achieved catalytically.

Reduced osmium supported on alumina or activated charcoal is said to be an excellent catalyst for reduction of $\alpha\beta$-unsaturated aldehydes to unsaturated alcohols. With this catalyst cinnamyl aldehyde gave cinnamyl alcohol in 95 per cent yield (Rylander and Steele, 1969) (7.15). The catalyst is apparently not effective for the selective reduction

$$\text{C}_6\text{H}_5\text{CH}\!=\!\text{CHCHO} \xrightarrow[\text{100°C, 30 atm}]{\text{H}_2\text{, Os–C}} \text{C}_6\text{H}_5\text{CH}\!=\!\text{CHCH}_2\text{OH} \quad (7.15)$$

of unsaturated ketones, however, for mesityl oxide was reduced to methyl isobutyl ketone. Selective reduction of a carbonyl group in presence of an aromatic nucleus is often possible with a copper chromite catalyst. Palladium is the most effective catalyst for the reduction of aromatic aldehydes and ketones, and excellent yields of the alcohols can be obtained if the reaction is interrupted after absorption of one mole of hydrogen. Prolonged reaction, particularly at elevated temperatures or in presence of acid, leads to hydrogenolysis with formation of the methyl or methylene compound and hydrogenation over a palladium catalyst in presence of acid is a convenient method for the reduction of aromatic ketones to methylene compounds.

When no other reducible groups are present, or when complete reduction of the carbonyl compound is required, hydrogenation over Raney nickel at high temperature and pressure is the most effective procedure for large-scale work.

Nitriles, oximes and nitro compounds. Functional groups with multiple bonds to nitrogen are also readily reduced by catalytic hydrogenation. Nitriles, oximes, azides and nitro compounds, for example, are all smoothly converted into primary amines. Reduction of nitro compounds takes place very easily and is generally faster than reduction of olefinic bonds or carbonyl groups. Raney nickel or any of the platinum metals can be used as catalyst, and the choice will be governed by the nature of other functional groups in the molecule. Thus β-phenylethylamines, useful for the synthesis of isoquinolines, are conveniently obtained by catalytic reduction of $\alpha\beta$-unsaturated nitro compounds (7.16).

$$C_6H_5CH{=}CHNO_2 \xrightarrow[C_2H_5OH,\ 25^\circ C,\ H_2SO_4]{H_2,\ Pd\text{-}C} C_6H_5CH_2CH_2NH_2 \qquad (7.16)$$

Nitriles are conveniently reduced with platinum or palladium at room temperature, or with Raney nickel under pressure. Unless precautions are taken, however, large amounts of secondary amines may be formed in a side reaction of the amine with the intermediate imine (7.17). With

$$RC{\equiv}N \xrightarrow{H_2} RCH{=}NH \xrightarrow{H_2} RCH_2NH_2$$

$$RCH_2NH_2 + RCH{=}NH \rightleftharpoons RCH_2NHCHR(NH_2) \underset{}{\overset{-NH_3}{\rightleftharpoons}} RCH_2N{=}CHR$$

$$RCH_2N{=}CHR \xrightarrow{H_2} RCH_2NHCH_2R \qquad (7.17)$$

the platinum metal catalysts this reaction can be suppressed by conducting the hydrogenation in acid solution or in acetic anhydride, which removes the amine from the equilibrium as its salt or as its acetate. For reactions with Raney nickel, where acid cannot be used, secondary amine formation is prevented by addition of ammonia (7.18). Hydrogenation of nitriles containing other functional groups may lead to cyclic compounds. Thus, hydrogenation of γ-cyanoesters over copper chromite in an alcohol as solvent has been used as a route to *N*-alkylpiperidines; the *N*-alkyl group is derived from the alcohol used as

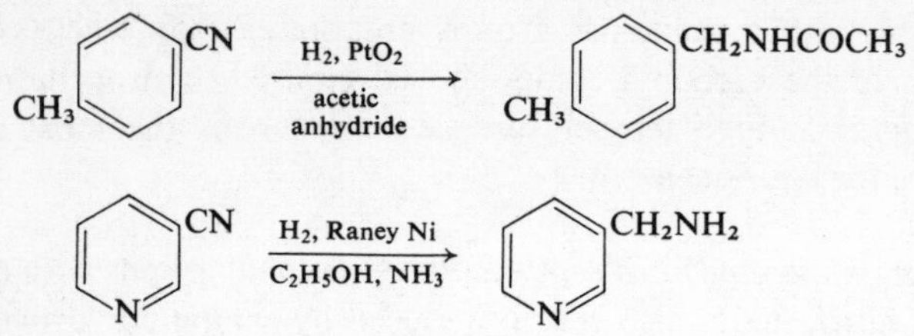

(7.18)

(7.19)

solvent. Suitably oriented aminonitriles may also afford cyclic products on hydrogenation. Thus indolizidine and quinolizidine derivatives have been obtained as shown in (7.19).

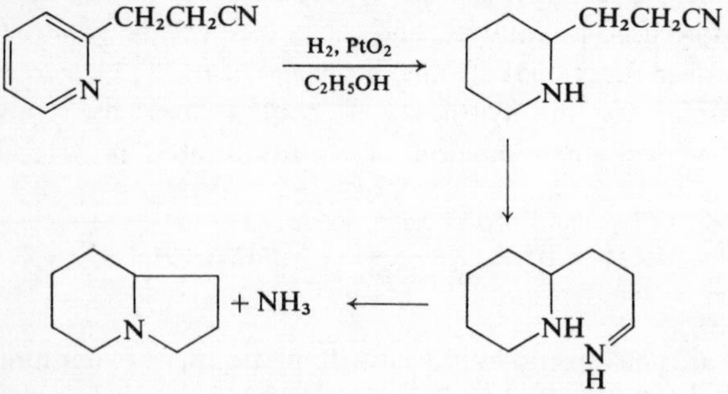

Reduction of oximes to primary amines takes place under conditions similar to those used for nitriles, with palladium or platinum in acid solution, or with Raney nickel under pressure.

Stereochemistry and mechanism. Hydrogenation of an unsaturated compound takes place by adsorption of the compound on to the catalyst surface, followed by transfer of hydrogen from the catalyst to the side of the molecule which is adsorbed on it. Adsorption on to the catalyst is largely controlled by steric factors, and it is found in general that hydrogenation takes place by *cis* addition of hydrogen atoms to the less hindered side of the unsaturated centre. However, it is not always easy to decide which is the less hindered side and in such cases it may be difficult to predict what the steric course of a hydrogenation will be.

Thus, hydrogenation of the *trans*-stilbene derivative (7) forms the (±)-dihydro compound (8) by *cis* addition of hydrogen, while the *cis*-stilbene (9) gives the *meso* isomer (10) (7.20). Hydrogenation of the pinene derivative (11) and of the trimethylcyclohexanone (12) gave products formed by *cis* addition of hydrogen to the more accessible side of the double bonds (7.21). Similarly in the hydrogenation of the diter-

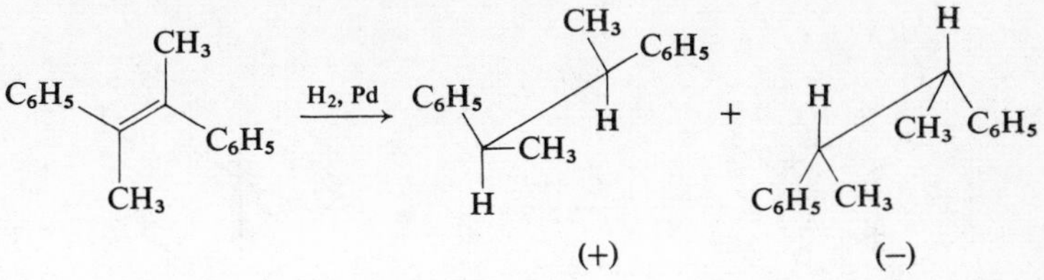

(7) $\xrightarrow{H_2,\ Pd}$ (8) (+) + (−) (7.20)

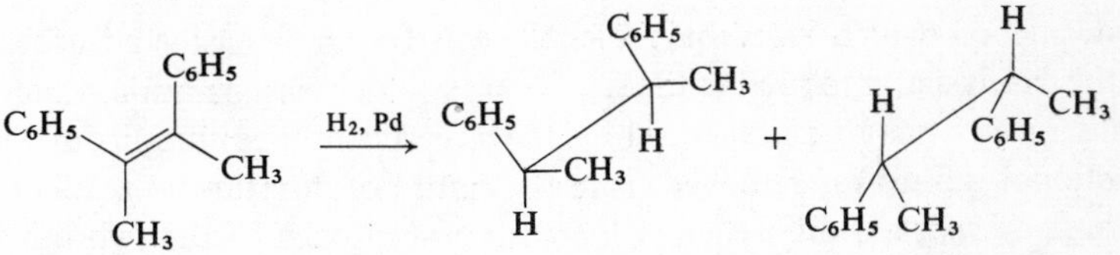

(9) $\xrightarrow{H_2,\ Pd}$ + *meso* (10)

pene derivative (13) addition of hydrogen takes place on the α-face of the molecule which is not hindered by the two axial methyl groups at C-4 and C-10 (7.22). In all these examples the molecule possesses a certain degree of rigidity and it is clear which is the less hindered face of the double bond. With more flexible molecules it may be more difficult

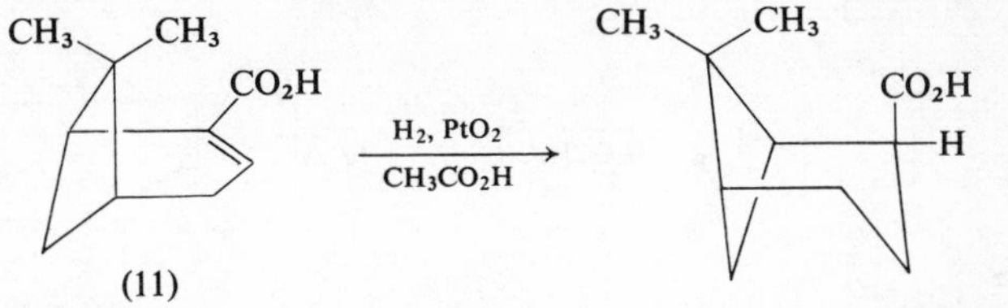

(11) $\xrightarrow[CH_3CO_2H]{H_2,\ PtO_2}$ (90%)

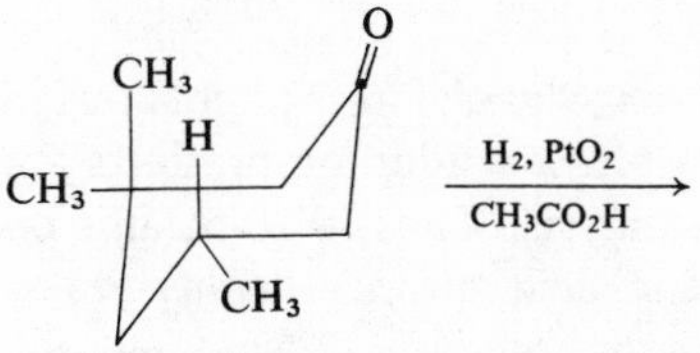

(12) $\xrightarrow[CH_3CO_2H]{H_2,\ PtO_2}$ (83%) + (17%) (7.21)

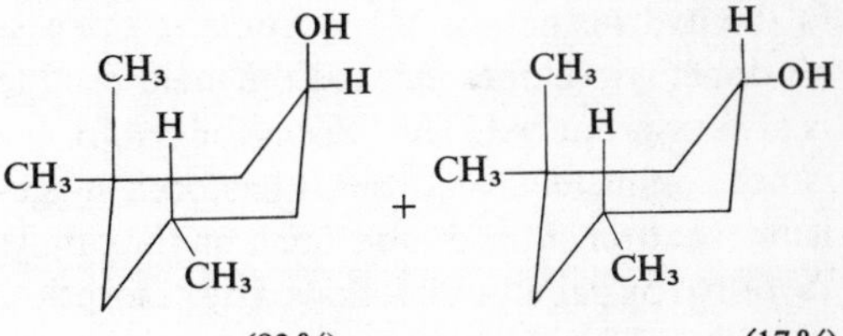

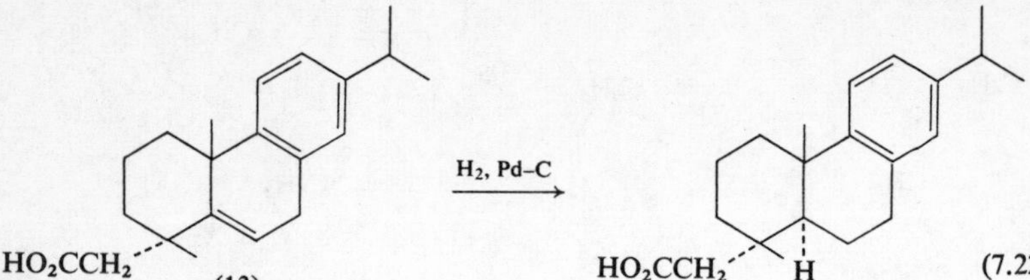

HO_2CCH_2 (13) $\xrightarrow{H_2,\ Pd\text{–}C}$ HO_2CCH_2 H (7.22)

to decide on which side the molecule will be more easily adsorbed on the catalyst and to predict the steric course of a hydrogenation which, in such cases, may be influenced by the experimental conditions or by functional groups far removed from the centre of unsaturation. Where there is no marked difference in the ease of approach to the two faces of the unsaturated centre mixtures of isomers are generally formed on hydrogenation. In some cases the affinity of a particular substituent group for the catalyst surface may induce addition of hydrogen from its own side of the molecule, irrespective of steric effects; the $—CH_2OH$ group is particularly effective in this respect. Thus, in the hydrogenation of the tetrahydrofluorene derivative in (7.23) there was 95 per cent *cis* addition of hydrogen when $R = CH_2OH$ but only 10 per cent *cis* addition when $R = CONH_2$.

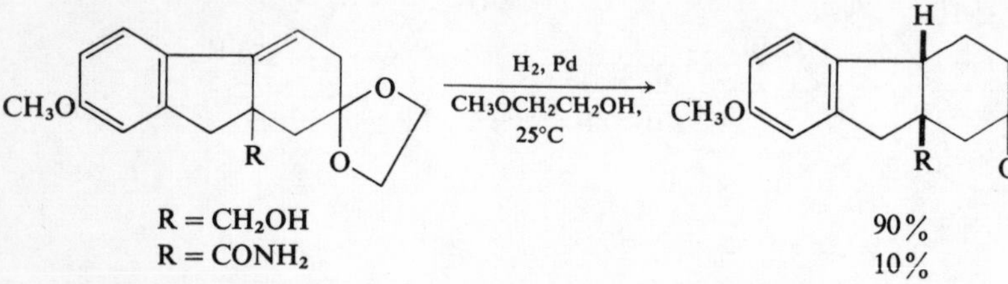

CH_3O R O O $\xrightarrow[CH_3OCH_2CH_2OH,\ 25°C]{H_2,\ Pd}$ CH_3O H R O O

$R = CH_2OH$ 90%
$R = CONH_2$ 10% (7.23)

The hydrogenation of substituted monocyclic olefins is anomalous in many cases in that substantial amounts of *trans* addition-products are formed, particularly with palladium catalysts. Thus, 9,10-octalin on hydrogenation over palladium in acetic acid affords mainly *trans*-decalin, and other examples are shown in the equations (7.24). Similarly, in the hydrogenation of the isomeric xylenes over platinum oxide, the *cis*-dimethylcyclohexanes are the main products, but some *trans* isomer is always produced. The reason for the formation of the *trans* products is not completely clear, but it has been suggested that it may be due to isomerisation of the double bond on the catalyst surface (which would be favoured by palladium) followed by desorption and random readsorption (Siegel and Smith, 1960).

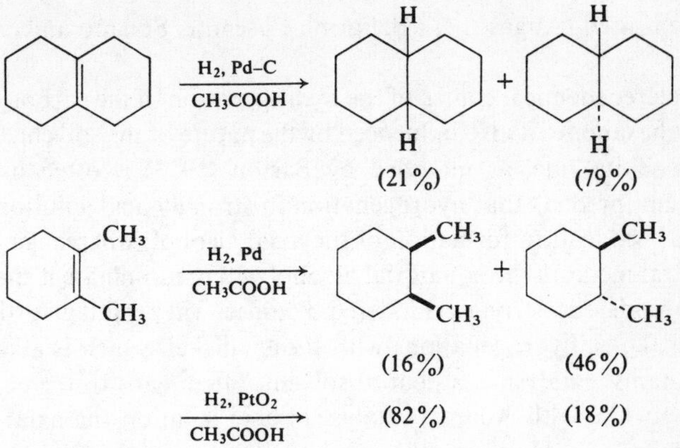

H_2, Pd–C / CH_3COOH → (21%) + (79%)

(7.24)

H_2, Pd / CH_3COOH → (16%) + (46%)

H_2, PtO_2 / CH_3COOH → (82%) (18%)

The rates and steric courses of many hydrogenations are altered by the presence of acid or base in the reaction mixtures. Hydrogenation of 4-cholestene in neutral solution affords coprostane while in acid solution cholestane is obtained. Similarly for 3-oxo-steroids with a double bond at C-4 the course of hydrogenation is appreciably influenced by the solvent (McQuillin, Ord and Simpson, 1963). Alkaline conditions and a hydroxylic solvent strongly favour β-addition of hydrogen leading to *cis* A/B ring fusion, whereas with acid in an aprotic less polar solvent more product of α-addition (*trans* A/B ring fusion) is formed (7.25). For the simpler

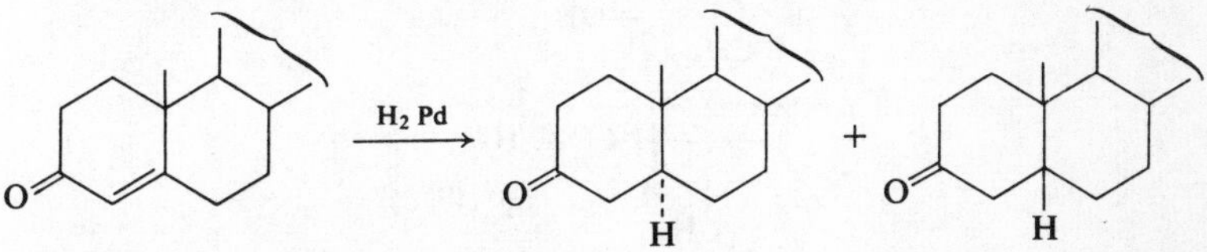

H_2 Pd →

(7.25)

unsaturated ketone in (7.26) the proportion of *cis* and *trans* products obtained on reduction varied with the solvent. In neutral medium the stereochemistry of the product was dependent on the polarity and type

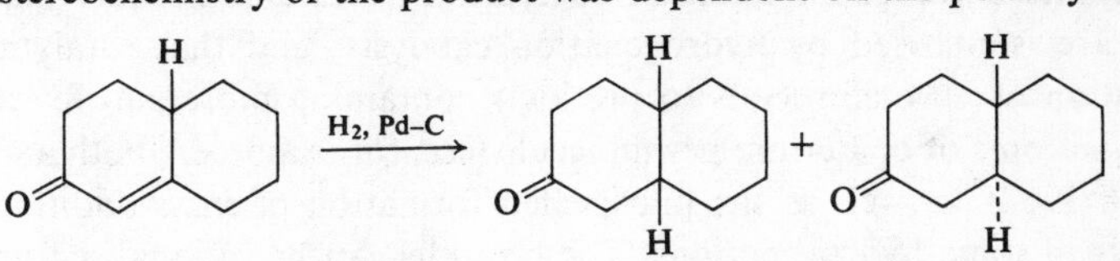

H_2, Pd–C → (7.26)

Solvent	Product composition (%)	
C_2H_5OH	53	47
$C_2H_5OH + 10\%$ HCl	93	7

of solvent used (Augustine, Migliorini, Foscante, Sodano and Sisbarro, 1969).

The stereochemical course of the hydrogenation of the carbonyl group of cyclohexanones is also influenced by the nature of the solvent. The von Auwers–Skita rule, as modified by Barton (1953) is often used as a guide, and predicts that hydrogenation in strongly acid solution, which is rapid, will lead to formation of the axial alcohol, whereas in alkaline or neutral medium the equatorial alcohol will predominate if the ketone is not hindered; strongly hindered ketones will again give the axial alcohol. Thus, hydrogenations with Raney nickel, which is alkaline, or with Adams' catalyst in a neutral solvent, often leads to the equatorial alcohol, while with Adams' catalyst in acid solution the axial alcohol is formed (7.27). However, the rule must be used with caution, and the

H_2, PtO_2, CH_3COOH → (95%)

Raney Ni, H_2, methanol → (95%)

(7.27)

effect of acid and alkali is not always as straightforward as the von Auwers–Skita rule suggests (Wicker, 1956).

A satisfactory mechanism for catalytic hydrogenation must explain not only the observed *cis* addition of hydrogen, but also the fact that olefins are isomerised by hydrogenation catalysts, and that catalytic deuteration of an olefin leads to products containing more and fewer than two atoms of deuterium per molecule (see, for example, Smith and Burwell, 1962), as well as the unexpected formation of *trans* addition products in some hydrogenations. These results can be rationalised on the basis of a mechanism which was first suggested by Horiuti and Polanyi (1934) in which transfer of hydrogen atoms from the catalyst to the adsorbed substrate is supposed to take place in a stepwise manner.

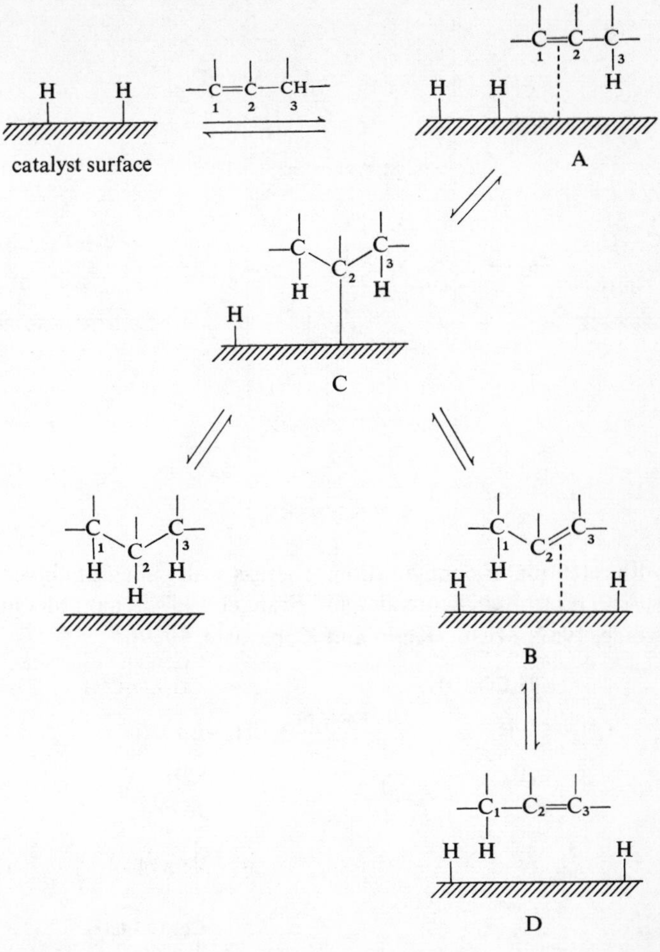

catalyst surface

A

C

B

D

(7.28)

The process is thought to involve equilibria between π-bonded forms A and B and a half hydrogenated form C which can either take up another atom of hydrogen or revert to starting material or to an isomeric olefin D (7.28). Another mechanism involving the π-allyl intermediate E has been postulated (Gault, Rooney and Kemball, 1962) (7.29). π-Allyl intermediates have also been invoked to explain the hydrogenolysis of benzyl alcohols and their derivatives. These reactions generally occur with high stereoselectivity, although the steric course depends to a large extent on the catalyst metal. Thus, hydrogenolysis of optically active methyl 3-hydroxy-3-phenylbutyrate over Raney nickel takes

$$-\underset{H}{C_1}=C_2-\underset{H}{C_3}- \rightleftharpoons \left\{ E \right\} \rightleftharpoons -\underset{H}{C_1}-C_2=C_3- \quad H + H \tag{7.29}$$

E

place with retention of configuration, whereas with palladium inversion occurs (7.30) (Garbisch, Schreader and Frankel, 1967; Khan, McQuillin and Jardine, 1967; Mitsui, Kudo and Kobayashi, 1969).

$$\underset{R(-)}{CH_3-C(CH_2CO_2CH_3)(C_6H_5)-H} \xleftarrow{H_2,\ \text{Raney Ni}} \underset{S(+)}{CH_3-C(CH_2CO_2CH_3)(C_6H_5)-OH} \xrightarrow{H_2,\ Pd} \underset{S(+)}{H-C(CH_2CO_2CH_3)(C_6H_5)-CH_3} \tag{7.30}$$

Homogeneous hydrogenation. Catalysts for heterogeneous hydrogenation of the types discussed above, although useful, have some disadvantages. They may show lack of selectivity when more than one unsaturated centre is present, they may cause double-bond migration and, in reactions with deuterium, they usually bring about allylic interchanges with deuterium. This, in conjunction with double-bond migration, results in unspecific labelling with, in many cases, introduction of more than two deuterium atoms. Again, a number of functional groups suffer easy hydrogenolysis over heterogeneous catalysts and this sometimes leads to complications in the reduction of other unsaturated groups in the molecule. The stereochemistry of reduction, despite a number of rules, is difficult to predict since it depends on chemisorption and not on reactions between molecules. Some of these difficulties have been overcome by the recent introduction of soluble catalysts which allow hydrogenation in homogeneous solution (Harmon, Gupta and Brown, 1973; McQuillin, 1973).

A number of soluble catalyst systems have been used, but the most effective found so far are the rhodium and ruthenium complexes tris(triphenylphosphine)chlororhodium $[(C_6H_5)_3P]_3RhCl$, and hydridochlorotris(triphenylphosphine)ruthenium $[(C_6H_5)_3P]_3RuClH$.

The rhodium complex is easily made by reaction of rhodium chloride with excess of triphenylphosphine in boiling ethanol (7.31). It is an

$$RhCl_3.3H_2O \xrightarrow[\text{boiling } C_2H_5OH]{\text{excess of } (C_6H_5)_3P} [(C_6H_5)_3P]_3RhCl \qquad (7.31)$$

extremely efficient catalyst for the homogeneous hydrogenation of non-conjugated olefins and acetylenes at ordinary temperature and pressure in benzene or similar solvents. Functional groups such as oxo, cyano, nitro, chloro, azo are not reduced under these conditions. Mono- and disubstituted double bonds are reduced much more rapidly than tri- or tetra-substituted ones (Hussey and Takeuchi, 1969), permitting

OH OH

H_2, $[(C_6H_5)_3P]_3RhCl$

C_6H_6

(14)

(7.32)

O O

H_2, $[(C_6H_5)_3P]_3RhCl$

benzene

(15)

the partial hydrogenation of compounds containing different kinds of double bonds (Birch and Walker, 1966; Osborn, Jardine, Young and Wilkinson, 1966; Biellmann, 1968). Thus, in the reduction of linalool (14) addition of hydrogen occurred selectively at the vinyl group, giving the dihydro compound in 90 per cent yield; carvone (15) was similarly converted into carvotanacetone (7.32). The selectivity of the catalyst is shown further by the remarkable reduction of ω-nitrostyrene to phenylnitroethane (7.33).

$$C_6H_5CH{=}CHNO_2 \xrightarrow[C_6H_6]{H_2,\ [(C_6H_5)_3P]_3RhCl} C_6H_5CH_2CH_2NO_2 \qquad (7.33)$$

Hydrogenations take place by *cis* addition to the double bond. This was shown by the catalysed reaction of deuterium with maleic acid to form *meso*-dideuterosuccinic acid, while fumaric acid gave the (±)-compound. Likewise, interrupted reduction of 2-hexyne gave a mixture of hexane and *cis*-2-hexene.

An important practical advantage of this catalyst in addition to its selectivity is that deuterium is introduced without scrambling; that is, only two deuterium atoms are added, at the site of the original double bond. Thus methyl oleate is converted into methyl 9,10-dideuterostearate in quantitative yield, and 1,4-androstadien-3,17-dione (16) gave the dideutero compound (17) by *cis* addition to the α-face of the disubstituted double bond (7.34).

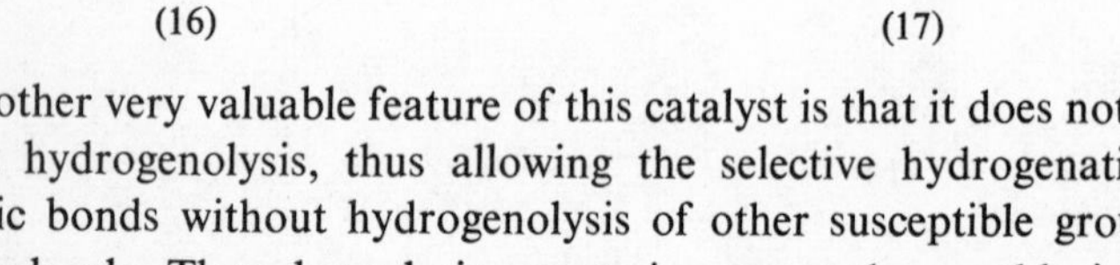

$$\xrightarrow[C_6H_6\text{–}C_2H_5OH]{{}^2H_2,\ ((C_6H_5)_3P)_3RhCl} \qquad (7.34)$$

(16) (17) (85%)

Another very valuable feature of this catalyst is that it does not bring about hydrogenolysis, thus allowing the selective hydrogenation of olefinic bonds without hydrogenolysis of other susceptible groups in the molecule. Thus, benzyl cinnamate is converted smoothly into the dihydro compound without attack on the benzyl ester group (7.35),

$$C_6H_5CH{=}CHCO_2CH_2C_6H_5 \longrightarrow C_6H_5CH_2CH_2CO_2CH_2C_6H_5 \qquad (7.35)$$

and allyl phenyl sulphide is reduced to phenyl propyl sulphide in 93 per cent yield (7.36).

$$CH_2{=}CHCH_2\text{—}S\text{—}C_6H_5 \longrightarrow C_3H_7\text{—}S\text{—}C_6H_5 \qquad (7.36)$$

Because of the strong affinity of $[(C_6H_5)_3P]_3RhCl$ for carbon monoxide it decarbonylates aldehydes, and olefinic compounds containing aldehyde groups cannot be hydrogenated with this catalyst under the usual conditions (Jardine and Wilkinson, 1967). Thus cinnamaldehyde is converted into styrene in 65 per cent yield, and benzoyl chloride gives chlorobenzene in 90 per cent yield. Acceptable yields of saturated aldehydes can be obtained under special experimental conditions. The iridium complex $IrH_3[(C_6H_5)_3P]_3$ in presence of acetic acid can be used for the homogeneous hydrogenation of aldehydes and some olefins at 50°C and atmospheric pressure (Coffey, 1967) (7.37).

$$C_3H_7CHO \xrightarrow[CH_3CO_2H,\ 50°C]{H_2,\ IrH_3[(C_6H_5)_3P]_3} C_3H_7CH_2OH \tag{7.37}$$

$[(C_6H_5)_3P]_3RhCl$ dissociates in solvents (S) to give the solvated species $[(C_6H_5)_3P]_2Rh(S)Cl$ which in presence of hydrogen is in equilibrium with the dihydrido complex $[(C_6H_5)_3P]_2Rh(S)ClH_2$ in which the hydrogen atoms are directly attached to the metal (7.38). Reduction is believed to

$$[(C_6H_5)_3P]_2Rh(S)Cl \underset{}{\overset{H_2}{\rightleftharpoons}} [(C_6H_5)_3P]_2Rh(S)ClH_2 \overset{RCH{=}CHR'}{\rightleftharpoons}$$

$$[(C_6H_5)_3P]_2RhCl(RCH{=}CHR')H_2 \longrightarrow RCH_2CH_2R' + [(C_6H_5)_3P]_2Rh(S)Cl \tag{7.38}$$

take place by displacement of solvent by the olefin followed by step-wise stereospecific *cis*-transfer of the two hydrogen atoms from the metal to the loosely co-ordinated olefin by way of an intermediate with a carbon–metal bond (McQuillin, 1974). Diffusion of the saturated substrate away from the transfer site leaves the complex again ready to combine with dissolved hydrogen and continue the reduction.

The ruthenium complex $[(C_6H_5)_3P]_3RuClH$, formed *in situ* from $[(C_6H_5)_3P]_3RuCl_2$ and molecular hydrogen in benzene in presence of a base such as triethylamine, is an even more efficient catalyst which is specific for the hydrogenation of monosubstituted alkenes $RCH{=}CH_2$. Rates of reduction for other types of alkenes are slower by a factor of at least 2×10^3 (Hallman, McGarvey and Wilkinson, 1968). Thus 1-heptene was rapidly converted into heptane with molecular hydrogen in benzene solution but 3-heptene was unaffected. Some isomerisation of alkenes is observed with this catalyst but the rate is slow compared with the rate of hydrogenation. 1-Octene, for example, gave 8 per cent of *cis*- and *trans*-2-octenes after 20 hours. Very similar behaviour is shown by the rhodium complex hydridocarbonyl-tris-(triphenylphosphine)rhodium(I) $[(C_6H_5)_3P]_3RhH(CO)$. Terminal acetylenes apparently react with the catalyst, but disubstituted acetylenes are converted into *cis* alkenes. Thus stearolic acid gave oleic acid and diphenylacetylene formed *cis*-stilbene (Jardine and McQuillin, 1966) (7.39).

$$CH_3(CH_2)_7C{\equiv}C(CH_2)_7CO_2H \xrightarrow[C_6H_6]{H_2,\ [(C_6H_5)_3P]_3RuCl_2} \underset{H\qquad\qquad H}{CH_3(CH_2)_7CH{=}CH(CH_2)_7CO_2H} \tag{7.39}$$

Hydrogenations with $[(C_6H_5)_3P]_3RuClH$ and $[(C_6H_5)_3P]_3RhH(CO)$ are two-step processes which proceed by the reversible formation of a metal–alkyl intermediate. The actual catalyst in the former case is thought to be the square monomer $[(C_6H_5)_3P]_2RuClH$ with *trans* $P(C_6H_5)_3$ groups, formed by dissociation in solution (7.40). The high selectivity for reduc-

$$[(C_6H_5)_3P]_2Ru\begin{smallmatrix}H\\Cl\end{smallmatrix} \xrightarrow{RCH{=}CH_2} \left[[(C_6H_5)_3P]_2Ru\begin{smallmatrix}H\\Cl\end{smallmatrix} \longleftarrow \begin{matrix}R\\|\\CH\\\|\\CH_2\end{matrix} \right] \tag{7.40}$$

$$[(C_6H_5)_3P]_2\underset{Cl}{Ru}{-}\underset{R}{C}HCH_3 \xrightarrow{H_2} RCH_2CH_3 + [(C_6H_5)_3P]_2Ru\begin{smallmatrix}H\\Cl\end{smallmatrix}$$

tion of terminal double bonds is attributed to steric hindrance by the bulky $P(C_6H_5)_3$ groups to the formation of the metal–alkyl intermediate with other types of olefins.

Induced asymmetry via homogeneous hydrogenation. An interesting development has been the use of soluble metal catalysts to effect asymmetric hydrogenations (Bogdanović, 1973). The catalysts used include a number of rhodium complexes containing as ligands chiral phosphines or amides. Trichlorotripyridylrhodium, treated with sodium borohydride in dimethylformamide, gives a complex $[py_2.dmf.RhCl_2(BH_4)]$ which is highly active for the hydrogenation of alkenes. By using asymmetric ligands related to dimethylformamide catalysts are obtained which can bring about asymmetric hydrogenations. Thus, methyl 3-phenyl-2-butenoate is hydrogenated at a rhodium complex formed in (+)- or (−)-1-phenylethylformamide to give (+)- or (−)-methyl 3-phenylbutanoate in better than 50 per cent optical yield (7.41) (Abley and McQuillin, 1971). Very high optical yields of α-amino acids have also been obtained

$$C_6H_5CCH_3{=}CHCO_2CH_3 \longrightarrow C_6H_5CHCH_3CH_2CO_2CH_3 \quad (7.41)$$

by hydrogenation of α-acylaminoacrylic acids using rhodium(I) complexes containing chiral phosphines (e.g. 18), or the cyclo-octadiene complex [Rh(1,5-cyclo-octadiene)Cl]$_2$. α-Acetylaminocinnamic acid, for example, gave *N*-acetylphenylalanine in 95 per cent yield and 85 per cent optical purity and, using the laevorotatory form of the ligand (18) atropic acid was converted into (*S*)-hydratropic acid with 64 per cent optical purity (7.42) (Kagan and Dang, 1972; Knowles, Sabacky

(18): 2,2-dimethyl-1,3-dioxolane with H and $CH_2P(C_6H_5)_2$ at C-4 and C-5

$$C_6H_5CH{=}C(NHCOCH_3)CO_2H \xrightarrow[H_2,\ C_6H_6{-}C_2H_5OH,\ 25°C]{RhCl[L^*]S} C_6H_5CH_2CH(NHCOCH_3)CO_2H$$

(85 % optical purity)

$$CH_2{=}C(C_6H_5)CO_2H \xrightarrow[H_2,\ C_6H_6,\ 25°C]{RhCl[L^*]S} CH_3CH(C_6H_5)CO_2H \quad (7.42)$$

(64 % optical purity)

and Vineyard, 1972; Knowles, Sabacky, Vineyard and Weinkauff, 1975). Remarkably, completely stereospecific hydrogenation of a number of αβ-unsaturated acids has been achieved using micro-organisms as catalysts. With the two allenic carboxylic acids (19) and (20) for example, hydrogenation in phosphate buffer with *Clostridium kluveri* gave (21)

$$C_6H_5(H)C{=}C{=}C(C_2H_5)CO_2H\ (19) \longrightarrow C_6H_5(H)C{=}C(H){-}CH(C_2H_5)CO_2H\ (21)$$

$$H(C_6H_5)C{=}C{=}C(C_2H_5)CO_2H\ (20) \longrightarrow H(C_6H_5)C{=}C(H){-}CH(C_2H_5)CO_2H\ (22) \quad (7.43)$$

and (22) in quantitative yield by *trans* addition of hydrogen (Rambeck and Simon, 1974). This procedure provides a convenient method for fashioning chiral methyl groups, as illustrated for the conversion of *cis*-dideuterioacrylic acid into the (3*S*)-[3-^{2}H,3-^{3}H]propionic acid shown in (7.44).

$$(H)(^2H)C{=}C(CO_2H)(^2H) \xrightarrow[C.\ kluyveri]{^3H_2,\ H^3HO} (H)(^3H)(^2H)C{-}C^2H^3HCO_2H \quad (7.44)$$

7.2. Reduction by dissolving metals

General. Chemical methods of reduction are of two main types: those which take place by addition of electrons to the unsaturated compound followed or accompanied by transfer of protons; and those which take place by addition of hydride ion followed in a separate step by protonation (Augustine, 1968).

Reductions which follow the first path are generally effected by a metal, the source of the electrons, and a proton donor which may be water, an alcohol or an acid. They can result either in the addition of hydrogen atoms to a multiple bond or in fission of a single bond between atoms, usually, in practice, a single bond between carbon and a hetero-atom. In these reactions an electron is transferred from the metal surface (or from the metal in solution) to the organic molecule being reduced, giving, in the case of addition to a multiple bond, an anion radical, which in many cases is immediately protonated. The resulting radical

$$A{-}B \xrightarrow{2e} A^- + B^-$$
$$A{-}B \xrightarrow{e} \bar{A}{-}B^\bullet \xrightarrow{e} A^- + B^- \quad (\text{or } A^\bullet + B^-)$$

$$A{=}B \xrightarrow{e} \left\{ \bar{A}{-}B^\bullet \longleftrightarrow A^\bullet{-}B^- \right\} \xrightarrow{e} \bar{A}{-}B^-$$

$$\left\{ \bar{A}{-}B^\bullet \longleftrightarrow A^\bullet{-}B^- \right\} \longrightarrow \begin{matrix} A{-}B^- \\ | \\ A{-}B^- \end{matrix}$$

$$\left\{ \bar{A}{-}B^\bullet \longleftrightarrow A^\bullet{-}B^- \right\} \xrightarrow{H^+} AH{-}B^\bullet \text{ or } A^\bullet{-}BH \xrightarrow{e} AH{-}B^- \text{ or } \bar{A}{-}BH \quad (7.45)$$

subsequently takes up another electron from the metal to form an anion which may be protonated immediately or remain as the anion until work-up. In the absence of a proton source dimerisation or polymerisation of the anion-radical may take place. In some cases a second electron may be added to the anion-radical to form a dianion, or two anions in the case of fission reactions (Birch, 1950; House, 1965). These transformations may be represented as shown in (7.45).

Thus, in the reduction of benzophenone with sodium in ether or liquid ammonia, the first product is the resonance-stabilised anion radical (23) which, in absence of a proton donor, dimerises to the pinacol. In presence of a proton source, however, protonation leads to the radical (24) which is subsequently converted into the anion and thence into benzhydrol (7.46). The presence in these anion radicals of an unpaired electron which interacts with the atoms in the conjugated system has been established by measurements of the electron spin resonance spectra of various anion radical solutions.

$$C_6H_5C(=O)C_6H_5 \underset{}{\overset{Na,\ liq.\ NH_3}{\rightleftharpoons}} C_6H_5\dot{C}(\text{:}\ddot{O}\text{:}^-)C_6H_5 \longrightarrow (C_6H_5)_2C(O^-)\text{—}C(O^-)(C_6H_5)_2 \xrightarrow{H^+} (C_6H_5)_2C(OH)\text{—}C(OH)(C_6H_5)_2$$

$$C_6H_5\dot{C}(\text{:}\ddot{O}\text{:}^-)C_6H_5 \longleftrightarrow C_6H_5.\ddot{C}^-.C_6H_5(\text{:}\ddot{O}\cdot)$$

(23)

$$(23) \xrightarrow{C_2H_5OH} C_6H_5CH(\text{:}\ddot{O}\cdot)C_6H_5 \; (24) \xrightarrow{Na} C_6H_5CH(\text{:}\ddot{O}\text{:}^-)C_6H_5 \xrightarrow{H^+} C_6H_5CHOHC_6H_5 \qquad (7.46)$$

The metals commonly employed in these reductions include the alkali metals, calcium, zinc, magnesium, tin and iron. The alkali metals are often used in solution in liquid ammonia (Birch reduction, see p. 443) or as suspensions in inert solvents such as ether or toluene, frequently with addition of an alcohol or water to act as a proton source. Many reductions are also effected by direct addition of sodium or, particularly, zinc, tin or iron, to a solution of the compound being reduced in a hydroxylic solvent, such as ethanol, acetic acid or an aqueous mineral acid.

Reduction with metal and acid. In the well-known Clemmensen reduction of the carbonyl group of aldehydes and ketones to methyl or methylene, a mixture of the carbonyl compound and amalgamated zinc is boiled with hydrochloric acid, sometimes in presence of a non-miscible solvent which serves to keep the concentration in the aqueous phase low and thus prevent bimolecular condensations at the metal surface (Martin, 1942; Asinger and Vogel, 1970). Amalgamation of the zinc raises its hydrogen overvoltage to the point where it survives as a reducing agent in the acid solution and is not consumed in reaction with the acid to give molecular hydrogen. The choice of acid is confined to the hydrogen halides, which appear to be the only strong acids whose anions are not reduced by zinc-amalgam. Thus, stearophenone is converted into octadecylbenzene in 88 per cent yield, and 4-bromoindanone gives 4-bromoindane in 70 per cent yield without removal of the bromine atom. Isolated olefinic double bonds are not reduced under Clemmensen conditions, but conjugated bonds may be. For example benzalacetone is converted into butylbenzene in 50 per cent yield. In many cases reduction of $\alpha\beta$-unsaturated ketones affords mixtures containing rearranged products formed by way of cyclopropanoid intermediates. With 1,3- and 1,4-dicarbonyl compounds, and with compounds in which there is a substituent on the α-carbon atom, reduction is often accompanied by bond cleavage, rearrangement or pinacol formation as in the examples (7.47) (Buchanan and Woodgate, 1969).

The mechanism of the Clemmensen reaction is uncertain. Alcohols are not normally reduced under Clemmensen conditions and it would appear that they are not intermediates in the reduction of the carbonyl compounds. The following mechanism (7.48) involving transfer of electrons from the metal surface to the carbonyl carbon atom has been proposed by Nakabayashi (1960). A related mechanism involving initial attack of zinc on the oxygen atom of the carbonyl group has been used by

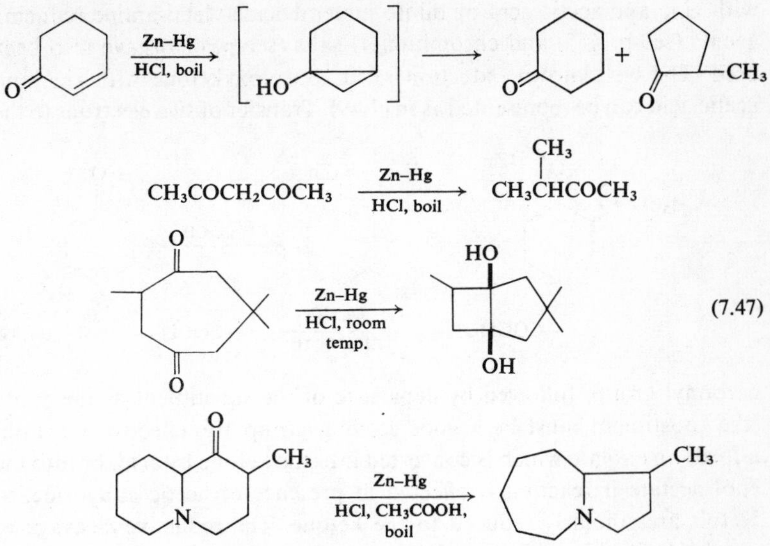

$$\xrightarrow[\text{HCl, boil}]{\text{Zn–Hg}}$$

$$CH_3COCH_2COCH_3 \xrightarrow[\text{HCl, boil}]{\text{Zn–Hg}} CH_3CH(CH_3)COCH_3$$

$$\xrightarrow[\text{HCl, room temp.}]{\text{Zn–Hg}} \qquad (7.47)$$

$$\xrightarrow[\text{HCl, } CH_3COOH\text{, boil}]{\text{Zn–Hg}}$$

$$R{-}C({=}O){-}R' \;(Zn{:}\; Cl^-) \rightleftharpoons R{-}C(O^-)(ZnCl){-}R' \underset{}{\overset{H^+}{\rightleftharpoons}} R{-}C(OH)(ZnCl){-}R' \xrightarrow[-H_2O]{H^+} R{-}\overset{+}{C}(ZnCl){-}R'$$

$$R{-}\overset{+}{C}(ZnCl){-}R' \xrightarrow{2Zn{:}} R{-}\bar{C}(ZnCl){-}R' + 2\dot{Z}n^{+}$$

$$R{-}\bar{C}(ZnCl){-}R' \xrightarrow{H^+} R{-}CH(ZnCl){-}R' \xrightarrow{H^+} R{-}CH_2{-}R' + \overset{+}{Zn}Cl \qquad (7.48)$$

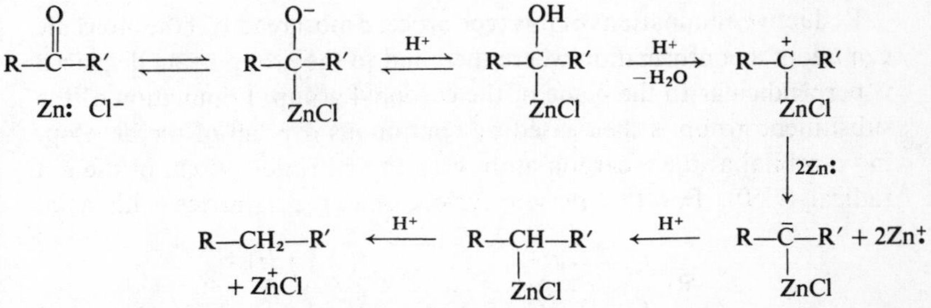

Buchanan and Woodgate (1969) to rationalise the rearrangements observed during Clemmensen reduction of difunctional ketones (see also Vedejs, 1975).

The Clemmensen reaction as ordinarily effected employs rather vigorous conditions and may not be suitable for the reduction of polyfunctional molecules. An alternative procedure which is often effective and which proceeds under mild conditions uses zinc dust and solutions of hydrogen chloride gas in aprotic organic solvents such as ether (compare Vedejs, 1975). Other methods for converting carbonyl groups into methyl or methylene are described on pp. 488–490.

Reductive cleavage of α-substituted ketones, such as α-halo, α-acyloxy- and α-hydroxy-ketones to the unsubstituted ketone is commonly effected with zinc and acetic acid or dilute mineral acid. Metal-amine reducing agents (see p. 455) and chromium(II) salts (see p. 460) have also been used. The well-known reduction of an α-acetoxyketone with zinc and acetic acid can be represented as in (7.49). Transfer of two electrons to the

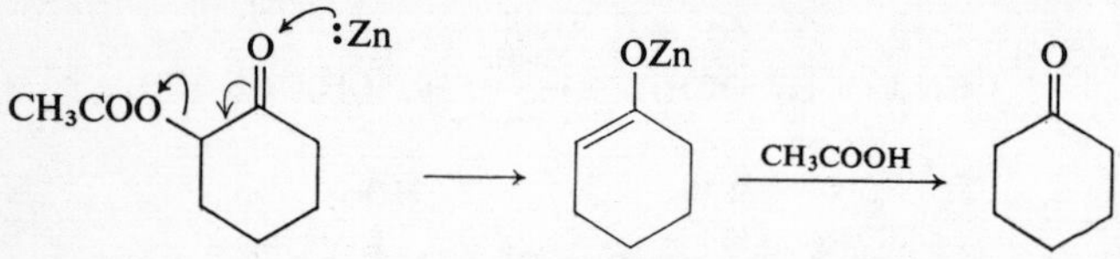

$$\text{—COCHOH—} \xrightarrow[\text{CH}_3\text{COOH}]{\text{Zn, HCl}} \text{—COCH}_2\text{—} \qquad (7.49)$$

carbonyl group, followed by departure of the substituent as the anion (the substituent must be a good leaving group for effective reaction) affords an enolate which is converted into the ketone by acid, or into the enol acetate if reaction is effected in presence of acetic anhydride. α-Ketols are similarly reduced to the ketone. The reductive cleavage of α-halo ketones may have a different mechanism.

Reductive eliminations of this type proceed most readily if the molecule can adopt a conformation where the bond to the group being displaced is perpendicular to the plane of the carbonyl group. Elimination of the substituent group is then eased by continuous overlap of the developing *p*-orbital at the α-carbon atom with the π-orbital system of the ion radical (7.50). For this reason cyclohexanone derivatives with axial

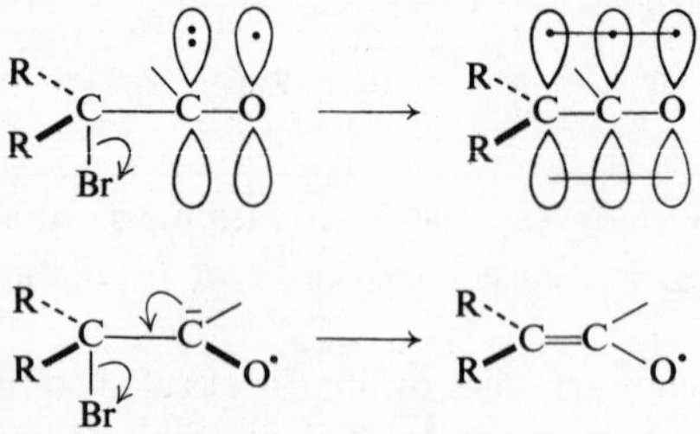

(7.50)

α-substituents are reductively cleaved more readily than their equatorial isomers. For example, the (axial) hydroxyl group of the 20-keto-20*a*-α-hydroxy-20*a*-β-methyl-D-homo-steroid derivative (25) was cleaved in high yield with zinc and acetic acid, whereas the 20*a*-β-hydroxy epimer was unaffected (7.51). Vinylogous α-substituted ketones are also reductively cleaved. Thus, the (axial) acetoxy compound (26) was smoothly converted into cholestenone and the epoxide (27) gave the

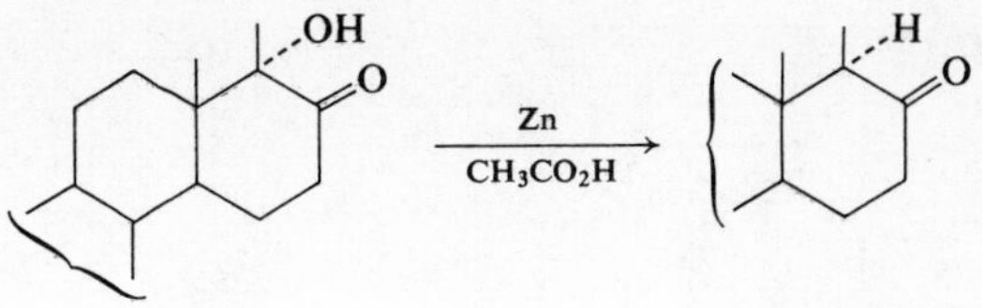

$$\xrightarrow[CH_3CO_2H]{Zn} \quad (7.51)$$

(25)

derivative (28). Again the stereochemistry of the substituent determines the ease of fission (7.52).

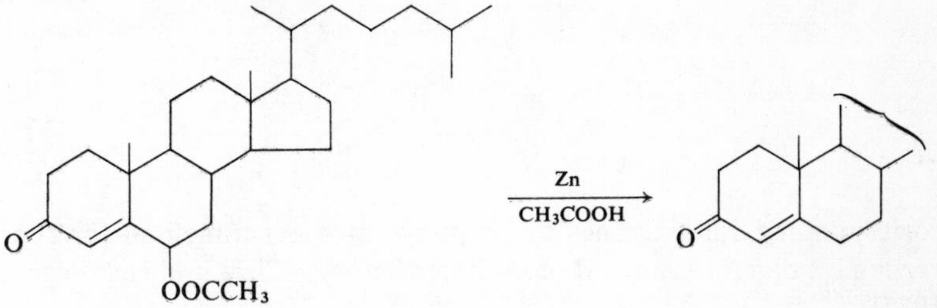

$$\xrightarrow[CH_3COOH]{Zn}$$

OOCCH$_3$

(26)

(7.52)

$$\xrightarrow[CH_3COOH]{Zn}$$

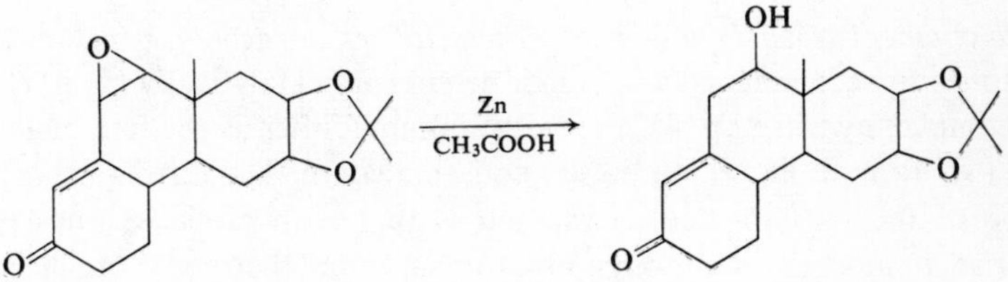

(27) (28)

Even carbon–carbon bonds may be broken in favourable circumstances. Thus, the cyclopropyl ketone (29) was converted into the bicyclic compound (30) with zinc–copper couple in acetic acid (7.53).

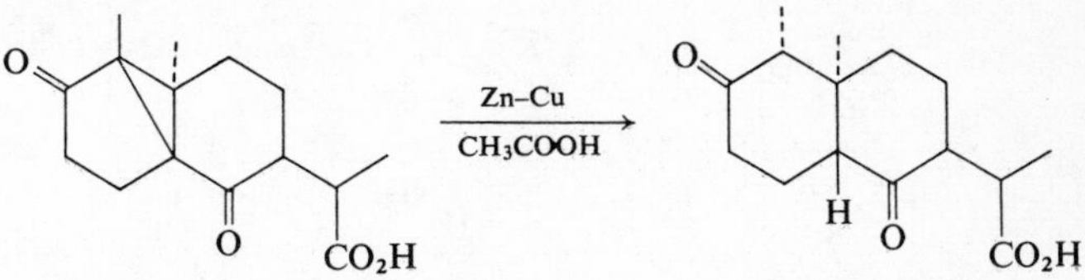

$$\xrightarrow[CH_3COOH]{Zn\text{–}Cu} \quad (7.53)$$

(29) (30)

Although alkyl halides are not normally cleaved by zinc and acid, benzylic halides are. For example, 1-chloromethylnaphthalene is readily converted into 1-methylnaphthalene. The more active of two halogen atoms can often be removed selectively, as in (7.54).

BrCH₂ H Br —Zn, HBr / CH_2Cl_2→ BrCH₂ O (7.54)

In a new reaction β-chloro-αβ-unsaturated ketones are reduced selectively to the halogen-free compounds with silver-promoted zinc dust in methanol (7.55). Since the β-chloroenones are readily obtained from the

—Zn(Ag) / CH_3OH, 25°C→ (81%) (7.55)

corresponding 1,3-diketones the sequence provides a method for converting 1,3-diketones into αβ-unsaturated ketones (Clark and Heathcock, 1973).

Reduction of carbonyl compounds with metal and an alcohol. Reduction of ketones to secondary alcohols can be effected catalytically (p. 417), with complex hydrides (p. 467) or with sodium (either as the free metal or as a solution in liquid ammonia) and an alcohol. The distinguishing feature of the sodium–alcohol method is that with cyclic ketones it gives rise, in most cases although not always, to the thermodynamically more stable alcohol either exclusively or in preponderant amount. Table 7.2 shows the proportions of more stable *trans* (equatorial) alcohol formed by reduction of 2-methylcyclohexanone with different reducing agents (7.56). Similarly 4-t-butylcyclohexanone gives the more

CH_3 → CH_3 H OH H + CH_3 H H OH (7.56)

stable *trans*-4-t-butylcyclohexanol almost exclusively on reduction with lithium and propanol in liquid ammonia, and numerous other experiments confirm that in the vast majority of cases the more stable alcohol is the main product.

Two main hypotheses have been put forward to account for the high proportion of the more stable epimer formed in these reductions with

TABLE 7.2 *Proportion of trans-2-methylcyclohexanol by reduction of 2-methylcyclohexanone with different reagents*

Reagent	Proportion *trans* alcohol (%)
Na–alcohol	99
Lithium aluminium hydride	82
Sodium borohydride	69
Aluminium isopropylate	42
Catalyst and hydrogen	7–35

sodium and an alcohol. In one due to Barton and Robinson (1954) it is supposed that a tetrahedral dianion is formed initially and adopts the more stable configuration with equatorial oxygen which, on protonation, affords the more stable arrangement of the alcohol.

A second suggestion is that the reaction proceeds by transfer of one electron from the metal to the carbonyl group, followed by protonation of the resulting anion radical from the less hindered side and further reduction to form the alkoxide. This alkoxide, once formed, can react with unchanged ketone by a mechanism similar to that of the Meerwein–Pondorff–Verley reduction to give an equilibrium mixture which will favour the more stable of the two alkoxides. Final acidification of the reaction mixture thus affords the more stable alcohol as the main product.

However, it appears that neither of these mechanisms is completely satisfactory, and modifications have been suggested by Huffman and Charles (1968) and by Taylor (1969). It may be that in the reduction of simple unhindered cyclohexanones protonation of the anion radical (Huffman and Charles) or the dianion (Taylor) takes place from an axial direction to lead directly to the more stable equatorial alcohol. House (1972) considers that reaction proceeds by donation of one electron from the metal to the carbonyl group followed by protonation on oxygen to give a free-radical intermediate which adopts the more stable configuration with an equatorial hydroxyl group. Further reduction and protonation of the anionic intermediate takes place with retention of configuration at the carbon atom to give the observed equatorial alcohol (7.57).

In the case of a few strained or sterically hindered ketones, reduction with sodium and alcohol has been found to give the less stable product. For example reduction of norcamphor with lithium and ethanol gave

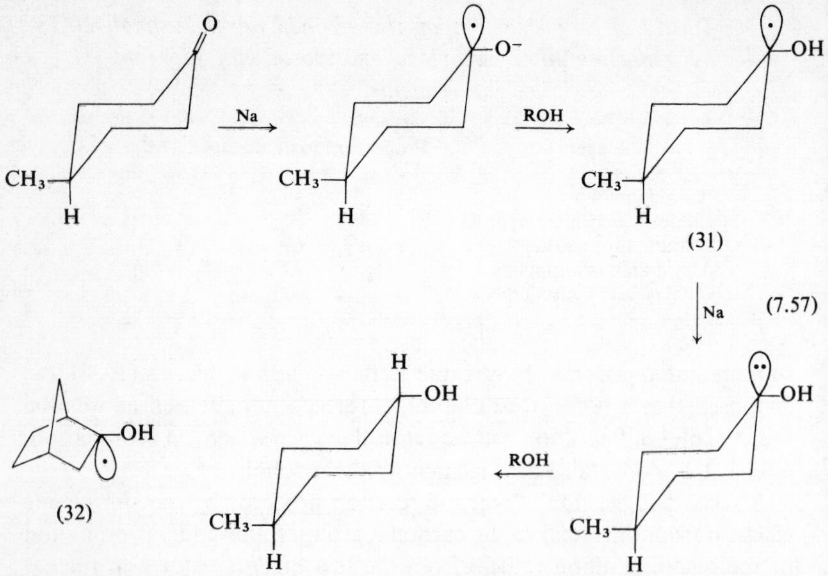

$$\text{(7.57)}$$

(31)

(32)

mainly the less stable *endo* alcohol. This may be because of torsional strain in the intermediate *exo* α-hydroxy radical (32) so that the *endo* isomer is preferred. Aldehydes can be reduced to primary alcohols with sodium and ethanol or aluminium amalgam, but in many cases better yields are obtained by catalytic hydrogenation (p. 417) or with hydride reducing agents (p. 467).

Reduction of ketones with dissolving metals in absence of a proton donor leads to the formation of bimolecular products. The usual reagents are magnesium, magnesium amalgam or aluminium amalgam. Thus, reduction of acetone to pinacol follows the course shown in (7.58).

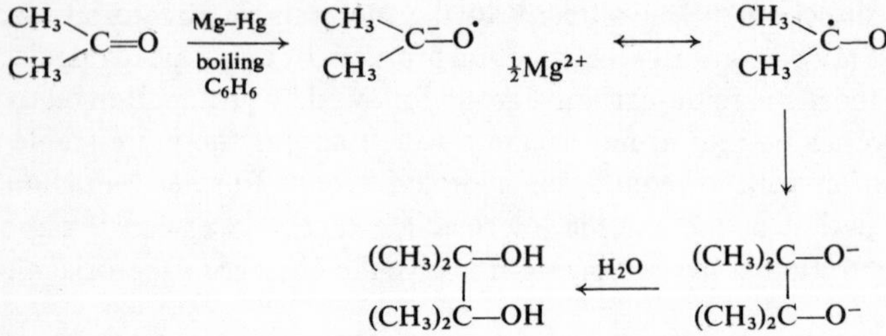

$$(CH_3)_2C{=}O \xrightarrow[\text{boiling } C_6H_6]{\text{Mg–Hg}} (CH_3)_2C{-}\dot{O}\ \ \tfrac{1}{2}Mg^{2+} \longleftrightarrow (CH_3)_2\dot{C}{-}O^-$$

$$\longrightarrow \begin{matrix}(CH_3)_2C{-}O^- \\ | \\ (CH_3)_2C{-}O^-\end{matrix} \xrightarrow{H_2O} \begin{matrix}(CH_3)_2C{-}OH \\ | \\ (CH_3)_2C{-}OH\end{matrix} \qquad (7.58)$$

Bimolecular reduction of this type is often a competing reaction in other reductions with dissolving metals.

The pinacol condensation has not been widely employed in synthesis up to now, but its potentiality has recently been greatly increased by the introduction of a number of new reagents derived from low-valent transition metal species which bring about both inter- and intra-molecular condensations in high yield. One of the most effective of these reagents is the Ti(II) species generated by reaction of titanium tetrachloride with amalgamated magnesium (Corey, Danheiser and Chandrasekaran, 1976). With this reagent cyclohexanone is converted into the corresponding pinacol in 93 per cent yield, and mixed pinacols are also readily obtained (7.59). Intramolecular condensations proceed equally well to

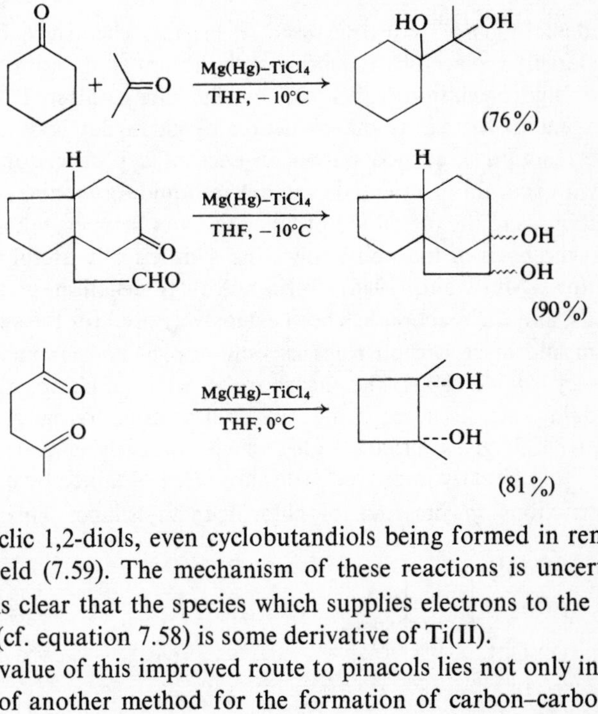

Mg(Hg)–$TiCl_4$, THF, – 10°C (76%)

Mg(Hg)–$TiCl_4$, THF, – 10°C (90%) (7.59)

Mg(Hg)–$TiCl_4$, THF, 0°C (81%)

give cyclic 1,2-diols, even cyclobutandiols being formed in remarkably high yield (7.59). The mechanism of these reactions is uncertain, but it seems clear that the species which supplies electrons to the carbonyl group (cf. equation 7.58) is some derivative of Ti(II).

The value of this improved route to pinacols lies not only in the provision of another method for the formation of carbon–carbon bonds, but also in the fact that the unsymmetrical pinacols which it makes readily available for the first time are themselves useful synthetic intermediates. Thus, treatment with methanesulphonyl chloride in triethylamine affords epoxides, and both the pinacols and the epoxides rearrange readily to carbonyl compounds under mild acidic conditions (7.60).

Esters are also reduced by sodium and alcohols to form primary alcohols. This, the Bouveault–Blanc reaction, is one of the oldest

$$\text{(1-hydroxy-1-methylethyl)cyclohexanol} \xrightarrow{BF_3.(C_2H_5)_2O,\ 0°C} \text{1-acetyl-1-methylcyclohexane}\ (88\%)$$

$$\xrightarrow{70\%\ HClO_4,\ -20°C} \text{2,2-dimethylcycloheptanone}\ (81\%) \qquad (7.60)$$

established methods of reduction used in organic chemistry, but has now been largely replaced by reduction with lithium aluminium hydride or catalytic hydrogenation with a copper chromite catalyst. It follows the same general course as the reduction of aldehydes and ketones. When the reaction is carried out in absence of a proton donor, for example with sodium in xylene or sodium in liquid ammonia, dimerisation, analogous to the formation of pinacols from ketones, takes place, and this is the basis of the well-known and synthetically useful acyloin condensation (McElvain, 1948). Intramolecular reaction gives ring compounds, and the reaction has been extensively used for the synthesis of medium and large carbon rings as well as five- and six-membered rings (Finley, 1964). Thus, the dicarboxylic ester (33) gives the ten-membered ring ketol (34), and, in the steroid series, dimethyl marrianolate methyl ether (35) gives the ketol (36) which is readily converted into oestrone (7.61). Greatly improved yields are often obtained by carrying out the reactions in presence of chlorotrimethylsilane. This serves

$$C_2H_5O_2C(CH_2)_8CO_2C_2H_5\ (33) \xrightarrow[\text{then } H_3O^+]{Na,\ \text{boiling xylene,}} \overline{C{=}O\ (CH_2)_8\ CHOH}\ (34) \qquad (7.61)$$

$$\text{(35) } [CH_3O\text{-aryl, } CO_2CH_3,\ CH_2CO_2CH_3] \xrightarrow{Na,\ liq.\ NH_3} \text{(36) } [OH,\ {=}O]$$

principally to remove alkoxide ion from the reaction medium, thus preventing wasteful base-catalysed side reactions such as β-eliminations and Claisen or Dieckmann condensations (see p. 312).

The acyloin reaction is generally considered to proceed by dimerisation of the anion radicals formed by electron transfer from the metal to the carbonyl group of the ester (see p. 433). Some doubt has recently been cast on this mechanism and an alternative which is said to account better for the anomalous products formed in some reactions has been proposed (Bloomfield, Owsley, Ainsworth and Robertson, 1975).

Reduction with metal and ammonia or an amine. Conjugated systems. Isolated carbon–carbon double bonds are not normally reduced by dissolving metal reducing agents because formation of the intermediate electron-addition products requires more energy than ordinary reagents can provide. Reduction is possible when the double bond is conjugated, because then the intermediate can be stabilised by resonance. By far the best reagent is a solution of an alkali metal in liquid ammonia, with or without addition of an alcohol – the so-called Birch reduction conditions. Under these conditions conjugated dienes, $\alpha\beta$-unsaturated ketones, styrene derivatives and even benzene rings can be reduced to dihydro derivatives (Birch and Smith, 1958; Birch and Subba Rao, 1972).

Birch reductions are usually effected with solutions of lithium, sodium or potassium in liquid ammonia. In some reactions an alcohol is added to act as a proton donor and as a buffer against the accumulation of the strongly basic amide ion; in other cases acidification with an alcohol or with ammonium chloride is effected at the end of the reaction. Solutions of alkali metals in liquid ammonia contain solvated metal cations and electrons (7.62), and part of the usefulness of these reagents arises

$$M \xrightarrow{\text{liq. } NH_3} M^+(NH_3) \cdots e^-(NH_3) \qquad (7.62)$$

from the small steric requirement of the electrons, which sometimes allows reactions which are difficult to achieve with other reducing agents, and in many cases leads to different stereochemical results. Wasteful reaction of the metal with ammonia to form hydrogen and amide ion is slow in the absence of catalysts; part of the superiority of lithium over sodium and potassium is due to the fact that its reaction with ammonia is catalysed to a lesser extent by colloidal iron present as an impurity in ordinary commercial liquid ammonia. Using distilled ammonia there is little to choose among the three metals (see Dryden, Webber, Burtner and Cella, 1961). Reactions are usually carried out at the boiling point of ammonia (−33°C), and since the solubility of many organic compounds in liquid ammonia is low at this temperature, co-solvents such as ether, tetrahydrofuran or dimethoxyethane are often added to aid solubility. A common technique is to add the metal to a stirred solution of the compound in liquid ammonia and the co-solvent, in presence of an alcohol where appropriate, but variations of this procedure are often used (Djerassi, 1963).

Isolated olefinic bonds are not reduced by metal–ammonia reagents alone. In the presence of an alcohol terminal olefinic bonds are reduced. Thus 1-hexene is converted into hexane by sodium and methanol in liquid ammonia, but 2-hexene is unaffected.

Conjugated dienes are readily reduced to the 1,4-dihydro derivatives with metal–ammonia reagents in the absence of added proton donors. Thus, isoprene is reduced to 2-methyl-2-butene by sodium in ammonia through the anion radical. The protons required to complete the reduction are supplied by the ammonia (7.63).

$$CH_2{=}C(CH_3){-}CH{=}CH_2 \xrightarrow[\text{liq. } NH_3]{Na} [CH_2{\cdots}C(CH_3){\cdots}CH{\cdots}CH_2]^{\cdot -} \longrightarrow CH_3{-}C(CH_3){=}CH{-}CH_3 \qquad (7.63)$$

Although it is generally true that isolated olefinic bonds are resistant to reduction by alkali metals and liquid ammonia, in some cases reduction of isolated double bonds can be assisted by stabilisation of the anion radical by means other than direct conjugation. Thus, norbornadiene is converted into norborene in 98 per cent yield by lithium in ammonia, possibly by way of the stabilised non-classical anion radical (37).

Li, NH_3 → (37) → (98%)

(7.64)

(38) → Na, NH_3, t-C_4H_9OH → (39)

Another remarkable example is provided by the compound (38) which is reduced by sodium and t-butanol in liquid ammonia to the tetrahydro compound (39); the olefinic bond is reduced first. In this case the benzene ring and the olefinic bond are too far apart to allow orbital interaction through space, and the results are attributed to orbital interaction through σ-bonds (Paddon-Row, Hartcher and Warrener, 1976).

Reduction of cyclic ketones with metal–ammonia–alcohol reagents leads to the alcohol with the thermodynamically more stable configuration. With $\alpha\beta$-unsaturated ketones the saturated ketone or the saturated alcohol can be obtained depending on the conditions. Reduction in

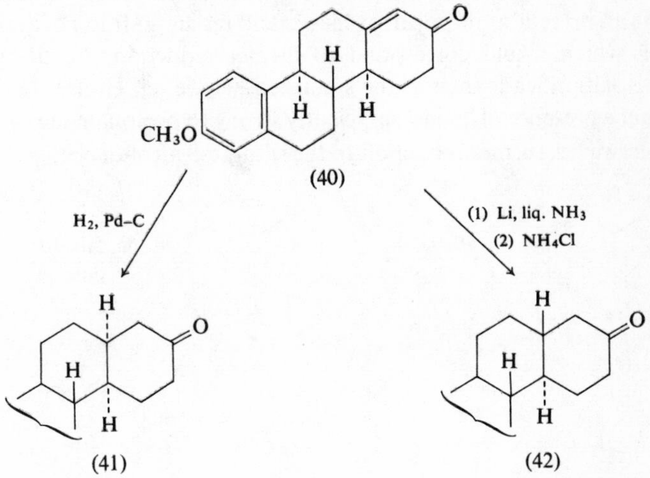

CH_3O H H H O

(40)

H_2, Pd–C

(1) Li, liq. NH_3
(2) NH_4Cl

H H O H

(41)

H H O H

(42)

(7.65)

O Li O⁻ NH_3 or ROH (1 equiv.) OH H H H

(43)

Li

H O NH_4Cl H O⁻ OH H H H

(45)

(44)

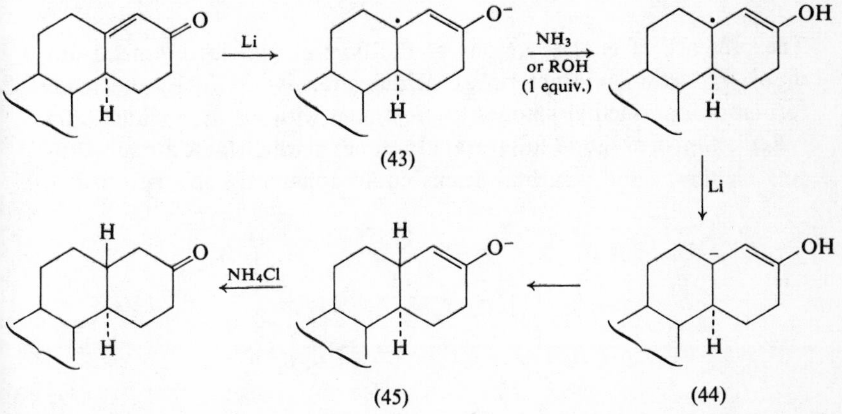

the presence of not more than one equivalent of a proton donor followed by acidification with ethanol or ammonium chloride leads to the saturated ketone. The same transformation can, of course, be effected by catalytic hydrogenation, but metal–ammonia reduction is more stereoselective and in many cases leads to a product with a different stereochemistry from that obtained by hydrogenation. Thus, the $\alpha\beta$-unsaturated ketone (40) on reduction with hydrogen and palladium affords the *cis* fused ketone (41), while with lithium in liquid ammonia the *trans* product (42) is obtained (7.65). The initially formed anion radical (43) abstracts a proton from the ammonia or from added alcohol to form, after addition of another electron the enolate anion (45). In the absence of a stronger acid this anion retains its negative charge and resists addition of another electron which would correspond to further reduction. Acidification during isolation leads then to the saturated ketone (cf. House, 1972, p. 175). In the presence of 'acids' sufficiently strong to protonate the enolate anion, however, further reduction to the saturated alcohol occurs (7.66).

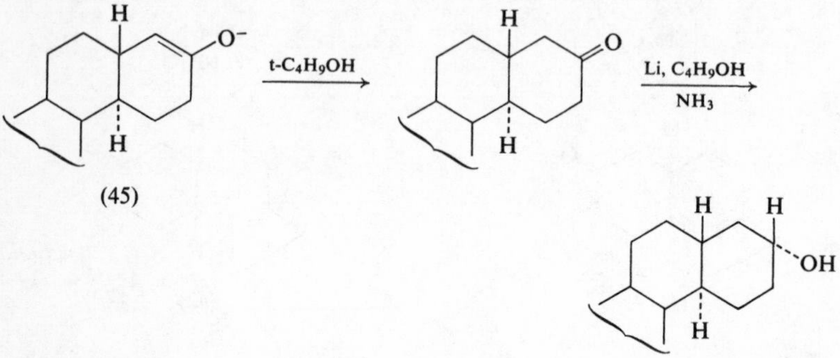

H O⁻ H (45) —t-C_4H_9OH→ H O H —Li, C_4H_9OH / NH_3→ H H OH H

(7.66)

The presence of enolate anions of the type of (45) in the metal–ammonia reduction product before acidification is shown by the ready formation of α-methyl ketones by treatment with methyl iodide (7.67).

Reduction of cyclic $\alpha\beta$-unsaturated ketones in which there are substituents on the β- and γ-carbon atoms could apparently give rise to two

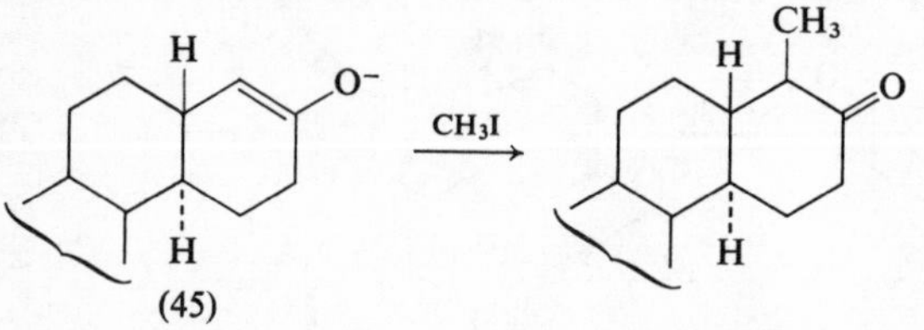

H O⁻ H (45) —CH_3I→ CH_3 H O H

(7.67)

epimeric products (7.68), but it is found in practice that in most cases only one isomer is formed, and that is generally the more stable of the two. The guiding principle appears to be that protonation of the inter-

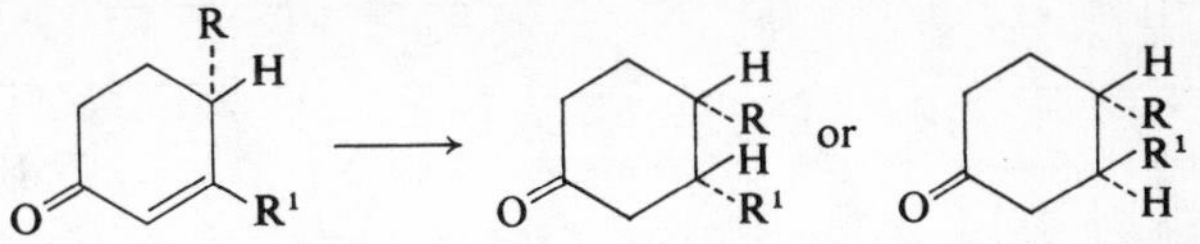

(7.68)

mediate allylic anion (as 44) takes place in a direction perpendicular to the double bond and to the most stable conformation/configuration of the carbanion which allows continuous overlap of the sp^3 orbital on the β-carbon atom with the π-orbital system of the enol (Stork and Darling, 1964). Thus, in the reduction of the octalone (46) to the decalone (47)

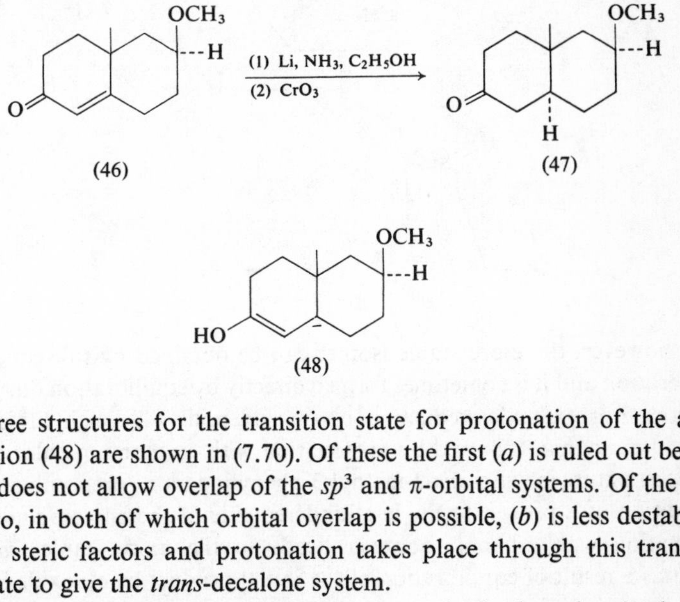

(1) Li, NH_3, C_2H_5OH
(2) CrO_3

(46) (47)

(7.69)

(48)

three structures for the transition state for protonation of the allylic anion (48) are shown in (7.70). Of these the first (*a*) is ruled out because it does not allow overlap of the sp^3 and π-orbital systems. Of the other two, in both of which orbital overlap is possible, (*b*) is less destabilised by steric factors and protonation takes place through this transition state to give the *trans*-decalone system.

After protonation at the β-carbon atom, the final product is obtained by protonation of the α-carbon atom during work-up (7.71). The stereochemistry of this process is not easy to predict. This protonation is kinetically controlled and usually takes place from the less hindered side of the enolate system, although other factors may influence the direction of addition (Zimmerman, 1961), to give a ketone which is not necessarily the thermodynamically more stable isomer. In many

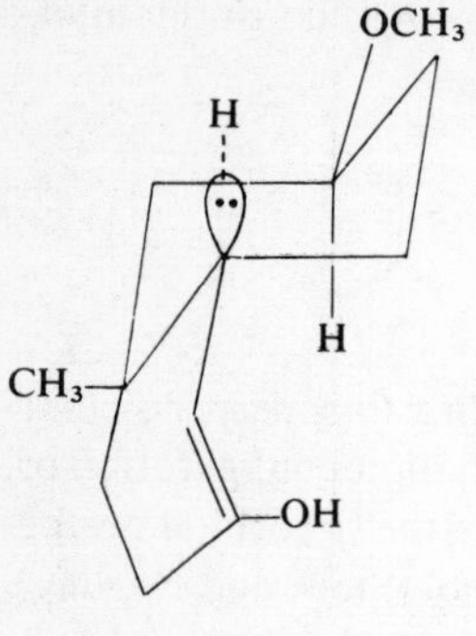

(*a*) (*b*) (*c*)

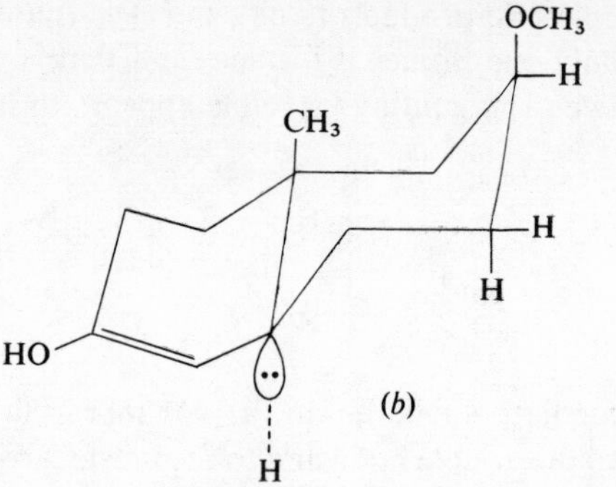

(7.70)

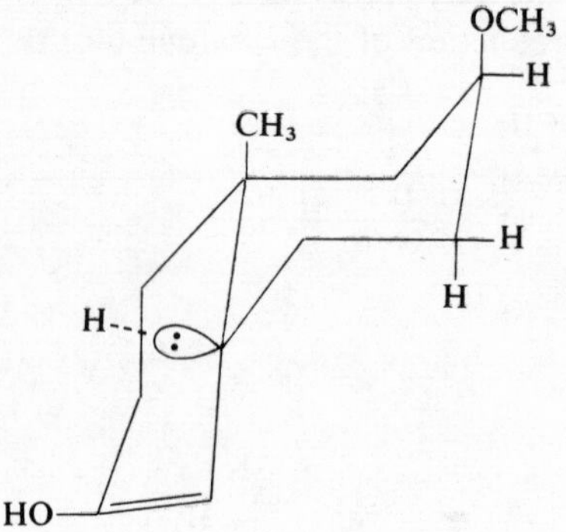

cases, however, the more stable isomer can be obtained by subsequent equilibration and it is sometimes formed directly by equilibration during isolation. For example, reduction of the indenone (49) with lithium in liquid ammonia, followed by protonation with ammonium chloride, affords a mixture in which *cis*-2-methyl-3-phenylindanone predominates, by addition of a proton to the less hindered side of the enolate anion. Protonation with ethanol, however, leads to the more stable *trans* isomer as a result of equilibration of the initial product by ethoxide ion

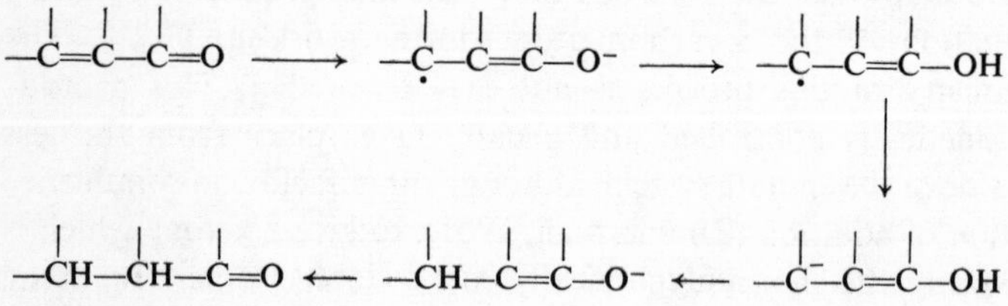

$$-\overset{|}{C}{=}\overset{|}{C}{-}\overset{|}{C}{=}O \longrightarrow -\underset{\bullet}{\overset{|}{C}}{-}\overset{|}{C}{=}\overset{|}{C}{-}O^{-} \longrightarrow -\underset{\bullet}{\overset{|}{C}}{-}\overset{|}{C}{=}\overset{|}{C}{-}OH$$

$$\downarrow \qquad (7.71)$$

$$-\overset{|}{C}H{-}\overset{|}{C}H{-}\overset{|}{C}{=}O \longleftarrow -\overset{|}{C}H{-}\overset{|}{C}{=}\overset{|}{C}{-}O^{-} \longleftarrow -\overset{|}{C}^{-}{-}\overset{|}{C}{=}\overset{|}{C}{-}OH$$

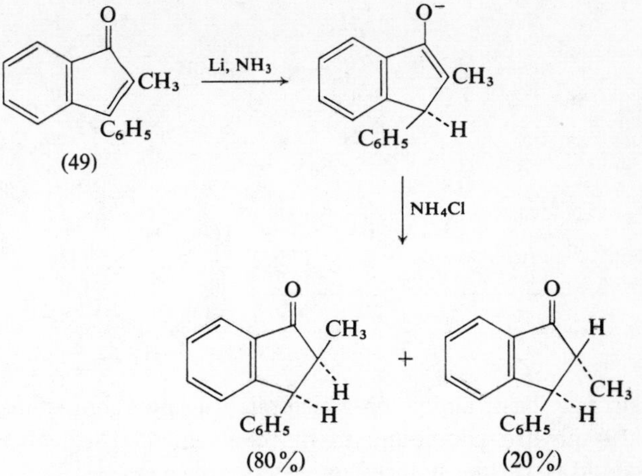

(49)

Li, NH_3

NH_4Cl

(80%) (20%)

(7.72)

(7.72). In the reduction of styrene derivatives also, the stereochemistry at the β-carbon atom in the dihydro derivative is that obtained by protonation of the most stable conformation/configuration of the β-anion. Thus, reduction of the steroid intermediate (50) with lithium in ammonia gave the more stable $8\beta,9\alpha$-isomer in major amount (7.73).

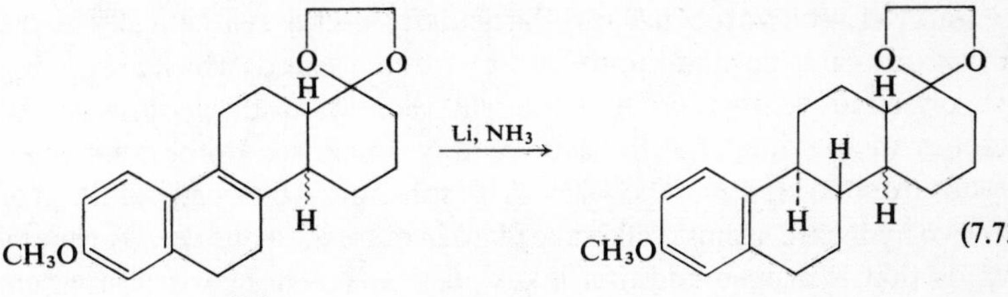

(50)

Li, NH_3

(7.73)

Aromatic compounds. One of the most useful synthetic applications of metal–ammonia–alcohol reducing agents is in the reduction of benzene rings to 1,4-dihydro derivatives. The reagents are powerful enough to reduce benzene rings, but specific enough to add only two hydrogen atoms. Benzene itself is reduced with lithium and ethanol in liquid ammonia to 1,4-dihydrobenzene by way of the anion radical (51) (Birch, 1950; Birch and Nasipuri, 1959) (7.74). The presence of an alcohol as a proton donor is necessary in these reactions, for the initial anion radical is an insufficiently strong base to abstract a proton from the ammonia. The alcohol also acts to prevent the accumulation

$$\text{C}_6\text{H}_6 \xrightleftharpoons{\text{Li, NH}_3} (51) \xrightarrow{\text{C}_2\text{H}_5\text{OH}} \xrightarrow{\text{Li, NH}_3} \xrightarrow{\text{C}_2\text{H}_5\text{OH}} \quad (7.74)$$

of the strongly basic amide ion which might bring about isomerisation of the 1,4-dihydro compound to the conjugated 1,2-dihydro isomer which would be further reduced to tetrahydrobenzene.

Naphthalene is converted into the 1,4,5,8-tetrahydro compound with sodium and alcohol in liquid ammonia (7.75). Alkylbenzenes are reduced

$$\xrightarrow{\text{Na, NH}_3\text{, C}_2\text{H}_5\text{OH}} \quad (7.75)$$

to 2-alkyl-1,4-dihydrobenzenes. Particularly useful synthetically is the reduction of anisoles and anilines to dihydro compounds which are readily hydrolysed to cyclohexenones. Under mild conditions the $\beta\gamma$-unsaturated ketones are obtained, but these are readily isomerised to the conjugated $\alpha\beta$-unsaturated compounds (7.76). With substituted benzenes addition of the two hydrogen atoms could take place in more than one way. A general rule is that hydrogen addition takes place at positions which are *para* to each other avoiding, if possible, carbon atoms to which electron-donating substituents are attached. The directive effect of $—OCH_3$ and $—N(CH_3)_2$ is stronger than that of alkyl groups so that 5-methoxytetralin gives (52) (7.77). On the other hand, reduction of benzene rings substituted by the electron-withdrawing carboxyl or carboxamide groups (the only electron-withdrawing groups which are not reduced before the benzene ring) gives derivatives of 1,4-dihydrobenzoic acid. These orientation effects are in line with molecular orbital calculations of the sites of highest electron density in the anion radicals (Zimmerman, 1961).

Selective reduction of a benzene ring in presence of another reducible group is possible if the other group can be protected by reversible conversion into a saturated derivative, or into a derivative containing only

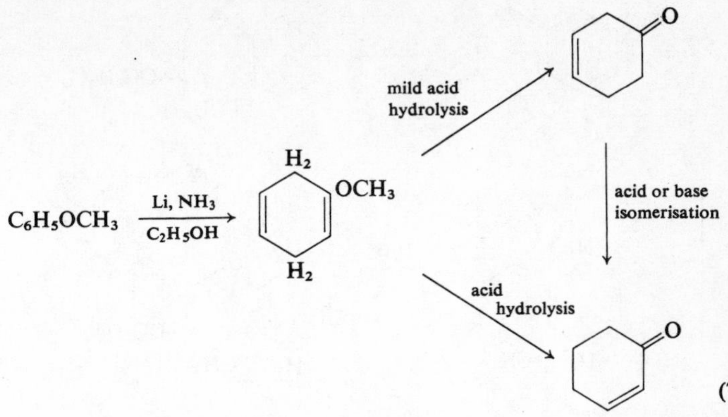
mild acid
hydrolysis
$C_6H_5OCH_3$
Li, NH_3
C_2H_5OH
H_2
OCH_3
H_2
acid or base
isomerisation
acid
hydrolysis
O
O

(7.76)

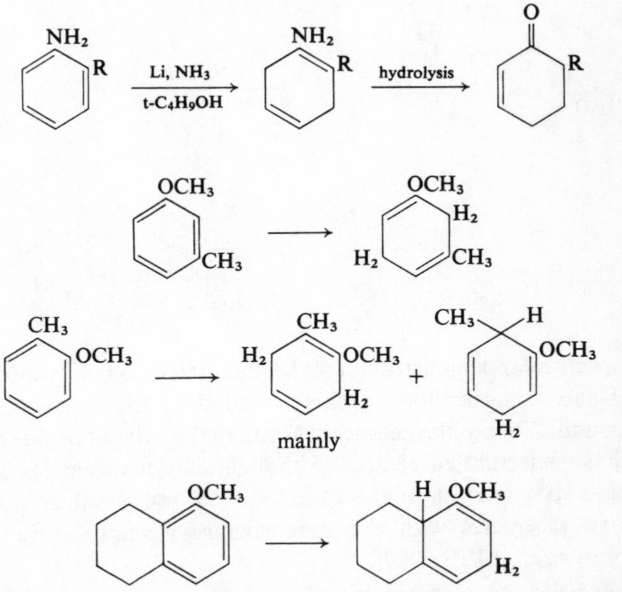
NH_2
R
Li, NH_3
t-C_4H_9OH
NH_2
R
hydrolysis
O
R
OCH_3
CH_3
OCH_3
H_2
H_2
CH_3
CH_3
OCH_3
CH_3
H_2
OCH_3
H_2
mainly
CH_3
H
OCH_3
H_2
+
OCH_3
H OCH_3
H_2

(7.77)

(52)

an isolated carbon–carbon double bond or into an anion by salt formation. Ketones, for example, may be converted into acetals or enol ethers, as shown in (7.78). But this is ineffective if the carbonyl group is conjugated with the benzene ring.

Conversely, reduction of benzene rings takes place only slowly in the absence of a proton donor and selective reduction of an $\alpha\beta$-unsaturated carbonyl system can be effected if no alcohol is added (7.79).

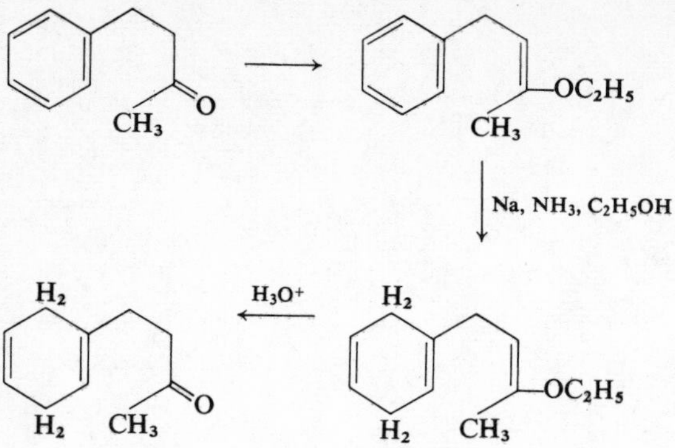

Na, NH_3, C_2H_5OH (7.78)

H_3O^+

(1) Li, NH_3
(2) NH_4Cl (7.79)

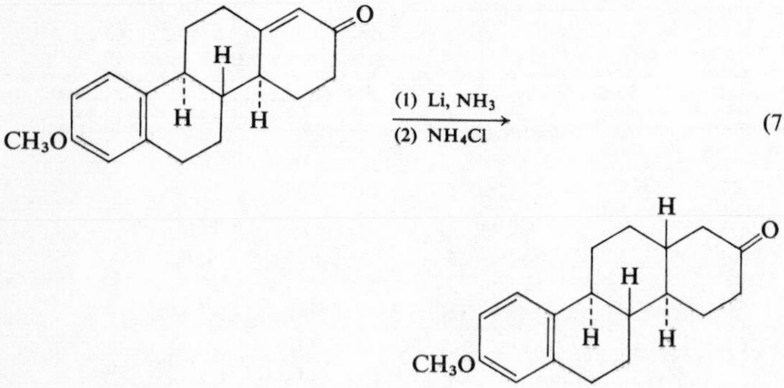

A spectacular demonstration of the high stereoselectivity shown by the metal–ammonia–alcohol reagents is found in the conversion of the compound (53) into the ketones (54) and (55) by simultaneous reduction of all the unsaturated centres. Although sixteen racemates of ketone (54) and sixty-four racemates of ketone (55) are possible, in fact only the two racemates with the stereochemistry shown were obtained (Johnson *et al.*, 1953) (7.80).

The applications in synthesis of the Birch reduction are manifold and go beyond the straightforward preparation of dihydrobenzenes or cyclohexenones. Equation (7.81) shows an example, taken from a synthesis of the alkaloid lycopine, in which a reduced anisole ring serves as a source of functionalised alkyl groups, and another in which oxidative decarboxylation of an (alkylated) dihydrobenzoic acid provides a method for alkylation of benzene derivatives with unusual orientation; in the example the pentyl group is introduced *meta* to the two methoxyl substituents (Birch and Slobbe, 1976).

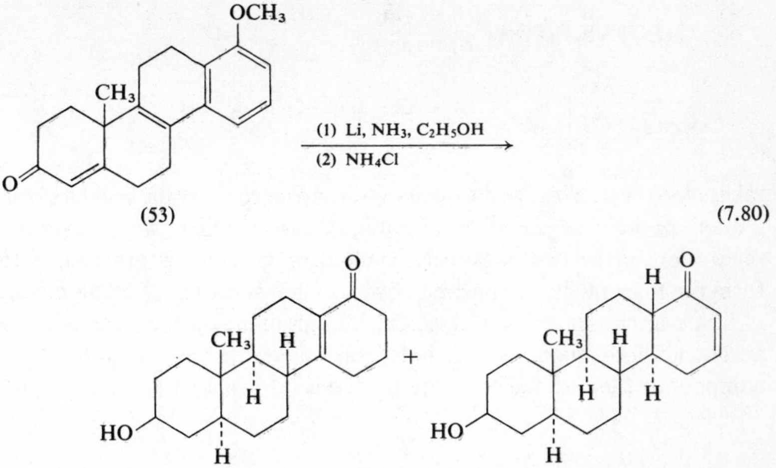

(53) —(1) Li, NH_3, C_2H_5OH; (2) NH_4Cl→ (54) + (55) (7.80)

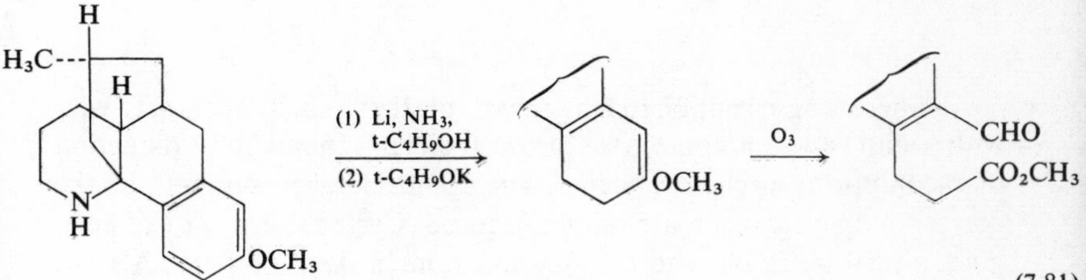

—(1) Li, NH_3, t-C_4H_9OH; (2) t-C_4H_9OK→ —O_3→ (7.81)

—(1) Na, NH_3; (2) $C_5H_{11}Br$→ —$Pb(OCOCH_3)_4$→

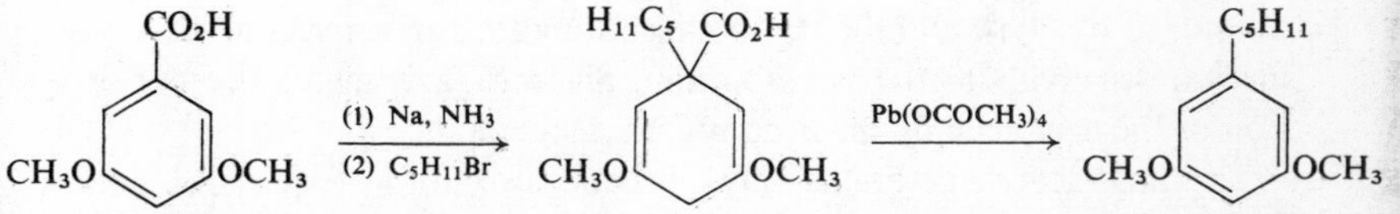

Reduction of acetylenes with metal and liquid ammonia. Although olefinic double bonds are not normally reduced by metal–ammonia reducing agents, the partial reduction of acetylenic bonds is very conveniently effected by these reagents. The procedure is highly selective and none of the saturated product is formed. Furthermore, the reduction is completely stereospecific and the only product from a disubstituted acetylene is the corresponding *trans* olefin (7.82). This method thus complements the formation of *cis* olefins by catalytic hydrogenation of acetylenes discussed on p. 415. It has been thought that the reaction

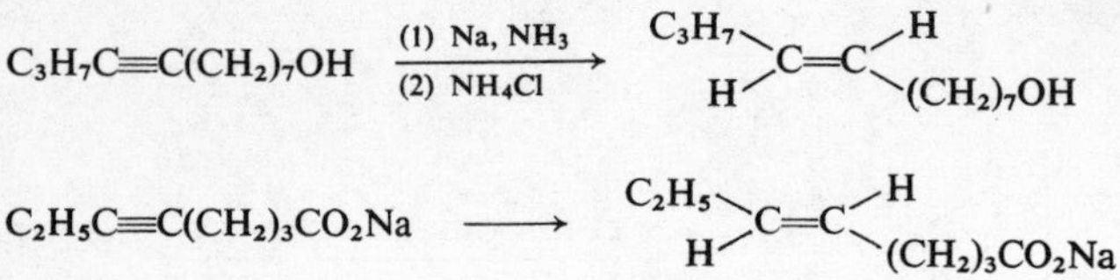

$$C_3H_7C{\equiv}C(CH_2)_7OH \xrightarrow[(2)\ NH_4Cl]{(1)\ Na,\ NH_3} \underset{H}{\overset{C_3H_7}{>}}C{=}C\underset{(CH_2)_7OH}{\overset{H}{<}}$$

$$C_2H_5C{\equiv}C(CH_2)_3CO_2Na \longrightarrow \underset{H}{\overset{C_2H_5}{>}}C{=}C\underset{(CH_2)_3CO_2Na}{\overset{H}{<}} \quad (7.82)$$

takes place by stepwise addition of two electrons to the triple bond to form a dianion, the more stable *trans* configuration of which, with maximum separation of the two negatively charged sp^2-orbitals, is protonated to form the *trans* olefin. It appears, however, that solutions of alkali metals in liquid ammonia are not powerful enough reducing agents to convert acetylenes into dianions and it is now suggested that organosodium compounds (56) are the true intermediates (House and Kinloch, 1974).

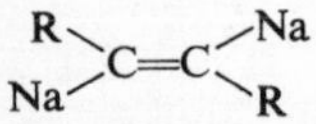

$$\underset{Na}{\overset{R}{>}}C{=}C\underset{R}{\overset{Na}{<}} \quad (56) \qquad\qquad CH{=}C{=}CH \text{ bridged by } (CH_2)_7 \quad (57) \quad (7.83)$$

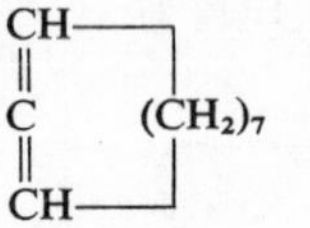

An interesting exception to the general rule that reduction of acetylenes with sodium and ammonia gives the *trans* olefin is found in the reduction of medium-ring cyclic acetylenes where considerable amounts of the corresponding *cis* olefin are often produced. Cyclodecyne, for example, gives a mixture of *cis*- and *trans*-cyclodecene in the ratio 47:1. A *trans* dianion, or disodium salt, is obviously formed with difficulty from compounds of this type and the *cis* olefin is thought to arise by an alternative mechanism involving the corresponding allene (57), formed by isomerisation of the acetylene by the accumulating sodamide.

Terminal acetylenic bonds may be reduced or not as required. Under ordinary conditions, reaction of the acetylene with the metal solution leads to the formation of the metallic acetylide which resists further reduction because of the negative charge on the ethynyl carbon atom. In presence of ammonium sulphate, however, the free ethynyl group is preserved and on reduction yields the corresponding olefin. Thus 1,6-heptadiyne is converted into 1,6-heptadiene in high yield under these conditions. This is a better method for preparing 1-alkenes than the alternative catalytic hydrogenation of 1-alkynes, which sometimes gives small amounts of the saturated hydrocarbons which may be difficult to separate from the olefin. On the other hand, reduction of a terminal aceylene can be suppressed by converting it into its sodium salt by reaction with sodamide, thus allowing the selective reduction of an internal triple bond in the same molecule. 1,7-Undecadiyne, for example, is directly converted into *trans*-7-undecen-1-yne in high yield (7.84).

$$CH_3(CH_2)_2C{\equiv}C(CH_2)_4C{\equiv}CH \xrightarrow[\text{liq. } NH_3]{NaNH_2} R{-}C{\equiv}C^-Na^+ \xrightarrow[(2)\ H^+]{(1)\ Na,\ NH_3} \underset{CH_3(CH_2)_2}{\overset{H}{>}}C{=}C\underset{H}{\overset{(CH_2)_4C{\equiv}CH}{<}} \qquad (7.84)$$

Reductive fission. Metal–amine reducing agents and other dissolving metal systems can bring about a variety of reductive fission reactions, some of which are useful in synthesis.

Most of these reactions probably proceed by direct addition of two electrons from the metal to the bond which is broken. The anions produced may be protonated by an acid in the reaction medium, or may survive until work-up. In some cases it may be desirable to maintain the fission products as anions to prevent further reduction. Reaction is

$$A{-}B \xrightarrow{2e} A^- + B^- \xrightarrow{2H^+} AH + BH$$

facilitated when the anions are stabilised by resonance or by an electronegative atom. As expected, therefore, bonds between hetero-atoms or between a hetero-atom and an unsaturated system which can stabilise a negative charge by resonance, are particularly easily cleaved. Thus allyl and benzyl halides, ethers and esters, and sometimes even the alcohols themselves, are readily cleaved by metal–amine systems and other dissolving metal reagents, as well as by catalytic hydrogenation (see p. 408). Fission reactions of this type, particularly with solutions of an alkali metal in liquid ammonia, have been widely used in structural studies, and also for the reductive removal of unsaturated groups used as protecting agents for amino, imino, hydroxyl and thiol groups. They are of particular value in peptide and nucleotide synthesis, and have an obvious general advantage over hydrolytic or catalytic methods for compounds which are labile or contain sulphur. Benzyl, *p*-toluenesulphonyl and benzyloxycarbonyl groups are all efficiently replaced by hydrogen, and cystinylpeptides are cleaved to cysteinyl derivatives (7.85). For example in the synthesis of the naturally occurring peptide, glutathione, *N*-benzyloxycarbonyl-γ-glutamyl-*S*-benzylcysteinylglycine was prepared as the

$$\begin{aligned} &RSCH_2C_6H_5 \longrightarrow RSH + CH_3C_6H_5 \\ &RNHSO_2C_6H_4CH_3\text{-}p \longrightarrow RNH_2 + HSC_6H_4CH_3\text{-}p \\ &RNHCO_2CH_2C_6H_5 \longrightarrow RNH_2 + CO_2 + CH_3C_6H_5 \\ &RSSR \longrightarrow 2RSH \\ &ROCH_2C_6H_5 \longrightarrow ROH + CH_3C_6H_5 \end{aligned} \qquad (7.85)$$

key intermediate in which the benzyloxycarbonyl group was used for protection of the amino group and the benzyl group for protection of the thiol. The protective groups were finally removed to yield glutathione by cleavage with sodium in liquid ammonia (7.86).

$$C_6H_5CH_2OCONHCH(CO_2H)CH_2CH_2CONHCH(CH_2SCH_2C_6H_5)CONHCH_2CO_2H \xrightarrow{Na,\ NH_3} H_2NCH(CO_2H)CH_2CH_2CONHCH(CH_2SH)CONHCH_2CO_2H \qquad (7.86)$$

Reductive fission has been of great assistance in the elucidation of the structures of a number of naturally occurring allyl and benzyl alcohols, ethers and esters. Thus, the structure of the alcohol lanceol was neatly confirmed by reduction with sodium and alcohol in liquid ammonia to the known sesquiterpene hydrocarbon bisabolene (7.87).

$$\text{lanceol } (CH_3, CH_2OH) \xrightarrow[C_2H_5OH]{Na,\ NH_3} \text{bisabolene } (CH_3) \qquad (7.87)$$

The antibiotic mycelianamide similarly gave methylgeraniolene and a derivative of *p*-hydroxybenzoic acid (7.88). In these examples the

$$\begin{array}{c}(CH_3)_2C{=}CHCH_2CH_2C(CH_3){=}CHCH_2OC_6H_4R \\ \downarrow \\ (CH_3)_2C{=}CHCH_2CH_2C(CH_3){=}CHCH_3 + HOC_6H_4R\end{array} \quad (7.88)$$

allylic groups are cleaved specifically without reduction of the olefinic double bonds.

The reductive fission of allyl or benzyl alcohols can be prevented, if necessary, by converting them into the corresponding alkoxide ions by reaction with sodamide or other strong base, as in (7.89).

$$C_6H_5CH{=}CH{-}C(CH_3)_2OH \xrightarrow[\text{(2) Na–liq. NH}_3]{\text{(1) NaNH}_2} C_6H_5CH_2CH_2C(CH_3)_2OH \quad (7.89)$$

In contrast to benzyl alcohols, benzylamines are not readily reduced with metal–ammonia reagents, because nitrogen is less electronegative than oxygen, but quaternary salts are easily cleaved as in the well-known Emde reaction of quaternary ammonium salts (7.90).

$$\text{[1-methyl-tetrahydroisoquinolinium, } {}^{+}N(CH_3)_2\text{, } CH_3] \xrightarrow{\text{Na, NH}_3} \text{[}o\text{-ethylphenethyl-}N(CH_3)_2\text{; } CH_2CH_3] \quad (7.90)$$

Solutions of sodium in liquid ammonia also cleave aryl ethers. Diaryl ethers react particularly readily. Thus phenyl *p*-tolyl ether gives a mixture of *p*-cresol and phenol in the ratio 3:1. Alkyl aryl ethers are cleaved only slowly, but aryl phosphonates, which are easily obtained from the phenol and diethyl phosphite or tetraethyl pyrophosphate, give the hydrocarbon in high yield, and this is one of the few available methods for removal of the oxygen from a phenolic ring. Enols are similarly converted into olefins. This has been exploited in a method, illustrated in (7.91), for converting $\alpha\beta$-unsaturated ketones into structurally specific olefins (Ireland, Muchmore and Hengartner, 1972).

Alcoholic hydroxyl groups can also be removed by reaction of the corresponding methanesulphonates with sodium and an alcohol in liquid ammonia, or of the diethyl phosphates with lithium and t-butanol in

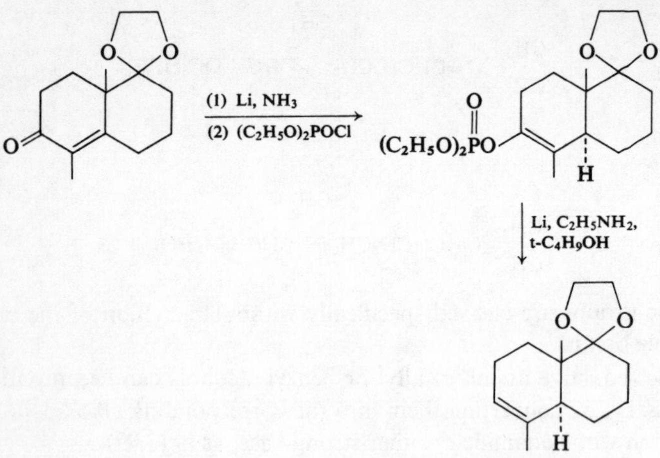

(1) Li, NH_3
(2) $(C_2H_5O)_2POCl$

$(C_2H_5O)_2PO$ H

Li, $C_2H_5NH_2$, t-C_4H_9OH (7.91)

H

ethylamine; even tertiary alcohols are readily deoxygenated by the latter route (Ireland, Muchmore and Hengartner, 1972).

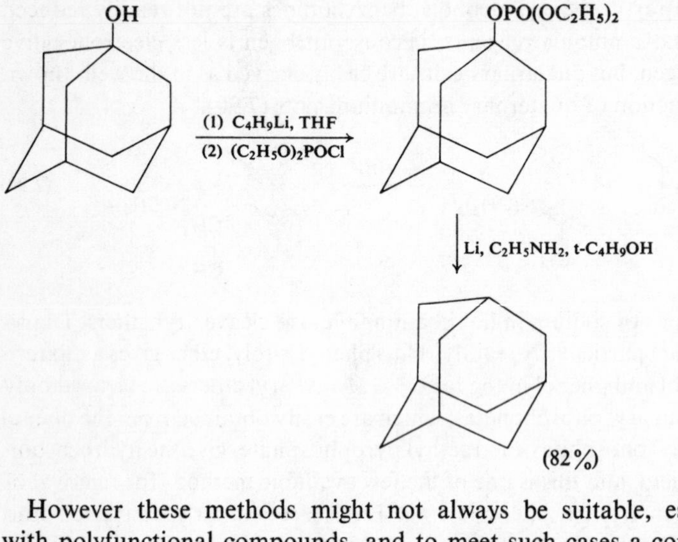

OH $OPO(OC_2H_5)_2$

(1) C_4H_9Li, THF
(2) $(C_2H_5O)_2POCl$

Li, $C_2H_5NH_2$, t-C_4H_9OH

(7.92)

(82%)

However these methods might not always be suitable, especially with polyfunctional compounds, and to meet such cases a convenient alternative procedure for the deoxygenation of secondary alcohols under neutral conditions has been described by Barton and McCombie (1975). Reaction of *O*-cycloalkyl thiobenzoates or *S*-methyldithiocarbonates with tributylstannane in refluxing toluene or xylene gives the corresponding hydrocarbon directly. The reaction proceeds by a radical

$$\mathrm{R{-}O{-}C({=}S){-}R' + (C_4H_9)_3\dot{S}n \longrightarrow R{-}O{-}\dot{C}(S{-}Sn(C_4H_9)_3){-}R' \longrightarrow R\cdot + O{=}C(S{-}Sn(C_4H_9)_3){-}R'} \qquad \mathrm{R\cdot \xrightarrow{(C_4H_9)_3SnH} RH} \tag{7.93}$$

mechanism and has the added advantage that rearrangements common in reactions involving carbonium ions are avoided. Lanosterol, for example, is converted into the oxygen-free compound without rearrangement and, in the carbohydrate series, the *S*-methyl dithiocarbonate (58) gives the 3-deoxy compound in 80–90 per cent yield.

$$\xrightarrow[\text{toluene}]{(C_4H_9)_3SnH} \tag{7.94}$$

$R = OC(=S){-}SCH_3$

(58)

(80–90%)

Organic halides can be converted into the halogen-free compounds with dissolving metal reagents, but, since the highly electropositive metals such as sodium induce Wurtz coupling, metal–amine systems are not often used. Good results are often obtained with magnesium or zinc in a protic solvent such as dilute mineral acid, acetic acid or an alcohol (7.95). The mechanism of these reactions is uncertain but it is thought

$$\mathrm{C_{15}H_{31}CH_2I \xrightarrow[CH_3COOH]{Zn,\ HCl} C_{15}H_{31}CH_3} \tag{7.95}$$

that they may proceed by way of short-lived organometallic intermediates. This is supported by the observation that if a substituent which can be lost as a stable anion (e.g. —OH, —OR, —OCOR, halogen) is present on the adjacent carbon atom elimination rather than simple reductive cleavage is observed (7.96).

$$\mathrm{C_3H_7CH(Br){-}CH(OCH_3)C_3H_7 \xrightarrow[C_2H_5OH]{Zn} C_3H_7CH{=}CHC_3H_7} \tag{7.96}$$

(55% *cis*, 45% *trans*)

Reductive cleavage of alkyl halides may or may not occur with retention of configuration. With vinyl halides cleavage with sodium in liquid ammonia usually gives mixtures of the *cis* and *trans* olefins formed by interconversion of the intermediate radicals or anions (7.97) (House and Kinloch, 1974).

$$(C_2H_5)HC{=}C(C_2H_5)Cl \xrightarrow[\text{THF}]{\text{Na, NH}_3} \underset{(52\%)}{(C_2H_5)HC{=}C(C_2H_5)H} + \underset{(48\%)}{(C_2H_5)HC{=}CH(C_2H_5)} \qquad (7.97)$$

Other useful reagents for reduction of carbon–halogen bonds are trialkyltin hydrides and chromous salts; other hydride reducing agents may also be employed (see pp. 482, 486). Reduction with trialkyltin hydrides is a general method and can be applied to both alkyl and aryl halides (Kuivila, 1968). It proceeds by a free-radical chain process. From the synthetic viewpoint, stepwise reduction of geminal dihalides is a useful application (7.98). As ordinarily employed the reagent is expensive

$$\text{CH}_3, \text{CH}_3, =\text{CH}_2, \text{Br}, \text{Br} \xrightarrow{(C_4H_9)_3SnH} \text{CH}_3, \text{CH}_3, =\text{CH}_2, \text{H}, \text{Br} \xrightarrow{(C_4H_9)_3SnH} \text{CH}_3, \text{CH}_3, =\text{CH}_2, \text{H}, \text{H} \qquad (7.98)$$

and isolation of the product may be troublesome. A new procedure using catalytic amounts of a trialkyltin halide with sodium borohydride has been reported by Corey and Suggs (1975*b*).

Reduction of simple alkyl or aralkyl halides with chromous salts usually effects replacement of halogen by hydrogen (Hanson and Premuzic, 1968; Hanson, 1974). Reaction is thought to proceed by two one-electron transfer steps as in (7.99). Compounds carrying a

$$\geq C{-}Br \xrightarrow{Cr^{2+}} \geq C\cdot \xrightarrow{Cr^{2+}} \geq C{-}Cr \;(\leftarrow H_2O) \longrightarrow \geq C{-}H \qquad (7.99)$$

vicinal substituent which can be eliminated as an anion (e.g. OH, OR, Cl) usually give olefins (7.100). Elimination can be circumvented by

$$CH_2BrCHBrCH_2Cl \xrightarrow[\text{aq. dimethylformamide}]{CrSO_4} CH_2{=}CHCH_2Cl \qquad (7.100)$$

effecting the reduction with chromous acetate in presence of a hydrogen atom transfer agent such as butanethiol, thus providing, for example, an excellent route from a bromohydrin to the corresponding alcohol as in the route to 11-β-hydroxysteroids (7.101).

HO, Br $\xrightarrow{Cr^{2+}}$ [HO, •] $\xrightarrow{RSH}$ HO, H (7.101)

Lithium in alkylamines. Solutions of lithium in aliphatic amines of low molecular weight are powerful reducing agents, but are less selective than metal–ammonia reagents (Kaiser, 1972). The amines have the advantage that they are better solvents for organic substances than is ammonia, and have higher boiling points which makes for easier handling. The higher working temperatures facilitate the initial stages of the reduction, but they also favour conjugation (and therefore further reduction) of the initial product, and unless precautions are taken reduction of an aromatic compound leads to the tetrahydro or even the hexahydro derivative (7.102). Under suitable conditions yields of 1,4-dihydro compounds comparable to those formed under Birch conditions can be obtained (Benkeser, Burrous, Hazdra and Kaiser, 1963; Benkeser, Agnihotri, Burrous, Kaiser, Mallan and Ryan, 1964; Kwart and Conley, 1973).

Solutions of lithium in alkylamines have been used successfully for reductive cleavage of allylic ethers and esters. Reduction of steroid

$$C_6H_5C_2H_5 \xrightarrow{Li,\ CH_3NH_2} \text{1-ethylcyclohexene-type product } (C_2H_5)\ (44\%) + \text{ethylcyclohexane } (C_2H_5)\ (24\%)$$

$$C_6H_5C_2H_5 \xrightarrow{Li,\ C_2H_5NH_2,\ C_2H_5OH} \text{ethylcyclohexadiene } (C_2H_5)\ (89\%) \qquad (7.102)$$

epoxides leads to the axial alcohols, and in this reaction they are more powerful and specific reagents than lithium aluminium hydride (Hallsworth and Henbest, 1957).

7.3. Reduction by hydride-transfer reagents

Reactions which proceed by transfer of hydride ions are widespread in organic chemistry (Deno, Peterson and Saines, 1960), and they are important also in biological systems. Reductions involving the reduced forms of coenzymes I and II, for example, are known to proceed by transfer of a hydride ion from a 1,4-dihydropyridine system to the substrate. In the laboratory the most useful reagents of this type in synthesis are aluminium isopropoxide and the various metal hydride reducing agents.

Aluminium alkoxides. The reduction of carbonyl compounds to alcohols with aluminium isopropoxide has long been known under the name of the Meerwein–Pondorff–Verley reduction (Wilds, 1944). The reaction is easily effected by heating the components together in solution in isopropanol. An equilibrium is set up and the product is obtained by using an excess of the reagent or by distilling off the acetone as it is formed. The reaction is thought to proceed by transfer of hydride ion from the isopropoxide to the carbonyl compound through a six-membered cyclic transition state (7.103). Aldehydes are reduced to primary alcohols, and ketones give secondary alcohols, often in high yield. The reaction owes its usefulness to the fact that olefinic double bonds and many other unsaturated groups are unaffected, thus allowing selective reduction of carbonyl groups. Cinnamaldehyde, for example, is con-

$$R^1R^2C{=}O + (CH_3)_2CH{-}O{-}Al(OPr^i)_2 \rightleftharpoons R^1R^2CH{-}O{-}Al< + (CH_3)_2C{=}O \xrightarrow{H_3O^+} R^1R^2CH{-}OH \quad (7.103)$$

verted into cinnamyl alcohol, *o*-nitrobenzaldehyde gives *o*-nitrobenzyl alcohol, and phenacyl bromide gives styrene bromohydrin.

Reductions of a similar type can be brought about by other metallic alkoxides, but aluminium alkoxide is particularly effective because it is soluble in both alcohols and hydrocarbons and, being a weak base, it shows little tendency to bring about wasteful condensation reactions of the carbonyl compounds.

Reduction of cyclohexanones by the Meerwein–Pondorff–Verley method usually produces a higher proportion of the axial alcohol than other chemical methods of reduction (see Table 7.2, p. 438). Prolonged reaction may lead to the establishment of equilibrium through continuous reduction and reoxidation, with gradual enrichment of the more stable equatorial alcohol.

Lithium aluminium hydride and sodium borohydride. A number of metal hydrides have been employed as reducing agents in organic chemistry, but the most commonly used are lithium aluminium hydride and sodium borohydride, both of which are commercially available. Another useful reagent, diborane, is prepared *in situ* as required (see p. 269).

The anions of the two complex hydrides can be regarded as derived from lithium or sodium hydride and either aluminium hydride or borane.

$$LiH + AlH_3 \longrightarrow Li^+\,AlH_4^-$$

$$NaH + BH_3 \longrightarrow Na^+\,BH_4^-$$

These anions are nucleophilic reagents and as such they normally attack polarised multiple bonds such as C=O or C≡N by transfer of hydride ion to the more positive atom. They do not usually react with isolated carbon–carbon double or triple bonds. Thus reduction of acetone with sodium borohydride proceeds as in (7.104). In aprotic

$$\mathrm{H{-}\overset{H}{\underset{H}{\overset{|}{\underset{|}{B^-}}}}{-}H} \quad \mathrm{\overset{CH_3}{\underset{CH_3}{C{=}O}}} \longrightarrow \mathrm{H{-}\overset{H}{\underset{H}{B}}} + \mathrm{H{-}\overset{CH_3}{\underset{CH_3}{C}}{-}O^-}$$

$$\xrightarrow[\text{CH}_3\text{OH}]{} \mathrm{H{-}\overset{H}{\underset{H}{B^-}}{-}OCH_3} + \mathrm{H{-}\overset{CH_3}{\underset{CH_3}{C}}{-}OH} \qquad \xrightarrow{\text{ether}} \mathrm{H{-}\overset{CH_3}{\underset{CH_3}{C}}{-}O\bar{B}H_3} \tag{7.104}$$

solvents the alkoxyborohydride is formed, but in hydroxylic solvents reaction with the solvent may occur to give the alcohol directly.

With both reagents all four hydrogen atoms may be used for reduction, being transferred in a stepwise process as illustrated for reduction of a ketone (7.105). For reductions with lithium aluminium hydride (but

$$AlH_4^- + RCOR^1 \longrightarrow H_3\bar{A}l{-}O{-}CH{<}^{R}_{R^1} \xrightarrow{RCOR^1} H_2\bar{A}l\left[O{-}CH{<}^{R}_{R^1}\right]_2 \longrightarrow H\bar{A}l\left[O{-}CH{<}^{R}_{R^1}\right]_3$$

$$\bar{A}l\left[O{-}CH{<}^{R}_{R^1}\right]_4 \tag{7.105}$$

not with sodium borohydride) each successive transfer of hydride ion takes place more slowly than the one before, and this has been exploited for the preparation of modified reagents which are less reactive and more selective than lithium aluminium hydride itself by replacement of two or three of the hydrogen atoms of the anion by alkoxy groups (see p. 475).

Lithium aluminium hydride is more reactive than sodium borohydride. It reacts readily with water and other compounds containing active hydrogen atoms, and must be used under anhydrous conditions in a non-hydroxylic solvent; ether and tetrahydrofuran are often employed. Sodium borohydride reacts only slowly with water and most alcohols at room temperature, and reductions with this reagent are often effected in ethanol solution at room temperature. Table 7.3, from House 1965, shows some common functional groups which are reduced with lithium aluminium hydride. Sodium borohydride, being less reactive, is more discriminating in its action than lithium

TABLE 7.3 *Common functional groups reduced by lithium aluminium hydride*

Functional group	Product
$>C{=}O$	$>CH{-}OH$
$-CO_2R$	$-CH_2OH + ROH$
$-CO_2H$	$-CH_2OH$
$-CONHR$	$-CH_2NHR$
$-CONR_2$	$-CH_2NR_2$ or $\left[-\underset{\mid}{\underset{OH}{CH}}-NR_2\right] \longrightarrow -CHO + R_2NH$
$-C{\equiv}N$	$-CH_2NH_2$ or $[-CH{=}NH] \xrightarrow{H_2O} -CHO$
$>C{=}NOH$	$>CH{-}NH_2$
$-\overset{\mid}{\underset{\mid}{C}}-NO_2$ (aliphatic)	$-\overset{\mid}{\underset{\mid}{C}}-NH_2$
$ArNO_2$	$ArNHNHAr$ or $ArN{=}NAr$
$-CH_2OSO_2C_6H_5$ or $-CH_2Br$	$-CH_3$
$>CHOSO_2C_6H_5$ or $>CH{-}Br$	$>CH_2$
$-CH{-}C<$ (epoxide, bridged by O)	$-CH_2-\overset{\mid}{\underset{OH}{C}}-$

H.O. House, *Modern synthetic reactions*, copyright 1965, 1972, W. A. Benjamin, Inc., Menlo Park, California.

aluminium hydride (see also Walker, 1976). It generally does not attack esters or amides, and it is normally possible to reduce aldehydes and ketones selectively with this reagent in presence of a variety of other functional groups. Some typical reductions are illustrated in the equations (7.106).

$$CH_2{=}CH{-}CH{=}CH{-}CHO \xrightarrow[\text{or NaBH}_4]{\text{LiAlH}_4} CH_2{=}CH{-}CH{=}CH{-}CH_2OH$$

$$\text{4-}(CO_2C_2H_5)\text{cyclohexanone} \xrightarrow[CH_3OH]{NaBH_4} \text{(H, OH; H, }CO_2C_2H_5\text{ cyclohexane)} + \text{(HO, H; H, }CO_2C_2H_5\text{ cyclohexane)} \quad (7.106)$$

$$(CH_2)_5\text{–}C(=O)\text{–}NCH_3 \text{ (ring)} \xrightarrow[\text{ether}]{LiAlH_4} (CH_2)_5\text{–}CH_2\text{–}NCH_3 \text{ (ring)}$$

$$CH_2{=}CHCH{-}CH_2 \text{ (epoxide, O)} \xrightarrow[\text{ether}]{LiAlH_4} CH_2{=}CHCHOHCH_3$$

$$NO_2CH_2CH_2CH_2CHO \xrightarrow[C_2H_5OH]{NaBH_4} NO_2CH_2CH_2CH_2CH_2OH$$

$$C_6H_5COCH_2Br \xrightarrow[C_2H_5OH]{NaBH_4} C_6H_5CHOHCH_2Br$$

Some exceptions to the general rule that olefinic double bonds are not attacked by hydride reducing agents have been noted in the reduction of β-aryl-$\alpha\beta$-unsaturated carbonyl compounds with lithium aluminium hydride. Even in these cases selective reduction of the carbonyl group can often be achieved by working at low temperatures and with short reaction times or by using sodium borohydride or aluminium hydride (p. 479) as reducing agent (7.107). This type of reduction of the double bond of allylic alcohols is thought to proceed through a cyclic organo-aluminium compound (59), for it is found experimentally that only

$$C_6H_5CH{=}CH{-}CHO \xrightarrow[35°C,\ \text{ether}]{LiAlH_4\ \text{excess}} C_6H_5CH_2CH_2CH_2OH$$

$$C_6H_5CH{=}CH{-}CHO \xrightarrow{NaBH_4\ \text{or}\ LiAlH_4,\ \text{ether},\ -10°C} C_6H_5CH{=}CHCH_2OH \quad (7.107)$$

one of the two hydrogen atoms added to the double bond is derived from the hydride, and acidification with a deuterated solvent leads to the deuterated alcohol shown in (7.108).

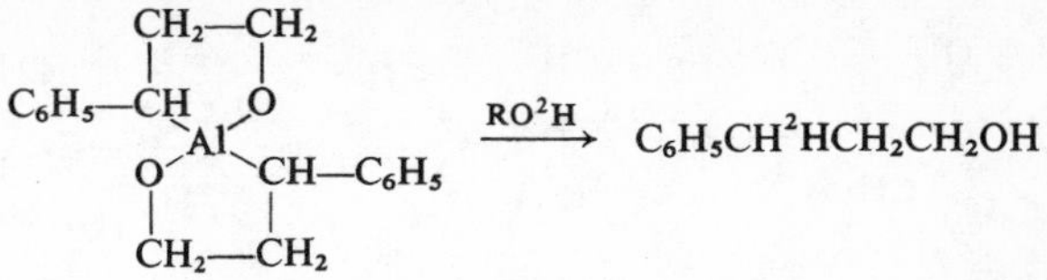

$$\underset{(59)}{\begin{array}{c} CH_2\text{—}CH_2 \\ | \quad\quad | \\ C_6H_5\text{—}CH \quad O \\ \diagdown Al \diagup \\ O \diagup \quad \diagdown CH\text{—}C_6H_5 \\ | \quad\quad | \\ CH_2\text{—}CH_2 \end{array}} \xrightarrow{RO^2H} C_6H_5CH^2HCH_2CH_2OH \qquad (7.108)$$

A similar type of aluminium compound is thought to be involved in the reduction of the triple bond of propargylic alcohols to a *trans* double bond with lithium aluminium hydride (cf. Grant and Djerassi, 1974). Only triple bonds flanked by hydroxyl groups are reduced, as illustrated in the example (7.109). The reaction is useful for the preparation of labelled allylic alcohols.

$$CH{\equiv}C(CH_2)_2C{\equiv}CCO_2C_2H_5 \xrightarrow{LiAlH_4} \underset{trans}{CH{\equiv}C(CH_2)_2CH{=}CHCH_2OH}$$

$$HC{\equiv}CCHOHC_4H_9 \xrightarrow[(2)\ ^2H_2O]{(1)\ LiAlH_4} \begin{array}{c} H \quad\quad\quad H \\ \diagdown C{=}C \diagup \\ ^2H \diagup \quad\quad \diagdown CHOHC_4H_9 \end{array} \qquad (7.109)$$

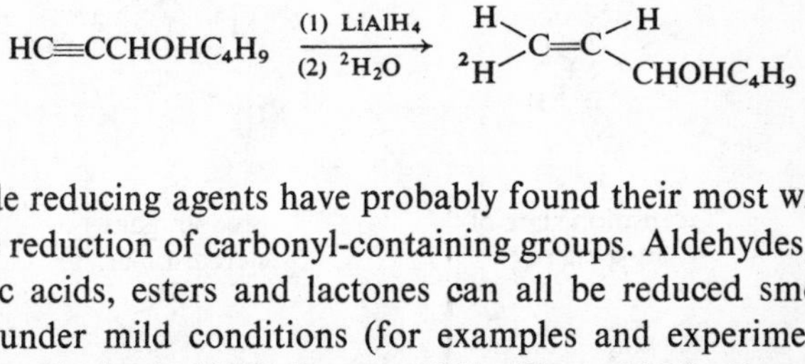

Hydride reducing agents have probably found their most widespread use in the reduction of carbonyl-containing groups. Aldehydes, ketones, carboxylic acids, esters and lactones can all be reduced smoothly to alcohols under mild conditions (for examples and experimental conditions see Gaylord, 1956; Brown, 1951; Pizey, 1974). Reaction with lithium aluminium hydride is the method of choice for the reduction of carboxylic esters to primary alcohols. Substituted amides are converted into amines or aldehydes, depending on the experimental conditions (see p. 476). When necessary, the carbonyl group of an aldehyde or ketone may be protected against reduction by conversion into the acetal. For example, acetoacetic ester can be converted into 4-hydroxy-2-butanone by reduction of the corresponding diethyl acetal with lithium aluminium hydride, followed by acid hydrolysis of the acetal group.

Reduction of unsymmetrical open-chain ketones such as ethyl methyl ketone leads, of course, to the racemic carbinol. With ketones containing a chiral centre, however, the two forms of the carbinol may not be formed in equal amount. Thus, in the reduction of the ketone (60) with lithium aluminium hydride, the *threo* form of the alcohol predominates in the product (7.110). The main product formed in these reactions can be

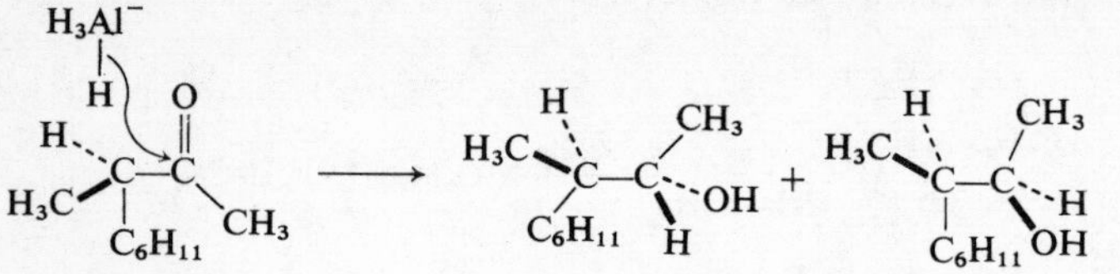

(60) → (72% of product) + (28% of product) (7.110)

predicted on the basis of Cram's rule (Cram and Abd Elhafez, 1952) according to which that diastereoisomer predominates in the product which is formed by approach of the reagent from the less hindered side of the carbonyl group when the rotational conformation of the molecule is such that the carbonyl group is flanked by the two least bulky groups on the adjacent chiral centre. This may be represented using Newman projection formulae, as (7.111), where S, M and L

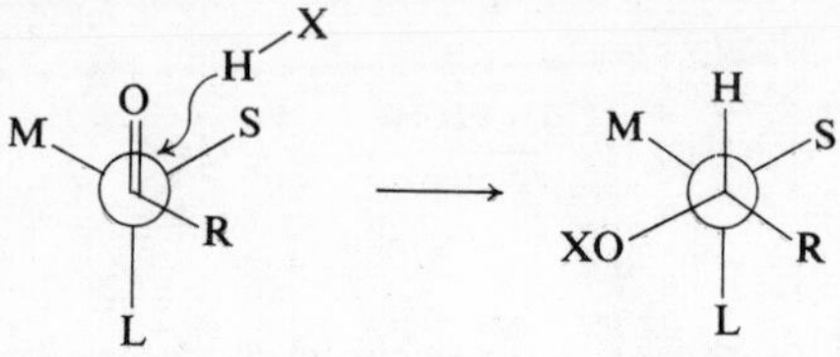

transition state of least energy → predominant stereoisomer

or as (7.111)

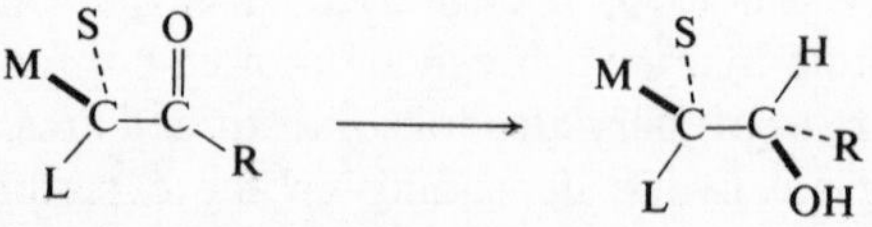

represent small, medium and large substituents. Thus, for the reduction of the ketone (60) the predominant *threo* carbinol arises by attack of the metal hydride anion on the less hindered side of the carbonyl group in the conformation shown. A somewhat different model for the transition state which is said to predict results more in agreement with experiment has been proposed by Cherest, Felkin and Prudent (1968) but the essential feature of Cram's hypothesis, involving approach of the hydride anion from the less hindered side of the carbonyl group is preserved.

The selectivity obtained in these reactions increases with the size of the reducing agent, and some remarkably stereoselective reactions have been achieved using complex hydride reducing agents specifically designed for the purpose (see p. 486). Thus, the prostaglandin precursor (61), in which the carbonyl group is several carbon atoms removed from a chiral centre, gave a product containing 92 per cent of the (*S*)-alcohol (62) on reduction with the chiral borohydride (63). When the reducing

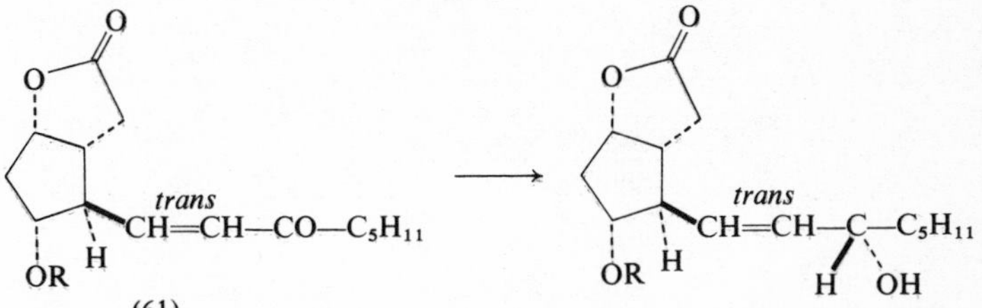

trans CH=CH—CO—C_5H_{11} →   *trans* CH=CH—C(H)(OH)—C_5H_{11}

(61)   (62)

H_3C, $\bar{B}$—H, $C(CH_3)_2CH(CH_3)_2$, Li^+, CH_3

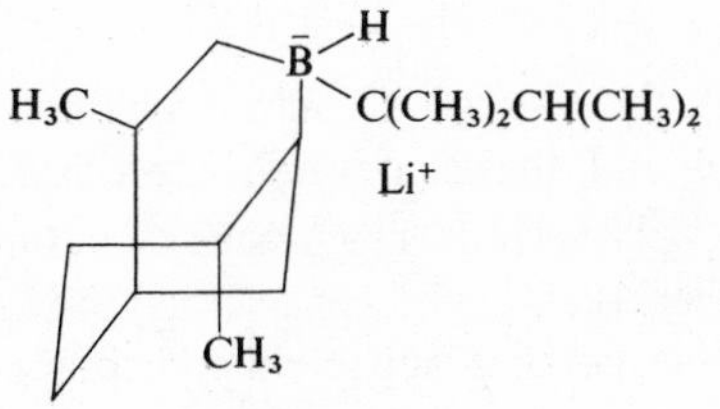

(63)   (7.112)

agent itself is chiral, asymmetric reduction of open-chain ketones can sometimes be achieved even when they themselves are achiral (see Morrison and Mosher, 1971). Thus, with the complex $LiAlH(OC_2H_5)(OR)_2$ obtained from lithium aluminium hydride and a monosaccharide in the presence of ethanol, carbinols of optical purity as high as 63 per cent have been obtained (Landor, Miller and Tatchell, 1967).

Enzymes are, of course, chiral reagents *par excellence*, and enzymic reduction of benzaldehyde derivatives using the enzyme liver alcohol dehydrogenase and the reduced form of its coenzyme, nicotinamide adenine dinucleotide (NADH), has been used recently to prepare a number of specifically labelled benzyl alcohols required for biosynthetic work. In these reactions transfer of hydride takes place from one face of NADH to the *re* face of the aldehyde, that is to the top face as it is written in (7.113). Thus, reaction of deuterated benzaldehyde (64) with liver alcohol dehydrogenase and NADH, generated *in situ* from NAD^+ and

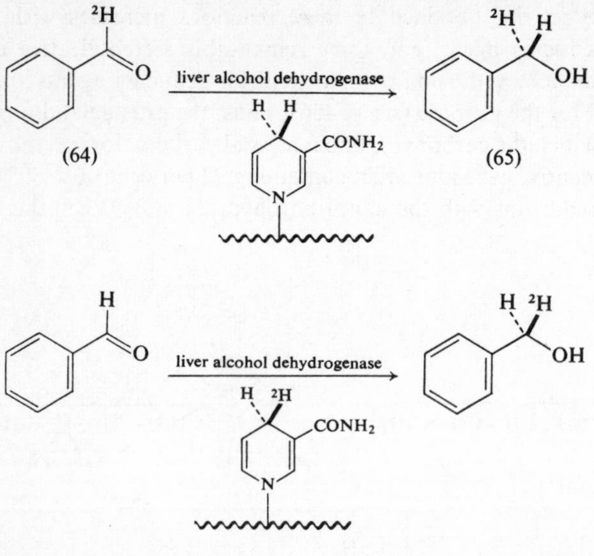

liver alcohol dehydrogenase

(64) (65)

(7.113)

liver alcohol dehydrogenase

ethanol, gives specifically the (*S*)-[1-^{2}H] benzyl alcohol (65), whereas using benzaldehyde itself and the deuterated coenzyme, prepared *in situ* from NAD^+ and $CH_3C^2H_2OH$, the (*R*)-alcohol is obtained (compare Battersby and Staunton, 1974).

In cases where there is a polar group on the carbon atom adjacent to the carbonyl group Cram's rule may not be followed, because the conformation of the transition state is no longer determined entirely by steric factors. In α-hydroxy and α-amino ketones for example, reaction is thought to proceed through a relatively rigid cyclic transition state of the type (66) with fixed conformation, and because of this reductions of α-hydroxy and α-amino ketones usually proceed with a comparatively high degree of stereoselectivity (7.114). Again, where an adjacent asymmetric carbon atom carries a chlorine substituent, the most reactive conformation of the molecule appears to be the one in which the polar halogen atom and the polar carbonyl group are *trans*, to minimise dipole–dipole repulsion. The predominant product is then formed by approach of the metal hydride anion to the less hindered side of this conformation (see p. 129).

With strongly hindered cyclic ketones the main product is again formed by approach of the reagent to the less hindered side of the carbonyl group. Thus, reduction of camphor with lithium aluminium hydride leads mainly to the *exo* alcohol (isoborneol), whereas nor-

$C_6H_5CH(OH)COC_6H_4CH_3$-*p* $\xrightarrow[\text{ether}]{LiAlH_4}$ (66) $\rightarrow$ $C_6H_5CH(OH)CH(OH)C_6H_4CH_3$-*p* (85% of product) (7,114)

camphor, in which approach of the hydride anion is now easier from the side of the methylene bridge, leads mainly to the *endo* alcohol.

The stereochemical course of the reduction of other less rigid cyclohexanones is not so easy to predict. In general, a mixture of products is obtained in which, with comparatively unhindered ketones, the more stable epimer predominates. But the relative proportion of the two isomers varies widely with the substitution of the ring (cf. Wigfield and Phelps, 1974).

It has been suggested (Dauben, Fonken and Noyce, 1956) that the steric course of these reactions is controlled by two usually opposing factors, a kinetic factor related to the degree of steric hindrance of the carbonyl group (steric approach control), and a thermodynamic factor related to the stability of the products (product development control). But it is now believed that the results in all cases can be accounted for largely by steric factors (Kirk, 1969; Cherest and Felkin, 1968; Eliel and Senda, 1970; Ashby and Laemmie, 1975). Richer (1965) has shown that, contrary to previous expectation, approach of a hydride ion to the carbonyl group of an unhindered cyclohexanone is easier from the axial side, because equatorial approach is hindered sterically by the axial hydrogen atoms or substituents in the 2- and 6-positions. The result is preferential formation of the more stable equatorial alcohol. But where the approaching group becomes larger, as with Grignard reagents, or where there are bulky axial substituents in the 3- and 5- positions, axial approach of the reagent is less easy and equatorial attack, leading to the axial alcohol, may be favoured. Thus, reduction of 4-t-butyl-2,2-dimethylcyclohexanone gives more equatorial alcohol than does the reduction of 4-t-butylcyclohexanone because the greater steric effect of the C-2 axial methyl group compared with that of an axial hydrogen atom, favours increased axial attack on the carbonyl group of the dimethyl compound. On the other hand, axial attack on

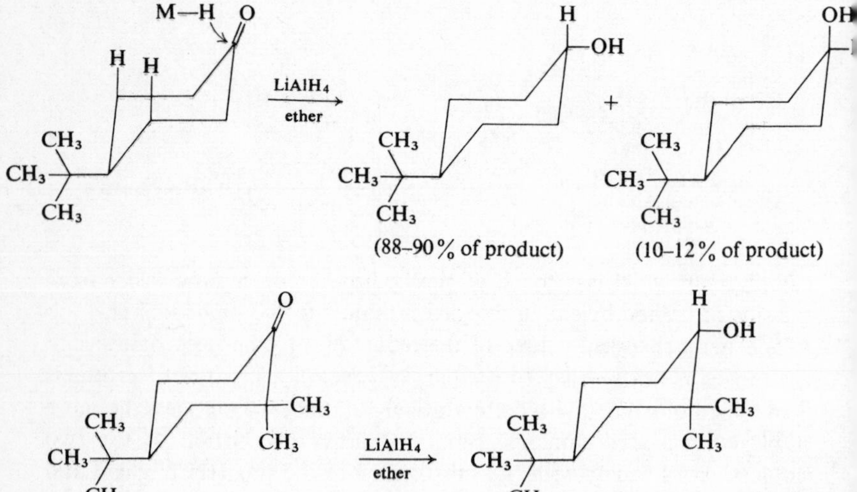

M—H O

H H

$\xrightarrow[\text{ether}]{LiAlH_4}$

(88–90% of product) + (10–12% of product)

$\xrightarrow[\text{ether}]{LiAlH_4}$ (95%)

(7.115)

$\xrightarrow[\text{ether}]{LiAlH_4}$ (45% of product) + (55% of product)

$LiAlH_4$, $AlCl_3$ (1:4), excess of ketone, ether

(only)

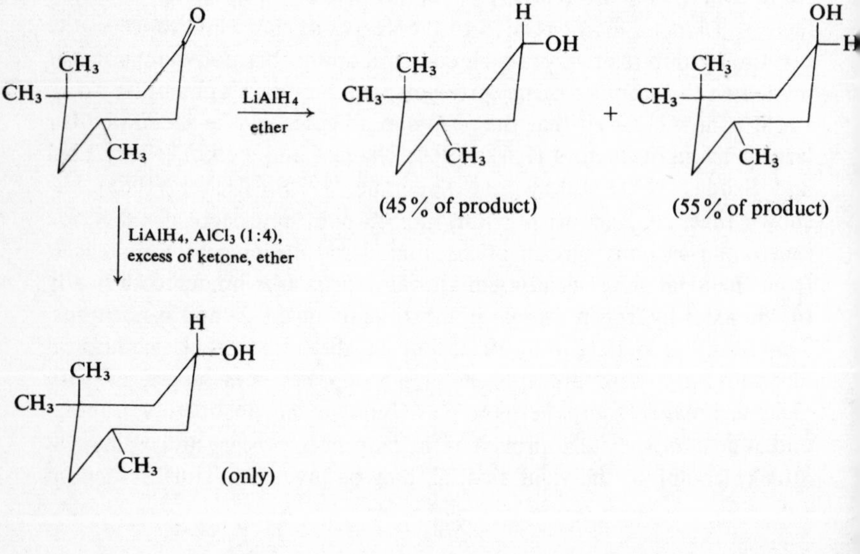

the carbonyl group of 3,3,5-trimethylcyclohexanone is hindered by the steric effect of the axial C-3 methyl group, and reduction of this compound leads to a mixture of isomeric alcohols containing approximately equal parts of each (7.115).

Improved yields of the more stable alcohol can often be obtained by effecting the reduction of a cyclohexanone with lithium aluminium hydride in presence of aluminium chloride and an excess of the ketone. Thus, 3,3,5-trimethylcyclohexanone is converted entirely into the more stable equatorial alcohol under these conditions. An equilibrium is set up between the aluminium alkoxide and the ketone similar to that found in the Meerwein–Pondorff–Verley reduction, resulting in gradual accumulation of the more stable alkoxide. The less stable axial alcohol can often be obtained by reduction with bulky hydride reagents (cf. p. 477).

Reductive fission of epoxides and of alkyl halides or sulphonates with lithium aluminium hydride proceeds by S_N2 substitution by hydride ion, with formation of a new C—H bond. Epoxides are thereby reduced to alcohols. With unsymmetrical epoxides reaction takes place at the less highly substituted carbon atom to give the more highly substituted alcohol. Thus, 3,4-epoxybutane yields a mixture in which the secondary alcohol predominates, epoxymethylenecyclohexane is converted into 1-methylcyclohexanol, and triphenylethylene oxide gives 1,1,2-triphenylethanol (7.116). In presence of 1/3 mole aluminium

$$CH_2{=}CHCH{-}CH_2 \text{ (epoxide, O bridging)} \xrightarrow[\text{ether}]{LiAlH_4} CH_2{=}CHCHOHCH_3 \text{ (mainly)}$$

$$(C_6H_5)_2C{-}CHC_6H_5 \text{ (epoxide, O bridging)} \xrightarrow[\text{ether}]{LiAlH_4} (C_6H_5)_2COHCH_2C_6H_5 \text{ only} \qquad (7.116)$$

$$\xrightarrow{LiAlH_4{-}AlCl_3\ (3:1)} (C_6H_5)_2CHCHOHC_6H_5 \text{ only}$$

chloride, the direction of ring opening in the last reaction is reversed, and high yields of the alternative alcohol are obtained, presumably because the reactive species is now the electrophilic aluminium hydride formed *in situ* from aluminium chloride and lithium aluminium hydride (Ashby and Prather, 1966).

$$AlCl_3 + 3LiAlH_4 \longrightarrow 4AlH_3 + 3LiCl$$

In accordance with the S_N2 type mechanism, reduction of epoxides with lithium aluminium hydride takes place with inversion of configuration at the carbon atom attacked. Thus 1,2-dimethylcyclohexene epoxide affords *trans*-1,2-dimethylcyclohexanol (7.117).

$$\text{(CH}_3\text{, O, CH}_3\text{)} \xrightarrow[\text{ether}]{LiAlH_4} \text{(CH}_3\text{, OH, H, CH}_3\text{)} \qquad (7.117)$$

Primary and secondary alkyl halides are reduced to the hydrocarbons with lithium aluminium hydride, although better results are obtained with sodium cyanoborohydride (p. 481) or with lithium triethylborohydride (p. 484). Tertiary halides react only slowly and give mostly olefins. Aryl iodides and bromides may be reduced to the hydrocarbons in boiling tetrahydrofuran (Brown and Krishnamurthy, 1969) and other supposedly inert compounds such as vinyl and bridgehead halides can also be converted into the parent hydrocarbons under appropriate conditions (Jefford, Kirkpatrick and Delay, 1972). α-Bromo-α-methylstyrene, for example, gives α-methylstyrene and 1-bromoadamantane is easily converted into adamantane. With cyclopropyl halides the halogens are replaced by hydrogen without rupture of the ring. 7,7-Dibromobicyclo[4.1.0]heptane, for example, gives the *syn*- and *anti*-monobromo compounds and, on further reduction, the parent hydrocarbon. Sulphonate esters of primary and secondary alcohols are also readily reduced with lithium aluminium hydride and this reaction has been widely used in synthesis to effect the replacement of an alcoholic hydroxyl group by hydrogen. Sodium borohydride in dimethyl sulphoxide or tetramethylene sulphone is an excellent reagent for reductive fission of benzylic and primary, secondary and, in certain cases, tertiary alkyl halides and tosylates in the presence of other reducible groups in the molecule, including carboxylic acid, ester or nitro groups (Hutchins, Hoke, Keogh and Koharski, 1969; Bell, Vanderslice and Spehars, 1969) (7.118).

$$p\text{-}NO_2C_6H_4CH_2Br \xrightarrow[\text{dimethyl sulphoxide}]{NaBH_4} p\text{-}NO_2C_6H_4CH_3 \quad (98\%) \qquad (7.118)$$

Another useful procedure for the formation of a methyl or methylene group is reduction of an acyl toluene-*p*-sulphonylhydrazide, or the toluene-*p*-sulphonylhydrazone of an aldehyde or ketone, with lithium aluminium hydride. Thus, 2-naphthaldehyde is converted into 2-methylnaphthalene and 1-naphthylacetic acid gives 1-ethylnaphthalene by this procedure (Cagliotti, 1972). Better results are obtained with sodium cyanoborohydride (p. 481).

Primary, secondary and tertiary bromides and tosylates are also said to be reduced rapidly to the corresponding hydrocarbons with the new reagent lithium butylcopperhydride, $LiCuHC_4H_9$, itself obtained by reaction of copper hydride with 2-butyl-lithium (Masamune, Bates and Georghiou, 1974). Thus, 1-bromoadamantane gave adamantane in 70 per cent yield and 3-cyclohexenyl tosylate was converted into cyclohexene in 75 per cent yield without elimination to give cyclohexadiene. Carboxylic ester groups are unaffected by the reagent, but aldehydes and ketones are reduced to the carbinol.

Lithium hydridoalkoxyaluminates. Lithium aluminium hydride itself is an extremely powerful and versatile reducing agent. More selective reagents can be obtained by modification of lithium aluminium hydride by treatment with alcohols or with aluminium chloride. One such reagent is the sterically bulky lithium hydridotri-t-butoxyaluminate, which is readily prepared by action of the stoichiometric amount of t-butyl alcohol on lithium aluminium hydride.

$$LiAlH_4 + 3ROH \longrightarrow LiAlH(OR)_3 + 3H_2$$

Analogous reagents are obtained in the same way from other alcohols, and by replacement of only one or two of the hydrogen atoms of the hydride by alkoxy groups, affording a range of reagents of graded activities (Rerick, 1968; Málek and Černý, 1972).

Lithium hydridotri-t-butoxyaluminate is a much milder reducing agent than lithium aluminium hydride itself (Brown, Weissman and Nung Min Yoon, 1966). Thus, although aldehydes and ketones are reduced normally to alcohols, carboxylic esters and epoxides react only slowly, and halides, nitriles and nitro groups are not attacked. Aldehydes and ketones can thus be selectively reduced in presence of these groups. A good example is the first reaction shown in equation (7.119).

One of the most useful applications of the alkoxy reagents is in the preparation of aldehydes from carboxylic acids by partial reduction of the acid chlorides or dimethylamides. Acid chlorides are readily reduced with lithium aluminium hydride itself or with sodium borohydride to the corresponding alcohols, but with one equivalent of the tri-t-butoxy compound high yields of the aldehyde can be obtained in the presence of a range of other functional groups (7.119).

$$\text{OHC}/\text{CH}_3\text{O}_2\text{C}\text{-bicyclic lactone} \xrightarrow[\text{THF, } -10^\circ\text{C}]{\text{LiAlH(OC}_4\text{H}_9\text{-t)}_3} \text{HOCH}_2/\text{CH}_3\text{O}_2\text{C}\text{-bicyclic lactone}$$

$$\text{N}\equiv\text{C}-\text{C}_6\text{H}_4-\text{COCl} \xrightarrow[\text{diglyme, } -78^\circ\text{C}]{\text{LiAlH(OC}_4\text{H}_9\text{-t)}_3} \text{N}\equiv\text{C}-\text{C}_6\text{H}_4-\text{CHO} \quad (80\%) \qquad (7.119)$$

Although esters, in general, are reduced only slowly, phenyl esters are converted into the aldehyde with $LiAlH(OC_4H_9)_3$. Thus, phenyl cyclohexanecarboxylate gives formylcyclohexane in 60 per cent yield.

Reduction of *N*-substituted amides with excess of lithium aluminium hydride affords the corresponding amines in good yield (7.120). Reaction

$$\text{C}_6\text{H}_{11}-\text{CON(CH}_3)_2 \xrightarrow[\text{ether}]{\text{LiAlH}_4} \text{C}_6\text{H}_{11}-\text{CH}_2\text{N(CH}_3)_2 \quad (88\%)$$

$$\downarrow \qquad\qquad \uparrow \qquad (7.120)$$

$$\underset{(67)}{\text{C}_6\text{H}_{11}-\text{CH(OAlH}_3)-\ddot{\text{N}}\text{(CH}_3)_2} \longrightarrow \text{C}_6\text{H}_{11}-\text{CH}=\overset{+}{\text{N}}\text{(CH}_3)_2$$

is believed to proceed through an aldehyde-ammonia derivative of the type (67). With the less active agents $LiAlH(OC_2H_5)_3$ or $LiAlH_2(OC_2H_5)_2$ (the tri-t-butoxy compound is ineffective in this case) reaction stops at the aldehyde–ammonia stage and hydrolysis of the product affords the corresponding aldehyde. Yields are usually high except in the case of $\alpha\beta$-unsaturated amides, and the reaction proceeds readily in presence of other reducible groups (7.121).

Similarly, reduction of nitriles with lithium aluminium hydride affords a primary amine by way of the imine salt (7.122). With lithium

$$p\text{-}O_2N\text{-}C_6H_4\text{-}CON(CH_3)_2 \xrightarrow{LiAlH(OC_2H_5)_3} p\text{-}O_2N\text{-}C_6H_4\text{-}CHO \ (89\%) \qquad (7.121)$$

$$CH_2{=}CH(CH_2)_8CON(CH_3)_2 \longrightarrow CH_2{=}CH(CH_2)_8CHO \ (85\%)$$

$$R{-}C{\equiv}N \xrightarrow[\text{ether}]{LiAlH_4} [R{-}CH{=}N{-}\bar{A}lH_3] \xrightarrow[(2)\ H_3O^+]{(1)\ LiAlH_4} RCH_2NH_2 \qquad (7.122)$$

hydridotriethoxyaluminate, however, reaction stops at the imine stage, and hydrolysis gives the aldehyde. By this procedure trimethylacetonitrile was converted into trimethylacetaldehyde in 75 per cent yield.

Sodium bis[2-methoxyethoxy] aluminium hydride, $NaH_2Al(OCH_2CH_2OCH_3)_2$, is a powerful reducing agent comparable in activity to lithium aluminium hydride. It is, therefore, of little value for selective reductions, but it has the advantage that it is more stable than lithium aluminium hydride and is soluble in aromatic solvents such as benzene and toluene (Málek and Černý, 1972).

Because of the steric effect of the alkoxy groups the hydridoalkoxyaluminates are more stereoselective in their action than is lithium aluminium hydride itself. But there is no very clear correlation between the size of the alkoxy groups and the selectivity of the reductions (cf. Haubenstock, 1973; Brown and Deck, 1965). The situation is complicated by the fact that some of the alkoxy compounds disproportionate readily, with regeneration of the reactive tetrahydroaluminate ion. Concen-

$$2H_3\bar{A}lOCH(CH_3)_2 \rightleftharpoons AlH_4^- + H_2\bar{A}l[OCH(CH_3)_2]_2$$

tration effects may also have to be considered. Thus, in the reduction of 3,3,5-trimethylcyclohexanone, lithium hydridotrimethoxyaluminate $LiAlH(OCH_3)_3$ was, as expected, more stereoselective than lithium aluminium hydride, but it was also more selective than the apparently more bulky tri-t-butoxy compound, $LiAlH(OC_4H_9\text{-t})_3$. The reason is that in tetrahydrofuran, the usual solvent, the trimethoxy compound is dimeric or trimeric, whereas the tri-t-butoxy compound is monomeric. The selectivity obtained with the trimethoxy compound depends on its concentration (compare Ashby, Sevenair and Dobbs, 1971).

TABLE 7.4 *Reduction of 3,3,5-trimethylcyclohexanone*

Reagent	cis-3,3,5-trimethylcyclohexanol (equatorial OH) (%)	trans-3,3,5-trimethylcyclohexanol (axial OH) (%)
$NaBH_4$, $(CH_3)_2CHOH$	36–45	55–64
$LiAlH_4$, ether	37–48	52–63
$LiAlH(OCH_3)_3$, THF	2–8	92–98
$LiAlH(OC_4H_9\text{-t})_3$, THF	4–12	88–96

Mixed lithium aluminium hydride–aluminium chloride reagents. Further useful modification of the properties of lithium aluminium hydride is achieved by addition of aluminium chloride in various proportions. This serves to release various mixed chloride-hydrides of aluminium as shown in (7.123). The general effect of the addition of

$$\begin{aligned} 3LiAlH_4 + AlCl_3 &\longrightarrow 3LiCl + 4AlH_3 \\ LiAlH_4 + AlCl_3 &\longrightarrow LiCl + 2AlH_2Cl \\ LiAlH_4 + 3AlCl_3 &\longrightarrow LiCl + 4AlHCl_2 \end{aligned} \qquad (7.123)$$

aluminium chloride is to lower the reducing power of lithium aluminium hydride and in consequence to produce reagents which are more specific for particular reactions (Eliel, 1961; Rerick, 1968). For example, the carbon–halogen bond is often inert to the mixed hydride reagents. Advantage is taken of this in the reduction of polyfunctional compounds in which retention of halogen is desired, as in the conversion of methyl 3-bromopropionate into 3-bromopropanol; lithium aluminium hydride alone produces propanol (7.124).

$$BrCH_2CH_2CO_2CH_3 \xrightarrow[\text{ether}]{LiAlH_4\text{–}AlCl_3\ (1:1)} BrCH_2CH_2CH_2OH \qquad (7.124)$$

Similarly, nitro groups are not so easily reduced as with lithium aluminium hydride itself, and *p*-nitrobenzaldehyde can be converted into *p*-nitrobenzyl alcohol in 75 per cent yield. Aldehydes and ketones are reduced to carbinols, and there is no advantage in the use of mixed hydrides in these cases, although it should be noted that the stereochemical result obtained in the reduction of cyclic ketones may not be the same as with lithium aluminium hydride itself. With diaryl ketones and with aryl alkyl ketones however the carbonyl group is reduced to methylene in high yield, and this procedure offers a useful alternative to the Clemmensen or Huang-Minlon methods for reduction of this type of ketone.

Reduction with lithium aluminium hydride–aluminium chloride (3:1) also provides an excellent route from $\alpha\beta$-unsaturated carbonyl compounds to unsaturated alcohols which are difficult to prepare with lithium aluminium hydride alone because of competing reduction of the olefinic double bond. The effective reagent is thought to be aluminium hydride formed *in situ* from lithium aluminium hydride and aluminium chloride (7.125).

$$C_6H_5CH{=}CHCO_2C_2H_5 \xrightarrow[\text{ether}]{3LiAlH_4\text{–}AlCl_3} \underset{(90\%)}{C_6H_5CH{=}CHCH_2OH} \qquad (7.125)$$

The most striking difference between lithium aluminium hydride and the mixed hydrides is seen in the reduction of epoxides, where two different modes of ring opening of the epoxide ring are observed (see p. 473) (Rerick and Eliel, 1962). Ethers in general are not attacked by either reagent. Acetals and ketals are reduced with the mixed reagent to ethers (7.126), probably through cleavage by excess of aluminium

$$C_6H_5C(CH_3)(OC_2H_5)_2 \xrightarrow[\text{ether}]{LiAlH_4\text{–}AlCl_3\ (1:4)} C_6H_5CH(CH_3)OC_2H_5 \qquad (7.126)$$

chloride to an oxonium salt which is then reduced by the aluminium hydride species present (7.127). Esters and lactones may also be reduced

$$C_6H_5C(CH_3)(OC_2H_5)(OC_2H_5 \cdot AlCl_3) \longrightarrow C_6H_5C(CH_3){=}\overset{+}{O}C_2H_5 \longrightarrow C_6H_5CH(CH_3)OC_2H_5 \qquad (7.127)$$

$$(68) \xrightarrow[\text{diglyme}]{NaBH_4,\ BF_3} (69)\ (86\%) \qquad (7.128)$$

CH₃ ; O ; H₃C C=O ; (68) → CH₃ ; O ; H₃C CH₂ ; (69) (86%)

to ethers with lithium aluminium hydride and aluminium chloride or sodium borohydride and boron trifluoride. Thus the lactone (68) readily affords the ether (69) (7.128). The actual reducing agent in the latter reaction is diborane, B_2H_6, formed by reaction of sodium borohydride and boron trifluoride. It is capable of reducing a number of functional groups but its most useful synthetic application is in the addition to carbon–carbon multiple bonds to form alkylboranes. These reactions are discussed in Chapter 5.

Di-isobutylaluminium hydride, DIBAL. This derivative of aluminium hydride is available commercially either as the neat liquid or as a solution in toluene. It finds its greatest use, probably, in the preparation of aldehydes. Lactones, esters, amides and nitriles are all reduced to aldehydes under mild conditions (e.g. Corey, Nicolaou and Toru, 1975) (7.129). It is also a useful reagent for the reduction of disubstituted

O O OH OH CHO

DIBAL

toluene, $-78°C$

(94%)

CH_3O OCH_3 CH_3O OCH_3

CH_3O_2C CO_2CH_3 DIBAL, 1·4 equiv. CH_2Cl_2, $-78°C$ CH_3O_2C CHO

50% (7.129)

O O O O

DIBAL

benzene, 25°C

CN CHO

(85%)

acetylenes to *cis* olefins, and for 1,2-reduction of $\alpha\beta$-unsaturated carbonyl compounds. Thus, in a synthesis of linoleic acid the diyne (70) was reduced to the *cis,cis*-diene in 82 per cent yield. The usual catalytic method might have cleaved the carbon–carbon bond as well.

$$CH_3(CH_2)_5C{\equiv}CCH_2C{\equiv}C(CH_2)_5CH_2Cl \xrightarrow[\text{(2) } CH_3OH]{\text{(1) DIBAL, 0°C}}$$

(70)

$$CH_3(CH_2)_5CH\overset{cis}{=}CHCH_2CH\overset{cis}{=}CH(CH_2)_5CH_2Cl$$

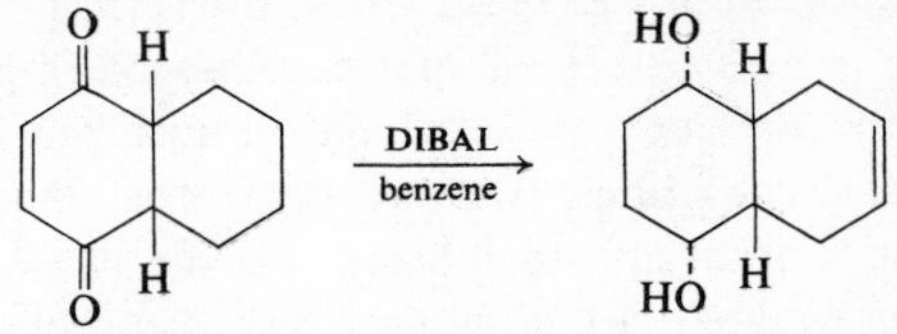

$$\xrightarrow[\text{benzene}]{\text{DIBAL}} \qquad (7.130)$$

(79%)

Mixed lithium aluminium hydride–copper reagents. Reagents prepared from lithium aluminium hydride and copper salts have also been used to bring about some selective reductions. Lithium butylcopper hydride is useful for the reductive fission of alkyl halides and tosylates (p. 475), and lithium aluminium hydride modified by addition of copper(I) iodide is an effective reagent for the reduction of open-chain conjugated enones to the saturated ketones (7.131). The active species is believed

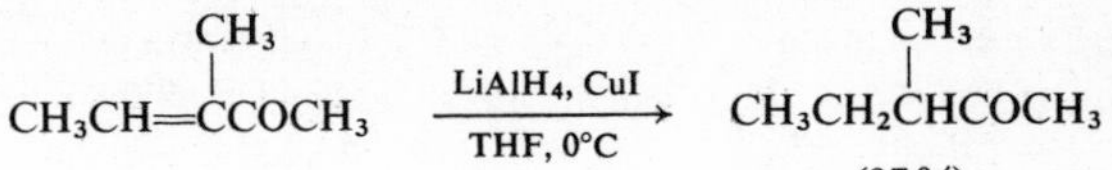

$$CH_3CH{=}C(CH_3)COCH_3 \xrightarrow[\text{THF, 0°C}]{LiAlH_4,\ CuI} CH_3CH_2CH(CH_3)COCH_3$$

(97%)

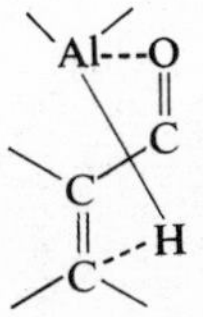

(7.131)

(71)

to be H_2AlI and a six-membered cyclic transition state (71) is proposed, explaining why cyclic enones do not react (Ashby and Lin, 1975). Reduction of $\alpha\beta$-unsaturated esters and ketones to the saturated compounds can also be effected with solutions of 'copper hydride' prepared by reaction of lithium hydridoalkoxyaluminates with copper(I) bromide (Semmelhack and Stauffer, 1975).

Sodium cyanoborohydride. A number of reagents derived from sodium borohydride by replacement of one or more of the hydrogen atoms by other groups has been developed in order to achieve more selective reduction. Among the most useful are sodium cyanoborohydride and some trialkylborohydrides (Walker, 1976).

Because of the electron-withdrawing effect of the cyano group, sodium cyanoborohydride is a weaker and more selective reagent than sodium borohydride itself (see Lane, 1975). It has the further advantage that it is stable in acid to pH = 3 and can thus be employed to effect reductions in the presence of functional groups which are sensitive to the more basic conditions of reduction with sodium borohydride. In neutral solution in hexamethylphosphoramide it is a useful reagent for reductive removal of iodo, bromo and tosyloxy groups. Under these conditions carbonyl groups and other reduceable groups are not affected. 1-Iododecane gives decane in 90 per cent yield, and 3-bromo-1,2-epoxy-1-phenylpropane is converted into the halogen-free compound without cleavage of the epoxide group (7.132).

$$C_6H_5\overset{O}{CH-CH}CH_2Br \xrightarrow[\text{hexamethylphosphoramide, 70°C}]{NaBH_3CN} C_6H_5\overset{O}{CH-CH}CH_3 \quad (63\%)$$

$$C_6H_5CHO + C_2H_5NH_2 \underset{}{\overset{pH6}{\rightleftharpoons}} [C_6H_5CH{=}\overset{+}{N}HC_2H_5] \xrightarrow{NaBH_3CN,\ CH_3OH} C_6H_5CH_2NHC_2H_5 \quad (80\%) \qquad (7.132)$$

$$C_6H_5CH_2COCO_2Na \xrightarrow[CH_3OH]{NH_4Br,\ NaBH_3CN} C_6H_5CH_2\underset{NH_2}{CH}CO_2H \quad (49\%)$$

Aldehydes and ketones are unaffected by sodium cyanoborohydride in neutral solution, but they are readily reduced to the corresponding alcohol at pH = 3–4 by way of the protonated carbonyl group (Borch, Bernstein and Durst, 1971). The reactions are conveniently effected in methanolic hydrogen chloride; under these conditions cyclohexanone is converted smoothly into cyclohexanol, and farnesal gives farnesol in 92 per cent yield. By previous exchange of the hydrogens of $^{-}BH_3CN$ for deuterium or tritium, by reaction with ^2H_2O or tritiated water, an efficient and economical route is available for deuteride or tritiide reduction of aldehydes and ketones.

$$NaBH_3CN + {}^2H_2O \longrightarrow NaB^2H_3CN \xrightarrow{C_6H_5CHO} C_6H_5CH^2HOH$$

Iminium groups are even more easily reduced than carbonyl groups in acid solution, and this has been exploited in a method for reductive amination of aldehydes and ketones by way of the iminium salts formed from the carbonyl compounds and a primary or secondary amine at pH = 6; at this pH the carbonyl compounds themselves are unaffected. The reaction has been adapted for the preparation of α-amino acids from α-keto acids (7.132).

Tetrabutylammonium cyanoborohydride, $(C_4H_9)_4\overset{+}{N}\,\overset{-}{B}H_3CN$, is said to be even more selective than the sodium compound. It reduces only primary iodides and some bromides and, with addition of acid can apparently be used for the selective reduction of aldehydes in the presence of ketones (Hutchins and Kandasamy, 1973).

A well-known method for the conversion of carbonyl compounds into the corresponding hydrocarbons involves reduction of the derived toluene-*p*-sulphonyl-(tosyl-) hydrazones with sodium borohydride (Cagliotti, 1972). Very much better yields are obtained using sodium cyanoborohydride in acidic dimethylformamide; the reactions are more selective and alkenes and other unwanted side products are not formed. The reaction is specific for aliphatic and alicyclic carbonyl compounds; aromatic compounds are unaffected. The tosylhydrazone need not be isolated and is prepared and reduced *in situ*.

$$CH_3(CH_2)_8CHO \xrightarrow[\text{(2) NaBH}_3\text{CN, dimethylformamide, 100°C}]{\text{(1) Tosylhydrazide}} CH_3(CH_2)_8CH_3 \quad (95\%)$$

(7.133)

NNHTos

NaBH₃CN

dimethylformamide, HCl, 100°C

(70%)

(72)

Tos = CH_3–C_6H_4–SO_2

With αβ-unsaturated carbonyl compounds reduction of the tosylhydrazone is accompanied by migration of the double bond to the carbon atom which originally carried the oxygen. For example, cinnamaldehyde tosylhydrazone gives 3-phenyl-1-propene in 98 per cent yield, and β-ionone is converted into the diene (72) with deconjugation of the double bonds. The double bond migrations are stereoselective, giving predominantly the *trans* olefin.

The proposed mechanism involves reduction of the iminium ion to the tosylhydrazine, elimination of *p*-toluenesulphinic acid and subsequent [1,5]-sigmatropic shift of hydrogen, with loss of nitrogen, to the rearranged alkene (Hutchins, Kacher and Rua, 1975). In agree-

(7.134)

ment with this mechanism reduction of the *cisoid* tosylhydrazone (73) with NaB^2H_3CN and HCl gave 6-[^{2}H]cholest-5-ene (74) while with NaB^2H_3CN and ^{2}HCl the 4α,6-bisdeuterated compound (75) was obtained (Taylor and Djerassi, 1976). In line with the requirement for a six-membered cyclic transition state in the hydrogen-transfer step the reaction proceeds particularly well with αβ-unsaturated ketones which have or can assume a *cisoid* conformation. For αβ-unsaturated tosylhydrazones which are restricted to a transoid conformation by structural factors, e.g. (76), reduction with sodium cyanoborohydride is more complex and usually gives mixtures containing the saturated alkane.

In an alternative procedure, tosylhydrazones may be reduced by reaction with catecholborane followed by decomposition of the reduction product with sodium acetate (Kabalka and Baker, 1975).

Trialkylborohydrides. The trialkylborohydrides shown in (7.136) have much the same functional group selectivity as sodium borohydride itself, but their reactions show special features which give them value in synthesis. Because of the inductive effect of the ethyl groups and possibly

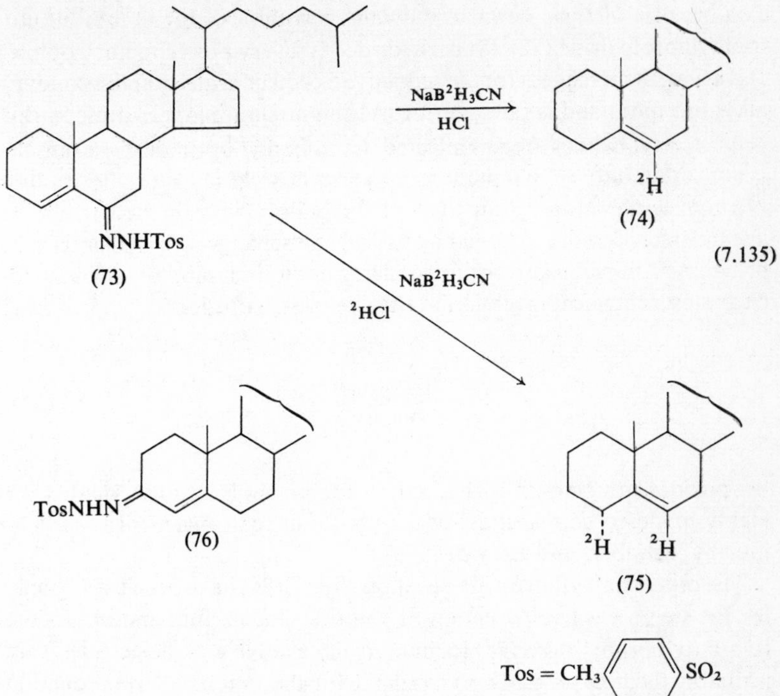

NNHTos

(73)

NaB^2H_3CN

HCl

^2H

(74)

(7.135)

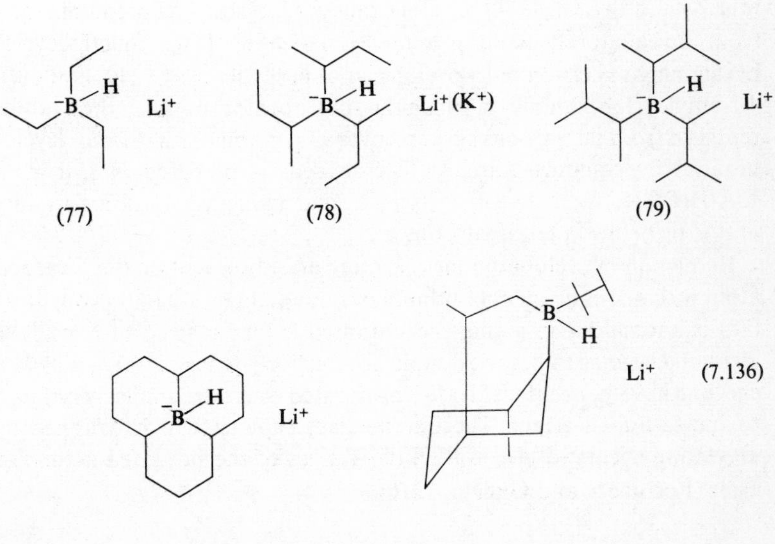

NaB^2H_3CN

^2HCl

TosNHN

(76)

^2H ^2H

(75)

Tos = CH_3 SO_2

^-B H Li^+

(77)

^-B H $Li^+(K^+)$

(78)

^-B H Li^+

(79)

$\bar{B}$ H

Li^+ (7.136)

^-B H Li^+

also because of their effect in reducing solvation of the anion, lithium triethylborohydride (77), 'Superhydride', is a very powerful nucleophile. It is an excellent reagent for the reductive fission of primary and secondary alkyl bromides and is far superior to lithium aluminium hydride in this respect. Aryl halides are not affected. Cycloheptyl bromide, for example, is converted into cycloheptane in 99 per cent yield in a smooth reaction at room temperature. Reduction of the halides with the deuterium reagent, which is easily obtained by hydrogen exchange in ^2H_2O at pH = 3, provides a simple means for introducing deuterium into an alkane, with clear stereochemical inversion at the site of substitution.

$$\xrightarrow[\text{THF, 65°C}]{Li(C_2H_5)_3BD} \qquad (7.137)$$

Br H H D

Epoxides are smoothly reduced to the alcohols in high yield, even highly hindered compounds reacting without rearrangement (Krishnamurthy, Schubert and Brown, 1973).

The other trialkylborohydrides shown in (7.136) have been used chiefly for the stereoselective reduction of ketones. The lithium and potassium tri-s-butylborohydrides (78) (commercially available as 'Selectrides') are probably the best reagents yet available for the stereoselective reduction of cyclic ketones (Brown and Krishnamurthy, 1972), although recently even better results have been claimed with the reagent (79) (Krishnamurthy and Brown, 1976). Cyclohexanones are attacked predominantly from the equatorial side to give the axial alcohol. Thus, 2-methylcyclohexanone gives *cis*-2-methylcyclohexanol in 99 per cent yield, and even 4-t-butylcyclohexanone, in which the substituent is several carbon atoms removed from the carbonyl group, gives 93 per cent of *cis*-4-t-butylcyclohexanol by equatorial attack. The reagent is prepared *in situ* from $LiAlH(OCH_3)_3$ and tri-s-butylborane, and reductions proceed rapidly at 0°C or below in tetrahydrofuran.

Reduction of cyclohexenones carrying no substituent on the β-carbon atom with one equivalent of lithium or potassium tri-s-butylborohydride (78) is anomalous in giving the saturated ketone instead of the allylic alcohol. Carvone (80), for example, is selectively reduced to 1,6-dihydrocarvone in 98 per cent yield. $\alpha\beta$-Unsaturated esters are similarly reduced to the saturated esters. The intermediate enolates can be trapped by alkylating agents to give α-alkyl derivatives of the saturated ketone or ester (Fortunato and Ganem, 1976).

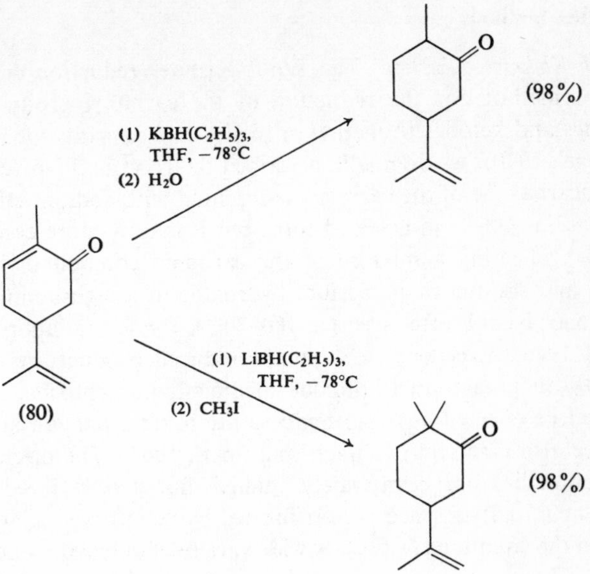

(1) $KBH(C_2H_5)_3$, THF, −78°C
(2) H_2O

(98%)

(7.138)

(80)

(1) $LiBH(C_2H_5)_3$, THF, −78°C
(2) CH_3I

(98%)

$^-B(C_4H_9)_2$ Li^+

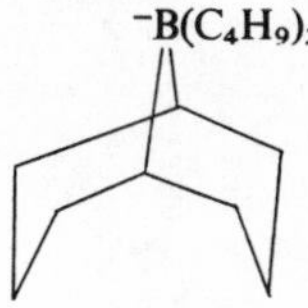

(81)

A promising new reagent is the dibutyl 'ate' complex of 9-borabicyclo-[3,3,1]nonane (81) which exhibits high regio- and stereo-selectivity in the reduction of carbonyl groups (reduction is effected by one of the bridge-head hydrogens) (Yamamoto, Toi, Sonoda and Murahashi, 1976). Usefully, the reagent distinguishes between ketonic carbonyl groups in different positions and appears to offer a method for the selective reduction of aldehydes in presence of ketones. Selective reduction of aldehydes can also be achieved with the modified reagent sodium triacetoxyborohydride, $NaBH(OOCCH_3)_3$, obtained by action of the stoichiometric amount of acetic acid on sodium borohydride in benzene. Thus a mixture of benzaldehyde and acetophenone on reaction with an excess of the reagent gave a product in which the aldehyde had been completely reduced to benzyl alcohol, but the ketone was scarcely affected (Gribble and Ferguson, 1975).

7.4. Other methods

Wolff–Kishner reduction. The Wolff–Kishner reduction provides an excellent method for the reduction of the carbonyl group of many aldehydes and ketones to methyl or methylene (Szmant, 1968; Asinger and Vogel, 1970). As originally described the reaction involved heating the semicarbazone of the carbonyl compound with sodium ethoxide or other base at 200°C in a sealed tube, but it is now more conveniently effected by heating a mixture of the carbonyl compound, hydrazine hydrate and sodium or potassium hydroxide in a high-boiling solvent (diethylene glycol is often used) at 180–200°C for several hours (Huang-Minlon, 1946). Excellent yields of the reduced product are often obtained. With potassium t-butoxide in dimethyl sulphoxide, reduction can often be effected at room temperature (Cram, Sahyun and Knox, 1962; see also Grundon, Henbest and Scott, 1963). The mechanism of the reaction has not been widely studied, but it is believed that the hydrazone initially formed is transformed by reactions similar to those shown in the equations (7.139). A wide variety of aldehydes and ketones

$$R'RC{=}O + H_2NNH_2 \rightleftharpoons R'RC{=}NNH_2 + H_2O$$

$$R'RC{=}NNH_2 \xrightarrow{OH^-} R'RC{=}\ddot{N}\ddot{N}H^- \longrightarrow R'RCH{-}N{=}\ddot{N}^- \longrightarrow R'RCH^- + N_2 \longrightarrow R'CH_2R \quad (7.139)$$

has been reduced by this method. Sterically hindered ketones, such as 11-keto steroids, are resistant, but they too can be reduced by use of anhydrous hydrazine and sodium metal as base.

Reduction of conjugated unsaturated ketones is sometimes accompanied by a shift in the position of the double bond (7.140). In other

$$\text{(2,6,6-trimethylcyclohex-1-ene-carbaldehyde, CHO)} \longrightarrow \text{(exocyclic } {=}CH_2\text{ product)} \quad (80\%) \quad (7.140)$$

cases pyrazoline derivatives may be formed which decompose yielding cyclopropanes isomeric with the expected olefin (7.141).

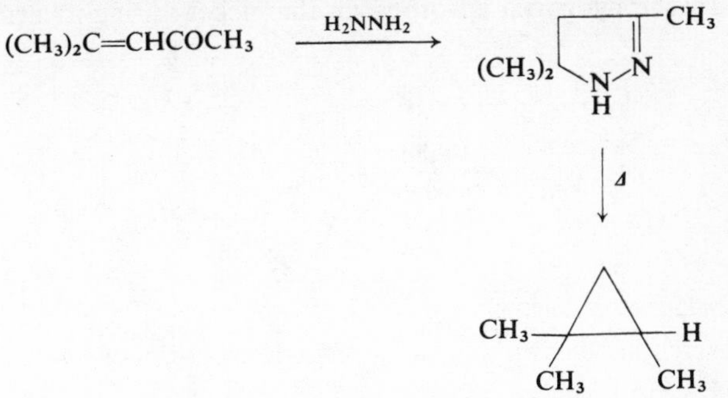

$$(CH_3)_2C{=}CHCOCH_3 \xrightarrow{H_2NNH_2} \text{pyrazoline} \xrightarrow{\Delta} \text{cyclopropane} \qquad (7.141)$$

With many α-substituted ketones elimination accompanies reduction. For example the sterol (82) affords the olefin (83) (7.142). In some cases

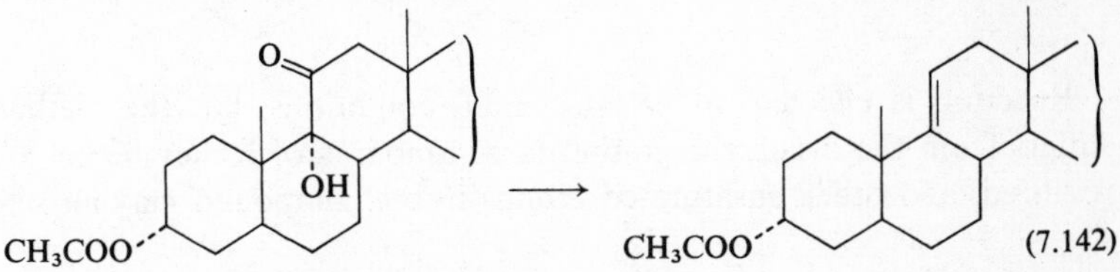

$$CH_3COO \ldots (82) \longrightarrow CH_3COO \ldots (83)\ (82\%) \qquad (7.142)$$

these eliminations take place under quite mild conditions. Of particular importance is the reductive opening of αβ-epoxyketones which are converted into allylic alcohols. This reaction has been exploited in a synthesis of linalool and other allylic alcohols (7.143).

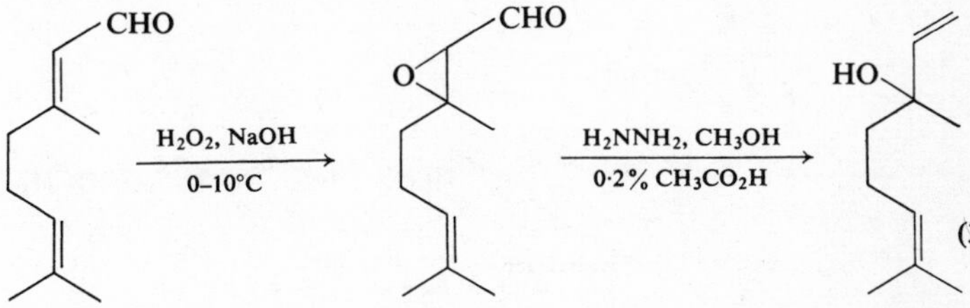

$$\text{citral} \xrightarrow[0\text{–}10°C]{H_2O_2,\ NaOH} \text{epoxide} \xrightarrow[0{\cdot}2\%\ CH_3CO_2H]{H_2NNH_2,\ CH_3OH} \text{linalool}\ (35\%) \qquad (7.143)$$

Raney nickel desulphurisation of thio-acetals. Another method for reduction of the carbonyl group of aldehydes and ketones to methyl or methylene which is useful on occasion is desulphurisation of the corresponding thio-acetals with Raney nickel in boiling ethanol (Pettit and van Tamelen, 1962; Asinger and Vogel, 1970). Hydrogenolysis is effected by the hydrogen adsorbed on the nickel during its preparation (7.144).

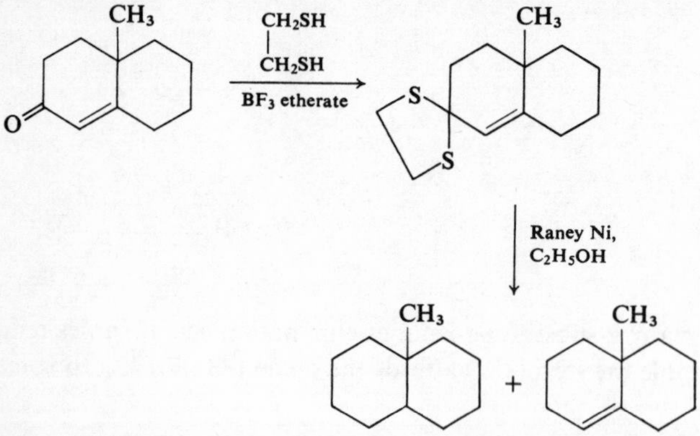

CH_3 CH_2SH | CH_2SH BF_3 etherate CH_3 S S O

Raney Ni, C_2H_5OH (7.144)

CH_3 CH_3 +

Reaction is effected under fairly mild conditions, but the method suffers from the disadvantage that large amounts of Raney nickel are required, and other unsaturated groups in the compound may also be

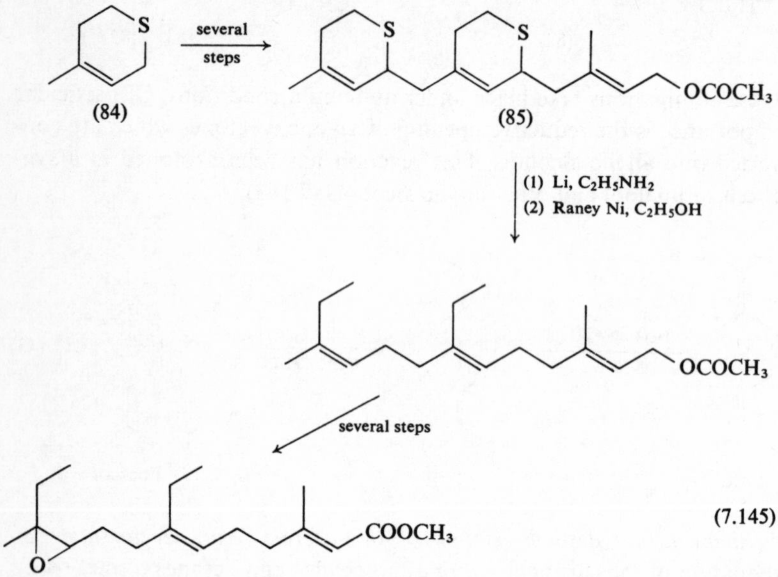

S several steps S S $OCOCH_3$

(84) (85)

(1) Li, $C_2H_5NH_2$
(2) Raney Ni, C_2H_5OH

$OCOCH_3$

several steps

O $COOCH_3$ (7.145)

(86)

reduced. Lithium and ethylamine may be used instead of Raney nickel, to bring about desulphurisation but again unsaturated groups in the molecule may be affected.

An illustration of the way in which desulphurisation can be used in the synthesis of a complex natural product is given in (7.145), which shows steps from a synthesis of *Cecropia* juvenile hormone (86). The dihydrothiopyran (84) is used to build up the precursor (85) in which the stereochemistry of two of the double bonds in the hormone is controlled by incorporating them in rings. Desulphurisation gives the open chain compound with the correct configuration of the double bonds, and at the same time unmasks the ethyl substituents, to give a product which is converted by further steps into the hormone.

Another method for effecting desulphurisation which is helpful in the formation of carbon–carbon double bonds is described on p. 123.

Reductions with di-imide. It has long been known that isolated olefinic bonds can be reduced with hydrazine in presence of oxygen or an oxidising agent. Thus, it was found as early as 1914 that oleic acid is reduced to stearic acid by this method. It has been suggested recently that the actual reducing agent in these reactions is in fact the highly active species di-imide, HN=NH, formed *in situ* by oxidation of hydrazine, and it has been found that this compound is a highly selective reducing agent which in many cases offers a useful alternative to catalytic hydrogenation for the reduction of carbon–carbon multiple bonds (Miller, 1965).

The reagent is not isolated but is prepared *in situ*, usually by oxidation of hydrazine with oxygen or an oxidising agent (e.g. hydrogen peroxide); by thermal decomposition of *p*-toluenesulphonylhydrazide; or from azodicarboxylic acid (7.146).

$$CH_3C_6H_4SO_2NHNH_2 \xrightarrow[\text{diglyme}]{\text{boiling}} CH_3C_6H_4SO_2H + HN{=}NH$$

$$HO_2CN{=}NCO_2H \longrightarrow HN{=}NH + CO_2 \qquad (7.146)$$

Di-imide is a highly selective reagent. In general, it is found that symmetrical double bonds such as C≡C, C=C, N=N, O=O are readily reduced, but unsymmetrical, more polar, bonds (C≡N, $N\!\overset{\nearrow O}{\searrow O}$,

C=N, S=O, S—S, C—S) are not. For example, under conditions where oleic acid and azobenzene were readily reduced, methyl cyanide, nitrobenzene and dibenzyl sulphide were unaffected, and the selectivity of the reagent is strikingly demonstrated by the reduction of diallyl disulphide to dipropyl disulphide in almost quantitative yield (7.147).

$$(CH_2{=}CHCH_2)_2S_2 \xrightarrow[\text{boiling glycol}]{\text{tosylhydrazide,}} (CH_3CH_2CH_2)_2S_2 \quad (93\text{–}100\%)$$

$$C_6H_5N{=}NC_6H_5 \xrightarrow[\text{boiling methanol}]{\text{azodicarboxylic acid,}} C_6H_5NHNHC_6H_5 \quad \text{quantitative} \qquad (7.147)$$

The reactions are highly stereospecific and take place by *cis* addition of hydrogen in all cases. Thus, reduction of fumaric acid with tetradeuterohydrazine or with potassium azodicarboxylate and deuterium oxide affords (±)-dideuterosuccinic acid exclusively, while maleic acid gives the *meso* isomer.

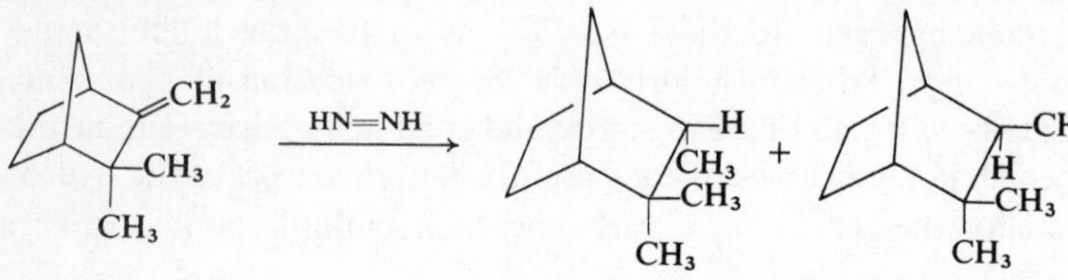

CH₂ / CH₃ / CH₃ $\xrightarrow{HN{=}NH}$ H / CH₃ / CH₃ / CH₃ + CH₃ / H / CH₃ / CH₃ (7.148)

(92% of product) (8% of product)

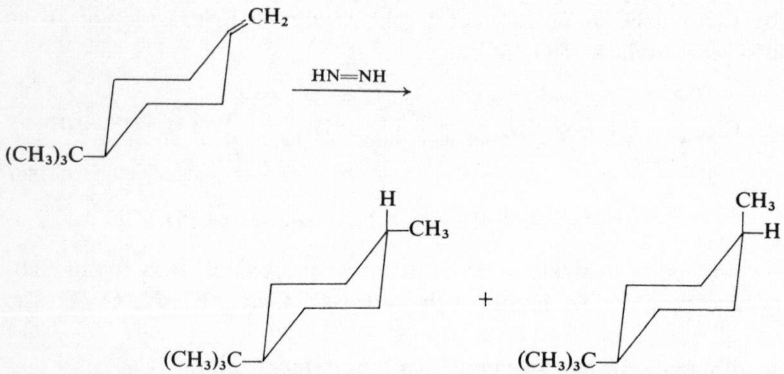

CH_2 / $(CH_3)_3C$ $\xrightarrow{HN{=}NH}$

H / CH_3 / $(CH_3)_3C$ + CH_3 / H / $(CH_3)_3C$

(51% of product) (49% of product)

Reduction of acetylenes is also a *cis* process. Thus, partial reduction of diphenylacetylene gave, besides starting material and diphenylethane, only *cis*-stilbene; no *trans*-stilbene was detected.

In sterically hindered molecules, addition takes place to the less hindered side of the double bond, but in examples where steric influences are moderate, much less stereochemical discrimination is observed (7.148).

The reactions are regarded as taking place by synchronous transfer of a pair of hydrogen atoms through a cyclic six-membered transition state (7.149). This mechanism explains the high stereospecificity of the reaction, and couples the driving force of nitrogen formation with the addition reaction. Concerted *cis* transfer of hydrogen is symmetry allowed for the ground-state reaction (Woodward and Hoffmann, 1969).

$$\begin{array}{c} >C{=}C< \\ HN{=}NH \end{array} \longrightarrow \begin{array}{c} >C{\cdots}C< \\ H \quad\quad H \\ N{\equiv}N \end{array} \longrightarrow \begin{array}{c} >C{-}C< \\ H \quad H \\ N{\equiv}N \end{array} \qquad (7.149)$$

Reduction with low-valent titanium species. Some synthetically useful reductions with low-valent titanium species have been reported (McMurry, 1974). Titanium(III), conveniently available as titanium chloride, is a mild reducing agent in aqueous solution.

$$Ti^{3+} + H_2O \longrightarrow TiO^{2+}\,(aq.) + 2H^+ + e^-; \qquad E = -0{\cdot}1\ V$$

The Table overleaf shows some reductions which can be effected with the reagent. One of the most useful reactions synthetically is the conversion of aliphatic nitro compounds into imines, which are directly hydrolysed to carbonyl compounds under the conditions of the reaction. Aliphatic —$\overset{|}{C}HNO_2$ groups thus become equivalent to carbonyl groups. This greatly extends the usefulness of nitro compounds in synthesis. They are easily converted into anions which serve, for example, as excellent donors in Michael addition to electrophilic olefins, and the sequence Michael addition followed by reaction with titanium chloride provides an excellent route to 1,4-dicarbonyl compounds and thence to cyclopentenones. *cis*-Jasmone, for example was readily obtained as shown in (7.150).

Some reductions effected with aqueous Ti(III)

Reactant	Product
$ArNO_2$	$ArNH_2$
R_2CHNO_2	$R_2CH{=}NH \rightarrow R_2C{=}O$
$R_2S{=}O$	R_2S
$R_3N{\rightarrow}O$	R_3N
2-bromocyclohexanone (Br)	cyclohexanone
$CO_2C_2H_5$ (unsaturated keto ester)	$CO_2C_2H_5$ (saturated keto ester)
ene-dione adduct	dione

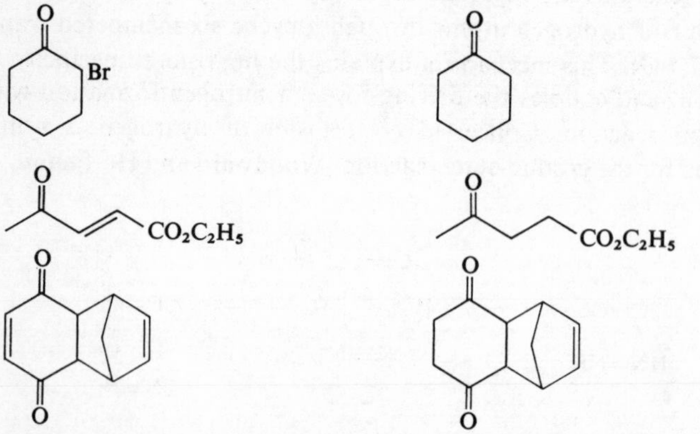

$$CH_3(CH_2)_4CH_2NO_2 \xrightarrow[\text{dimethoxyethane}]{TiCl_3,\ H_2O} CH_3(CH_2)_4CHO \quad (74\%)$$

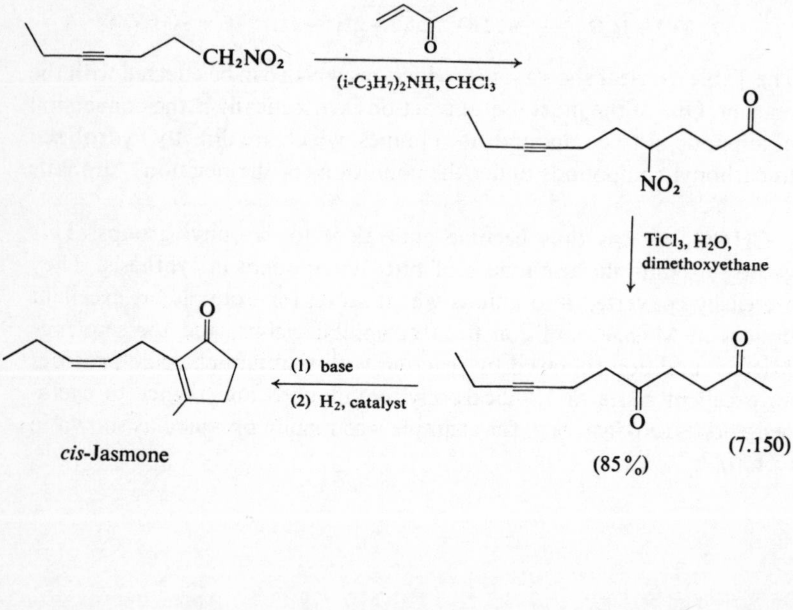

CH_2NO_2 → ($(i\text{-}C_3H_7)_2NH$, $CHCl_3$) → NO_2 → ($TiCl_3$, H_2O, dimethoxyethane) → (85%) → (1) base (2) H_2, catalyst → *cis*-Jasmone

(7.150)

The reagent is also very effective for reduction of α-halo ketones to the halogen-free compounds, and is more convenient than chromium(II) salts for this, and for the reduction of ene-diones to the saturated diones.

Another useful reagent of indefinite structure, but believed to be a titanium(II) species, is obtained from titanium(III) chloride and lithium aluminium hydride. It can be used for converting epoxides into olefins, although unfortunately deoxygenation is accompanied by isomerisation so that mixtures of the *cis* and *trans* olefins arise when that is possible. *cis*-5-Decene oxide, for example, gives a 70 per cent yield of 5-decene containing a preponderance of the *trans* isomer (McMurry and Fleming, 1975). The reagent is remarkably effective also for reductive coupling of carbonyl compounds to give olefins. Benzophenone, for example, gave tetraphenylethylene in 95 per cent yield, and the highly hindered tetra-isopropylethylene was obtained from di-isopropyl ketone. Even the highly unsaturated and reactive *β*-carotene was readily obtained by this method, from vitamin A aldehyde.

It may be that in these reactions the active species is active titanium metal and not a derivative of titanium(II) as originally supposed. In agreement, even better conversions of carbonyl compounds into olefins are obtained using an active titanium metal powder prepared by reduction of titanium(III) chloride with potassium (McMurry and Fleming, 1976). The mechanism of the reactions is uncertain but it is believed that it may involve the pinacol and a five-membered cyclic intermediate containing titanium. Pinacols also are readily converted into olefins.

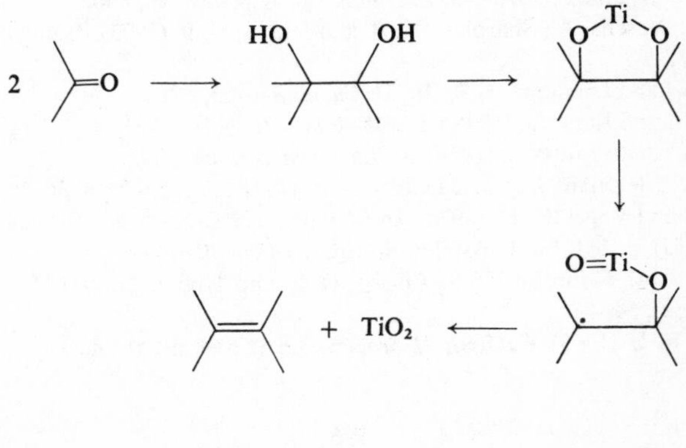

Ti^0 ($TiCl_3$, K), THF (40%)

2 ⟶ HO OH ⟶ Ti (7.151)

+ TiO_2 ⟵ O=Ti

References

Abley, P. and McQuillin, F. J. (1971). *J. chem. Soc.* (*C*), 844.
Abramovitch, R. A. and Davis, B. A. (1964). *Chem. Rev.* **64**, 149.
Acott, B. and Beckwith, A. L. J. (1964). *Austral. J. Chem.* **14**, 1342.
Adam, W., Baeza, J. and Liu, J. C. (1972). *J. Am. chem. Soc.* **94**, 2000.
Adams, W. R. (1971). In *Oxidation*, vol. 2. Ed. R. L. Augustine and D. J. Trecker. (New York: Marcel Dekker.)
Akhtar, M. (1962). In *Advances in Photochemistry*, vol. 2, p. 263. Ed. W. A. Noyes, G. S. Hammond and J. N. Pitts (London: Interscience.)
Akhtar, A. and Barton, D. H. R. (1961). *J. Am. chem. Soc.* **83**, 2213.
Akhtar, A. and Barton, D. H. R. (1964). *J. Am. chem. Soc.* **86**, 1528.
Akhtar, M. and Pechet, M. M. (1964). *J. Am. chem. Soc.* **86**, 265.
Albright, J. D. (1974). *J. org. Chem.* **39**, 1977.
Alder, K. (1948). In *Newer Methods of Preparative Organic Chemistry*, vol. 1. (London: Interscience.)
Alder, K. and von Brachel, H. (1962). *Liebigs Ann.* **651**, 141.
Alder, K. and Schumacher, M. (1953). In *Fortschritte der Chemie organischer Naturstoffe*, vol. **10**, p. 69. Ed. L. Zechmeister. (Vienna: Springer Verlag.)
Allen, D. W., Heatley, P., Hutley, B. G. and Mellors, M. T. J. (1974). *Tetrahedron Lett.* 1787.
Allen, J. A., Boar, R. B., McGhie, J. F. and Barton, D. H. R. (1973). *J. Chem. Soc. Perkin I*, 2402.
Alston, P. V. and Ottenbrite, R. M. (1974). *J. org. Chem.* **39**, 1584.
Alston, P. V. and Ottenbrite, R. M. (1975). *J. org. Chem.* **40**, 111.
Alston, P. V., Ottenbrite, R. M. and Shillady, D. D. (1973). *J. org. Chem.* **38**, 4075.
Anh, N. T. and Seyden-Penne, J. (1973). *Tetrahedron*, **29**, 3227.
Ansell, M. J. *et al.* (1971). *J. chem. Soc.* (*C*), 1401, 1414, 1423, 1429.
ApSimon, J. W. and Edwards, O. E. (1962). *Can. J. Chem.* **40**, 896.
Arigoni, D., Vasella, A., Sharpless, K. B. and Jensen, H. P. (1973). *J. Am. chem. Soc.* **95**, 7917.
Ashby, E. C. and Laemmie, J. T. (1975). *Chem. Rev.* **75**, 521.
Ashby, E. C. and Lin, J. J. (1975). *Tetrahedron Lett.* 4453.
Ashby, E. C. and Prather, J. (1966). *J. Am. chem. Soc.* **88**, 729.
Ashby, E. C., Sevenair, J. P. and Dobbs, F. R. (1971). *J. org. Chem.* **36**, 197.
Asinger, F. and Vogel, H. H. (1970). In *Methoden der Organischen Chemie*, 4th edn, vol. 5(1a), p. 244. Ed. E. Müller (Stuttgart *Thieme Verlag*).
Auerbach, R. A., Crumrine, D. S., Ellison, D. L. and House, H. O. (1974). *Organic Syntheses*, **54**, 49.
Augustine, R. L. (1965). *Catalytic Hydrogenation*. (London: Arnold.)

Augustine, R. L. (1968). *Reduction.* (London: Arnold.)

Augustine, R. L., Migliorini, D. C., Foscante, R. E., Sodano, C. S. and Sisbarro, M. J. (1969). *J. org. Chem.* **34**, 1075.

Bacha, J. D. and Kochi, J. K. (1968). *Tetrahedron*, **24**, 2215.

Bachman, W. E. and Struve, W. S. (1942). *Organic Reactions*, vol. 1, p. 38. (New York: Wiley.)

Back, T. G., Barton, D. H. R., Britten-Kelly, M. R. and Guziec, F. S. (1975). *J. Chem. Soc. Chem. Commun.* 539.

Bacon, R. G. R. and Hill, H. A. O. (1965). *Quart. Rev. chem. Soc. Lond.* **19**, 95.

Badger, G. M. (1954). *The Structures and Reactions of Aromatic Compounds.* (Cambridge: Cambridge University Press.)

Bailey, D. S. and Saunders, W. H. (1970). *J. Am. chem. Soc.* **92**, 6904, 6911.

Bailey, P. S. (1958). *Chem. Rev.* **58**, 925.

Baker, R. (1973). *Chem. Rev.* **73**, 487.

Baldwin, J. E., Barton, D. H. R., Dainis, I. and Pereira, J. L. C. (1968). *J. chem. Soc.* (*C*), 2283.

Baldwin, J. E., Höfle, G. A. and Lever, O. W. (1974). *J. Am. chem. Soc.* **90**, 7125.

Baldwin, J. E. and Walker, J. A. (1972). *J. chem. Soc. Chem. Commun.* 354.

Balogh, V., Fétizon, M. and Golfier, M. (1971). *J. org. Chem.* **36**, 1339.

Barlow, M. G., Haszeldine, R. N. and Hubbard, R. (1969). *Chem. Commun.* 301.

Bartlett, P. D. (1968). *Science*, **159**, 833.

Bartlett, P. D. (1970). *Quart. Rev. chem. Soc. Lond.* **24**, 473.

Bartlett, P. D. (1971). *Pure and appl. Chem.* **27**, 597.

Barton, D. H. R. (1953). *J. chem. Soc.* 1027, n. 23.

Barton, D. H. R. (1959). *Helv. chim. Acta*, **42**, 2604.

Barton, D. H. R., Beaton, J. M., Geller, L. E. and Pechet, M. M. (1961). *J. Am. chem. Soc.* **83**, 4076.

Barton, D. H. R., Beckwith, A. L. J. and Goosen, A. (1965). *J. chem. Soc.* 181.

Barton, D. H. R. and Forbes, C. P. (1975). *J. chem. Soc. Perkin I*, 1614.

Barton, D. H. R., Guziec, F. S. and Shahak, I. (1974). *J. chem. Soc. Perkin I*, 1794.

Barton, D. H. R., Haynes, R. K., Leclerc, G., Magnus, P. D. and Menzies, I. D. (1975). *J. chem. Soc. Perkin I*, 2055.

Barton, D. H. R. and McCombie, S. W. (1975). *J. chem. Soc. Perkin I*, 1574.

Barton, D. H. R. and Morgan, L. R. (1962). *J. chem. Soc.* 622.

Barton, D. H. R. and Robinson, C. H. (1954). *J. chem. Soc.* 3045.

Barton, D. H. R. and Starratt, A. N. (1965). *J. chem. Soc.* 2445.

Battersby, A. R. and Staunton, J. (1974). *Tetrahedron*, **30**, 1707.

Beckwith, A. L. J., Cross, R. T. and Gream, G. E. (1974). *Austral. J. Chem.* **27**, 1693.

Beckwith, A. L. J. and Goodrich, J. E. (1965). *Austral. J. Chem.* **18**, 747.

Behan, J. M., Johnstone, R. A. W. and Wright, M. J. (1975). *J. chem. Soc. Perkin I*, 1216.

Bell, H. M., Vanderslice, C. W. and Spehars, A. (1969). *J. org. Chem.* **34**, 3923.

Belluš, D., Mez, H-C. and Rihs, G. (1974). *J. chem. Soc. Perkin II*, 884.

Belluš, D., von Bredow, K., Sauter, H. and Weis, C. D. (1973). *Helv. chim. Acta*, **56**, 3004.

Benkeser, R. A., Agnihotri, R., Burrous, M. L., Kaiser, E. M., Mallan, J. M. and Ryan, P. W. (1964). *J. org. Chem.* **29**, 1313.

Benkeser, R. A., Burrous, M. L., Hazdra, J. J. and Kaiser, E. M. (1963). *J. org. Chem.* **28**, 1094.
Benkley, E. S. and Heathcock, C. H. (1975). *J. org. Chem.* **40**, 2156.
Berson, J. A., Hamlet, Z. and Mueller, W. A. (1962). *J. Am. chem. Soc.* **84**, 297.
Berson, J. A. and Olin, S. S. (1969). *J. Am. chem. Soc.* **91**, 777.
Berson, J. A. and Remanick, A. (1961). *J. Am. chem. Soc.* **83**, 4947.
Berson, J. A., Wall, R. G. and Perlmutter, H. D. (1966). *J. Am. chem. Soc.* **88**, 187.
Berti, G. (1973). In *Topics in Stereochemistry*, vol. 7. Ed. N. L. Allinger and E. L. Eliel. (London: Wiley.)
Bestman, H. J., Armsen, R. and Wagner, H. (1969). *Chem. Ber.* **102**, 2259.
Bethell, D. (1969). In *Advances in Physical Organic Chemistry*, vol. 7, p. 153. Ed. V. Gold. (London: Academic Press.)
Bhalerao, U. T. and Rapaport, H. (1971). *J. Am. chem. Soc.* **93**, 4835, 5311.
Biellmann, J. F. (1968). *Bull. Soc. chim. Fr.* 3055.
Biellmann, J. F. and Ducep, J. B. (1971). *Tetrahedron*, **27**, 5861.
Birch, A. J. (1950). *Quart. Rev. chem. Soc. Lond.* **4**, 69.
Birch, A. J. and Nasipuri, D. (1959). *Tetrahedron*, **6**, 148.
Birch, A. J. and Slobbe, J. (1976). *Tetrahedron Lett.* 2079.
Birch, A. J. and Smith, H. (1958). *Quart. Rev. chem. Soc. Lond.* **12**, 17.
Birch, A. J. and Subba Rao, G. (1972). In *Advances in Organic Chemistry: Methods and Results*, vol. 8, p. 1. Ed. E. C. Taylor. (New York: Wiley-Interscience.)
Birch, A. J. and Walker, K. A. M. (1966). *J. chem. Soc.* (*C*) 1894.
Bird, C. L., Frey, H. M. and Stevens, I. D. R. (1967). *Chem. Commun.* 707.
Birkowitz, L. M. and Rylander, P. N. (1958). *J. Am. chem. Soc.* **80**, 6683.
Blackburn, E. V. and Timmons, C. J. (1969). *Quart. Rev. chem. Soc. Lond.* **23**, 482.
Blackwood, J. E., Gladys, C. L., Loening, K. L., Petrara, A. E. and Rush, J. E. (1968). *J. Am. chem. Soc.* **90**, 509.
Blatcher, P., Grayson, J. I. and Warren, S. (1976). *J. chem. Soc. Chem. Commun.* 547.
Bloomfield, J. J. (1968). *Tetrahedron Lett.* 587.
Bloomfield, J. J., Owsley, D. C., Ainsworth, C. and Robertson, R. E. (1975). *J. org. Chem.* **40**, 393.
Boeckmann, R. K. (1974). *J. Am. chem. Soc.* **96**, 6179.
Boeckman, R. K. and Bruza, K. T. (1974). *Tetrahedron Lett.* 3365.
Boeckman, R. E., Bruza, K. J., Baldwin, J. E. and Lever, O. W. (1975). *J. chem. Soc. Chem. Commun.* 519.
Boeckman, R. K. and Ganem, B. (1974). *Tetrahedron Lett.* 913.
Bogdanović, B. (1973). *Angew. Chem. internat. Edit.* **12**, 954.
Bogdanowicz, M. J. and Trost, B. M. (1974). *Organic Syntheses*, **54**, 27.
Bohlmann, F. and Zdero, C. (1973). *Chem. Ber.* **106**, 3779.
Borch, R. F., Bernstein, M. D. and Durst, H. D. (1971). *J. Am. chem. Soc.* **93**, 2897.
Bordwell, F. G. (1970). *Acc. chem. Res.* **3**, 281.
Bošnjak, J., Andrejević, V., Čeković, Z. and Mihailović, M. Lj. (1972). *Tetrahedron Lett.* 6031.
Bosworth, N. and Magnus, P. D. (1972). *J. chem. Soc. Perkin I*, 943.
Boutagy, J. and Thomas, R. (1974). *Chem. Rev.* **74**, 87.
Bowers, A., Halsall, T. G., Jones, E. R. H. and Lemin, A. J. (1953). *J. chem. Soc.* 2548.
Boyd, J., Epstein, W. and Frater, G. (1976). *J. chem. Soc. Chem. Commun.* 380.

Brady, S. F., Ilton, M. A. and Johnson, W. S. (1968). *J. Am. chem. Soc.* **90**, 2882.
Breslow, R. (1972). *Chem. Soc. Rev.* **1**, 553.
Breslow, R., Baldwin, S., Flechtner, T., Kalicky, P., Liu, S. and Washburn, W. (1973). *J. Am. chem. Soc.* **95**, 3251.
Breslow, R., Corcoran, R., Dale, J. A., Liu, S. and Kalicky, P. (1974). *J. Am. chem. Soc.* **96**, 1973.
Breslow, R., Corcoran, R. J. and Snider, B. B. (1974). *J. Am. chem. Soc.* **96**, 6791.
Breslow, R. and Winnik, M. A. (1969). *J. Am. chem. Soc.* **91**, 3083.
Bridges, A. J. and Whitham, G. H. (1974). *J. chem. Soc. Chem. Commun.* 142.
Brieger, G. and Nestrick, T. J. (1974). *Chem. Rev.* **74**, 567.
Brimacombe, J. S. (1969). *Angew. Chem. internat. Edit.* **8**, 401.
Brown, C. A. (1969). *J. Am. chem. Soc.* **91**, 5901.
Brown, H. C. (1962). *Hydroboration.* (New York: Benjamin.)
Brown, H. C. (1969). *Accounts chem. Res.* **2**, 65.
Brown, H. C. (1972). *Boranes in Organic Chemistry.* (Ithaca: Cornell University Press.)
Brown, H. C. and Bhatt, M. V. (1966). *J. Am. chem. Soc.* **88**, 1440.
Brown, H. C. and Bigley, D. (1961). *J. Am. chem. Soc.* **83**, 3166.
Brown, H. C., Bigley, D. B., Arora, S. K. and Nung Min Yoon. (1970). *J. Am. chem. Soc.* **92**, 7161.
Brown, H. C. and Brown, C. A. (1962). *J. Am. chem. Soc.* **84**, 2827.
Brown, H. C. and Brown, C. A. (1963). *J. Am. chem. Soc.* **85**, 1003.
Brown, H. C. and Coleman, R. A. (1969). *J. Am. chem. Soc.* **91**, 4606.
Brown, H. C. and Deck, H. R. (1965). *J. Am. chem. Soc.* **87**, 5620.
Brown, H. C. and Dickason, W. C. (1970). *J. Am. chem. Soc.* **92**, 709.
Brown, H. C. and Gallivan, R. M. (1968). *J. Am. chem. Soc.* **90**, 2906.
Brown, H. C. and Garg, C. P. (1961). *J. Am. chem. Soc.* **83**, 2951.
Brown, H. C., Garg, C. P. and Kwang-Ting Liu. (1971). *J. org. Chem.* **36**, 387.
Brown, H. C. and Gupta, S. K. (1971*a*). *J. Am. chem. Soc.* **93**, 4062.
Brown, H. C. and Gupta, S. K. (1971*b*). *J. Am. chem. Soc.* **93**, 1818.
Brown, H. C. and Gupta, S. K. (1972). *J. Am. chem. Soc.* **93**, 4370, 4371.
Brown, H. C. and Gupta, S. K. (1975). *J. Am. chem. Soc.* **97**, 5249.
Brown, H. C., Hamaoka, T. and Ravindran, N. (1973). *J. Am. chem. Soc.* **95**, 6456.
Brown, H. C., Heim, P. and Nung Min Yoon. (1972). *J. org. Chem.* **37**, 2942.
Brown, H. C., Heydkamp, W. R., Breuer, E. and Murphy, W. S. (1964). *J. Am. chem. Soc.* **86**, 3565.
Brown, H. C. and Kabalka, G. W. (1970). *J. Am. chem. Soc.* **92**, 712, 714.
Brown, H. C., Katz, J-J. and Carlson, B. A. (1973). *J. org. Chem.* **38**, 3968.
Brown, H. C., Katz, J-J., Lane, C. F. and Negishi, E. (1975). *J. Am. chem. Soc.* **97**, 2799.
Brown, H. C. and Keblys, K. A. (1964). *J. Am. chem. Soc.* **86**, 1791, 1795.
Brown, H. C., Knights, E. F. and Coleman, R. A. (1969). *J. Am. chem. Soc.* **91**, 2144.
Brown, H. C., Knights, E. F. and Scouten, C. G. (1974). *J. Am. chem. Soc.* **96**, 7765.
Brown, H. C., Kramer, G. W., Levy, A. B. and Midland, M. (1974). *Organic Syntheses via Boranes.* (New York: Wiley Interscience.)
Brown, H. C. and Krishnamurthy, S. (1969). *J. org. Chem.* **34**, 3918.
Brown, H. C. and Krishnamurthy, S. (1972). *J. Am. chem. Soc.* **94**, 7159.
Brown, H. C., Krishnamurthy, S. and Nung Min Yoon (1976). *J. org. Chem.* **41**, 1778.

Brown, H. C. and Lane, C. F. (1970). *J. Am. chem. Soc.* **92**, 6660.
Brown, H. C. and Midland, M. M. (1971). *J. Am. chem. Soc.* **93**, 4078.
Brown, H. C. and Midland, M. M. (1972). *Angew. Chem. internat. Edit.* **11**, 692.
Brown, H. C., Midland, M. M. and Levy, A. B. (1972). *J. Am. chem. Soc.* **94**, 2114.
Brown, H. C., Midland, M. M. and Levy, A. B. (1973). *J. Am. chem. Soc.* **95**, 2394.
Brown, H. C. and Murray, K. J. (1961). *J. org. Chem.* **26**, 631.
Brown, H. C., Murray, K. J., Müller, H. and Zweifel, G. (1966). *J. Am. chem. Soc.* **88**, 1443.
Brown, H. C., Nambu, H. and Rogic, M. (1969). *J. Am. chem. Soc.* **91**, 6852.
Brown, H. C. and Negishi, E. (1967). *J. Am. chem. Soc.* **89** (*a*) 5285; (*b*) 5477.
Brown, H. C. and Negishi, E. (1968). *Chem. Comm.* 594.
Brown, H. C. and Negishi, E. (1972). *Pure and appl. Chem.* **29**, 527.
Brown, H. C. and Nung Min Yoon (1968). (*a*) *J. Am. chem. Soc.* **90**, 2686; (*b*) *Chem. Comm.* 1549.
Brown, H. C. and Rathke, M. W. (1967). *J. Am. chem. Soc.* **89**, 2737.
Brown, H. C., Rathke, M. W. and Rogic, M. M. (1968). *J. Am. chem. Soc.* **90**, 5038.
Brown, H. C. and Ravindran, N. (1972). *J. Am. chem. Soc.* **94**, 2112.
Brown, H. C. and Ravindran, N. (1973*a*). *J. org. Chem.* **38**, 1617.
Brown, H. C. and Ravindran, N. (1973*b*). *J. organometal. Chem.* **61**, C5.
Brown, H. C. and Ravindran, N. (1973*c*). *J. Am. chem. Soc.* **95**, 2396.
Brown, H. C. and Rhodes, S. P. (1969). *J. Am. chem. Soc.* **91**, 2149, 4306.
Brown, H. C. and Rogic, M. M. (1969). *J. Am. chem. Soc.* **91**, 2146, 4304.
Brown, H. C., Rogic, M. M., Rathke, M. W. and Kabalka, G. W. (1967). *J. Am. chem. Soc.* **89**, 5709.
Brown, H. C., Rogic, M. M., Rathke, M. W. and Kabalka, G. W. (1969). *J. Am. chem. Soc.* **91**, 2150.
Brown, H. C. and Sharp, R. L. (1966). *J. Am. chem. Soc.* **88**, 5851.
Brown, H. C. and Sharp, R. L. (1968). *J. Am. chem. Soc.* **90**, 2915.
Brown, H. C. and Subba Rao, B. C. (1960). *J. Am. chem. Soc.* **82**, 681.
Brown, H. C. and Varma, V. (1974). *J. org. Chem.* **39**, 1631.
Brown, H. C., Verbrugge, C. and Snyder, C. H. (1961). *J. Am. chem. Soc.* **83**, 1001.
Brown, H. C., Weissman, P. M. and Nung Min Yoon (1966). *J. Am. chem. Soc.* **88**, 1458.
Brown, W. G. (1951). *Organic Reactions*, **6**, 469.
Brun, P. and Waegell, B. (1976). *Tetrahedron*, **32**, 517.
Bryce-Smith, D. (ed.) (1974). *Chemical Society Specialist Periodical Reports. Photochemistry*, vol. 5. (See also vols. 1, 2, 3 and 4.)
Bryce-Smith, D., Deshpande, R. R. and Gilbert, A. (1975). *Tetrahedron Lett.* 1627.
Buchanan, J. G. St C. and Woodgate, P. D. (1969). *Quart. Rev. chem. Soc. Lond.* **23**, 522.
Büchi, G. and Wüest, H. (1971). *Helv. chim. Acta*, **54**, 1767.
Bunnett, J. F. (1969). In *Survey of Progress in Chemistry*, vol. 5, p. 53. Ed. A. F. Scott. (Academic Press: London.)
Butsugan, Y., Yoshida, S., Muto, M. and Bito, T. (1971). *Tetrahedron Lett.* 1129.
Butz, L. W. and Rytina, A. W. (1949). *Organic Reactions*, **5**, 136.
Cadogan, J. I. G. (1968). *Quart. Rev. chem. Soc. Lond.* **22**, 222.
Cadogan, J. I. G. and Mackie, R. K. (1974). *Chem. Soc. Rev.* **3**, 102.
Cagliotti, L. (1972). *Organic Syntheses*, **52**, 122.

Cain, E. N., Vukov, R. and Masamune, S. (1969). *Chem. Commun.* 98.
Cambie, R. C., Hayward, R. C., Roberts, J. L. and Rutledge, P. S. (1974). *J. Chem. Soc. Perkin I*, 1858.
Cantello, B. C. C., Mellor, J. M. and Webb. C. F. (1974). *J. chem. Soc. Perkin II*, 22.
Cardin, D. J., Cetinkaya, B., Doyle, M. J. and Lappert, M. F. (1973). *Chem. Soc. Rev.* **2**, 99.
Carey, F. A. and Court, A. S. (1972). *J. org. Chem.* **37**, 1926.
Carlson, R. M. and Hill, R. K. (1970). *Organic Syntheses*, **50**, 24.
Carruthers, W. (1973). *Chem. Ind.* (*London*), 931.
Carter, M. J. and Fleming, I. (1976). *J. chem. Soc. Chem. Commun.* 679.
Čeković, Z. and Green, M. M. (1974). *J. Am. chem. Soc.* **96**, 3000.
Čeković, Z. and Srnic, T. (1976). *Tetrahedron Lett.* 561.
Challand, B. D., Hikino, H., Kornis, G., Lange, G. and de Mayo, P. (1969). *J. org. Chem.* **34**, 794.
Chan, Th. and Chang, E. (1974). *J. org. Chem.* **39**, 3264.
Chan, T. H. and Massuda, D. (1975). *Tetrahedron Lett.* 3383.
Chavdarian, C. G. and Heathcock, C. H. (1975). *J. Am. chem. Soc.* **97**, 3822.
Cherest, M. and Felkin, H. (1968). *Tetrahedron Lett.* 2205.
Cherest, M., Felkin, H. and Prudent, N. (1968). *Tetrahedron Lett.* 2199.
Chow, Y. L., Tam, J. N. S., Colón, C. J. and Pillay, K. S. (1973). *Can. J. Chem.* **51**, 2469.
Ciganek, E. (1967). *Tetrahedron Lett.* 3321.
Cimarusti, C. M. and Wolinsky, J. (1968). *J. Am. chem. Soc.* **90**, 113.
Clark, R. D. and Heathcock, C. H. (1973). *J. org. Chem.* **38**, 3658.
Clark, R. D. and Heathcock, C. H. (1976). *J. org. Chem.* **41**, 1396.
Clayton, R. B., Henbest, H. B. and Smith, M. (1957). *J. Chem. Soc.* 1982.
Clive, D. L. J. and Denyer, C. V. (1973). *J. chem. Soc. chem. Commun.* 253.
Coates, A. M., Pigott, H. D. and Ollinger, J. (1974). *Tetrahedron Lett.* 3955.
Coates, R. M. and Sandefur, L. O. (1974). *J. org. Chem.* **39**, 275.
Coates, R. M., Sandefur, L. O. and Smillie, R. D. (1975). *J. Am. chem. Soc.* **97**, 1619.
Coates, R. M. and Sowerby, R. L. (1971). *J. Am. chem. Soc.* **93**, 1027.
Coffey, R. S. (1967). *Chem. Commun.* 923.
Cohen, T., Bennett, D. A. and Mura, A. J. (1976). *J. org. Chem.* **41**, 2506.
Collins, J. C. and Hess, W. W. (1972). *Organic Syntheses*, **52**, 5.
Collins, J. C., Hess, W. W. and Frank, F. J. (1968). *Tetrahedron Lett.* 3363.
Colonge, J. and Descotes, G. (1967). In 1,4-*Cycloaddition Reactions*. Ed. J. Hamer. (New York: Academic Press.)
Colonna, F. P., Fattuta, S., Risaliti, A. and Russo, C. (1970). *J. chem. Soc.* (*C*) 2377.
Conia, J-M. (1963). *Rec. chem. Prog.* **24**, 43.
Conia, J-M. and Girard, C. (1973). *Tetrahedron Lett.* 3327.
Conia, J-M. and Le Perchec, P. (1975). *Synthesis*, 1.
Conia, J-M. and Leriverend, P. (1960). *C. r. hebd. Séanc. Acad. Sci., Paris*, **250**, 1078.
Cook, A. G. (1968). (ed.) *Enamines: their Synthesis, Structure and Reactions.* (Marcel Dekker, New York.)
Cookson, R. C. and Rogers, N. R. (1973). *J. chem. Soc. Perkin I*, 2738, 2741.
Cookson, R. C., Stevens, I. D. R. and Watts, C. T. (1966). *Chem. Commun.* 744.
Cope, A. C. and Trumbull, E. R. (1960). *Organic Reactions*, **11**, 317.

Corey, E. J. and Achiwa, K. (1969). *Tetrahedron Lett.* 3257.
Corey, E. J., Bass, J. D., LeMahieu, R. and Mitra, R. B. (1964). *J. Am. chem. Soc.* **86**, 5570.
Corey, E. J. and Beames, D. J. (1972). *J. Am. chem. Soc.* **94**, 7210.
Corey, E. J. and Cane, D. E. (1969). *J. org. Chem.* **34**, 3053.
Corey, E. J., Carey, F. A. and Winter, R. A. E. (1965). *J. Am. chem. Soc.* **87**, 935.
Corey, E. J. and Chen, R. H. K. (1973). *Tetrahedron Lett.* 1611, 3817.
Corey, E. J., Chow, S. W. and Scherrer, R. A. (1957). *J. Am. chem. Soc.* **79**, 5773.
Corey, E. J., Crouse, D. N. and Anderson, T. E. (1975). *J. org. Chem.* **40**, 2140.
Corey, E. J., Danheiser, R. L. and Chandrasekaran, S. (1976). *J. org. Chem.* **41**, 260.
Corey, E. J. and Enders, D. (1976). *Tetrahedron Lett.* **3**, 11.
Corey, E. J., Erickson, B. W. and Noyori, R. (1971). *J. Am. chem. Soc.* **93**, 1724.
Corey, E. J. and Fleet, G. W. J. (1973). *Tetrahedron Lett.* 4499.
Corey, E. J., Gilman, N. W. and Ganem, B. E. (1968). *J. Am. chem. Soc.* **90**, 5616.
Corey, E. J. and Hamanaka, E. (1967). *J. Am. chem. Soc.* **89**, 2758.
Corey, E. J. and Hertler, W. R. (1959). *J. Am. chem. Soc.* **81**, 5209.
Corey, E. J. and Hertler, W. R. (1960). *J. Am. chem. Soc.* **82**, 1657.
Corey, E. J. and Hortmann, A. G. (1963). *J. Am. chem. Soc.* **85**, 4033.
Corey, E. J. and Katzenellenbogen, J. A. (1969). *J. Am. chem. Soc.* **91**, 1851.
Corey, E. J., Katzenellenbogen, J. A., Gilman, N. W., Roman, S. A. and Erickson, B. W. (1968). *J. Am. chem. Soc.* **90**, 5618.
Corey, E. J., Katzenellenbogen, J. A. and Posner, G. H. (1967). *J. Am. chem. Soc.* **89**, 4245.
Corey, E. J. and Kim, C. U. (1972). *J. Am. chem. Soc.* **94**, 7587.
Corey, E. J. and Kim, C. U. (1973). *Tetrahedron Lett.* 919.
Corey, E. J. and Kim, C. U. (1974). *Tetrahedron Lett.* 287.
Corey, E. J., Kim, C. U., Chen, R. H. K. and Takeda, M. (1972). *J. Am. chem. Soc.* **94**, 4395.
Corey, E. J. and Kirst, H. A. (1972). *J. Am. chem. Soc.* **94**, 667.
Corey, E. J. and Kozikowsky, A. P. (1975). *Tetrahedron Lett.* 925, 2389.
Corey, E. J. and Kwiatkowsky, G. T. (1968). *J. Am. chem. Soc.* **90**, 6816.
Corey, E. J., Mitra, R. B. and Uda, H. (1964). *J. Am. chem. Soc.* **86**, 485.
Corey, E. J., Nicolaou, K. C. and Toru, T. (1975). *J. Am. chem. Soc.* **97**, 2287.
Corey, E. J. and Ravindranathan, T. (1972). *J. Am. chem. Soc.* **94**, 4013.
Corey, E. J., Ravindranathan, T. and Terashima, S. (1971). *J. Am. chem. Soc.* **93**, 4326.
Corey, E. J. and Seebach, D. (1965). *Angew. Chem. internat. Edit.* **4**, 1075, 1077.
Corey, E. J. and Semmelhack, M. F. (1967). *J. Am. chem. Soc.* **89**, 2755.
Corey, E. J., Semmelhack, M. F. and Hegedus, L. S. (1968). *J. Am. chem. Soc.* **90**, 2416.
Corey, E. J. and Suggs, J. W. (1975*a*). *Tetrahedron Lett.* 2644.
Corey, E. J. and Suggs, J. W. (1975*b*). *J. org. Chem.* **40**, 2554.
Corey, E. J. and Taylor, W. C. (1964). *J. Am. chem. Soc.* **86**, 3881.
Corey, E. J. and Venkateswarlu, A. (1972). *J. Am. chem. Soc.* **94**, 6190.
Corey, E. J. and Walinsky, S. W. (1972). *J. Am. chem. Soc.* **94**, 8932.
Corey, E. J. and Watt, D. S. (1973). *J. Am. chem. Soc.* **95**, 2303.
Cornforth, J. W., Cornforth, R. H. and Mathew, K. K. (1959). *J. chem. Soc.* 112.
Cornforth, J. W., Milborrow, B. V. and Ryback, G. (1965). *Nature, Lond.* **206**, 715.

Cotton, F. A. and Wilkinson, G. (1972). In *Advanced Inorganic Chemistry*, 3rd edn, pp. 113, 115. (London: Interscience.)
Cowell, G. W. and Ledwith, A. (1970). *Quart. Rev. chem. Soc. Lond.* **24**, 119.
Coyle, J. D. and Carless, H. A. J. (1972). *Chem. Soc. Rev.* **1**, 471.
Cragg, G. L. (1973). *Organoboranes in Organic Synthesis.* (New York: Marcel Dekker.)
Cram, D. J. and Abd Elhafez, F. A. (1952). *J. Am. chem. Soc.* **74**, 4828.
Cram, D. J., Sahyun, M. R. V. and Knox, G. R. (1962). *J. Am. chem. Soc.* **84**, 1734.
Crandall, J. K. and Crawley, L. C. (1973). *Organic Syntheses*, **53**, 17.
Crandall, J. K. and Luan-Ho Chang (1967). *J. org. Chem.* **32**, 435.
Crawford, R. J., Erman, W. F. and Broaddus, C. D. (1972). *J. Am. chem. Soc.* **94**, 4298.
Criegee, R. (1965). *Oxidation in Organic Chemistry*, p. 297. Ed. K. B. Wiberg. (New York: Academic Press.)
Criegee, R. (1975). *Angew. Chem. internat. edn.* **14**, 745.
Criegee, R. and Gunther, P. (1963). *Chem. Ber.* **96**, 1564.
Cuppen, Th. J. H. M. and Laarhoven, W. H. (1972). *J. Am. chem. Soc.* **94**, 5914.
Danishefsky, S. and Kitahara, T. (1974). *J. Am. chem. Soc.* **96**, 7807.
Danishefsky, S. and Kitahara, T. (1975). *J. org. Chem.* **40**, 538.
Dauben, W. G., Beasley, G. H., Broadhurst, M. D., Muller, B., Peppard, D. J., Pesnelle, P. and Suter, C. (1975). *J. Am. chem. Soc.* **97**, 4973.
Dauben, W. G., Fonken, G. J. and Noyce, D. S. (1956). *J. Am. chem. Soc.* **78**, 2579.
Dauben, W. G. and Ipaktschi, J. (1973). *J. Am. chem. Soc.* **95**, 5088.
Dauben, W. G., Lorber, M. and Fullerton, D. S. (1969). *J. org. Chem.* **34**, 3587.
Dauben, W. G., Rivers, G. T., Zimmerman, W. T., Yang, N. C., Kim, B. and Yang, J. (1976). *Tetrahedron Lett.* 2951.
Dave, V. and Warnhoff, E. W. (1970). *Organic Reactions*, **18**, 217.
Davidson, A. H. and Warren, S. (1976). *J. Chem. Soc. Perkin I*, 639.
Deem, M. L. (1972). *Synthesis*, 675.
Dehmlow, E. V. (1974). *Angew. Chem. internat. Edit.* **13**, 170.
De Los, F. de Tar and Relyea, D. I. (1956). *J. Am. chem. Soc.* **78**, 4302.
de Mayo, P. (1971). *Accounts chem. Res.* **4**, 41.
Demoulin, A., Gorissen, H., Hesbain-Frisque, A-M. and Ghosez, L. (1975). *J. Am. chem. Soc.* **97**, 4409.
Denis, J. M. and Conia, J. M. (1972). *Tetrahedron Lett.* 4593.
Denny, R. W. and Nickon, A. (1973). *Organic Reactions*, **20**, 133.
Deno, N. C., Billups, W. E., Kramer, K. E. and Lastomirsky, R. R. (1970). *J. org. Chem.* **35**, 3080.
Deno, N. C., Peterson, H. J. and Saines, G. S. (1960). *Chem. Rev.* **60**, 7.
DePuy, C. H. and King, R. W. (1960). *Chem. Rev.* **60**, 431.
DePuy, C. H., Naylor, C. G. and Beckman, J. A. (1970). *J. org. Chem.* **35**, 2750.
Dervan, P. B. and Shippey, M. A. (1976). *J. Am. chem. Soc.* **98**, 1265.
Dewar, M. J. S. (1971). *Angew. Chem. internat. Edit.* **10**, 761.
Dewar, M. J. S., Griffin, A. C. and Kirschner, S. (1974). *J. Am. chem. Soc.* **96**, 6225.
Dilling, W. L. (1966). *Chem. Rev.* **66**, 373.
Dittius, G. (1965). *Methoden der organischen Chemie*, vol. 6(3), p. 434. Ed. E. Müller. (Stuttgart: Georg Thieme Verlag.)
Djerassi, C. (1951). *Organic Reactions*, **6**, 207.

Djerassi, C. (1963). *Steroid Reactions*, p. 267. (San Francisco: Holden-Day, Inc.)
Dockx, J. (1973). *Synthesis*, 441.
Dodd, G. H., Golding, B. T. and Ionannu, P. V. (1975). *J. chem. Soc. Chem. Commun.* 249.
Doering, W. E. von and Dorfman, E. (1953). *J. Am. chem. Soc.* **75**, 5595.
Doering, W. E. von, Franck-Neumann, M., Hasselmann, D. and Kaye, R. L. (1972). *J. Am. chem. Soc.* **94**, 3833.
Dopper, J. H., Oudman, D. and Wynberg, H. (1973). *J. Am. chem. Soc.* **95**, 3692.
Dryden, H. L., Webber, G. M., Burtner, R. R. and Cella, J. A. (1961). *J. org. Chem.* **26**, 3237.
Dumont, W. and Krief, A. (1976). *Angew. Chem. internat. Edit.* **15**, 161.
Dust, L. A. and Gill, E. W. (1970). *J. chem. Soc.* (*C*), 1630.
Eaborn, C. (1973). *J. chem. Soc. Chem. Commun.* 1255.
Eaborn, C., Pant, B. C., Peeling, E. R. A. and Taylor, S. C. (1969). *J. chem. Soc.* (*C*), 2823.
Eastwood, F. W., Harrington, K. J., Josan, J. S. and Pura, J. L. (1970). *Tetrahedron Lett.* 5223.
Eaton, P. E. (1968). *Accounts chem. Res.* **1**, 50.
Eglinton, G. and McCrae, W. (1963). *Advances in Organic Chemistry: Methods and Results*, vol. 4, p. 225. Ed. R. A. Raphael, E. C. Taylor and H. Wynberg. (London: Interscience.)
Eisch, J. J. and Trainor, J. T. (1963). *J. org. Chem.* **28**, 2870.
Eliel, E. L. (1961). *Rec. chem. Prog.* **22**, 129.
Eliel, E. L. and Senda, Y. (1970). *Tetrahedron*, **26**, 2411.
Epstein, W. W. and Ollinger, J. (1970). *Chem. Commun.* 1338.
Epstein, W. W. and Sweat, F. W. (1967). *Chem. Rev.* **67**, 247.
Erickson, B. W. (1974). *Organic Syntheses*, **54**, 19,
Eschenmoser, A. (1970). *Quart. Rev. chem. Soc. Lond.* **24**, 366.
Evans, D. A. and Andrews, G. C. (1974). *Accounts chem. Res.* **7**, 147.
Evans, D. A., Andrews, G. C. and Buckwalter, B. (1974). *J. Am. chem. Soc.* **96**, 5560.
Evans, D. A., Bryan, C. A. and Wahl, G. M. (1970). *J. org. Chem.* **35**, 4122.
Evans, D. A., Carroll, G. L. and Truesdale, L. K. (1974). *J. org. Chem.* **39**, 914.
Evans, D. A. and Domeier, L. A. (1974). *Organic Syntheses*, **54**, 93.
Evans, D. A. and Golob, A. M. (1975). *J. Am. chem. Soc.* **97**, 4765.
Evans, D. A., Grimm, K. G. and Truesdale, L. K. (1975). *J. Am. chem. Soc.* **97**, 3229.
Evans, D. A. and Hoffman, J. M. (1976). *J. Am. chem. Soc.* **98**, 1983.
Evans, D. A., Scott, W. L. and Truesdale, L. K. (1972). *Tetrahedron Lett.* 121.
Evans, D. A., Thomas, R. C. and Walker, J. A. (1976). *Tetrahedron Lett.* 1427.
Evans, D. A., Truesdale, L. K., Grimm, K. G. and Nesbitt, S. L. (1977). *J. Am. chem. Soc.* **99**, 5009.
Evans, R. M. (1959). *Quart. Rev. chem. Soc. Lond.* **13**, 61.
Farmer, R. F. and Hamer, J. (1966). *J. org. Chem.* **31**, 2418.
Faulkner, D. J. and Petersen, M. R. (1973). *J. Am. chem. Soc.* **95**, 553.
Fétizon, M. and Golfier, M. (1968). *C. r. hebd. Séanc. Acad. Sci., Paris*, **267**(*C*), 900.
Fétizon, M., Golfier, M. and Louis, J. M. (1975). *Tetrahedron*, **31**, 171.
Fétizon, M., Golfier, M., Milcent, R. and Papadakis, I. (1975). *Tetrahedron*, **31**, 165.

Fétizon, M., Golfier, M. and Mourgues, P. (1972). *Tetrahedron Lett.* 4445.
Fieser, L. F. and Fieser, M. (1967). *Reagents for Organic Synthesis*, Vol. 1, p. 759. (London: Wiley.)
Filler, R. (1963). *Chem. Rev.* **63**, 21.
Fleming, I. and Kargar, M. H. (1967). *J. chem. Soc.* (*C*) 226.
Fleming, I. and Harley-Mason, J. (1964). *J. chem. Soc.* 2165.
Fleming, I. and Pearce, A. (1975). *J. chem. Soc. Chem. Commun.* 633.
Finley, K. T. (1964). *Chem. Rev.* **64**, 573.
Fliszar, S. and Carles, J. (1969). *J. Am. chem. Soc.* **81**, 2637.
Foote, C. S. (1971). *Pure appl. Chem.* **27**, 635.
Foote, C. S. (1968). *Accounts chem. Res.* **1**, 104.
Foote, C. S. and Wexler, S. (1964). *J. Am. chem. Soc.* **86**, 3879, 3880.
Fortunato, J. M. and Ganem, B. (1976). *J. org. Chem.* **41**, 2194.
Fráter, G. (1975). *Helv. chim. Acta*, **58**, 442.
Freeman, F. and Yamachika, N. J. (1972). *J. Am. chem. Soc.* **94**, 1214.
Fuks, R. and Viehe, H. G. (1969). In *Chemistry of Acetylenes*, Ed. H. G. Viehe, p. 477. (New York: Marcel Dekker.)
Ganem, B. and Boeckman, R. K. (1974). *Tetrahedron Lett.* 917.
Garbisch, E. W., Schreader, L. and Frankel, J. J. (1967). *J. Am. chem. Soc.* **89**, 4233.
Gardner, J. N., Carlon, F. E. and Gnoj, O. (1968). *J. org. Chem.* **33**, 3294.
Garsky, V., Koster, D. F. and Arnold, R. T. (1974). *J. Am. chem. Soc.* **96**, 4207.
Gassman, P. G. and Richmond, G. D. (1966). *J. org. Chem.* **31**, 2355.
Gassman, P. G. *et al.* (1974). *J. Am. chem. Soc.* **96**, 3002, 5487, 5495, 5512.
Gault, F. G., Rooney, J. J. and Kemball, C. (1962). *J. Catalysis*, **1**, 255.
Gaylord, N. G. (1956). *Reduction with Complex Metal Hydrides.* (New York: Wiley.)
Geiss, K., Seuring, B., Pieter, R. and Seebach, D. (1974). *Angew. Chem. internat. Edit.* **13**, 479.
George, M. V. and Balachandran, K. S. (1975). *Chem. Rev.* **75**, 491.
Geschwend, H. W. (1973). *Helv. chim. Acta* **56**, 1763.
Gibson, T. W. and Erman, W. F. (1967). *Tetrahedron Lett.* 905.
Giering, W. P., Rosenblum, M. and Tancrede, J. (1972). *J. Am. chem. Soc.* **94**, 7170.
Gilchrist, T. L. and Rees, C. W. (1969). *Carbenes, Nitrenes and Arynes.* (London: Nelson.)
Gilchrist, T. L. and Storr, R. C. (1972). *Organic Reactions and Orbital Symmetry.* (Cambridge: Cambridge University Press.)
Gillis, B. T. (1967). In 1,4-*Cycloaddition Reactions.* Ed. J. Hamer. (New York: Academic Press.)
Girard, C., Amice, P., Barnier, J. P. and Conia, J. M. (1974). *Tetrahedron Lett.* 3329.
Girard, C. and Conia, J. M. (1974). *Tetrahedron Lett.* 3327.
Gollnick, K. (1968). *Adv. Photochem.* **6**, 1.
Gollnick, K. and Schenck, O. (1967). In 1,4-*Cycloaddition Reactions.* Ed. J. Hamer. (New York: Academic Press.)
Gopal, H., Adams, T. and Moriarty, R. M. (1972). *Tetrahedron*, **28**, 4259.
Gorlier, J-P., Hamon, L., Levisalles, J. and Wagnon, J. (1973). *J. chem. Soc. Chem. Commun.* 88.
Gorman, A. (1973). In *Chemical Society Specialist Periodical Report 'Photochemistry'*, vol. 4, p. 734.
Grant, B. and Djerassi, C. (1974). *J. org. Chem.* **39**, 968.

Grewe, R. and Hinrich, I. (1964). *Chem. Ber.* **97**, 443.
Gribble, G. W. and Ferguson, D. C. (1975). *J. chem. Soc. Chem. Commun.* 535.
Grieco, P. A., Chia-Lin, J. Wang and Majetich, G. (1976). *J. org. Chem.* **41**, 726.
Grieco, P. A. and Miyashita, M. (1974). *J. org. Chem.* **39**, 120.
Grieco, P. A. and Pogonowski, C. S. (1974). *J. org. Chem.* **39**, 732.
Grieco, P. A. and Pogonowski, C. S. (1975). *J. chem. Soc. Chem. Commun.* 79.
Grieco, P. A., Pogonowski, C. S. and Burke, S. (1975). *J. org. Chem.* **40**, 542.
Grob, C. A. and Kammüller, H. (1957). *Helv. chim. Acta*, **40**, 2139.
Grob, C. A. and Schiess, P. W. (1967). *Angew. Chem. internat. Edit.* **6**, 1.
Gröbel, B-T. and Seebach, D. (1974). *Angew. Chem. internat. Edit.* **13**, 83.
Groves, J. T. and Van der Puy, M. (1974). *J. Am. chem. Soc.* **96**, 5274.
Grundon, M. F., Henbest, H. B. and Scott, M. D. (1963). *J. chem. Soc.* 1855.
Gunstone, F. D. (1960). In *Advances in Organic Chemistry: Methods and Results.* vol. 1, p. 103. Ed. R. A. Raphael, E. C. Taylor and H. Wynberg. (New York: Wiley Interscience.)
Gutsche, C. D. and Redmore, D. (1968). *Advances in Alicyclic Chemistry.* Supplement 1. *Carbocyclic Ring Expansion Reactions*, p. 111. (London: Academic Press.)
Gygax, P., Das Gupta, T. K. and Eschenmoser, A. (1972). *Helv. chim. Acta*, **55**, 2205.
Hallman, P. S., McGarvey, B. R. and Wilkinson, G. (1968). *J. chem. Soc.* (*A*), 3143.
Hallsworth, A. S. and Henbest, H. B. (1957). *J. chem. Soc.* 4604.
Hamer, J. and Ahmad, M. (1967). In *1,4-Cycloaddition Reactions.* Ed. J. Hamer. (New York: Academic Press.)
Hammond, G. S., Turro, N. J. and Liu, R. S. H. (1963). *J. org. Chem.* **28**, 3297.
Hanson, J. R. (1974). *Synthesis*, 1.
Hanson, J. R. and Premuzic, E. (1968). *Angew. Chem. internat. Edit.* **7**, 247.
Harmon, R. E., Gupta, S. K. and Brown, D. J. (1973). *Chem. Rev.* **73**, 21.
Harris, T. M. and Harris, C. M. (1969). *Organic Reactions*, **17**, 155.
Harrison, C. R. and Hodge, R. (1976). *J. Chem. Soc. Perkin I*, 605.
Hartmann, W. G., Heine, H. G. and Schrader, L. (1974). *Tetrahedron Lett.* 883, 3101.
Hartzell, S. L., Sullivan, D. F. and Rathke, M. W. (1974). *Tetrahedron Lett.* 1403.
Hassall, C. H. (1957). *Organic Reactions*, **9**, 73.
Haubenstock, H. (1973). *J. org. Chem.* **38**, 1765.
Hegedus, L. S. and Stiverson, R. K. (1974). *J. Am. chem. Soc.* **96**, 3250.
Henbest, H. B. (1965). *Chem. Soc. Spec. Publ.* No. 19, p. 83.
Herrmann, J. L., Kieczykowski, G. R., Romaret, R. F. and Schlessinger, R. H. (1973). *Tetrahedron Lett.* 4715.
Herz, J. E. and Gonzalez, E. (1969). *Chem. Commun.* 1395.
Heusler, K. and Kalvoda, J. (1964). *Angew. Chem. internat. Edit.* **3**, 525.
Heyns, K. and Paulsen, H. (1963). In *Newer Methods of Preparative Organic Chemistry*, vol. 2, p. 303. Ed. W. Foerst. (New York: Academic Press.)
Heyns, K., Soldat, W-D. and Köll, P. (1975). *Chem. Ber.* **108**, 3611, 3619.
Hill, R. K. and Carlson, R. M. (1965). *J. org. Chem.* **30**, 2414.
Hill, R. K., Morgan, J. W., Shetty, R. V. and Synerholm, M. E. (1974). *J. Am. chem. Soc.* **96**, 4201.
Hill, R. K. and Rabinovitz, M. (1964). *J. Am. chem. Soc.* **86**, 965.
Hine, J. (1964). *Divalent Carbon.* (New York: Ronald Press.)

Hines, J. N., Peagram, M. J., Thomas, E. J. and Whitham, G. H. (1973). *J. chem. Soc. Perkin I*, 2332.
Hiyama, T. and Nozaki, H. (1973). *Bull. chem. Soc. Japan*, **46**, 2248.
Hoffmann, H. M. R. (1969). *Angew. Chem. internat. Edit.* **8**, 556.
Hoffmann, H. M. R. (1973). *Angew. Chem. internat. Edit.* **12**, 819, 831.
Hoffmann, R. and Woodward, R. B. (1965). *J. Am. chem. Soc.* **87**, (*a*) 2046; (*b*) 4388.
Hoffmann, R. W. (1967). *Dehydrobenzene and Cycloalkynes.* (New York: Academic Press.)
Holmes, H. L. (1948). *Organic Reactions*, **4**, 60.
Hooz, J., Gunn, D. M. and Kono, H. (1971). *Can. J. Chem.* **49**, 2371.
Hooz, J. and Smith, J. (1972). *J. org. Chem.* **37**, 4200.
Horiuti, I. and Polanyi, M. (1934). *Trans. Faraday Soc.* **30**, 1164.
Houk, K. N. (1973). *J. Am. chem. Soc.* **95**, 4092.
Houk, K. N. (1975). *Accounts chem. Res.* **8**, 361.
Houk, K. N. and Luskus, L. J. (1971). *J. Am. chem. Soc.* **93**, 4606.
Houk, K. N. and Strozier, R. W. (1973). *J. Am. chem. Soc.* **95**, 4094.
House, H. O. (1965). *Modern Synthetic Reactions.* (New York: Benjamin.)
House, H. O. (1967). *Rec. chem. Prog.* **28**, 99.
House, H. O. (1972). *Modern Synthetic Reactions*, 2nd edn, p. 175. (Benjamin: Menlo Park, California.)
House, H. O. and Cronin, T. H. (1965). *J. org. Chem.* **30**, 1061.
House, H. O., Czuba, L. J., Gall, M. and Olmstead, H. D. (1969). *J. org. Chem.* **34**, 2324.
House, H. O., Gall, M. and Olmstead, H. D. (1971). *J. org. Chem.* **36**, 2361.
House, H. O. and Kinloch, E. F. (1974). *J. org. Chem.* **39**, 747.
House, H. O. and Larson, J. K. (1968). *J. org. Chem.* **33**, 61.
House, H. O. and Trost, B. M. (1965). *J. org. Chem.* **30**, 2502.
House, H. O. and Umen, M. J. (1972). *J. Am. chem. Soc.* **94**, 5495.
House, H. O. and Umen, M. J. (1973). *J. org. Chem.* **38**, 1000.
House, H. O. and Weeks, P. D. (1975). *J. Am. chem. Soc.* **97**, 2770.
Huang-Minlon (1946). *J. Am. chem. Soc.* **68**, 2487.
Huckin, S. N. and Weiler, L. (1974). *J. Am. chem. Soc.* **90**, 1082.
Hudrlik, P. F. and Peterson, D. (1975). *J. Am. chem. Soc.* **97**, 1464.
Hudrlik, P. F., Peterson, D. and Rona, R. J. (1975). *J. org. Chem.* **40**, 2263.
Huffman, J. W. and Charles, J. T. (1968). *J. Am. chem. Soc.* **90**, 6486.
Hughes, R. J., Pelter, A. and Smith, K. (1974). *J. chem. Soc. Chem. Commun.* 863.
Hughes, R. J., Pelter, A., Smith, K., Negishi, E. and Yoshida, T. (1976). *Tetrahedron Lett.* 87.
Huisgen, R. (1968). *Angew. Chem. internat. Edit.* **7**, 321.
Huisgen, R., Grashey, R. and Sauer, J. (1964). In *The Chemistry of Alkenes*, p. 739. Ed. S. Patai. (London: Interscience.)
Hussey, A. S. and Takeuchi, Y. (1969). *J. Am. chem. Soc.* **91**, 672.
Hutchins, R. O., Hoke, D., Keogh, J. and Koharski, D. (1969). *Tetrahedron Lett.* 3495.
Hutchins, R. O., Kacher, M. and Rua, L. (1975). *J. org. Chem.* **40**, 923.
Hutchins, R. O. and Kandasamy, D. (1973). *J. Am. chem. Soc.* **95**, 6131.
Inukai, T. and Kojima, T. (1966). *J. org. Chem.* **31**, 2032.

Inukai, T. and Kojima, T. (1970). *J. org. Chem.* **35**, 1342; (1971). *J. org. Chem.* **36**, 924.
Ireland, R. E., Muchmore, D. C. and Hengartner, U. (1972). *J. Am. chem. Soc.* **94**, 5098.
Ireland, R. E., Mueller, R. H. and Willard, A. K. (1976). *J. Am. chem. Soc.* **98**, 2868.
Jackson, W. R. and Strauss, J. H. G. (1975). *Tetrahedron Lett.* 2591.
Jardine, F. H. and Wilkinson, G. (1967). *J. chem. Soc.* (*C*), 270.
Jardine, I. and McQuillin, F. J. (1966). *Tetrahedron Lett.* 4871.
Jefford, C. W., Kirkpatrick, D. and Delay, F. (1972). *J. Am. chem. Soc.* **94**, 8905.
Jenner, E. L. (1962). *J. org. Chem.* **27**, 1031.
Jerussi, R. A. (1970). In *Selective Organic Transformations*, vol. 1. Ed. B. S. Thyagarajan. (London: Wiley-Interscience.)
Johnson, A. W. (1966). *Ylid Chemistry*. (London: Academic Press.)
Johnson, C. R. (1973). *Accounts chem. Res.* **6**, 341.
Johnson, C. R., Schroeck, C. W. and Shanklin, J. R. (1973). *J. Am. chem. Soc.* **95**, 7424.
Johnson, W. S., Bannister, B., Bloom, B. M., Kemp, A. D., Pappo, R., Rogier, E. R. and Szmuszkovicz, J. (1953). *J. Am. chem. Soc.* **75**, 2275.
Johnson, W. S., Brocksom, T. J., Loew, P., Rich, D. H., Werthemann, L., Arnold, R. A., Tsung-tee Li and Faulkner, D. J. (1970). *J. Am. chem. Soc.* **92**, 4463.
Johnson, W. S., Tsung-tee Li, Faulkner, D. J. and Campbell, S. F. (1968). *J. Am. chem. Soc.* **90**, 6225.
Johnson, W. S., Werthemann, L., Bartlett, W. R., Brocksom, T. J., Tsung-tee Li, Faulkner, D. J. and Petersen, M. R. (1970). *J. Am. chem. Soc.* **92**, 741.
Jones, M. and Moss, R. A. (1972). *Carbenes*, vol. 1. (London: Wiley-Interscience.)
Jones, P. F., Lappert, M. F. and Szary, A. C. (1973). *J. chem. Soc. Perkin I*, 2272.
Jones, R. L. and Rees, C. W. (1969). *J. chem. Soc.* (*C*), 2249.
Julia, M., Julia, S. and Tchen, S-Y. (1961). *Bull. Soc. chim. Fr.* 1849.
Jung, M. E. and McCombs, C. A. (1976). *Tetrahedron Lett.* 2935.
Kabalka, G. W. and Baker, J. D. (1975). *J. org. Chem.* **40**, 1834.
Kabalka, G. W., Brown, H. C., Suzuki, A., Honma, S., Arase, A. and Itoh, M. (1970). *J. Am. chem. Soc.* **92**, 710.
Kabalka, G. W. and Hedgecock, H. C. (1975). *J. org. Chem.* **40**, 1776.
Kabasakalian, P., Townley, E. R. and Yudis, M. D. (1962). *J. Am. chem. Soc.* **84**, 2718.
Kagan, H. B. and Dang, T-P. (1972). *J. Am. chem. Soc.* **94**, 6429.
Kaiser, E. M. (1972). *Synthesis*, 391.
Kalvoda, J. (1970). *J. chem. Soc. Chem. Commun.* 1002.
Kalvoda, J. and Heusler, K. (1971). *Synthesis*, 501.
Kametani, T., Shibuya, S., Nakano, T. and Fukumoto, K. (1971). *J. chem. Soc.* (*C*), 3818, 3976.
Katzenellenbogen, J. A. and Christy, K. J. (1974). *J. org. Chem.* **39**, 3315.
Kauffmann, T. (1974). *Angew. Chem. internat. Edit.* **13**, 627.
Kaufman, G., Cook, F., Schechter, H., Bayless, J. and Friedman, L. (1967). *J. Am. chem. Soc.* **89**, 5736.
Kearns, D. R. (1971). *Chem. Rev.* **71**, 395.
Kearns, D. R., Hollins, R. A., Khan, A. U., Chambers, R. W. and Radlick, P. (1967). *J. Am. chem. Soc.* **89**, 5455, 5456.

Keung, E. C. and Alper, H. (1972). *J. chem. Educ.* **49**, 97.
Khan, A. M., McQuillin, F. J. and Jardine, I. (1967). *J. chem. Soc.* (*C*), 136.
Kharasch, M. S. and Tawney, P. O. (1941). *J. Am. chem. Soc.* **63**, 2308.
Kimmer, W. (1955). In *Methoden der Organischen Chemie* (*Houben-Weyl*), vol. 4(2). Ed. E. Müller. (Stuttgart: Georg Thieme Verlag.)
Kingsbury, C. A. and Cram, D. J. (1960). *J. Am. chem. Soc.* **82**, 1810.
Kirby, G. W. and Sweeny, J. G. (1973). *J. chem. Soc. Chem. Commun.* 704.
Kirk, D. N. (1969). *Tetrahedron Lett.* 1727.
Kirmse, W. (1964). *Carbene Chemistry*. (London: Academic Press.)
Kishi, Y., Aratani, M., Tanino, H., Fukuyama, T. and Goto, T. (1972). *J. chem. Soc. Chem. Commun.* 64.
Klebe, J. F. (1972). In *Advances in Organic Chemistry*, vol. 8, p. 97. Ed. E. C. Taylor. (London: Wiley-Interscience.)
Kloetzel, M. C. (1948). *Organic Reactions*, **4**, 1.
Klopman, G. and Joiner, C. M. (1975). *J. Am. chem. Soc.* **97**, 5287.
Knowles, W. S., Sabacky, M. J. and Vineyard, B. D. (1972). *J. chem. Soc. chem. Commun.* 10.
Knowles, W. S., Sabacky, M. J., Vineyard, B. D. and Weinkauff, D. J. (1975). *J. Am. chem. Soc.* **97**, 2567.
Kobrich, G. (1972). *Angew. Chem. internat. Edit.* **11**, 473.
Kobuke, Y., Fueno, T. and Furukawa, J. (1970). *J. Am. chem. Soc.* **92**, 6548.
Koenig, K. E. and Weber, W. P. (1973). (*a*) *J. Am. chem. Soc.* **95**, 3416; (*b*) *Tetrahedron Lett.* 2533.
Kofron, W. G. and Yeh, M-K. (1976). *J. org. Chem.* **41**, 439.
Konaka, R., Terabe, S. and Kuruma, K. (1969). *J. org. Chem.* **34**, 1334.
Kondo, K. and Matsumoto, M. (1972). *Chem. Commun.* 1332.
Kornblum, N., Jones, W. J. and Anderson, G. J. (1959). *J. Am. chem. Soc.* **81**, 4113.
Köster, R. (1964). *Angew. Chem. internat. Edit.* **3**, 174.
Krantz, A. and Lin, C. Y. (1973). *J. Am. chem. Soc.*, **95**, 5662.
Kreiser, W., Haumesser, W. and Thomas, A. F. (1974). *Helv. chim. Acta*, **57**, 164.
Kreiser, W. and Wurziger, H. (1975). *Tetrahedron Lett.* 1669.
Krespan, C. G., McKusick, B. C. and Cairns, T. L. (1961). *J. Am. chem. Soc.* **83**, 3428.
Krishnamurthy, S. and Brown, H. C. (1976). *J. Am. chem. Soc.* **98**, 3383.
Krishnamurthy, S., Schubert, R. M. and Brown, H. C. (1973). *J. Am. chem. Soc.* **95**, 8486.
Kuhn, H. J. and Gollnick, K. (1972). *Tetrahedron Lett.* 1909.
Kuivila, H. G. (1968). *Accounts chem. Res.* **1**, 299.
Kupchan, S. M. and Wormsey, H. C. (1965). *Tetrahedron Lett.* 359.
Kuwajima, I. and Nakamura, E. (1975). *J. Am. chem. Soc.* **97**, 3257.
Kuwajima, I., Sato, T., Arai, M. and Minami, N. (1976). *Tetrahedron Lett.* 1817.
Kwart, H. and Conley, R. A. (1973). *J. org. Chem.* **38**, 2011.
Kwart, H. and King, K. (1968). *Chem. Rev.* **68**, 415.
Laarhoven, W. H., Cuppen, Th. J. H. M. and Nivard, R. F. J. (1970). *Tetrahedron*, **26**, 4865.
Lambert, J. and Napoli, J. J. (1973). *J. Am. chem. Soc.* **95**, 294.
Landini, D., Montanari, F. and Pirisi, F. M. (1974). *J. chem. Soc. Chem. Commun.* 879.
Landor, S. R., Miller, B. J. and Tatchell, A. R. (1967). *J. chem. Soc.* (*C*), 197.

Lane, C. F. (1975). *Synthesis*, 135.

Lane, C. F. and Kabalka, G. W. (1976). *Tetrahedron*, **32**, 981.

Lee, J. B. and Uff, B. C. (1967). *Quart. Rev. chem. Soc. Lond.* **21**, 429.

Lee, R. A., McAndrews, C., Patel, K. M. and Reusch, W. (1973). *Tetrahedron Lett.* 965.

Lemieux, R. U. and von Rudloff, E. (1955). *Can. J. Chem.* **33**, 1701.

Leyendecker, F., Drouin, J. and Conia, J. M. (1974). *Tetrahedron Lett.* 2931.

Lutz, E. F. and Bailey, G. M. (1964). *J. Am. chem. Soc.* **86**, 3899.

Lwowski, W. (1967). *Angew. Chem. internat. Edit.* **6**, 897.

McCormick, J. P. (1974). *Tetrahedron Lett.* 1701.

McElvain, S. M. (1948). *Organic Reactions*, **4**, 256.

McKillop, A., Hunt, J. D., Kienzle, F., Bigham, E. and Taylor, E. C. (1973). *J. Am. chem. Soc.* **95**, 3635.

McKillop, A., Oldenziel, O. H., Swann, B. P., Taylor, E. C. and Robey, R. L. (1973). *J. Am. chem. Soc.* **95**, 1296.

McKillop, A., Swann, B. P. and Taylor, E. C. (1973). *J. Am. chem. Soc.* **95**, 3340.

McKillop, A. and Taylor, E. C. (1973). *Chem. Britain*, **9**, 4.

McMurry, J. E. (1974). *Accounts chem. Res.* **7**, 281.

McMurry, J. E. and Fleming, M. P. (1975). *J. org. Chem.* **40**, 2555.

McMurry, J. E. and Fleming, M. P. (1976). *J. org. Chem.* **41**, 896.

McQuillin, F. J. (1973). In *Progress in Organic Chemistry*, vol. 8, p. 314. Eds. W. Carruthers and J. K. Sutherland. (London: Butterworths.)

McQuillin, F. J. (1974). *Tetrahedron*, **30**, 1661.

McQuillin, F. J., Ord, W. O. and Simpson, P. L. (1963). *J. chem. Soc.* 5996.

Maercker, A. (1965). *Organic Reactions*, **14**, 270.

Makosza, M. and Jończyk, A. (1976). *Organic Syntheses*, **55**, 91.

Málek, J. and Černý, M. (1972). *Synthesis*, 217.

Marchand, A. P. and Brockway, N. M. (1974). *Chem. Rev.* **74**, 431.

Marino, J. P. and Mesbergen, W. B. (1974). *J. Am. chem. Soc.* **96**, 4050.

Marshall, J. A. (1969). *Rec. chem. Prog.* **30**, 3.

Marshall, J. A. (1971). *Synthesis*, 229.

Marshall, J. A. and Babler, J. H. (1970). *Tetrahedron Lett.* 3861.

Marshall, J. A. and Seitz, D. E. (1974). *J. org. Chem.* **39**, 1814.

Martin, E. L. (1942). *Organic Reactions*, **1**, 155.

Martin, J. G. and Hill, R. K. (1961). *Chem. Rev.* **61**, 537.

Martin, R. H. (1974). *Angew. Chem. internat. Edit.* **13**, 649.

Masamune, S., Bates, G. S. and Georghiou, P. E. (1974). *J. Am. chem. Soc.* **96**, 3688.

Mauldin, C. H., Biehl, E. R. and Reeves, P. C. (1972). *Tetrahedron Lett.* 2955.

Meinwald, J. and Frauenglass, E. (1960). *J. Am. chem. Soc.* **82**, 5235.

Meinwald, J. and Meinwald, Y. C. (1966). In *Advances in Alicyclic Chemistry*, vol. 1, p. 1. Ed. H. Hart and G. J. Karabatsos. (New York: Academic Press.)

Mellor, J. M. and Webb, C. F. (1974). *J. Chem. Soc. Perkin II*, **17**, 26.

Meyers, A. I. *et al.* (1973). *J. org. Chem.* **38**, 36.

Meyers, A. I. and Durandetta, J. L. (1975). *J. org. Chem.* **40**, 2021.

Meyers, A. I., Knaus, G., Kamata, K. and Ford, M. E. (1976). *J. Am. chem. Soc.* **98**, 567.

Meyers, A. I. and Mihelich, E. D. (1976). *Angew. Chem. internat. Edit.* **15**, 270.

Meyers, A. I. and Smith, E. M. (1970). *J. Am. chem. Soc.* **92**, 1084.

Meyers, A. I., Temple, D. L., Haidukewych, D. and Mihelich, E. D. (1974). *J. org. Chem.* **39**, 2787.
Meyers, A. I., Temple, D. L., Nolen, R. L. and Mihelich, E. D. (1974). *J. org. Chem.* **39**, 2778.
Meyers, A. J. and Whitten, C. E. (1975). *J. Am. chem. Soc.* **97**, 6266.
Meyers, A. I., Williams, D. R. and Druelinger, M. (1976). *J. Am. chem. Soc.* **98**, 3032.
Midland, M. M. and Brown, H. C. (1975). *J. org. Chem.* **40**, 2845.
Midland, M. M., Sinclair, J. A. and Brown, H. C. (1974). *J. org. Chem.* **39**, 731.
Mihailović, Lj., Čeković, Z. and Stanković, J. (1969). *Chem. Commun.* 981.
Miles, D. H., Loew, P., Johnson, W. S., Kluge, A. F. and Meinwald, J. (1972). *Tetrahedron Lett.* 3019.
Miller, C. E. (1965). *J. chem. Educ.* **42**, 254.
Mitchell, R. H. and Boekelheide, V. (1974). *J. Am. chem. Soc.* **96**, 1547, 1558.
Mitsui, S., Kudo, Y. and Kobayashi, M. (1969). *Tetrahedron*, **25**, 1921.
Miyamoto, N., Isayama, S., Utimoto, K. and Nozaki, H. (1971). *Tetrahedron Lett.* 4597.
Moffatt, J. G. (1971). *J. org. Chem.* **36**, 1909.
Moore, W. M., Morgan, D. D. and Stermitz, F. R. (1963). *J. Am. chem. Soc.* **85**, 829.
Morrison, J. D. and Mosher, H. S. (1971). *Asymmetric Organic Reactions*, p. 202. (Englewood Cliffs, New Jersey: Prentice-Hall.)
Moser, W. R. (1969). *J. Am. chem. Soc.* **91**, 1135, 1141.
Mukaiyama, T., Banno, K. and Narasaka, K. (1974). *J. Am. chem. Soc.* **96**, 7503.
Müller, E., Kessler, H. and Zeeh, B. (1966). *Fortschr. chem. Forsch.* **7**, 128.
Munch-Petersen, J. (1958). *Acta chem. scand.* **12**, 2007.
Murray, R. W. (1968). *Accounts chem. Res.* **1**, 313.
Nace, H. R. (1962). *Organic Reactions*, **12**, 57.
Näf, F., Decorzant, R. and Thommen, W. (1975). *Helv. chim. Acta*, **58**, 1808.
Näf, F. and Degen, P. (1971). *Helv. chim. Acta*, **54**, 1939.
Näf, F. and Ohloff, G. (1974). *Helv. chim. Acta*, **57**, 1868.
Nakabayashi, T. (1960). *J. Am. chem. Soc.* **82**, 3900, 3906, 3909.
Nakagawa, K., Konaka, R. and Nakata, T. (1962). *J. org. Chem.* **27**, 1597.
Nakamura, H., Yamamoto, H. and Nozaki, H. (1973). *Tetrahedron Lett.* 111.
Neckers, D. C. (1967). *Mechanistic Organic Photochemistry*. (New York: Reinhold.)
Nedelec, J. Y. and Lefort, D. (1972). *Tetrahedron Lett.* 5073.
Nedelec, L., Gasc, J. C. and Bucourt, R. (1974). *Tetrahedron*, **30**, 3263.
Needleman, S. B. and Chang Kuo, M. C. (1962). *Chem. Rev.* **62**, 405.
Negishi, E., Lew, G. and Yoshida, T. (1973). *J. chem. Soc. Chem. Commun.* 874.
Negishi, E. and Yoshida, T. (1973). *J. chem. Soc. Chem. Commun.* 606.
Negishi, E., Yoshida, T., Silveira, A. and Chiou, B. L. (1975). *J. org. Chem.* **40**, 814.
Nielson, A. T. and Houlihan, W. J. (1968). *Organic Reactions*, **16**, 1.
Ninomiya, I., Naito, T. and Kiguchi, T. (1973). *J. chem. Soc. Perkin I*, 2257.
Ninomiya, I., Naito, T. and Takasugi, H. (1975). *J. chem. Soc. Perkin I*, 1791.
Nishimuru, J., Kawabata, N. and Furukawa, J. (1969). *Tetrahedron*, **25**, 2647.
Nozake, H., Kato, H. and Noyori, R. (1969). *Tetrahedron*, **25**, 1661.
Nussbaum, A. L. and Robinson, C. H. (1962). *Tetrahedron*, **17**, 35.
Ogura, K., Yamashita, M., Suzuki, M. and Tsuchihashi, G. (1974). *Tetrahedron Lett.* 3653.

Olofson, R. A. and Dougherty, C. M. (1973). *J. Am. chem. Soc.* **95**, 582.
Onischenko, A. S. (1964). *Diene Synthesis.* (New York: Davey.)
Oppolzer, W. (1973). *Helv. chim. Acta*, **56**, 1812.
Oppolzer, W. (1974). *Tetrahedron Lett.* 1001.
Oppolzer, W., Pfenninger, E. and Keller, K. (1973). *Helv. chim. Acta*, **56**, 1807.
Orban, I., Schaffner, K. and Jeger, O. (1963). *J. Am. chem. Soc.* **85**, 3033.
Osborn, J. A., Jardine, F. H., Young, J. F. and Wilkinson, G. (1966). *J. chem. Soc.* (*A*), 1711.
Oshima, K., Shimoji, K., Takahashi, H., Yamamoto, H. and Nozaki, H. (1973). *J. Am. chem. Soc.* **95**, 2694.
Oshima, K., Takahashi, H., Yamamoto, H. and Nozaki, H. (1973). *J. Am. chem. Soc.* **95**, 2693.
Paddon-Row, M. N., Hartcher, R. and Warrener, R. N. (1976). *J. chem. Soc. Chem. Commun.* 305.
Padwa, A. and Brodsky, L. (1974). *J. org. Chem.* **39**, 1318.
Pappas, J. J. and Keaveney, W. P. (1966). *Tetrahedron Lett.* 4273.
Paquette, L. A. (1968). *Accounts chem. Res.* **1**, 209.
Paquette, L. A., Itoh, I. and Farnham, W. B. (1975). *J. Am. chem. Soc.* **97**, 7280.
Parham, W. E. and Schweizer, E. E. (1963). *Organic Reactions*, **13**, 55.
Parker, A. J. (1962). *Quart. Rev. chem. Soc. Lond.* **16**, 163.
Parker, R. E. and Isaacs, N. S. (1959). *Chem. Rev.* **59**, 737.
Partch, E. (1964). *Tetrahedron Lett.* 3071.
Partridge, J. J., Chadha, N. K. and Uskoković, M. R. (1973). *J. Am. chem. Soc.* **95**, 532.
Payne, G. B. (1962). *Tetrahedron*, **18**, 763.
Pearson, D. E. and Buehler, C. A. (1974). *Chem. Rev.* **74**, 45.
Pelter, A., Harrison, C. R. and Kirkpatrick, D. (1973). *J. chem. Soc. Chem. Commun.* 544.
Pelter, A., Hutchings, M. G., Smith, K. and Williams, D. J. (1975). *J. chem. Soc. Perkin I*, 145.
Pelter, A., Rowe, K. and Smith, K. (1975). *J. chem. Soc. Chem. Commun.* 532.
Pelter, A., Smith, K., Hutchings, M. G. and Rowe, K. (1975). *J. chem. Soc. Perkin I*, 129, 138.
Pelter, A., Smith, K. and Tabata, M. (1975). *J. chem. Soc. Chem. Commun.* 857.
Pelter, A., Subramanyan, C., Laub, R. J., Gould, K. J. and Harrison, C. R. (1975). *Tetrahedron Lett.* 1633. (See also Pelter, A., Gould, K. J. and Harrison, C. R. (1975). *Tetrahedron Lett.* 3327.)
Peterson, D. J. (1968). *J. org. Chem.* **33**, 781.
Pettit, G. R. and van Tamelen, E. E. (1962). *Organic Reactions*, **12**, 356 (London: Wiley).
Pfau, M. and Ribière, C. (1970). *Chem. Commun.* 66.
Piatak, D. M., Herbst, G., Wicha, J. and Caspi, E. (1969). *J. org. Chem.* **34**, 112, 116.
Piers, E. and Nakamura, I. (1975). *J. org. Chem.* **40**, 2694.
Pillot, J-P., Dunogues, J. and Calas, R. (1976). *Tetrahedron Lett.* 1871.
Pizey, S. S. (1974). *Synthetic Reagents*, vol. 1, p. 101. (London: Wiley.)
Politzer, I. R. and Meyers, A. I. (1971). *Organic Syntheses*, **51**, 24.
Poos, G. I., Arth, G. E., Beyler, R. E. and Sarett, L. H. (1953). *J. Am. chem. Soc.* **75**, 422.

Posner, G. H. (1975*a*). *Organic Reactions*, **19**, 1.
Posner, G. H. (1975*b*). *Organic Reactions*, **22**, 253.
Posner, G. H., Sterling, J. J., Whitten, C. E., Lentz, C. M. and Brunelle, D. J. (1975). *J. Am. chem. Soc.* **97**, 107.
Posner, G. H. and Whitten, C. E. (1976). *Organic Syntheses*, **55**, 122.
Posner, G. H., Whitten, C. E. and Sterling, J. J. (1973). *J. Am. chem. Soc.* **95**, 7788.
Poulter, C. D., Friedrich, E. C. and Winstein, S. (1969). *J. Am. chem. Soc.* **91**, 6893.
Prager, R. H. and Reece, P. A. (1975). *Austral. J. Chem.* **28**, 1775.
Rabjohn, N. (1949). *Organic Reactions*, **5**, 331.
Radlick, P., Klem, R., Spurlock, S., Sims, J. J., van Tamelen, E. E. and Whitesides, T. (1968). *Tetrahedron Lett.* 5117.
Rambeck, B. and Simon, H. (1974). *Angew. Chem. internat. Edit.* **13**, 608.
Ratcliffe, R. W. (1976). *Organic Syntheses*, **55**, 84.
Ratcliffe, R. and Rodehorst, R. (1970). *J. org. Chem.* **35**, 4000.
Rees, C. W. and Storr, R. C. (1969). *J. chem. Soc.* (*C*), 1474.
Reich, H. J. (1975). *J. org. Chem.* **40**, 2571.
Reich, H. J. and Chow, F. (1975). *J. chem. Soc. chem. Commun.* 790.
Reich, H. J., Reich, I. L. and Renga, J. M. (1973). *J. Am. Chem. Soc.* **95**, 5813.
Reich, H. J., Renga, J. M. and Reich, I. L. (1974). *J. org. Chem.* **39**, 2133.
Reich, H. J. and Shah, S. K. (1975). *J. Am. chem. Soc.* **97**, 3250.
Rémion, J., Dumont, W. and Krief, A. (1976). *Tetrahedron Lett.* 1385.
Rerick, M. N. (1968). In *Reduction*, p. 1. Ed. R. L. Augustine. (London: Arnold.)
Rerick, M. N. and Eliel, E. L. (1962). *J. Am. chem. Soc.* **84**, 2356.
Reucroft, J. and Sammes, P. G. (1971). *Quart. Rev. chem. Soc. Lond.* **25**, 135.
Rhoads, S. J. (1963). In *Molecular Rearrangements*, vol. 1, p. 655. Ed. P. de Mayo. (London: Interscience.)
Rhoads, S. J. and Raulins, N. R. (1975). *Organic Reactions*, **22**, 1.
Richer, J. C. (1965). *J. org. Chem.* **30**, 324.
Richman, J. E., Herrmann, J. L. and Schlessinger, R. H. (1973). *Tetrahedron Lett.* 3267, 3271.
Rickborn, B. and Gerkin, M. (1968). *J. Am. chem. Soc.* **90**, 4193.
Rickborn, B. and Thummel, R. P. (1969). *J. org. Chem.* **34**, 3583.
Ringold, H. J. and Malhotra, S. K. (1962). *J. Am. chem. Soc.* **84**, 3402.
Robinson, R. (1916). *J. Chem. Soc.* **109**, 1038.
Roth, von W. R. (1966). *Chimia*, **20**, 229.
Ruden, R. A. and Bonjouklian, R. (1974). *Tetrahedron Lett.* 2095.
Rühlmann, K., Seefluth, H. and Becker, H. (1967). *Chem. Ber.* **100**, 3820.
Russell, G. A. and Ochrymowycz, L. A. (1969). *J. org. Chem.* **34**, 3624.
Rylander, P. N. (1967). *Catalytic Hydrogenation over Platinum Metals.* (London: Academic Press.)
Rylander, P. N. and Steele, D. R. (1969). *Tetrahedron Lett.* 1579.
Sam, D. J. and Simmons, H. E. (1972). *J. Am. chem. Soc.* **94**, 4024.
Sammes, P. G. (1970). *Quart. Rev. chem. Soc. Lond.* **24**, 37.
Sample, T. E. and Hatch, L. F. (1970). *Organic Syntheses*, **50**, 43.
Sato, K., Inoue, S., Ota, S. and Fujita, Y. (1972). *J. org. Chem.* **37**, 462.
Sauer, J. (1966). *Angew. Chem. internat. Edit.* **5**, 211.
Sauer, J. (1967). *Angew. Chem. internat. Edit.* **6**, 16.
Sauer, J. and Kredel, J. (1966). *Tetrahedron Lett.* (*a*) 731; (*b*) 6359.

Sauers, R. R. and Ahearn, G. P. (1961). *J. Am. chem. Soc.* **83**, 2759.
Saunders, W. H. (1964). In *The Chemistry of Alkenes*, p. 149. Ed. S. Patai. (London: Interscience.)
Saunders, W. H. and Cockerill, A. F. (1973). *Mechanisms of Elimination Reactions*, p. 105. (London: Wiley-Interscience.)
Schaffner, K., Arigoni, D. and Jeger, O. (1960). *Experientia*, **16**, 169.
Schatzmiller, S. and Eschenmoser, A. (1973). *Helv. chim. Acta*, **56**, 2975.
Schenck, G. O. (1957). *Angew. Chem.* **69**, 579.
Schiller, G. (1955). In Houben-Weyl, *Methoden der organischen Chemie*, vol. 4(2), p. 241. Ed. E. Müller. (Stuttgart: Georg Thieme Verlag.)
Schlosser, M. (1970). In *Topics in Stereochemistry*, vol. 5, p. 1. Ed. E. L. Eliel and N. L. Allinger. (New York: Interscience.)
Schlosser, M. (1972). In Houben-Weyl, *Methoden der organischen Chemie*, vol. 5(1*b*), p. 134. Ed. E. Müller. (Stuttgart: Georg Thieme.)
Schlosser, M. and Christmann, K. F. (1966). *Angew. Chem. internat. Edit.* **5**, 126.
Schlosser, M., Hartmann, J. and David, V. (1974). *Helv. chim. Acta*, **57**, 1567.
Schmidt, R. R. (1973). *Angew. Chem. internat. Edit.* **12**, 212.
Schmitz, E. and Murawski, Z. (1966). *Chem. Ber.* **99**, 1493.
Schönberg, A. (1968). *Preparative Organic Photochemistry*. (Berlin: Springer Verlag.)
Schulte-Elte, K. H. and Ohloff, G. (1964). *Tetrahedron Lett.* 1143.
Seebach, D. (1969). *Synthesis*, **1**, 17.
Seebach, D. and Beck, A. K. (1971). *Organic Syntheses*, **51**, 76.
Seebach, D. and Corey, E. J. (1975). *J. org. Chem.* **40**, 231.
Seebach, D., Jones, N. R. and Corey, E. J. (1968). *J. org. Chem.* **33**, 300.
Seebach, D., Kolb, M. and Gröbel, B-T. (1974). *Tetrahedron Lett.* 3171.
Semmelhack, M. F. (1972). *Organic Reactions*, **19**, 115.
Semmelhack, M. F. and Stauffer, R. D. (1973). *Tetrahedron Lett.* 2667.
Semmelhack, M. F. and Stauffer, R. D. (1975). *J. org. Chem.* **40**, 3619.
Shapiro, R. H. and Heath, M. J. (1967). *J. Am. chem. Soc.* **89**, 5734.
Sharpless, K. B. and Akashi, K. (1975). *J. Am. chem. Soc.* **97**, 5927.
Sharpless, K. B. and Akashi, K. (1976). *J. Am. chem. Soc.* **98**, 1986.
Sharpless, K. B., Chong, A. O. and Oshima, K. (1976). *J. org. Chem.* **41**, 177.
Sharpless, K. B. and Gordon, K. M. (1976). *J. Am. chem. Soc.* **98**, 300.
Sharpless, K. B. and Hori, T. (1976). *J. org. Chem.* **41**, 176.
Sharpless, K. B. and Lauer, R. F. (1973). *J. Am. chem. Soc.* **95**, 2697.
Sharpless, K. B., Lauer, R. F. and Teranishi, A. Y. (1973). *J. Am. chem. Soc.* **95**, 6137.
Sharpless, K. B. and Michaelson, R. C. (1973). *J. Am. chem. Soc.* **95**, 6136.
Sharpless, K. B., Umbreit, M. A., Nieh, M. T. and Flood, T. C. (1972). *J. Am. chem. Soc.* **94**, 6538.
Sharpless, K. B. and Young, M. W. (1975). *J. org. Chem.* **40**, 947.
Sharpless, K. B., Young, M. W. and Lauer, R. F. (1973). *Tetrahedron Lett.* 1979.
Sheldon, R. A. and Kochi, J. K. (1972). *Organic Reactions*, **19**, 279.
Shen-Hong Dai and Dolbier, W. R. (1972). *J. Am. chem. Soc.* **94**, 3953.
Shimoji, K., Taguchi, H., Oshima, K., Yamamoto, H. and Nozaki, H. (1974). *J. Am. chem. Soc.* **96**, 1620.
Sicher, J. (1972). *Angew. Chem. internat. Edit.* **11**, 200.

Siddall, J. B., Biskup, M. and Fried, J. H. (1969). *J. Am. chem. Soc.* **91**, 1853.
Siegel, S. and Smith, G. V. (1960). *J. Am. chem. Soc.* **82**, 6082, 6087.
Simmons, H. E., Cairns, T. L., Vladuchick, S. A. and Hoiness, C. M. (1973). *Organic Reactions*, **20**, 1.
Smissman, E. E., Suh, J. T., Oxman, M. and Daniels, R. (1962). *J. Am. chem. Soc.* **84**, 1040.
Smith, C. W. and Holm, R. T. (1957). *J. org. Chem.* **22**, 747.
Smith, G. V. and Burwell, R. L. (1962). *J. Am. chem. Soc.* **84**, 925.
Smith, K. (1974). *Chem. Soc. Rev.* **3**, 443.
Smith, P. A. S. (1963). In *Molecular Rearrangements*, vol. 1, p. 571. Ed. P. de Mayo. (London: Wiley-Interscience.)
Smolinsky, G. and Feuer, B. I. (1964). *J. Am. chem. Soc.* **86**, 3085.
Sneen, R. A. and Matheny, N. P. (1964). *J. Am. chem. Soc.* **86**, 5503.
Starks, C. M. (1971). *J. Am. chem. Soc.* **93**, 195.
Stechl, H.-H. (1975). In Houben-Weyl, *Methoden der organischen Chemie*, vol. 4(1*b*), 877. Ed. E. Müller. (Stuttgart: Georg Thieme Verlag.)
Stemke, J. E. and Bond, F. T. (1975). *Tetrahedron Lett.* 1815.
Stephenson, L. M., McClure, D. E. and Sysak, P. K. (1973). *J. Am. chem. Soc.* **95**, 7888.
Stetter, H. and Kuhlmann, H. (1974). *Angew. Chem. internat. Edit.* **13**, 539.
Stevens, R. V. and Wentland, M. P. (1968). *J. Am. Soc.* **90**, 5580.
Stewart, R. (1965). In *Oxidation in Organic Chemistry*, part A, p. 42. Ed. K. B. Wiberg. (London: Academic Press.)
Still, W. C. and Macdonald, T. L. (1974). *J. Am. chem. Soc.* **96**, 5561.
Still, W. C. and Macdonald, T. L. (1976). *Tetrahedron Lett.* 2659.
Stork, G. and Benaim, J. (1971). *J. Am. chem. Soc.* **93**, 5938.
Stork, G., Brizzolana, A., Landesman, H., Szmuszkovicz, J. and Terrell, R. (1963). *J. Am. chem. Soc.* **85**, 207.
Stork, G. and Colvin, E. (1971). *J. Am. chem. Soc.* **93**, 2080.
Stork, G. and Danheiser, R. L. (1973). *J. org. Chem.* **38**, 1775.
Stork, G. and Darling, S. D. (1964). *J. Am. chem. Soc.* **86**, 1761.
Stork, G. and Dowd, S. R. (1963). *J. Am. chem. Soc.* **85**, 2178.
Stork, G. and Dowd, S. R. (1974). *Organic Syntheses*, **54**, 46.
Stork, G. and Ganem, B. (1973). *J. Am. chem. Soc.* **95**, 6152.
Stork, G. and Hudrlik, P. F. (1968). *J. Am. chem. Soc.* **90**, 4462, 4464.
Stork, G. and Jung, M. E. (1974). *J. Am. chem. Soc.* **96**, 3682.
Stork, G., Kraus, G. A. and Garcia, G. A. (1974). *J. org. Chem.* **39**, 3459.
Stork, G. and Macdonald, T. L. (1975). *J. Am. chem. Soc.* **97**, 1265.
Stork, G. and Maldonado, L. (1974). *J. Am. chem. Soc.* **96**, 5272.
Stork, G., Meisels, A. and Davies, J. E. (1963). *J. Am. chem. Soc.* **85**, 3419.
Stork, G., Rosen, P., Golman, N., Coombs, R. V. and Tsuji, J. (1965). *J. Am. chem. Soc.* **87**, 275.
Stork, G. and Singh, J. (1974). *J. Am. chem. Soc.* **96**, 6181.
Stotter, P. L. and Hill, K. A. (1973). *J. org. Chem.* **38**, 2576.
Streitwieser, A. and Williams, J. E. (1975). *J. Am. chem. Soc.* **97**, 191.
Sustmann, R. (1974). *Pure appl. Chem.* **40**, 569.
Suzuki, A., Miyaura, N., Abiko, S., Itoh, M., Brown, H. C., Sinclair, J. A. and Midland, M. M. (1973). *J. Am. chem. Soc.* **95**, 3080.

Suzuki, A., Nozawa, S., Itoh, M., Brown, H. C., Kabalka, G. W. and Holland, G. W. (1970). *J. Am. chem. Soc.* **92**, 3503.
Swern, D. (1953). *Organic Reactions*, **7**, 378.
Syper, L. (1966). *Tetrahedron Lett.* 4493.
Szmant, H. H. (1968). *Angew. Chem. internat. Edit.* **7**, 120.
Szmuszkovicz, J. (1963). In *Advances in Organic Chemistry: Methods and Results*, vol. 4, p. 1. Eds. R. A. Raphael, E. C. Taylor and H. Wynberg. (London: Wiley-Interscience.)
Tagaki, W. and Hara, H. (1973). *Chem. Commun.* 89.
Takahashi, H., Oshima, K., Yamamoto, H. and Nozaki, H. (1973). *J. Am. chem. Soc.* **95**, 5803.
Tam, J. N. S., Mojelsky, T., Hanaya, K. and Chow, Y. L. (1975). *Tetrahedron*, **31**, 1123.
Tamao, K., Sumitani, K. and Kumada, M. (1972). *J. Am. chem. Soc.* **94**, 4374.
Tanaka, S., Yamamoto, H., Nozaki, H., Sharpless, K. B., Michaelson, R. C. and Cutting, J. D. (1974). *J. Am. chem. Soc.* **96**, 5254.
Taylor, D. A. H. (1969). *Chem. Commun.* 476.
Taylor, E. J. and Djerassi, C. (1976). *J. Am. chem. Soc.* **98**, 2275.
Thies, R. W. (1972). *J. Am. chem. Soc.* **94**, 7074.
Thies, R. W. and Billigmeier, J. E. (1974). *J. Am. chem. Soc.* **96**, 200.
Thomas, A. F. (1969). *J. Am. chem. Soc.* **91**, 3281.
Thomas, A. F. and Ohloff, G. (1970). *Helv. chim. Acta* **53**, 1145.
Thomas, A. F. and Ozainne, M. (1970). *J. chem. Soc.* (*C*), 220.
Thummel, R. P. and Rickborn, B. (1970). *J. Am. chem. Soc.* **92**, 2064.
Tinnemanns, A. H. A. and Laarhoven, W. H. (1974). *J. Am. chem. Soc.* **96**, 4611, 4617.
Trippett, S. (1963). *Quart. Rev. chem. Soc. Lond.* **17**, 406.
Trost, B. M. (1974). *Accounts chem. Res.* **7**, 85.
Trost, B. M. (1975). *Pure appl. Chem.* **43**, 563.
Trost, B. M., Bogdanowicz, M. J. and Kern, J. (1975). *J. Am. chem. Soc.* **97**, 2218.
Trost, B. M. and Bridges, A. J. (1976). *J. Am. chem. Soc.* **98**, 5017.
Trost, B. M. and Fullerton, T. J. (1973). *J. Am. chem. Soc.* **95**, 292.
Trost, B. M. and Leung, K. K. (1975). *Tetrahedron Lett.* 4197.
Trost, B. M. and Melvin, L. S. (1975). *Sulfur Ylides.* (New York: Academic Press.)
Trost, B. M., Preckel, M. and Leichter, L. M. (1975). *J. Am. chem. Soc.* **97**, 2224.
Trost, B. M. and Salzmann, T. N. (1975). *J. org. Chem.* **40**, 149. (See also Trost, B. M. and Salzmann, T. N. (1973). *J. Am. chem. Soc.* **95**, 6840.)
Trost, B. M. and Stanton, J. L. (1975). *J. Am. chem. Soc.* **97**, 4018.
Trost, B. M. and Tamaru, Y. (1975). *J. Am. chem. Soc.* **97**, 3528.
Turro, N. J. (1969). *Acc. chem. Res.* **2**, 25.
van Rheenen, V., Kelly, R. C. and Cha, D. Y. (1976). *Tetrahedron Lett.* 1973.
van Tamelen, E. E. and Sharpless, K. B. (1967). *Tetrahedron Lett.* 2655.
van Tamelen, E. E., Pappas, S. P. and Kirk, K. L. (1971). *J. Am. chem. Soc.* **93**, 6092.
Vedejs, E. (1974). *J. Am. chem. Soc.* **96**, 5944.
Vedejs, E. (1975). *Organic Reactions*, **22**, 401.
Vogel, E., Grimme, W. and Korte, S. (1963). *Tetrahedron Lett.* 3625.
Wagner, P. J. and Hammond, G. S. (1968). In *Advances in Photochemistry*, vol. 5, p. 136. Ed. W. A. Noyes, G. S. Hammond and J. N. Pitts. (London: Interscience.)

Walker, E. R. H. (1976). *Chem. Soc. Rev.* **5**, 23.
Walling, C. (1957). *Free Radicals in Solution.* (London: Wiley.)
Walling, C. and Padwa, A. (1961). *J. Am. chem. Soc.* **83**, 2207.
Walling, C. and Padwa, A. (1963). *J. Am. chem. Soc.* **85**, 1597.
Wasserman, H. H. and Terao, S. (1975). *Tetrahedron Lett.* 1975, 1735.
Wasserman, H. H. and Lipshutz, B. H. (1975). *Tetrahedron Lett.* 1731.
Wassermann, A. (1965). *Diels–Alder Reactions.* (London: Elsevier.)
Watt, D. S. (1976). *J. Am. chem. Soc.* **98**, 271.
Wawzonek, S., Nelson, M. F. and Thelen, P. J. (1951). *J. Am. chem. Soc.* **73**, 2806.
Wawzonek, S. and Thelen, P. J. (1950). *J. Am. chem. Soc.* **72**, 2118.
Wendisch, D. (1971). In Houben-Weyl, *Methoden der organischen Chemie*, vol. 4(3), p. 126. Ed. E. Müller. (Stuttgart: Georg Thieme Verlag.)
Wendler, N. L., Taub, D. and Kuo, H. (1960). *J. Am. chem. Soc.* **82**, 5701.
Wenkert, E. (1968). *Acc. chem. Res.* **1**, 78.
Wiberg, K. B. (1965). In *Oxidation in Organic Chemistry*, part A, p. 69. Ed. K. B. Wiberg. (London: Academic Press.)
Wicker, R. J. (1956). *J. chem. Soc.* 2165.
Wigfield, D. C. and Phelps, D. J. (1974). *J. Am. chem. Soc.* **96**, 543.
Wilds, A. L. (1944). *Organic Reactions*, **2**, 178.
Wilke, G. and Heimback, P. (1962). *Liebigs Ann.* **652**, 7.
Windaus, A. and Brunken, J. (1928). *Liebigs Ann.* **460**, 225.
Wittig, G. (1962). *Angew. Chem. internat. Edit.* **1**, 415.
Wolfe, S., Hasan, S. K. and Campbell, J. R. (1970). *Chem. Commun.* 1420.
Wolff, M. E. (1963). *Chem. Rev.* **63**, 55.
Woodward, R. B. (1963). *The Harvey Lectures* vol. 31. (Compare Fleming, I. (1973) *Selected Organic Synthesis* p. 202, (London: Wiley.)
Woodward, R. B., Freary, E. R. J., Nestler, G., Raman, H., Sitrin, R., Suter, Ch. and Whitesell, J. K. (1973). *J. Am. chem. Soc.* **95**, 6853.
Woodward, R. B. and Hoffmann, R. (1969). *Angew. Chem. internat. Edit.* **8**, 781.
Woodward, R. B. and Hoffmann, R. (1970). *The Conservation of Orbital Symmetry.* (New York: Academic Press.)
Woodward, R. B. and Katz, T. J. (1959). *Tetrahedron* **5**, 70.
Woodward, R. B., Pachter, I. J. and Scheinbaum, M. L. (1974). *Organic Syntheses*, **54**, 39.
Yamamoto, Y. and Brown, H. C. (1973). *J. chem. Soc. Chem. Commun.* 801.
Yamamoto, Y., Toi, H., Sonoda, A. and Murahashi, S-I. (1976). *J. Am. chem. Soc.* **98**, 1965.
Yang, N. C. and Yang, D. H. (1958). *J. Am. chem. Soc.* **80**, 2913.
Yasuda, A., Tanaka, S., Oshima, K., Yamamoto, H. and Nozaki, H. (1974). *J. Am. chem. Soc.* **96**, 6513.
Yasuda, A., Yamamoto, H. and Nozaki, H. (1976). *Tetrahedron Lett.* 2621.
Yates, P. and Eaton, P. (1960). *J. Am. chem. Soc.* **82**, 4436.
Ziegler, F. E. and Spitzer, E. B. (1970). *J. Am. chem. Soc.* **92**, 3492.
Zimmerman, H. E. (1961). *Tetrahedron*, **16**, 169.
Zimmerman, H. E. (1963). In *Molecular Rearrangements*, vol. 1, p. 345. Ed. P. de Mayo. (London: Wiley-Interscience.)
Zurflüh, R., Wall, E. N., Siddall, J. B. and Edwards, J. A. (1968). *J. Am. chem. Soc.* **90**, 6224.

Zweifel, G. and Arzoumanian, H. (1967). *J. Am. chem. Soc.* **89**, 5086.
Zweifel, G., Arzoumanian, H. and Whitney, C. C. (1967). *J. Am. chem. Soc.* **89**, 5086.
Zweifel, G. and Brown, H. C. (1963). *Organic Reactions*, **13**, 1.
Zweifel, G., Fisher, R. P., Snow, J. T. and Whitney, C. C. (1971). *J. Am. chem. Soc.* **93**, 6309.
Zweifel, G. and Polston, N. L. (1970). *J. Am. chem. Soc.* **92**, 4068.
Zweifel, G., Polston, N. L. and Whitney, C. C. (1968). *J. Am. chem. Soc.* **90**, 6243.
Zweifel, G. and Steele, R. B. (1967). *J. Am. chem. Soc.* **89**, (*a*) 2754, (*b*) 5085.

Index